The Agricultural Potential of the Middle East

THE MIDDLE EAST
Economic and Political Problems and Prospects

Studies from a research program of
The Rand Corporation
and
Resources for the Future, Inc.
Sidney S. Alexander, *Program Director*

Published:

Marion Clawson, Hans H. Landsberg, *and* Lyle T. Alexander
The Agricultural Potential of the Middle East

Sam H. Schurr *and* Paul T. Homan
with Joel Darmstadter, Helmut Frank, John J. Schanz, Jr.,
Thomas R. Stauffer, *and* Henry Steele
Middle Eastern Oil and the Western World:
Prospects and Problems

In Preparation:

Sidney S. Alexander
The Economics and Politics of the Middle East

Charles A. Cooper *and* Sidney S. Alexander *(eds.)*
Economic Development and Population Growth
in the Middle East

Paul Y. Hammond *and* Sidney S. Alexander *(eds.)*
Political Dynamics in the Middle East

The Agricultural Potential of the Middle East

Marion Clawson
Director of the Land Use and Management Studies
Resources for the Future, Inc.
Washington, D.C.

Hans H. Landsberg
Director of Resources Appraisal Studies
Resources for the Future, Inc.
Washington, D.C.

Lyle T. Alexander
Consultant, Resources for the Future, Inc.
Retired chief of the Soil Survey Laboratories
of the U.S. Department of Agriculture

With 2 four-color maps, tables, and illustrations

American Elsevier Publishing Company, Inc.
NEW YORK · 1971

AMERICAN ELSEVIER PUBLISHING COMPANY, INC.
52 Vanderbilt Avenue, New York, N.Y. 10017

ELSEVIER PUBLISHING COMPANY, LTD.
Barking, Essex, England

ELSEVIER PUBLISHING COMPANY
335 Jan Van Galenstraat, P.O. Box 211
Amsterdam, The Netherlands

International Standard Book Number 0-444-00093-3

Library of Congress Card Number 79-135058

Manufactured in the United States of America

CONTENTS

ILLUSTRATIONS

TABLES

Foreword

In the Middle East, except for the oil-rich countries, agriculture is and for some few decades ahead will continue to be the mainstay of economic life. In the countries examined in this book—stretching from Iraq on the east through to Egypt on the west—the agricultural population totals twenty-eight million, just over 50 percent of the total population of the region. Many other people in the region depend for their livelihood on the processing and handling of agricultural products. Thus it is hard to imagine sustainable economic progress apart from improvements in agriculture. An examination of recent history and a consideration of future potentialities both affirm this conclusion, as do the theoretical and model-building approaches of the scholars of economic development.

The studies reported in this book aim to delineate the economic development potential of agriculture in the Middle East, the principal problems involved, and the measures required to solve the problems and achieve the potential. The analysis is based upon the physical resources of the region; upon the technology available for using those resources; and upon economic analysis of costs, markets, returns, and the like. Stress is laid upon both imaginative projection of the development possibilities of the region, and a hardheaded analysis of the difficulties of achieving them.

This book is directed at everyone concerned with the welfare of the rural population of the Middle East, and because of the dominant role of agriculture in most of the countries studied, the welfare of the area as a whole. It is written to provide perspective for national and international programs, whether public or private, and for improvement in the living conditions of rural people. In the book substantial use is made of scientific and technical information, but at the same time the style of writing and the terms employed are such as to make the analysis and findings readily understandable by any interested reader who is neither an agricultural specialist nor an economist.

Among other things, the study presents informal projections of Middle East agricultural outputs to a point roughly a generation into the future, perhaps as far ahead as the year 2000. The principal input requirements are also indicated; these include such items as drainage and irrigation works, fertilizers, new varieties of crops, and pesticides, as well as soil and water. Attention is also given to markets, manpower requirements, foreign exchange, assistance from more developed countries, governmental stability, administrative capacity, education, and the other elements of social infrastructure. The authors make no claims to discovering the "best" path for economic development; they present no blueprint for the future. More realistically they argue that no one solution is to be found for the age-old, intractable problems of agricultural improvement in the Middle East. Whole "packages" of measures must be pursued in interrelated sequences if significant gains are to be made.

But despite numerous difficulties, most of which will have to be surmounted by the people in the region functioning under their own motivations and through their own institutions, a central conclusion of the study stands out: that the region has the natural resource potentialities of soil, water, and climate on the basis of which rapid agricultural development can be achieved, provided the required effort can be organized and population increase moderated. Despite all manner of difficulties which lie ahead, this is a heartening view.

This study is one of several being undertaken jointly by The RAND Corporation and Resources for the Future, Inc., in response to a request from, and with the support of, the Ford Foundation. The studies aim to delineate the main economic development and related potentialities and problems of the Middle East region. In addition to its role as co-sponsor, RFF took particular responsibility for two of the principal component studies: this one on *The Agricultural Potential of the Middle East* and another on *Middle Eastern Oil and the Western World: Prospects and Problems*. Although quite different in most respects, these are the two great industries in the Middle East: agriculture because it is essential to livelihood and employs such a large number of people, and oil because of its overwhelming importance as a source of government revenue and foreign exchange.

Each of the three authors brought to the task of preparing this study not only expertise in several relevant aspects of agriculture, but also the advantages of considerable experience in the Middle East region. Furthermore, as the acknowledgments indicate, research contributions were received from others whose professional careers are centrally focused on the Middle East either as residents or scholars from the outside. Both Mr. Clawson and Mr. Landsberg have had extensive experience in agricultural, economic, and development research and in the related field of water resources, while Mr. Alexander has had many years of distinguished work in soil chemistry and related subjects.

This study stands on its own feet as a treatment of the potentials and problems of Middle East agricultural development; at the same time it forms a unit in the larger RAND-RFF program of Middle East studies, being prepared under the overall direction of Sidney Alexander, consultant to RAND and professor of economics at the Massachusetts Institute of Technology. It is hoped that leaders concerned not only with agricultural matters in the Middle East, but also with agricultural and economic development of less developed regions in general will profit from this book.

Washington, D. C.
January 1970

JOSEPH L. FISHER
President, Resources for the Future, Inc.

Preface

The purpose of this book is to consider the economic development potential of agriculture in the Middle East, the problems involved, and the measures required to achieve that potential. The term "Middle East" as used in this study includes Egypt, Israel, Jordan, Lebanon, Syria, Iraq, and Saudi Arabia, the last-named dealt with primarily in a brief Appendix. The data and analyses use the national boundaries as they existed before June, 1967.

Although considerable differences exist within the region, it has many common features of agricultural resources, problems, and opportunities. In the degree to which it brings together and contrasts data for the various countries, on such matters as cropping patterns, crop yields, economic output per unit of land, and the like, the study is, to our knowledge, unique.

Informal projection of output and main inputs was made to the end of the present century; then attention was directed toward the problems and possibilities of moving toward or to the goals or end points which it would be technically possible to reach. The report is not intended as a blueprint for the future, however. The extent to which the technically feasible potentials can be realized is subject to great uncertainty; the purpose of this report is to focus on these possibilities and to provide a general background for programs which may be undertaken to reach them.

Several major assumptions underlie the report:

(1) Agriculture will continue to be the source of employment for a large proportion of the total population in these countries for a generation ahead, even if industrialization and other economic development proceed rapidly. The welfare of farm and other rural people is therefore a matter of continuing importance.

(2) The traditional main resources for agricultural production—climate, soil, and water—will continue to remain highly important, although they may be used in new ways or in new combinations in the future.

(3) Chemical synthesis of foods or hydroponics will not become established on a scale sufficient to displace in any significant degree the growing of crops on the land. Chemical synthesis of food supplements, of fertilizers, and of other chemicals for use in agricultural production will continue and expand, but to supplement rather than to replace reliance upon soil.

(4) Agricultural technology throughout the world is likely to advance slowly in the future, no matter what public or private measures are taken; but the rate of development and of adoption of agricultural technology new to the Middle East is susceptible to acceleration sufficient to bring about a virtual agricultural revolution within a generation.

Data for this report have been obtained from many sources. Full use has been made of the published reports of international organizations such as the Food and Agriculture Organization, of U.S. sources such as the Department of Agriculture, and of published reports of the countries in the region. In addition, use has been made of many unpublished reports or materials, in the region and elsewhere; in some instances it has not been feasible to cite the source, a circumstance that any scholar will regard with regret, but which we decided could not rule out the use of important information that was easily accessible but technically not in the public domain. In some cases, where reliable data were missing or where available data were contradictory, we have had to make our best judgment as to what the facts are. As a result there are instances in which no source gives exactly the data we present. Whenever possible, we have tried to give the reader an assessment of the quality of the data used.

Our preference has been for data from official sources; and in no case have we revised or corrected such official data. We have accepted the data as given, on the theory that any attempt at revision would usually create more problems than it would solve. For that reason, in making monetary comparisons between countries, we have used official exchange rates in all cases, in spite of the fact that such rates are sometimes quite unrealistic.

Since this book is likely to have predominantly an American audience, we have included some analyses and illustrations which should be particularly helpful to that audience. Also, we have in the text used mainly U. S. measuring units. But for the benefit of the international audience, units of measure in Appendix C, Statistical Tables, are those widely used in international comparisons.

The report consists of three major parts: Part I, Natural Resources of the Middle East as They Affect Agriculture; Part II, Present Organization of Agriculture in the Middle East; and Part III, Development Potential of Middle Eastern Agriculture. They are preceded by a summary chapter which strives to bring the highlights of the report to the attention of one who wants a quick picture without the fullness of supporting evidence or detail.

Because data are too scanty to justify the detailed treatment given the other countries, Saudi Arabia is dealt with in a brief appendix (D).

Two other appendices (A and B) contain reports on soils and water resources, respectively, that are greatly more detailed than Chapters 2 and 3, which are based on them. Their inclusion was decided upon as a help to those who wish to go more deeply into what are, when all is said and done, the basic resources for agriculture. It is, of course, not possible to connect, especially, the appendix material on soils closely to the rest of the report in the sense that this or that piece is tied to the output potential, in area and tons, of a given crop. But in a broader sense the basic information on soils and water does underlie the quantification of the potential and should be of significant help to those who will proceed from the generalities of this study to particularized plans and programs. If its language is too technical for some, it will strike soil experts as not technical enough. Our aim has been to bend the language toward the nontechnical wherever this was possible without sacrifice of needed precision.

Two folded maps on a scale of 1:2,000,000 are included in a pocket at the back of the book; these show major soil groups of the region and are to be used with the soils appendix.

An appendix of statistical tables (C) contains supporting statistics and other data that, while not essential in the text, are of value to one concerned with the region in detail. This report has assembled a great deal of information that previously was not easily available or not available in a form to facilitate comparisons between countries; we publish this particular appendix therefore as a service to professional workers interested in this region. Appendix C also contains the data on which the text charts are based.

In preparing the report we have had the assistance of many professionals, not all of whom can be fully acknowledged. Of particular assistance has been Professor Salim W. Macksoud of the American University of Beirut, an irrigation specialist, who provided us with memoranda on irrigation, drainage, and land use in Egypt, Lebanon, Jordan, Syria, and Iraq, and with citations to numerous professional reports. In addition, we gratefully acknowledge the assistance of Dr. George S. Medawar, formerly of the American University of Beirut and now consultant economist in Beirut, who prepared for us reports, invaluable in our analysis, on the economic aspects of agriculture in those same countries.

Dean K. Fuhriman of Brigham Young University assembled a great deal of information from numerous sources on the general matter of water supply use and potential, by river basins, prepared Appendix B, and wrote other memoranda on which we have drawn heavily.

William J. Vaughan, a graduate student in economics at Georgetown University, assembled statistical information on crop acreages, output, and yield; on livestock numbers and production; on agricultural inputs, farm numbers, and the like, from various published reports. Laura Lee gave invaluable service in checking both text and appendix statistical tables, in updating many of them, and in helping to put together the final version of the report.

Various officials at the Food and Agriculture Organ-

ization in Rome gave us liberally of their time and expertise and provided us with relevant published information. In particular, we wish to acknowledge the assistance of K. L. Bachman, D. Luis Bramao, David Burdon, J. J. Doyle, R. Dudal, M. R. El Ghonemi, K. R. Ellinger, L. Kadry, L. B. Kristjanson, J. de Meredieu, B. L. Nestel, P. A. Oran, W. H. Pauley, R. A. Peterson, R. Poduval, M. Thielebein, B. G. West, and J. Wolf, and of Oris V. Wells, Deputy Director General, for arranging a review of the first draft. Various officials at the U.S. Department of Agriculture in Washington performed a similar service for us, and drew upon their personal experience in the region for our information. In particular, we wish to acknowledge the help of Charles A. Gibbons, Lyle Moe, Afif Tannous, H. Charles Treakle, and Cline J. Warren.

Staff members of the International Land Development Consultants, of Arnhem, the Netherlands, who had personal experience of the region (and experts from other organizations associated with their work in the region) were also very helpful to us. Meetings with P. Buringh, J. Schilstra, P. Snethlage, J. van Assen, J. S. Veeenbos, and J. Westerhout, and above all P. M. van der Sluis, who made these possible, introduced us to and clarified many important issues.

Thanks are due to Raanan Weitz, Director of the Settlement Study Centre of the National and University Institute of Agriculture, Rehovot, Israel, who assembled a group of experts to meet with us at the Centre in a discussion of Israel agriculture and allied issues and generally assisted us in those parts of the study dealing with Israel.

During a week's stay of the authors in Lebanon, Hugh Walker and Owen Brough of the Ford Foundation in Beirut were particularly helpful in arranging for us to meet with people familiar with agriculture in that country and in the region in general. On a subsequent visit to Washington, Owen Brough made many helpful suggestions for improving the manuscript.

At an early stage in our study, a group of persons with personal experience in one or more countries of the region met with us for a day, and offered comments and advice. That group included (organizations listed for identification only, since each came as an individual):

Albert Y. Badre, University of Southern Illinois

Maurice Bart, World Bank (IBRD)

Ralph W. Cummings, Rockefeller Foundation (now with North Carolina State University)

B. Delworth Gardner, Utah State University

Robert M. Hagan, University of California (Davis)

Bent Hansen, University of California (Berkeley)

Morris A. Huberman, United Nations Development Programme

Donald L. McCune, Tennessee Valley Authority

Wyn F. Owen, University of Colorado
Dean F. Peterson, Utah State University
Gabriel Saab, L'Union Catholique des Cultivateurs (Montréal)
A. Robert Sadove, World Bank (IBRD)
B. Abdel-Sayed, Michigan State University
Roger Woodworth, Tennessee Valley Authority

as well as others previously acknowledged.

A review draft of this report was circulated in early 1969 to a considerable number of agricultural experts and others, in the Middle East and elsewhere; and valuable comments were received from Roscoe Bell, Consultant; D. R. Campbell, University of Toronto; Leonard Rosenberg, AID, U.S Department of State; William D. Romig, AID, U.S. Department of State; Lawrence Witt, Michigan State University; and others named previously in the Preface. Not all of their comments could be, or were, incorporated in this final report, nor all the lacunae filled that they called to our attention.

Our thanks are due also to eight Arab graduate students from the Middle East who in 1969 attended departments of agricultural economics at U.S. universities, and spent a week with us reviewing the manuscript. They called our attention to many points that needed clarification or correction; and they will find much of their effort reflected in the pages of the final version.

We wish to express our thanks and to acknowledge our debt to all these persons, and to any whom we may have inadvertently omitted. To the extent that this report ultimately helps to improve the living conditions of rural people in the region, the latter will be in no small way indebted to these many persons who have assisted us. At the same time, of course, only the authors are responsible for what is said herein.

MARION CLAWSON
HANS H. LANDSBERG
LYLE T. ALEXANDER

CONVERSION TABLES

Factors Used in This Report

To convert Column 1 into Column 2, multiply by	Column 1	Column 2	To convert Column 2 into Column 1, multiply by
		Length	
0.621	kilometer, km.	mile, mi.	1.609
1.094	meter, m.	yard, yd.	0.914
3.282	meter, m.	foot, ft.	0.305
0.394	centimeter, cm.	inch, in.	2.540
0.0394	millimeter, mm.	inch, in.	0.254
		Area	
0.386	kilometer^2, km.^{2}	mile^2, mi.^{2}	2.590
247.1	kilometer^2, km.^{2}	acre	0.00405
2.471	hectare, ha.−0.01 km.^{2}	acre	0.405
1.038	feddan (Egypt)	acre	0.963
0.2471	dunum (Israel, Lebanon, Jordan, Syria)−0.10 ha.	acre	4.05
0.618	meshara[a] (Iraq)	acre	1.620
		Volume	
0.0008107	meter^3, m.^{3}	acre-foot, ac. ft.	1,233.5
326,000	acre-foot, ac. ft.	gallon, gal.	0.000,003,067
2.838	hectoliter, hl.	bushel, bu.	0.352
1.057	liter, l.	quart, qt.	0.946

(1.0 cubic foot per second, cusec., flowing continuously for 12.1 hours equals 1.0 acre-foot; 1,000 m.^{3} per second, flowing continuously for 278 hours equals 1.0 billion [milliard] m.^{3})

To convert Column 1 into Column 2, multiply by	Column 1	Column 2	To convert Column 2 into Column 1, multiply by
		Mass	
1.102	ton (metric)	ton (short)	0.9072
1,000.0	ton (metric)	kilogram, kg.	0.001
220.5	quintal, q.	pound, lb.	0.00454
2.205	kilogram, kg.	pound, lb.	0.454
44.928	cantar (Egypt)	kilogram, kg.	0.0223
935 (approx.)	dariba (rice, Egypt)	kilogram, kg.	0.00107
		Rates and Yields	
0.446	ton (metric)/hectare (ha.)	ton (short)/acre	2.242
0.892	kg./ha.	lb./acre	1.12
0.892	quintal/ha.	hundredweight/acre	1.12
0.0149	kg./ha.	bushel (wheat)/acre	67.253
0.0186	kg./ha.	bushel (barley)/acre	53.52
0.0168	kg./ha.	bushel (sorghum, maize)/acre	59.406
0.0198	hg./ha.	bushel (rice)/acre	50.440

Crop Weights per Unit of Volume

bushel, bu.   = wheat, 60 pounds; barley, 48 pounds; sorghum, corn (maize) shelled, 56 pounds; rice, 45 pounds
bale   = 500 pounds (gross) cotton lint
ardeb (Egypt) = 198 liters; wheat = 150 kg. (320 lbs.); barley = 120 kg. (264 lbs.); maize = 140 kg. (309 lbs.)

[a] Sometimes listed as donom.

Official Foreign Exchange Rates, Middle Eastern Countries, 1948–68

1. (in units of foreign currency per U.S. $)

Year	United Arab Republic (Egyptian pound)	Israel (Israel pound)	Lebanon (Lebanon pound)	Jordan (Jordan dinar)	Syria (Syrian pound)	Iraq (Iraqi dinar)	Saudi Arabia (Riyal)
1968	.435	3.50	3.15	.357	3.82	.357	4.50
1967	.435	3.00	3.13	.357	3.82	.357	4.50
1966	.435	3.00	3.17	.357	3.82	.357	4.50
1965	.435	3.00	3.07	.357	3.82	.357	4.50
1964	.435	3.00	3.08	.357	3.82	.357	4.50
1963	.435	3.00	3.08	.357	3.82	.357	4.50
1962	.352	3.00	3.06	.357	3.82	.357	4.50
1961	.352	1.80	3.02	.357	3.58	.357	4.50
1960	.352	1.80		.357	3.58	.357	
1959	.352	1.80		.357	3.58	.357	
1958	.352	1.80		.357	3.58	.357	
1957	.352	1.80		.357	3.58	.357	
1956	.352	1.80		.357	3.58	.357	
1955	.349	1.80		.357	3.58	.357	
1954	.349	1.80		.357	3.58	.357	
1953	.349	1.80		.357	2.19	.357	
1952	.349	1.80		.357	2.19	.357	
1951	.349	.36		.357	2.19	.357	
1950	.349	.36		.357	2.19	.357	
1949	.349	.36		.357	2.19	.357	
1948	.242	.25		.248	2.19	.248	

2. (in U.S. $ per unit of foreign currency)

Year	United Arab Republic (Egyptian pound)	Israel (Israel pound)	Lebanon (Lebanon pound)	Jordan (Jordan dinar)	Syria (Syrian pound)	Iraq (Iraqi dinar)	Saudi Arabia (Riyal)
1968	2.30	0.29	0.317	2.80	0.262	2.80	0.222
1967	2.30	0.33	0.319	2.80	0.262	2.80	0.222
1966	2.30	0.33	0.315	2.80	0.262	2.80	0.222
1965	2.30	0.33	0.326	2.80	0.262	2.80	0.222
1964	2.30	0.33	0.325	2.80	0.262	2.80	0.222
1963	2.30	0.33	0.325	2.80	0.262	2.80	0.222
1962	2.30	0.33	0.327	2.80	0.262	2.80	0.222
1961	2.838	0.56	0.331	2.80	0.279	2.80	0.222
1960	2.838	0.56		2.80	0.279	2.80	
1959	2.838	0.56		2.80	0.279	2.80	
1958	2.838	0.56		2.80	0.279	2.80	
1957	2.838	0.56		2.80	0.279	2.80	
1956	2.838	0.56		2.80	0.279	2.80	
1955	2.862	0.56		2.80	0.279	2.80	
1954	2.862	0.56		2.80	0.279	2.80	
1953	2.862	0.56		2.80	0.457	2.80	
1952	2.862	0.56		2.80	0.457	2.80	
1951	2.862	2.78		2.80	0.457	2.80	
1950	2.862	2.78		2.80	0.457	2.80	
1949	2.862	2.78		2.80	0.457	2.80	
1948	4.127	4.00		4.03	0.457	4.03	

Source: International Monetary Fund.

The Potential of Middle East Agriculture—A Summary

AGRICULTURE is a major sector of the economy and the social structure of Middle Eastern countries. Except in Jordan and Israel, half or more of the people gain their living in agriculture, and only in Israel is the rural smaller than the urban population. The potential of agricultural resources, in combination with truly modern agricultural technology, offers these countries an opportunity to escape within one generation the old vicious circle of poverty, excessive dependence on agriculture, high birth and high death rates, low capital accumulation—and poverty. If the development potential of agriculture is seized upon and exploited rapidly, this cycle can be broken; new economies, new social structure, new urbanism, and new national destiny can be achieved. But another generation of unrestrained population increase and slow agricultural development might well lead to a situation from which escape would be far more difficult.

The obstacles to achievement of sustained and rapid increases in agricultural output are great; and many are lodged deeply in the social and political fabric. Above all, little can be accomplished without governments that are stable without being rigid, capable of orderly transfer of power from one set of national leaders to another, responsive to the needs of its people, and—as a result—capable of carrying to completion resource development and agriculture-improvement projects over a series of years and on an adequate scale. A continuously improving general infrastructure—transportation, communication, education, financial institutions—is essential to both achieving the potential levels of production and exploiting them profitably. And, of course, improvements within agriculture itself are the indispensable basis for progress. Finally, the new attitudes toward knowledge, new working relationships with one's friends and with others, and a willingness to modify custom and inherited practices that are no longer functionally efficient are not only necessary, but are most difficult of all to achieve and even to measure.

Nevertheless great as are the problems on the road toward rural comfort, well-being, and human dignity—all rooted in greatly increased agricultural output—they are within human control, in the Middle East no less than in South Asia or Latin America. A realistic appraisal of both the opportunity and the problems is an essential step toward achievement.

The Record

Over the centuries, man has shown great ingenuity in using the climate, soil, and other agricultural resources of the Middle East. The Nile Valley has been used for millennia; the river overflowed, wetting the valley soils, depositing a thin layer of fertile silt, and flushing out the salts; winter crops were raised, especially grain; and in the valley, a complex and productive society evolved. The Mesopotamian Plain was a great garden spot, with irrigation, vast green areas, hanging gardens, and all the other attributes of a productive area. The hill country, not only from Dan to Beersheba but much farther, was terraced and planted to deep-rooted vines and olives or used to grow annual winter crops such as barley and wheat, each in order to utilize the modest winter rainfall.

The soil, the winter grain, and the exploitable water resources provided the basis for a crop agriculture. Other men grazed their flocks and herds of camels, cattle, horses, donkeys, sheep, and goats on the natural pasture and grazing lands, following the grass wherever it might be found. This contrast between settled crop farmer and nomadic herdsman has persisted to the present and severely impedes the emergence of a modern livestock enterprise.

Repeatedly fought over by contending forces native to the region and by Persian, Greek, Roman, and other conquerors, the Middle East has never been a land of enduring peace. Farmers have repeatedly been dispossessed of their land, and their buildings and movable property destroyed. Their lives have been taken or their liberty ended by slavery. The wonder is not that agriculture has sometimes retrogressed sharply, but rather that it survived at all.

In more modern times, but still for several hundred years, the countries of the Middle East as here considered were part of the Ottoman Empire—an Empire which began and expanded with vigor and skill, but gradually became old and sick. In the first half of this century, two World Wars produced drastic political changes in the Middle East: the first saw transition from the provinces of the Ottoman Empire to the colonies of European powers—Britain and France; the second led to independence. Egypt followed a somewhat different route, breaking away from the Ottoman Empire earlier, but to the same essentially colonial status. The point is that as independent nations with present boundaries, the countries of the Middle East are very young.

Long before these political changes, the once highly productive irrigated areas now generally referred to as the Mesopotamian Plain fell into an agricultural decay from which they have scarcely begun to emerge. Soil erosion from unstable and overgrazed surrounding lands covered ancient soils with one to several feet of overburden; today, these ancient soil horizons and numerous intermediate but less fully developed ones can be uncovered by excavation on part of the Mesopotamian Plain—incontrovertible evidence of past land use and misuse. In the presence of naturally inadequate drainage, these developments have led to accumulation of almost unbelievable levels of soluble salts, deposited as the flood and irrigation waters evaporated. In many areas, the soil surface is white with salt. A meager crop of salt-tolerant barley (ten bushels per acre) can be grown under irrigation in alternate years only. The salts are temporarily flushed down into the soil and the crop is planted, sparingly irrigated, and harvested. Then the land lies idle with a growth of native weeds while it dries out in preparation for another meager round. Crop production can scarcely be more marginal and yet survive.

Agriculture in the Nile Valley began to change in the nineteenth century. Cotton, introduced earlier, became a major summer crop during and after the American Civil War, as world cotton supplies shrank and prices rose. This necessitated summer irrigation and led to the construction of several low dams or barrages. Keeping water in the canals all year and applying some irrigation water throughout the naturally dry seasons steadily saturated the soil, the water table throughout most of the valley and delta rose, until today it is within one meter (39 inches) of the surface in extensive areas, and within two meters (78 inches) for a very large portion of the irrigable area. The latest and most dramatic development on the Nile is the High Aswan Dam, just now coming into operation. Because of the dam, valley flooding will no longer occur, seasonal flood flows will be stored not only for the low flow season within the year but for the occasional dry year, and enough irrigation water will be available for every season. Fully controlled, the river can be of better service. But the ancient fertilizing effect of silt deposition from flood waters and the even more important salt-flushing action of the floods will also cease. Moreover, river waters no longer laden with silt may pick up new loads of sediments from downriver channels and create serious hydrologic and hydraulic problems as the river adjusts its slopes and cross sections to the new conditions; some fear that perennial irrigation will render the control of bilharzia—Egypt's waterborne, debilitating rural disease—still harder. Thus, full harnessing of the Nile has increased its capabilities for good, but raised new problems; the premium on skilled water and land management rises sharply on both accounts.

Agriculture in the Middle East today is a curious and complex mixture of ancient and fully modern technology. Tractors plow and combines reap much of the land planted to grain—but women still pull meager barley crops from the soil, harvesting roots as well as stalks for livestock feed, and grain for human use. Statistics indicate that in Egypt and Israel chemicals for control of insects, plant diseases, and weeds are used far more heavily in relation to crop acreage than they are in the United States. But over large areas weeds choke out grains and chemicals have yet to make their appearance. Specialized, intensively operated dairies, usually with the familiar large black and white Holstein cows one sees in U. S. dairies, have appeared in various areas, especially to provide milk for the larger cities. Specialized broiler and egg chicken production also is not uncommon. At the same time, nomads continue to drive their herds of sheep and goat across sparse grassland, much as they have done in the remote past.

By and large, the present agriculture of the Middle East is not highly productive, particularly in terms of the opportunities its natural resources afford. Winter grains (wheat

and barley) on the rainfed lands on an alternate crop-fallow basis produce meager crops—rarely above fifteen bushels per acre—even in relatively wet years, and in the all-too-frequent dry years yields are sharply lower. This need not be so. In the Pacific Northwest of the United States on areas closely comparable in soil and climate, farmers using generally similar equipment get yields three or more times larger than these low yields. The irrigated winter wheat of Egypt does yield about forty bushels per acre, which is indeed high compared to the national average yield of nearly all countries; but in comparable irrigated areas in the United States—and perhaps in other areas where water is also not a limiting factor—yields are about double those of Egypt. Likewise Egyptian cotton yields are just above a bale per acre, which is also high compared with national average yields in most countries. But irrigated cotton in the American Southwest averages two bales and more per acre; so does irrigated cotton in neighboring Israel.

Such crop comparisons could be extended. In some instances, average yields in the Middle East have trended upward. But in other instances—irrigated corn in Egypt is a prime example—the Middle Eastern yield is not only low but shows no upward trend. Some writers have stated that Egypt's crop yields are high; indeed, they often are when compared with quite dissimilar areas where natural features are much less favorable. But comparisons should be made with areas elsewhere in the world where climate, soil, water availability, and other conditioning factors are reasonably similar.

Livestock too is relatively unproductive in the Middle East, for a variety of reasons. Most livestock is not fed or pastured on crop farms, but grazes natural pastures and grain aftermath the year around. Without exception, far too many livestock are kept in relation to the grazing capacity. The grasses and other forage plants are eaten literally into the ground; this greatly reduces the capacity of such areas to produce feed another year and accelerates runoff of precipitation. Most of the food the animals obtain is required for their maintenance, leaving little margin for growth or fattening. As long as native pastures are open to grazing by any man's livestock, no one has any incentive to conserve forage; any that he left ungrazed would be taken by someone else. Another obstacle to progress is that native breeds are perhaps well-adapted to the harsh conditions under which they live, but lack the capacity to grow rapidly and to multiply prolifically even when conditions are more favorable. Finally, many animals are diseased.

As a result of factors such as these, the returns to human effort expended in agriculture are low in most Middle Eastern countries. The value added, per agricultural worker, is about $300 in Egypt, and is not greatly different in other Arab countries. The cost of farm inputs in these countries is 25 percent of the gross value of output—in the United States the comparable figure is nearly 70 percent. The Arab farmer is efficient in the sense that he expends little on his output; but he is extremely inefficient because, spending little, he produces little. And it is the size of the margin between cost of input and value of output that finally sets his standard of living.

Living conditions in most Middle Eastern countries are what they can be at GNP levels that, with the exception of Israel, range between $150 and $400 per capita. Information on rural as compared to general conditions is sparse. Most of it is episodic rather than statistical. Even at lower income levels rural people are probably better fed, but worse off in such matters as electricity, sanitary water supplies, safe disposal of human wastes, and other basic conveniences of modern life. Ill health is common: bilharzia, malaria, other pest-borne diseases, and blindness are widespread and difficult to control. These diseases not only are a burden on those suffering and those having to cure them, but also greatly affect the capability of the labor force. However, substantial progress in health and education has been made in recent years—there are even some encouraging research results on control of bilharzia. But one cannot help being impressed by the likelihood that fundamental change is so much part and parcel of improvement of the efficiency of agriculture as a whole that it is difficult to foresee any one segment of the general problem moving far out ahead of the others.

Closely associated with health conditions is nutrition. No data have come to our attention that deal specifically with nutrition in rural areas; but on a national average, diets are marginal in terms of calories, badly deficient in protein intake, and, one must suppose, also other nutrients. Since national averages show these deficiencies, and since some strata of the population must be assumed to exceed this by differing margins, it follows that a significant segment of the population in most of the countries subsists on a sub-marginal diet. Apart from its immediate adverse effect on health and performance, this condition also sets the stage for a large absorptive capacity for additional food output in the future. But the degree to which such increases will in fact occur depends not only on increased output but also on increased incomes, so that domestic demand can compete successfully with export demand. This is especially important as governments can by and large be expected to favor exports in order to strengthen their foreign exchange position.

This favoring of export demand over domestic demand is an import consideration. At present all countries reviewed are in deficit on their food accounts. Grains are a sizeable import item; so in lesser degree are fats and oils, meats, and some other products. For countries like Iraq—and even more so Saudi Arabia—that have large incomes from oil this is not a serious matter. But continuing large outlays for basic food stuffs to feed a rapidly rising population hobble development efforts elsewhere. Nor is this much offset by exports of farm products. Cotton from Egypt, citrus from Israel, dates from Iraq, and miscellaneous fruits and vegetables from Lebanon are the principal agricultural export from the region. Of these, only cotton, citrus, and dates have substantial markets outside the region itself, and in value they are a long way from covering the cost of food imports, especially when one adds the cost of supplies in support of domestic agriculture, such as fertilizer, pesticides, machinery, etc.

One is on safe ground in saying that if organized to pro-duce efficiently, agriculture in each of the Arab countries of the region could be carried on with a greatly reduced number of workers. It is also certain that even under current conditions only a continuing flow of rural manpower to the towns and cities is keeping rural unemployment, hard as it is to ascertain and measure, from becoming a major problem.

As discussed in Chapter 6, there is a wide spectrum of opinion regarding the volume of unemployment, open or disguised, in Egypt. In other countries similar arguments could probably be advanced, but the statistical base in manpower matters is so thin that none have arisen.

Where differences of opinion emerge is in the appraisal of the magnitude of the problem. We tend to agree with those who believe that with present farm practices and structure it would be difficult, at least in Egypt, to withdraw sizeable numbers of agricultural workers without creating at least seasonal shortages, and that this would be the more pronounced the greater the desirable elimination of young children and of women from the labor force.

One must at the same time concede that in none of the countries considered will agriculture be in a position to offer employment opportunities to the natural increase in agricultural manpower; and the limits to employment opportunities would narrow with the spread of modern practices.

The foregoing description of Middle Eastern rural life and circumstances applies to all the Arab countries more or less equally; Israel is an exception to much of this description. Culturally and politically, for reasons associated with its origin and brief history, Israel stands apart from the rest of the Middle East; so does its agricultural technology and its economy. Although crop yields and livestock production might well be improved, in important respects Israel agriculture compares very favorably with that of the United States. Particularly, value added per agricultural worker is high—over $2,100—or more than seven times that of Egypt and within 20 percent of that of the United States. Rural living conditions are generally good in Israel, birth rates low, and life expectancy long, even compared with the United States. Why then, one might ask, include Israel in this regional study? The answer is that it furnished a great many clues to the magnitude of the area's potential and serves that purpose better than do experimental results or faraway places, however comparable in other respects.

The Potential

Two popular opinions are widespread about agriculture in the Middle East. One is that a region that was once the granary of the ancient world has played out its usefulness forever, exhausted its soils, denuded its forests, and generally worn out its resources beyond repair or, as in Egypt, reached its natural limits. The other looks to water as the rejuvenator that can in short order restore the region to its former productiveness.

In studying the region's resources we have come to the conclusion that both judgments are myths. As for the first, the region is not played out. With appropriate modern farm technology, practices and managment, and freed from restrictions, inhibitions, and rigidities that lie in the political, social, and institutional field, the natural resources of the Middle East could permit production on a vastly higher level. In a matter of two to three decades, more or less, agricultural output could be doubled in Egypt, multiplied tenfold in Iraq, and increased to intermediate degrees in the other countries except Israel.

As to the second, water alone is not the key. It is only one among many necessary inputs, and it is not now and will not for a considerable time be a limiting factor in most of the region. Indeed, neither water nor any other single input is the magic wand that will quickly and painlessly produce agricultural plenty and prosperity. Instead, there is a whole bundle, or package, of separate, but closely interrelated programs which will lead to sustained advances in agriculture. An improved variety of wheat, grown in adequately irrigated and properly drained soil, with good land tillage and uniform seeding, nourished by enough fertilizers of the right kind, protected against competition by weeds and infestations of diseases and of insects, may be tremendously productive. The same wheat in an unimproved environment may produce little more than the old variety; the fertilizer applied to the old variety might increase yields little or none; uncontrolled weeds might capture all added water and fertilizer. Several improvements, properly coordinated, form the productive package; each part, alone, has limited value. Under some conditions it might even be a handicap. Nor will all of them suffice unless the prices offered for the commodities produced can be counted upon to leave a reasonable margin of profit; unless there are ways of obtaining necessary inputs at prices that permit this profit margin; unless the products can be easily moved to market, etc.

In this context, water is an important ingredient. So are suitable soils to receive it. And suitable soils are probably scarcer in the Middle East than water. To make the package analogy does not mean that the significance of one component vis-a-vis another will not vary from country to country, or that one remains powerless to move unless and until all one's pieces are nicely lined up. It does mean, however, that sustained rapid agricultural progress will require much hard work, sound planning, and competent execution of plans over a period of years. And it also means that nearly all of the ingredients for agricultural success in the Middle East must come from *within* the region. To be sure, outside help will be of value. Especially in the initial phase, new technology and new plant and livestock material from abroad will be indispensable; capital limitations may be eased by foreign grants or loans; and technical expertise may fill in shortages in locally available manpower. These aids are not to be underestimated. But basically, agricultural progress will either capture the imagination of the government and people of the country itself, or it will not take place.

What we refer to as the potential is more easily defined in terms of what it is not than what it is. It is not a forecast or projection for a given target date. Even less is it a program or a project or a group of projects. Instead, it is a racking up of output levels that could be achieved over the span of a generation or so on the basis of natural resource endowment, best practices now known and carried out elsewhere, and a favorable infrastructure that furnishes agriculture with what these practices require if they are to be effective in raising output.

So interpreted, potential grain production, perhaps the single most significant measure for a region in which grain is the staff of life, could rise from its present 11 million tons to 50. At that rate of output, the region would be a net exporter of grain in a volume that raises questions of export outlets. Production of other crops could rise similarly, in some instances perhaps even more sharply.

Another measure is gross income of agriculture, and here rough calculations lead us to estimate the possibility of a rise from $3,500 million to $9,100 million a year. The degree to which such an increase would result in higher per capita income is of course dependent upon the rate of population growth. In general, one can judge that if present rates were to continue unchecked on the farm, much if not most of the estimated growth in output would be dissipated with little increase in per capita income.

Having given two broad dimensions of the potential, we now look at some of the prerequisities to achievement in the specific context of the region, and begin with the package of physical production inputs.

Drainage of irrigated land is basic for Iraq, will be highly productive in Egypt, and will be useful in local situations elsewhere. Surface or buried drains in irrigated areas of Iraq are necessary for flushing out the high concentration of salts. Fortunately, the salts are soluble, the soils have adequate internal drainage, and the costs are of an order of magnitude that local budgets or foreign borrowing can be expected to support if full advantage is subsequently taken of the improved productive capacity. For several years drainage water will be highly saline, with salt concentration roughly ten times that of sea water. Disposal of these wastes presents serious problems; at first they may be dumped in nearby desert depressions, to evaporate there; but, over the longer run, conduits to the sea must be provided, to bypass the rich agricultural areas along the Shatt Al Arab. Nor is it enough to construct drains; they must be maintained, especially the surface ones.

In Egypt, buried tile or plastic drains will greatly improve soil-water relationships, thereby making possible increases in crop yields of up to 50 percent. Buried drains to replace present surface drains will save enough land (12 percent of the area and perhaps more) to pay for themselves; here, as in Iraq, drainage presents no great technical problems, and costs are not excessive. Until the present restricted drainage of the Nile Valley is improved, other efforts at improving crop yields will be severely limited in effect.

Improvement of irrigated land by leveling that allows uniform water application and by functional canals can further help to raise crop yields.

Moisture conservation or rainfed crop lands is another important measure. The present crop-fallow rotation, with the land largely growing up to weeds in the fallow year, does not conserve well either moisture or fertility. If the drier and poorer lands were seeded to permanent grasses, and if annual cropping were instituted on the better soils and wetter areas, more forage for livestock and more grain for human consumption could be produced. Cultural practice could conserve more of the moisture which falls. On those soils—unfortunately not very common—where depth is adequate to store most of two years' rainfall, the crop-fallow system could be made much more productive by stricter control of weeds during the fallow year and by tillage practices designed to retain moisture.

Crop rotation on rainfed lands would be changed by the improvements noted above. Some land would no longer be used for annual crops, other areas would grow a crop every year (sometimes with pulses or other legumes alternating with grain to facilitate weed and disease control); and those areas remaining in crop-fallow would have a greatly different fallow operation. On irrigated lands in Egypt the present crops would largely continue, but quicker maturing winter grains would permit earlier seeding or corn; and more summer irrigation water should lead to much improved corn yields. Together with other cropping changes in Egypt this could bring the ratio of cropped to cultivated area from 1.6 at present to 2.0. In Iraq, the present crop-fallow system on irrigated land could be dropped entirely; not only could a crop of wheat be harvested each year but a good crop of sorghum or corn could be harvested from the same land later in the season. Other irrigated land in Iraq could grow summer crops such as cotton.

Fertilizer would have to be applied in far greater amounts than at present if the foregoing measures were to achieve their potential. Although requirements will vary with different varieties and cultural practices, approximately 2.7 pounds of nitrogen are needed for each bushel of wheat for most of the normal yield range. Some of this can come from the soil, the amount varying by areas; but most must be added in the shape of chemical fertilizer. Much more phosphate and potash fertilizer than currently used will be necessary, if yields are to be increased. Most importantly the capacity of crops to utilize fertilizer must be raised, basically through replacing native strains of crops.

Improved crop varieties will be highly productive if drainage is provided, if moisture control results in better soil moisture conditions, if land is properly prepared and seeds placed uniformly and at proper depth and density, and if fertilizer is added to adequate levels. The new short-stem (Mexipak, etc.) wheats are very much higher yielding than old varieties, in large part because they can efficiently utilize far more nitrogen and other fertilizer, as well as water, and respond to improved seeding practices. The same is true for rice, which is important in Egypt and of some significance in Iraq. Improved corn varieties, whether open pollinated or hybrid, can likewise produce more under similarly favorable conditions. But yields of nearly every crop can be raised, in part by cultural and fertility measures, in part by new varieties, in part by improved drainage. Some new varieties may be introduced from other parts of the world; some may be developed within the region through use of plant breeding methods well tested elsewhere.

Weed control on rainfed and irrigated crop lands is essential if the added moisture and added fertility are not to be merely the basis of a more glorious weed crop. Most crop land in the Middle East is now heavily weed infested. Weed control is possible by chemical or cultural means or both. The cost of spraying may be prohibitive for a 10-bushel grain crop, but reasonable for a 40-or 60-bushel crop. Returns to spraying have been known to be ten or fifteen times its cost. Once weeds are brought under control for considerable areas, continued control is less onerous. The initial control thus basically calls for a capital outlay.

Livestock production, no less than crop production, requires a package approach. Animals require better feed and better disease control, and improved breeding by selection of native stock and imports of foreign animals can be as productive for animals as for crops. As matters stand now, most of what the animals are fed goes into maintenance. But profit lies solely in what the animal produces above maintenance. There is a quick increase in profits once the maintenance level is crossed; below it nothing is accomplished. Though not a physical input, regulating the use of nonprivate grazing lands is a sine qua non of improvement of such areas, and thus of better livestock management and production.

Additional water, which in an essentially arid country means additional *irrigation*, is frequently cited as a major means of increasing crop production. We have intentionally placed it last, for two reasons: (1) because irrigation alone, without the other measures discussed, would produce relatively little; and (2) because there remain only moderate amounts of unused water in the region. The Nile Valley is now fully irrigated from the High Aswan Dam; indeed, we judge that there is now more water than there is good land to put it on. On the Tigris-Euphrates system additional dams could develop additional water, but as long as totally inadequate drainage and other reasons keep cropping practices in the Mesopotamian Plain at their present primitive level, the productivity of more irrigation water is very low. Iraq or any other country can employ foreign contractors to build big dams; there is no lack of interested and competent construction firms. But water so developed would have extremely limited value, unless it were part of the package of measures here described. Moreover, it would seem unwise to undertake water development on the Euphrates River, and to a lesser extent on the Tigris, until treaties or other arrangements define the amounts of water available to each country, including Turkey, through which these rivers flow. The above comments are pertinent to new works, not to improvement and adjustment of existing ones.

Complete development of all natural water supplies in the Middle East would require investments of the general order of $1 billion. Complete drainage of all lands requiring it, both in the Nile Valley and in the Mesopotamian Plain (and for the latter conduits for conducting saline drainage water to the sea), would require investments of the general order of $1.5 billion. These are large investments for these countries, but they are not out of reach. In recent years, Iraq has allocated $100 million annually to investment in irrigation, drainage, and agriculture. Its oil revenue could probably support a higher rate of investment. Syria and Egypt lack large oil revenues, but capital could probably be found abroad if potential lenders could have clear evidence of its productivity. A factor easing the capital problem, in any event, is that drainage would require some years for completion and further irrigation development should await adequate drainage.

Additional costs would arise in the development of groundwater—such as exists in Egypt's western desert, for example; in Saudi Arabia; in parts of Lebanon and Syria; and probably elsewhere. But information on the characteristics of these sources is very scanty, especially as it bears upon the likelihood of a sustained yield over a long period of time. And since we consider that for some time to come lack of suitable soil rather than of water will be the greater constraint on output, we have not included in the above accounting any costs for groundwater development of as yet unexploited sources.

The use of nuclear power to desalt sea water on a large scale as a basis for commercial agriculture has in recent years attracted a good deal of attention, at least in the United States. Such proposals appear on inspection to be completely uneconomic, in the sense that few if any crops can tolerate the high costs of water that nuclear desalting involves. Even if proposed but untested methods are as successful in engineering terms as their proponents estimate, cost of water delivered to the land on the time schedule the farmer wants will be at least ten times the value of the water acceptable for the staple crops that are the only ones grown on a sufficiently large acreage to absorb water on the scale suggested. Water costs decline sharply as the size of the nuclear desalting plant increases, so that the volume of water produced would be far too large to be confined to such high-value crops as out-of-season horticultural products which might be able to absorb high-cost water. The magic wand of desalted water applied to arid coastal zones resulting in the production of specialties sold at fancy prices in European capitals at Christmas time is largely, if not wholly, a product of wishful thinking.

Moreover, there is not much good irrigable land within reasonable distance of the sea which does not already have enough irrigation water, except perhaps in the Northern Negev and the northern Sinai. But for those areas desalted water, whether by nuclear or any other type of energy, is not likely to be the answer at present cost levels and probably not even in the amounts that could be most cheaply produced. In summary, desalting sea water does not offer a path to agricultural development of the Middle East. By diverting attention—and perhaps capital—to infeasible proposals, it may serve to delay it.

Achievement of what we judge to be the potential of Middle Eastern agricultural resources might require additional annual outlays of roughly $900 million for fertilizer; of roughly $500 million for farm mcahinery including tractors; and of perhaps $450 million for other current physical inputs, above all pesticides and the like. Much, perhaps most, of the fertilizer, machinery, and other inputs could be produced within the region. The advantages of a regional organization of agricultural supply are obvious in terms of lower cost and assured access. On such a basis the volume of tractor manufacture, for instance, might readily

attain economic levels. Development of irrigation and drainage, over some twenty-five years or so, would cost in the neighborhood of $300 million per year, including cost of capital, operation, and maintenance. Then, the additional costs here identified would run in the order of $2,150 million per year. Compared with an estimated increase in gross output of about $5,500 million, this would not appear to be an infeasible proposition.

Having dealt with the physical aspects of output and input, we are left with the intangibles as well as the development costs associated not with agriculture, or with agriculture alone, but with the economy as a whole—transportation, education, communication, etc. These we have made no attempt to estimate, either in terms of dollars or likelihood of achievement over a specified time horizon.

However, mention must be made of at least two of the hard-to-quantify areas that are close to the heart of agricultural development: markets and manpower.

The prospects for marketing the potential output, estimated to exceed current production by about 150 percent, are reasonably good. That is, population and income growth in the region itself can be instrumental in creating a market for most of it, with two principal exceptions: grain and cotton. A *grain* surplus, greatly exceeding regional needs as expressed through the market for grain as such, would have to look to feedstuff markets, in connection with the ample room there exists for improvement in the diet of large strata of the population by way of greater consumption of livestock products, or to foreign outlets. The latter might not be a simple matter, especially for a newcomer in a highly competitive market, but needs for foreign exchange to procure newly necessary inputs for a modernized agriculture might give such exports added importance, except in the oil-rich countries. *Cotton's* future in the world is uncertain. Its share of the world's textile fiber market (measured in weight which overstates cotton's share since man-made fibers go farther pound for pound) has fallen from just under up 70 percent in 1960 to just over 56 percent by 1968, and continues to fall. Competition has increased with the entry of many new producers. And the relative advantage of long-staple fibers—Egypt's forte—may be eroded in a textile world dominated by synthetics. At the least, cotton prices, as well as those of grain, cannot be expected to show much buoyancy and are more likely to enter a period of long-run decline rather than rise.

Of the rest of agricultural output, there may be problems for citrus, and perhaps other fruits, if these crops are raised in the absence of market information that at the moment does not exist in sufficient precision to permit judgment. Since there is a lag of several years between planting and the first harvest, and abandonment of productive capacity is correspondingly costly, long-term market knowledge here is especially important.

Obviously, no market can absorb, at reasonable prices, unlimited increases in agricultural output. The task of agricultural statesmanship will be to foster increased output at the most reasonable cost and in such directions and at such rates as the market can absorb.

The employment prospect in agriculture is less optimistic. Except for Israel, the countries of the Middle East today are half or more agricultural in employment; the annual surplus of young workers reaching employment age is substantial in comparison with agriculture's replacement needs for its aging workers. In sharp contrast, the technological measures necessary to increase agricultural output will almost certainly *lower* the need for human labor, perhaps sharply so; if not at once, then soon. Nor can it be taken for granted that the annual surplus labor force from agriculture can readily be absorbed into urban employment. Unless highly labor-intensive industries are introduced—an unlikely development—any displacement of agricultural workers would exacerbate the present situation. Except in Israel, where further declines from an already relatively low level of agricultural employment are unlikely to have much of an impact on the labor market, Middle Eastern countries will face a serious dilemma—technological efficiency versus welfare of the rural population.

Establishment of cottage-type industries in the countryside, of repair and maintenance stations, and perhaps of processing industries for farm products could reduce the size of the problem. But they could hardly be expected to solve it.

The Road Ahead

The technical problems of doubling agricultural output in two or three decades are severe but by no means impossible of solution. The basic resources are there, the expertise can be acquired, the practices adapted to local conditions, and so on down the line. Moreover, with allowance for basic differences in other respects, Israel is endowed with basically similar resources and constitutes in a sense a living laboratory that lends realism to the proposition. As long as present political tensions continue, the Arab countries cannot be expected to directly utilize the Israel agricultural experience in their own country. But the same sources of scientific methodology, of agricultural technology, and of genetic materials in plants and animals are available to the Arab countries. Moreover, the technical and economic aspects of Israeli agriculture have been described in books, pamphlets, and scientific journal articles which are readily available.

But the Israeli agricultural experiments—and similar ones in Mexico, Pakistan, and Taiwan—direct attention not only to the similarities in climate and other natural resources but also to the differences in levels of capital investment in agriculture, absence or successful removal of the fetters of time-honored farm practices, size and quality of the supporting research and information structure, availability of marketing and processing facilities, and other factors, and perhaps foremost of all, the express design on the part of society and government to carve out a dominant role for agriculture.

If these latter prerequistes were easy to come by there would be no reason why similar rapid increases in agricultural output could not be achieved elsewhere in the region. But they are not. There are no well-identified roads to progress or even methods of measuring it. Yet without them exploitation of the physical opportunities is bound to be slow, uncertain, and the result far short of the potential.

Present population levels and growth rates exacerbate the agricultural problem, especially in Egypt, and to a lesser extent in other Arab countries. With 30,000,000 people and 6,700,000 acres of agricultural land, Egypt has an overall average density of over four persons per acre, or over 2,500 per square mile. This is about the same density for the entire country as that of a moderately high-income residential subdivision in the United States. In another generation, with another doubling of total population and with very little real opportunity to expand irrigated acreage, average population density for the nation will rise to that of an average-income American residential subdivision. Agricultural progress may provide food and other agricultural commodities for the population growth of *one* more generation; it is now difficult to see how it could do so for the population doubling of still another generation of rapid growth. Unless great progress can be achieved in the next generation in population control or industrial development, it may be simply too late for agricultural progress, no matter how rapid, to provide the necessary economic support for the nation.

Fortunately, population densities are not as high, and population pressures on land are not as great, in the other countries, although rates of population increase are high. Moreover, agricultural development prospects are better in some other countries, not in terms of what can be expected to happen in the coming years, but in the sense that the potential is much higher in relation to present agricultural output.

The opportunity to use natural resources and modern technology to more than double agricultural output, the necessity to overcome severely inhibiting institutional and cultural barriers and political uncertainties if the potential is to be realized, and the compelling need to accelerate economic growth before rapidly growing population makes such growth well-nigh impossible, all combine to present the Arab countries of the Middle East with a great challenge, no matter whether it be a doubling of output in Egypt or a rise of many hundred percent in Iraq or Syria.

To translate that challenge into a series of concrete steps is the task of administrators and planners who are intimately acquainted with a given country. But there is a useful checklist of things to keep in mind. First, in the absence of unlimited means, concentration of effort is likely to bear better fruit than dispersal of effort. A given locality, a given product, a given group of farmers, or any other well-identified sector that can be made to demonstrate dramatic improvement—such as in yield—will serve as a focus of attention, stimulating interest and then emulation. Second, a price structure that will allow farmers to make a good profit will lead them to abandon traditional ways and to assume what they will otherwise consider undue risks. Third, only a government convinced that it must put its agricultural house in order before it can embark on great forward drives elsewhere will be able to cope successfully with the many-sided problems associated with agricultural development. Fourth, agriculture is unlikely to advance as the result of pushing one particular input, be that water or fertilizer or the extension service. Only a convergence of programs on many fronts, though by no means at an equal pace, will bring development.

Indeed, some of the steps require gradual progress at a proper sequence. An example will help. Improved wheat varieties may be introduced in relatively favorable natural conditions; weed control, perhaps by spraying, is likely to be both necessary and profitable, so that still other varieties may become attractive, or use of available varieties may spread, or both. The use of each input strengthens the other; gradual improvement in one makes improvement in others more profitable. In one sense, drainage is an exception to the gradualistic approach; wherever installed, it must be fully adequate and unless the other parts of the

package are quickly added on the drained areas, drainage will have little or no effect on productivity. From a national standpoint, though, its installation can be gradual, proceeding from one area to another. Continuous changes in agricultural technology, with each specific change opening up new horizons for other changes, must be the overall strategy.

In the chapters that follow, we have of necessity accorded the components of a successful approach separate treatment. Beginning with the natural resource endowments and proceeding to management and infrastructure, the important parts of the package are discussed and each is treated as though it was making an independent contribution and was unhampered in its effect by shortcomings elsewhere. They are not brought together until Chapter 19, where the ring opened in Chapter 1 is closed.

PART I

Natural Resources of the Middle East as They Affect Agriculture

ONE assumption basic to this whole study is that natural resource endowment affects the possibilities of economic development—that it is easier to develop "good" than to develop "poor" natural resources. The differentiation of good and poor in turn depends upon many factors, not the least of which is the technology available to the users of resources in the region. What can be done with a given soil type, or with a water supply of specified characteristics, depends in large part upon the science, engineering, and general technological development of the area. Such economic factors as the cost of inputs and the prices of outputs will largely determine the profitability of a particular resource development or use. The social, institutional, and managerial framework or competence of the region also exerts a major influence. Thus, the kind and degree of resource use is a resultant of many factors and forces; nevertheless, the basic characteristics of the resources themselves are important.

This part of the study considers some of the basic natural resources of the Middle East; in so doing, it necessarily depends to a large extent upon the knowledge accumulated as a result of the economic use of these resources over a period of many years—in some cases, literally, of centuries. The use of these resources is considered in later parts of this study. To the extent that our discussion of the resources themselves draws upon the knowledge which experience in their use has provided, this part depends upon later parts; but in turn, they depend upon this part for knowledge of the resource endowment.

Part I is the briefest possible consideration of the essential facts about the agriculturally useful natural resources of the Middle East. Some additional supporting information is found in Appendices A and B, for those interested in somewhat more detail. Chapter 2 is concerned with climate; it makes no pretense of originality, but presents only a very brief summary of the major climatic characteristics of the region. Chapter 3, Soils, is somewhat more extensive, and is backed up by a modest but nevertheless substantial discussion of soils in Appendix A and by a semi-detailed soils map of the region included in the pocket at the back of this report. Appendix A and the soil map bring together a great deal of information previously available only in widely scattered sources, some of which were not readily available in the United States. Although this study of soils is perhaps less detailed and less thorough than would ideally be desirable, yet it does represent a real step ahead in the compilation and interpretation of soils information for this region, and we venture to assert that it will prove a significant original contribution of this study. Chapter 4, Water, briefly summarizes the economically most important facts about the water resources of the region; it is based upon the somewhat more detailed Appendix B. Both are based upon various engineering studies made over a period of years, and neither claims to be a compilation of original data nor a new type of analysis. However, the bringing together in comparative form of data from different reports, relating to different countries, may in itself be a useful contribution.

Part I describes the salient features of the natural resources, based upon knowledge gained through man's experience with these resources. Choice of information about these resources has been guided by a consideration of those characteristics most important for their economic use and development. However, the actual patterns of resource use are considered in Part II, and the possibilities of resource use in the future are discussed in Part III.

Part I omits from consideration some important natural resources of the region, because they are not directly used by agriculture or because their importance to other spheres of economic activity overshadows their importance for agriculture. A notable example of natural resources omitted from our consideration are the highly important petroleum resources of the region. National income from petroleum development may be critical in providing capital for agricultural development, but such capital might be supplied from other sources; and natural gas or other petroleum resource might be used as the basis for a fertilizer industry, although fertilizers might come from other sources as well. The degree of direct involvement of petroleum in agriculture is relatively slight, and we have thus not considered the petroleum resources in this study. Likewise, the rather pleasant winter climate of much of the area, its rich resources of antiquities, and other factors might well become the base for a much more extensive tourist industry in the region, but the direct involvement of agriculture is slight and hence these resources are not considered here.

CHAPTER 1

Climate[1]

THE Middle East as defined in this study, is mostly arid; and all of it has a Mediterranean type of climate. That is to say, it has relatively mild temperatures; precipitation in the winter time; and hot, dry rainless summers. This climatic type, also found elsewhere in the world, dominates the region from which it takes its name. However, some differences within the region exist, and these importantly affect agriculture.

Egypt is the hottest and driest of the countries where agriculture is significant (Saudi Arabia is hotter and almost as dry but has little agriculture, cf. Appendix D). Annual rainfall at Cairo is only slightly more than one inch, and there is virtually no precipitation at all at Aswan. With very little cloud cover to impede it at any season, sunshine falls upon the ground surface well over 90 percent of the maximum possible time. As one goes northward and toward the east, the climate gradually shifts toward a continental type, characterized by greater extremes of temperature daily and seasonally and a different rainfall pattern. However, the climate of by far the greater part of the region is dominantly Mediterranean.

Within the region, local climates are considerably affected by differences in elevation. Examples of these differences are the Judean Hills, the Galilee, Mount Hermon, the Lebanon mountains, the Anti-Lebanon mountains, and some differences in elevation along the coast of the eastern Mediterranean, in Israel, Lebanon, Jordan, and Syria. These exert a significant influence upon total precipitation and temperature. Along the northern border of the region, in the high mountains of Iraq, this effect is even greater. Other less marked topographical features also exert important local influences.

In this connection, American readers should realize that the study region extends from about latitude 12 at the southern tip of Saudi Arabia (about the latitude of northern Costa Rica) to about latitude 37 in northern Iraq (about the latitude of Norfolk, Virginia). It is thus, compared to the United States, a relatively southern region. The parts of the region immediately bordering the Mediterranean are warmer in winter and cooler in summer than further inland, but the moderating influence of the Mediterranean extends some distance inland. The larger portion of the region, however, is not exposed to the moderating

influence of any ocean current or other large body of water.

By far the greater part of the region has less than 10 inches rainfall annually, with some parts, as noted, having very close to none at all. In the mountain areas bordering the region to the north and on the higher mountain areas within the region, precipitation ranges above 30 inches, even above 40 inches. Many stretches, historically important for nonirrigated agriculture, have annual precipitation well above 10 inches but generally below 20 inches. As in all low rainfall areas of the world, precipitation is highly variable from one year to another, and in evaluating rainfall, variability is often a more important element than averages. Relatively large areas may get as much as 16 to 20 inches of rain some years, and as little as 8 or 10 in others; the higher range makes good dryland grain production possible, the latter almost surely leads to crop failure. The boundary lines between rainfall zones shift, and this has great implications for agricultural practice and production, and for the kind of living farm people may be able to make from such areas, a matter further considered in Chapters 2 and 4. But variability in rainfall affects also the strictly grazing areas, where it results in great year-to-year variations in usable forage production.

The frost-free season is long for most of the region. Production of dates, which is highly sensitive to temperature, has been a basic part of the agriculture in parts of the region for centuries, especially in Iraq. Citrus is less sensitive to temperature, although intolerant of frost; its production on a commercial scale is more recent and also more widespread, since extensive areas can produce citrus which cannot grow dates. The relatively cool winters—but with temperatures high enough to permit plant growth—combined with winter precipitation, favor production without irrigation of winter grains, particularly wheat and barley. Production of summer crops without irrigation is difficult in much of the region. Such crops are therefore generally limited to deep-rooted tree and vine crops, olives and grapes particularly, which can draw upon moisture from a large volume of soil where it has accumulated during the winter rainfall months.

The region's water supply is dominated by rivers which rise outside of it and flow across it, to the sea (see Chapter 3). Much of the rain which does come, comes in rather heavy downpours. This is particularly true in the drier desert parts; for instance, a six-inch annual rainfall may consist of two rather heavy storms. With such low and infrequent precipitation, vegetative growth is sparse and soil development very limited. When the rain falls upon the bare and unprotected soil surface, it results not only in severe erosion down to resistant rock in large areas, but also in runoff which often is a large portion of the rainfall;

---

[1] This discussion is drawn primarily from three sources: (1) *Agricultural Geography of Europe and the Near East* (Miscellaneous Publication No. 665, Office Foreign Agricultural Relations, U. S. Department of Agriculture [Washington, D. C., June 1948]); (2) FAO Mediterranean Development Project, *The Integrated Development of Mediterranean Agriculture and Forestry in Relation to Economic Growth* (Rome: Food and Agriculture Organization of the United Nations, 1959); and (3) *Technical Report on a Study of Agroclimatology in Semi-arid and Arid Zones of the Near East* (Geneva: FAO, WMO, OMM; Paris: UNESCO, July 1962).

frequently it does not reach a major stream. As a result, flash floods along normally dry channels are common; even the larger rivers are characterized by sharp seasonal peaks and by great year-to-year variation in flow. Although soil development is limited, these soils do have developed pans of calcium carbonate and calcium sulfate that have been there a long time.

In these respects, the Middle East has considerable similarities to other dry or desert regions elsewhere in the world, such as the drier parts of the Southwest in the United States and the drier parts of northern Mexico. Although the specific problems differ, and hence the measure for dealing with them, the scientific knowledge and experience gained elsewhere have a large degree of applicability to this region.

The relatively high temperatures, the high percentage of sunshine, the variability and general infrequency of precipitation, and the sparseness of vegetation all make this a region of high evaporation. Hot, drying winds are also common, especially in the spring, with consequent severe damage to growing crops. The high natural evaporation leads to severe salt accumulation in the soil wherever water is applied and allowed to evaporate. The consequences of salt accumulation for agricultural development and practice will be explored in some detail below, especially in Chapters 3 and 4.

CHAPTER 2

Soils

THE countries with which we are concerned, excluding Saudi Arabia, have a total land area of about 435 million acres. Of this we estimate that only 44 million acres have appreciable potential for either rainfed or irrigated agricultural production (see Chapter 4). Saudi Arabia alone has a land area of about 560 million acres; but of this no more than perhaps 950,000 acres may eventually be cultivated. We assume that desalted water—discussed in Chapter 13—will not be an appreciable source for agriculture in the time horizon here considered and that, therefore, agriculture must get its water from rainfall and streams that now exist.

A great deal of effort is now being expended in the Middle East on a search for sources of groundwater. Some sources have been found. They may be very important in some of the countries as domestic supplies for man and animals; and perhaps for a limited development of irrigation for gardens and local feed supplies. We do not believe that these new groundwater sources will make a major contribution to the production of the Middle East as a whole (see also Chapter 13).

Chemical Characteristics

Soils of the area are relatively well supplied with nutrient elements other than nitrogen. The low content of nitrogen reflects the very low level of organic matter in these soils except for the recent alluvium such as in the Nile Valley and Delta.

There are no acid-leached soils such as are usually found in comparable latitudes where rainfall is high and the period of precipitation extends over a longer period of time each year. The soils are all well supplied—in fact nearly always oversupplied with calcium and magnesium due to the presence of much calcium and magnesium carbonate in all size fractions of the soils. This excess of limestone in the soils serves not only as a diluent for the active soils components, but also leads to a chlorotic condition in some crops—such as citrus.

Potassium is in good supply in most of the soils; and only with long-continued production of high potassium-demanding crops—such as alfalfa, cotton and corn—will the supplying of potassium from manufactured fertilizers be a likely major item of expense.

The supply of phosphorus, while not particularly low, is not adequate for sustained high production of most food and feed crops.

As mentioned before, nitrogen is not adequate for sustained *high* production in any of the soils, including the naturally very fertile soils of the Nile Valley and Delta. Hence Egypt and Israel with relatively high production of a number of crops now use large quantities of nitrogen fertilizers. They also use considerable quantities of phosphatic fertilizers.

Physical Characteristics

The physical characteristics of the soils of the Middle East area are not of the best. Neither are they the worst possible. Most of the soils have fairly high contents of clay-size particles. This gives soils that are not easy to work especially with hand tools or with animal power. Generally they are sticky when wet and crack deeply when dry.

In some areas such as the Jezira in Syria, they tend to puddle at the surface during the heavy rainfall in mid-winter, giving a condition that is unfavorable for the early period of growth of late planted winter grains and also one that encourages loss by runoff of the water that should soak into the deeper layers of soil for use by the crop during maturity time in the spring. Management of crop residues in the surface horizon of the cultivated soils could be of material help in reducing the breakdown of structure and in encouraging the infiltration of water during the wet season.

Many of the soils, especially in adjoining areas of Syria and Iraq are shallow—generally less than 40 inches thick—and many are less than 24 inches. The ones with no more depth than 40 inches are too shallow for efficient carry-over of moisture from one crop year to another in a dry fallow farming system such as that used, for example, in central Washington in the United States. These soils should, however, grow good crops on an annual basis in years of reasonable rainfall.

The soils with a rooting depth of little more than 24 inches cannot store enough moisture during the winter to mature a good crop in the early summer. When irrigated they must be watered too frequently. Another limiting factor in the use of these shallow soils in both Syria and Iraq is their generally high content of both calcium carbonate and gypsum. Too, they are frequently underlain by material that has gypsum as the major component. Under irrigated agriculture the gypsum is dissolved more in one place than another due to unequal penetration of water. This leads to the development of an uneven surface topography with the consequent difficulty of getting even distribution of irrigation water unless sprinkle equipment is used. This high content of gypsum also makes the lining of the canals and ditches a necessity to prevent unequal subsidence and excessive loss of water. A considerable area of soil in Syria that might otherwise be watered from the Euphrates River has this high content of gypsum beneath the thin depth of soil that can be cultivated. The development of a sustained high level of agricultural production on these thin gypseous soils is not a realistic goal, considering that roots do not penetrate a layer having more than about 25 percent of gypsum.

Considerable attention has been given to some of the limitations of the soils of the region. But they have favorable characteristics in some cases that are not being fully utilized. The residual soils from limestone and marls in the higher rainfall area from mid-Israel north to the Turkish border are favorable for the growth of winter grain crops and can do remarkably well in carrying crops into early and midsummer without irrigation or with only limited supplemental water.

The area is to a considerable extent hilly and much of the surface is exposed rock. But where there is sufficient depth of soil and where rock terraces can be built, crops can be and are now being grown. The area of cultivation could be expanded and present farm management practices could be greatly improved.

Rainfall as a Limiting Factor

The vast desert areas of Egypt, Jordan, Syria, and Iraq and the smaller desert area in Israel contribute some pasturage for sheep, goats, and camels. Because of the vagaries of the unevenness and irregular spatial occurrence of the year-to-year winter precipitation, the amount of forage produced on the soils of the deserts is highly variable however. In the truly desert areas little can be done to stabilize the production, but much could be done to improve the utilization by adequate control of the rate of grazing. The overgrazed areas produce less forage when the rains do come, and such overgrazed areas are more subject to erosion by wind and water than areas on which more vegetation is allowed to remain.

The soils at the edge of the desert—where the Sierozems get too dry for production of winter barley—could produce much more forage with more certainty from year to year if better managed. Runoff could be reduced by simple water-spreading structures of stone or earth that soak more water into the ground during the rainy season. This extra water in the soils that have adequate depth leads to more animal feed each year whether the forage plants are native or planted. A stabilization of feed production by water spreading in areas having too little rainfall for regular production of winter grains could very materially improve the meat production throughout the area, but especially in Jordan.

Soils of Egypt

Agriculture in Egypt is conducted primarily in the Nile Valley and the Nile Delta on calcareous Alluvial soils. The soils of these areas are formed from relatively recent calcareous alluvium laid several meters thick over alluvial gravels of an older age. This calcareous alluvium has been deposited at the rate of about 1 millimeter per year or 1 meter (3 feet) for each 1000 years. The source of the Nile "silt" or "mud" has been primarily the highland of Ethiopia. Because of the prevalence of the volcanic and other recent plutonic rocks of these highlands, the alluvial soils of the Nile Valley and Delta are rich in unweathered minerals that give most of the mineral nutrients that plants need. Thus the cultivated soils of Egypt are naturally fertile in contrast to many soils of the older tropical areas, especially those with high rainfall.

Although the soils of the Nile Valley and Delta are naturally rather productive when given water, they do have a high clay content that makes them very sticky when wet, hard and difficult to crush when dry, and rather slowly permeable to water. Thus they are burdensome to cultivate, especially when the source of effort is man and animal power only. When properly provided with drains and not overwatered these soils are among the most productive in the world. However in recent times multiple cropping and the use of excessive amounts of irrigation water has resulted in high water tables, especially in the Delta area, that are severely limiting yields of most crops. Balls has given a good picture of the historical development of these high water tables and has presented data on the effects on crop yield, particularly for cotton.[1]

The area of the good Alluvial soils of Egypt is rather sharply delimited to the flood plain and delta of the Nile River. Generally the sands and gravelly soils abut rather sharply on either side. There is little uncultivated soil of quality comparable to that now in use. To the west and south of Alexandria there are some very sandy soils that can be cultivated with irrigation by sprinkling systems. But the soils are rapidly permeable and thus have low water-holding capacities. It is unlikely that extensive use of these soils under sprinkler irrigation will be economically feasible for general crops such as cotton, wheat, and corn. Improving the drainage on the Delta soils with high water tables would yield a better return on any investment.

In recent years much effort has gone into the development of lands in the Western Desert. The large depressional oases in this desert thought to have been formed by the wind do have fossil water that can be used to produce crops. Indeed, this has been done to some extent since Roman times. However, most of the soils of these depressions are not very suitable for farming. Many of them are salty, some are very heavy clays, and in all cases there is a potential problem of drainage since these depressions are below surrounding higher ground and hence have no outlets. Moreover, the continued supply of large amounts of the fossil water from the underlying Nubian sandstone is at best very doubtful and would have to be buttressed by future investigation.

The Desert soils of Egypt produce a very scanty grazing potential and that only along the coast west of Alexandria. The total contribution to Egypt's meat production is small. The production of forage in the mountains to the east of the Nile Valley is negligible.

That population pressure may force the country into using such soils not withstanding their basically poor suitability is another matter and is further considered in Chapters 4 and 13. But there seems little doubt that such attempts would involve heavy outlays for soil improvement, such as, for example, the admixture of clayey material from the Nile Delta. These are considerations that go far beyond the description and evaluation of soils that is the subject of this chapter.

Changes in the area under cultivation are best considered

---

[1] W. Laurence Balls, *The Yields of a Crop* (London: E. & F. N. Spon Limited, 1953).

in three time segments: up to 1960, 1961–65 (First Five-Year Plan) and 1965–68.

It is generally agreed that additions were very slow up to the beginning of the First Five-Year Plan, totaling since the turn of the century some 600,000 acres, of which about 80,000 were added between 1952 and 1960, almost all of it in the Nile Valley and the desert edges of the delta. Additions in desert regions proper amounted to only 5,000 acres during the 1950's.

The pace of land reclamation picked up greatly in the 1960's. The plan called for 540,000 acres to be reclaimed in the valley and just over 200,000 acres outside it, between 1960–61 and 1964–65. Results did not quite match the plan, especially outside the valley, where only 40 percent of the target was achieved—i.e., about 80,000 acres, as against nearly 90 percent—or about 470,000 acres in the valley. The low performance rate outside the valley has generally been ascribed to lack of water, of which more below. The bulk of the newly-reclaimed valley lands was west and north of the delta, with small acreage added in Upper Egypt and east of the Delta. Of the 80,000 acres outside the valley, one-half was reclaimed in the so-called New Valley in the Western Desert, with the remainder spread out east and west of the delta in the arid coastal lands towards Sinai and Marsa Matrouh.

Thus by 1965, the land area had risen to about 6.7 million acres. Government plans call for additions, by 1980 or so, of some 1.3–1.5 million acres (these figures are not firm, depending on the date and originator of the plan). About 470,000 acres were slated for reclamation in the three years 1965–66 to 1967–68; but only a little over 200,000 acres are reported to have actually been reclaimed during the period, and those at a steeply declining rate.[2] In 1967–68 the rate was down to 35,000 acres. Thus the reclaimed area by the end of 1968 might be put at 6.9 million acres, though this would include considerable acreage not as yet cultivated. The cultivated area is probably closer to 6.7 million acres.

In the light of the record, what is one to assume as the area basis for calculating the potential of Egyptian agriculture? The limit on cultivated lands if all plans are 100 percent executed would appear to be 8 million acres of cultivated land, a little above the 7.8 million often found in the literature[3] as the goal for the mid-eighties.

Several factors make one reluctant to accept these figures. One is the difference between estimated and actual cost per unit of land reclaimed during the First Five-Year Plan: it jumped from £E 197/feddan estimated to £E 310/feddan actual.[2] Secondly, the achievements outside the valley have turned out to be very disappointing. The specific reference here is to the New Valley. According to

recent reports,[4] all work has been halted, since the wells have flowed at declining rates—some falling to as little as 27 percent of their original flow—and of the 45,000 feddans (47,000 acres) to be cultivated, 60 percent are totally without water supply, while the remaining 40 percent have insufficient water for irrigation and must wait for installation of pumps before work can be resumed. Thirdly, according to the same source, sandy soils have yielded such poor results that their reclamation is to be given bottom priority. Fourthly, new crop rotations have to be worked out capable of covering at least the cost of reclamation. Fifthly, yields have so far been found to be significantly lower on newly added land. Finally, and of substantial bearing, there are the results of the Food and Agriculture Organization/United Nations Special Fund soils investigation.[5] They show a poor availability of good soil generally.

In summary, Egypt has about 7.0 million acres of excellent land for irrigated agriculture if properly drained and managed with the best technology. Until one has a clearer idea of the costs involved in rendering additional soils suitable for cultivation and has watched the pace of additions for a few years after completion of the High Dam and until water resources of yet unproven yield are better defined, it seems prudent to us to stay substantially below the 7.8 million acres that government planners believe can be put under the plow by the mid-1980's. Instead, increased production must—and can—come from better management and multiple cropping on the suggested, smaller acreage.

Soils of Israel

In the main all of Israel's arable land is now in use: 600,000 acres under rainfed and 400,000 acres under irrigated farming. A recent government study estimates that by 1971 418,000 acres or 4.6 percent more than is now irrigated would be under irrigation.[6] Much of Israel's irrigated land is well-watered during the winter by natural rainfall. It is during the dry summer months that the irrigation water is required for trees (mostly citrus) and summer crops.

From Rehovot north the annual rainfall is 500 mm. (20 in.) or more. The soils from hard limestones are classed as Terra Rossa and those from soft limestones or marls are mostly Rendzinas. The Terra Rossas, as implied by the name, are red while the Rendzinas are black or gray. Both are calcareous, have high percentages of clay, good water-holding capacities, and are good for agriculture. But neither is particularly easy to cultivate, and the Rendzina especially is very sticky and cracks deeply when dried during the period without rain. Most of the Terra Rossa-Rendzina complex soil area in Israel is hilly or mountainous, with

[2] Mohamed G. Abou El Dahab, *Horizontal Expansion in UAR Agricultural Land Resources* (Seminar on Perspective Planning, Warsaw, May 20–25, 1968, organized jointly by Central School of Planning and Statistics, Warsaw, and the Institute of National Planning, Cairo.)

[3] It is shown, for example, as "utilizable land" in a tabulation forming part of Item 1 of the *Agenda* of the Land and Water Use Seminar for the Near East, Beirut, Lebanon, September 25–30, 1967.

[4] *Étude Mensuelle sur l'Économie et les Finances des Pays Arabes, XII, No. 2 (February 1969)*.

[5] *High Dam Soil Survey* (United Arab Republic, General Report, FAO/SF: 16/UAR [Rome: FAO, undated—after 1966]). See Chapter 4, n. 6.

[6] *Israel Economic Development* (Jerusalem: Israel Economic Planning Authority 1968).

much bare rock and many denuded areas or areas where soils are very thin. Only in positions of accumulation are the soils deep enough to have the water storage capacity needed for carrying crops well into the summer after rainfall has ceased for the season. Many of the steeper areas have been rock-terraced for production of olives. Other areas of thin soils or areas too small to use otherwise produce forage that, at present, is not adequately grazed.

Reddish Brown soils are found in the Plain of Esdraelon and the Vale of Jezreel in the northern part of Israel. These are on weakly dissected terraces. The parent materials of the soils are a mixture of material of basaltic and limestone origin in the hills to the east. These are among the most intensively cultivated soils in Israel. In the past they have been used primarily for winter grain. Now, however, they are mostly irrigated and are used for intensive cultivation during the summer. Rainfall is more than 500 mm. (20 in.).

Further south where the Coastal Plain is rather wide is a belt of Noncalcic Brown soils that are sandy. As their name implies they do not generally contain free calcium carbonate but they are near neutral in reaction. There are some hills but slopes are not generally too steep for cultivation. They developed on sandy coastal plain materials that had been cemented by calcium carbonate. The carbonate has now been weathered and leached from the soil profiles. This area is intensively used for production of citrus and vegetables under irrigation. Winter type rainfall is 500 mm. (20 in.) or more.

To the north of Lake Tiberias is an area of poorly drained Alluvial soils including an area of about 12 thousand acres of peat that has been drained and reclaimed for agricultural use. Rainfall is about as high as is found in Israel—up to 750 mm. (30 in.) or more. As elsewhere in this country all falls in the winter season.

To the south of the Terra Rossa and Noncalcic Brown soil areas is an area of windblown—loessial—materials several feet thick that have a lesser rainfall—120-300 mm. (4—12 in.). These soils, classified as Sierozems, have a steppe vegetation. The area is gently rolling with some deeply cut stream channels. But as a whole the area which is from 500 to 1000 feet above sea level is for the most part smooth. The soils are calcareous but generally nonsaline. To the south the windblown materials are sandier and to the east they thin out against the rising hills.

If water were plentiful some of these loessial areas not now producing could be productive under irrigation.

The deserts in the south of Israel are of little or no agricultural value. Some are rocky, some are windswept with gravelly surfaces. Still others are sandy. They furnish very little grazing. Most of the forage grazed by animals is in the higher rainfall areas. The nonarable land of Israel has the capacity—if properly used—to produce about one million sheep-months of grazing.

Soils of Jordan

Jordan has a total area of about 24 million acres, of which the land area is 22.3 million acres; but only about 2.7 million can be used for crops.[7] About 75,000 acres are now used for irrigated crops, but only about 85 percent of this is fully used each year. Perhaps as much as 150—160,000 acres could be irrigated if water were more plentiful. Also Jordan has areas between the deserts and the rainfed farming areas that are fairly good for grazing. These total perhaps 1.5 million acres. Thus Jordan has some 18 million acres of wasteland and desert that at best furnish some scanty rough grazing for animals belonging to the Nomads.

The soils of the higher rainfall areas of the West Bank of Jordan are the same as in corresponding topographic and rainfall areas of Israel. They are a complex of Terra Rossa and Rendzina. The uses of the soils and the crops grown are similar to those in Israel, but there is more room for improvement in agricultural management practices. These are good soils, where of sufficient depth. Some of them produce reasonable crops of tomatoes and tobacco when planted after the last rain in the spring or early summer. A handicap is the lateness of arrival of warm weather in the spring, particularly at the higher elevations. The olives are not generally well cared for, with resultant low production.

Toward the Jordan Valley, the Terra Rossa-Rendzina complex gives way to steep rocky Lithosols that have some grazing value but are of limited value for crop production. In some of the valleys soils are deep and there is some water for irrigation. These areas are cultivated and the available water used.

Soils of the Jordan Valley are mostly on two levels of terraces above the present flood plain. The parent materials of most of these soils is marl. Many of the soils are salty and require considerable leaching prior to agricultural production. The valley is very dry and hot, being several hundred feet below sea level.

Calcareous Alluvial soils are found in the north portion of the valley south of Lake Tiberias. Both north of and south of the Dead Sea the soils are Solonchak—salty soils. The development of irrigation agriculture on the East Bank of the Jordan has gone forward rapidly during the last few years. Plans have been made for bringing irrigation water from the East Bank to the west side but so far this has not been accomplished. About one-third of the irrigable land of the Jordan Valley is on the West Bank.

The main agricultural area of East Jordan centers in the high plateau country to the north and south of Amman. Although the rainfall at the higher elevations reaches 600 mm. (24 in.) or a bit more, the area with rainfall above 500 mm. (20 in.) is quite small. We classify the soils of the plateau as Reddish Brown rather than as Terra Rossa, as in the corresponding areas of the West Bank. Most of these soils are not quite as red as the typical Terra Rossa. However, in the highest rainfall areas of the East Bank of Jordan, some areas of the true Terra Rossa are found. These areas are not distinguishable with certainty on our soil map. They would perhaps be separated and shown on a more detailed one.

---

[7] See Chap. 4, p. 32, for differences in estimates of cultivated areas. Figure given here is at high end of range of estimates.

The distinction is stressed because, where deep, the Reddish Brown soils are excellent crop land. In most of the East Jordan area annual rainfall is enough for reasonable crops and in years of high rainfall grain crops—mostly wheat—are good by standards of the area. However, as pointed out repeatedly in this study, the high degree of variability of yields from year to year is a severe handicap; and although, as described below, there are ways of making better use of the rainfall that does come, there will always be more variability in year-to-year grain production on these soils than is desirable.

These Reddish Brown soils are nearly all highly calcareous with a general high degree of base saturation. As with the Terra Rossa of the West Bank and Israel, they are low in soil organic matter and hence deficient in nitrogen for sustained high production of all food crops. In a few places a layer of calcium carbonate has been formed within the soil profile. This shows that the calcium carbonate has dissolved in the percolating rainwater and then reprecipitated at a lower depth. These carbonated layers—called *croute calcaire* by the French—can interfere with root penetration, especially with tree crops.

Some of the Reddish Brown soils are derived mainly from basaltic materials. However, they too are calcareous from carbonates formed in place from weathering products of the rocks and carbon dioxide from the air. Some areas in northern Jordan are so covered with basalt boulders that they cannot be cultivated. They furnish pasture but may be difficult to transverse.

Between the Reddish Brown soils of the higher rainfall areas and the drier deserts to the east and south are soils of the Sierozem kind. These are characterized by a steppe vegetation of grasses and shrubs, whereas the Reddish Brown soil areas originally had oak and pine forests, particularly in the highest rainfall sites. An area of the Sierozem soils is also found on the west side of the Reddish Brown soils at the edge of the Jordan Valley. This area is narrow and rather steep and rough. On our soil maps it is shown as Lithosol (shallow soil).

The most extensive area of Jordan is to be found in the deserts. There are red sandy deserts, gravelly windswept deserts, and rocky deserts. All have in common such a low level of precipitation that little vegetation is produced. In these deserts there are some depressions where there is some crop production from groundwater. They may be of importance as occupied points in the desert, but are not a factor of consequence to the agriculture as a whole.

The grazing potential of Jordan's nonarable lands is approximately one million sheep months if well managed. At present the land is badly overstocked and overgrazed.

Soils of Lebanon

The soils of Lebanon have many characteristics in common with those of Israel and Jordan. Physiographically the coastal plain of Lebanon is a much narrower continuation of the one in Israel. The hill country and mountains of Israel and the West Bank of Jordan continue northward into Lebanon as the Lebanon Mountains. There, however,

the elevation is much greater. In the south the Lebanon range has some elevation near 2000 meters (6500 feet) and in the north some heights exceed 2500 meters (8000 feet). Snow is on these mountain heights during much of the year.

To the east of the Lebanon range of mountains is the Anti-Lebanon range, which is a continuation of the highland of East Jordan. Between these two mountain ranges is the Bekaa Valley, a part of the same great rift as the Jordan Valley and the Wadi Araba in Israel and Jordan. The southern portion of this valley is drained by the Litani River, which flows south then cuts through the Lebanon Mountains to the Mediterranean Sea. The northern portion is drained by the Orontes River, which flows north into Syria before cutting through the mountains to the sea.

The total area of Lebanon is 2.6 million acres, of which about 740,000 acres are cultivable.[8] In this country the use of very steep mountain land by means of rock terraces is very highly developed.

Relative to most of the areas of the Middle East, Lebanon is very well watered, though here, too, the rain comes during the winter season. The summers are dry. Along the coast the annual precipitation is about 900 mm. (35 in.), while in the mountains near the coast it is much higher. In much of the area of the Lebanon Mountains it exceeds 1000 mm. (40 in.).

The soils of Western Lebanon are derived primarily from limestones, mostly hard but some soft. Thus, as in Israel and Jordan, the soils are primarily Terra Rossa-Rendzina complexes at the lower elevations. At intermediate elevations the soils from hard limestones are not so red. Temperatures are not so high, precipitation is greater, and we find Brown Forest soils rather than Terra Rossa. These soils, also, have a high-base status and are nearly, if not always, calcareous throughout. At the highest elevation, even at 3000 meters (10,000 feet), the soils are not acid in reaction, have low carbon to nitrogen ratios, and a general high-base status. This is to say that these soils are young and relatively fertile even at the highest elevations. Terra Rossa, Rendzina, and Brown Forest soil areas are found in the rough mountain land that is primarily rock surface.

Soils of the Bekaa Valley are primarily of Alluvial or Colluvial origin. The material from which they were formed was either deposited by streams or was moved in mass down the slopes of the hills and mountains. These soils are calcareous throughout, and some have a relatively high content of organic matter. Thus they are naturally fertile soils. Some of the colluvial soils have lesser amounts of organic matter, and hence of native nitrogen. As with most of the limestone-derived soils in Lebanon, textures are from medium to heavy. Loams and heavier soils predominate. Rainfall is much lower in the Bekaa Valley than in the mountains to the west and is somewhat lower than in the Anti-Lebanon Mountains to the east of it.

In the Anti-Lebanon, in Lebanon, and on the western slopes of Mt. Hermon, the soils are similar to the soils on the western side of the valley. Precipitation levels, however,

---

[8] See Chap. 4, p. 33, for difficulties in ascertaining area of cultivable land; also Chap. 15, p. 132.

are less. Most of the soils are classified as Terra Rossa-Rendzina complexes.

The nonarable land of Lebanon has the capacity to produce about two million sheep mounths of grazing under good management of the pasture lands. As elsewhere in the area, the pasture lands of Lebanon are overstocked. The variability of production from year to year is perhaps less than in the other countries because of the greater amount of precipitation.

Soils of Syria

Almost one-half of Syria's 46 million acres of land surface consists of mountain rockland and deserts that give very little grazing. The rainfed cropland is usually cited as being between 15 and 16 million acres, but much of it is so deficient in rainfall as to barely qualify as such and a figure of 10 million acres would seem more realistic (see p. 000). Irrigated land and potentially irrigable land that is not now in use amount to 1.3 million acres, although some of this is not very good for irrigation. Thus, one-third or one-fourth of the country's area, depending on definition, can be cultivated. Some 12 million acres of land—or more if one adopts a restrictive definition of cultivable land—are useful for grazing animals. Much of this grazing land is in the Sierozem soil belt between the desert and the rainfed cropland. Too, it is intermediate in rainfall: too little moisture and too much uncertainty for reliable crop yields exists, and yet too much rainfall for it to be a desert.

The westernmost portion of Syria, next to the Mediterranean Coast, is a northward extension of the very narrow coastal plain and high mountain belt of Lebanon. From Lebanon to the north, the mountains decrease in height and precipitation falls off. However, at Tartus on the coast the annual rainfall is about 1000 mm. (39 in.). In this area, the soils are mostly Terra Rossa-Rendzina complexes. There are some areas of Terra Rossa soils that are developed on colluvial outwash from the mountains. At the higher elevations some Brown Forest soils are shown. These are intermingled with Terra Rossa and Rendzina soils.

To the east of the mountains is a lowland that is composed of an extension of the Bekaa Valley and the Ghab. The Orontes River flows through these depressions north across the Turkish border and then cuts west to the coast by the ancient city of Antioch in Turkey. The Ghab had been a marshy area due to a basalt sill across the Orontes River near the Turkish border. The sill, however, has recently been cut and the marsh areas drained to make more irrigable land. The Ghab depression was then filled with deep alluvium.

To the east of the Bekaa Valley and the Ghab depression is a belt of Terra-Rossa-Rendzina soils that extends south to the vicinity of Mt. Hermon. From there on south, Reddish Brown soils—primarily from basalt—are found. Precipitation of about 500 mm. (20 in.) prevails in much of the area, except that it occurs more on the higher elevations of Mt. Hermon. The soils are Reddish Brown, similar to those in northeastern Jordan. A great deal of the area is covered by relatively unweathered basalt flows. On these, there is no cultivation, but on some of the flows there is some vegetation for sheep grazing.

The eastern slopes of the Anti-Lebanon Mountains have Reddish Brown soils, except where the soils are classed as rough mountain areas or as Lithosols. Although the mountainous areas are mostly rough and steep, there are areas of land that are level enough for cultivation; and some areas are cultivated even though they are not level enough to permit this. All of the areas not cultivated are badly overgrazed.

As rainfall and elevation drop to the east the soils become Sierozems. One example is in the Damascus basin that has been under irrigation for thousands of years. Here the soils are much darker than in the original state. They have accumulated considerable organic matter from plants grown under irrigation and from the application of organic manures. This is a very productive basin, watered by the small Barada River and by the springs that derive their water from the mountains to the west.

Along the northern border of Syria the precipitation decreases from about 900 mm. (35 in.) at the Mediterranean Sea coast to 330 mm. (13 in.) at Jerablus. East from there rainfall increases to about 700 mm. (28 in.) at the northeastern tip of the country. The soils along this border are characteristically Reddish Brown. The parent materials are limestones in the west, unconsolidated materials at the foot of the mountains in the middle, and basaltic materials in the east. This is Syria's excellent dry-farm area, a gently rolling plains country. Primarily, wheat is grown in alternate years with a weed-fallow between the grain crop.

To the south, rainfall decreases rather rapidly. Sierozem soils are to be found on the more level areas, forming an extensive belt of soils on the border between areas of cultivation and noncultivation. Hassakeh is near the boundary between the Reddish Brown soils and the Sierozems in the Jezireh. It has average annual precipitation of about 300 mm. (12 in.). South of here some barley is grown, with very erratic year-to-year yields. W. J. van Liere has given a good discussion of the soils and agriculture of the Jezira (the area between the Tigris and Euphrates Rivers).[9] Grazing is perhaps the best present use of these soils.

South of the Sierozems are the Desert soils. As in other countries of the region the deserts furnish little grazing. Also as in other countries there are a number of kinds of deserts, but all of these are dry and formidable. Average rainfall is perhaps about 150 mm. (6 in.).

The one-fourth of Syria that is potentially good grazing land has the capability of producing about 18 million sheep months of grazing. However, at present the stocking rate is far too high for maximum production.

Soils of Iraq

The total area of Iraq is about 110 million acres. Of this at most 8 million acres is suitable for rainfed crop production. An equal area, approximately, is mostly too dry for

[9] W. J. van Liere, *Report to the Government of Syria on the Classification and Rational Utilization of Soils,* (Report No. 2075 [Rome: FAO, 1965]).

rainfed crop production but is suitable for irrigated crop production. In addition to these, about 23 million acres, 21 percent, are suitable for some permanent grazing. The greater portion of the irrigable land comprises the flood plain of the Tigris-Euphrates River system. This has a desert climate with about 150 mm. (6 in.) average annual precipitation. The remaining 70 million acres is mostly desert with an average annual precipitation of about 150 mm. (6 in.). A portion, however, consists of rockland and high mountain areas that do not produce usable grazing.

The Alluvial soils of the Mesopotamian Plain in Iraq are primarily a product of man's activity during the last six thousand years. Most of this plain has been covered to a depth of several feet with sediments brought in suspension by irrigation water. Thus the soils are not pedologically developed in the usual sense. Beneath these sediments, which are the result of man's activity, are great thicknesses of deposits not thus caused that are similar in composition but usually somewhat coarser. It has been said that these deposits are as much as three kilometers thick.

These Alluvial soils have a consistently high (usually 20–30 percent) content of calcium carbonate and a small percentage of gypsum. They are reasonably permeable both laterally and vertically. Textures are generally fine silty, fine loamy, or fine clayey. Near the rivers they are coarser than this.

The major impediment to the use of these basically good crop soils of the Mesopotamian Plain is salt that has accumulated in them during their use over the last six thousand years. Most of the soils have been occupied and abandoned more than once because of the salt. Although the water of the Tigris and Euphrates Rivers are relatively low in soluble salts, the repeated wetting and drying out of the soils many times has left a salt content that reduces or prohibits plant growth in a majority of the area. A systematic, planned removal of salts from irrigated land is a necessary part of any successful irrigation scheme. In the case of Iraq none has been provided, though the problem has been well identified. Consequently much of the land, especially south of Baghdad, is so salty that it can be used for cropping only in alternate years. Even then yields are very low. With proper leaching, drainage, and good management, these soils could be among the most productive in the world. Without these improvements they can only get saltier and further decline in their already low level of productivity.

The rainfed agricultural area of Iraq is in the northeast. In the high mountains on the border with Iran there are small areas of perpetual snow and ice. Annual precipitation is as high as 1300 mm. (50 in.) in this area of parallel mountain ranges. The valleys are used for crop production, both for rainfed winter grain and for irrigated summer crops. Where not too steep and rugged the lower slopes are used for grazing. The middle slopes have been and some are still forested. The forest is mostly open and used for grazing. The upper slopes and alpine meadows furnish grazing for sheep in the summer season.

In the mountain area the soils are primarily Terra Rossa-Rendzina complexes and Terra Rossa-Brown Forest soils vertically zoned. Above these the land is mapped as Rough Mountainous land of Terra Rossa-Brown Forest vertical soil zones. In the drier situations in these mountains there are some occurrences of soils that are similar to Chernozems and Chestnuts, as pointed out by Buringh.[10]

The foothills area on the southwest side of the mountain area is rather rugged but does have good rainfall during the winter. Grazing is the principal use of the land. The soils are Terra Rossa or Brown Forest.

The Kirkuk-Erbil-Mosul Area is the most important dry-farm area in Iraq. It has annual precipitation of about 400 mm. (16 in.). This was one of the early areas of grain production in the world. The soils are Reddish Brown. They are calcareous with a zone of carbonate accumulation at depths of from 12 to 20 inches. These excellent rainfed soils are not salty and could be made to produce several times their present output.

Southward from the Reddish Brown soil area is a zone of Sierozem soils. The rainfall is less here and it is more uncertain from year to year. Too, these soils frequently have a very high content of gypsum as well as of calcium carbonate. Many are shallow. This area, like the corresponding area in Syria, is very marginal for cultivated crops such as wheat and barley. It can be extensively farmed but the risk of failure is great. Well-managed grazing would be a better present use than cultivation.

The remaining portion of Iraq is primarily desert. Some vegetation is produced in the wadi bottoms, but in many locations drinking water for both animal and man is not plentiful. Little or nothing can be done with these desert areas other than their present scant use.

The grazing land of Iraq is estimated to be capable of producing about 19 million sheep months of grazing. As in the rest of our area the grazing land is badly overstocked, with resultant reduction of output of meat, milk, and fiber.

---

[10] P. Buringh, *Soils and Soil Conditions in Iraq* (Baghdad: Iraq Ministry of Agriculture, 1960).

CHAPTER 3

Water Resources[1]

THE water resources of the Middle East are considered here in two broad main categories: (1) that water which falls on the region in the form of rain or snow, a minor part of which later shows up in the flow of streams and springs, and (2) that water carried into the region by major rivers which arise outside of the region or which originate along its extreme northern border. In addition, there may be significant flows of groundwater into the region, as for instance in the Nubian sandstones of Egypt's western desert, or in the vast stretches of the Arabian peninsula. This supply of groundwater is dealt with briefly at the end of the chapter. As for desalted seawater, this cannot currently—and as we point out in Chapter 13 cannot for some time into the future—be considered as a significant source of water for agriculture and it thus is not included in this chapter.

The seven countries of Egypt, Israel, Jordan, Lebanon, Syria, Iraq, and Saudi Arabia contain almost one billion acres of land, or slightly more than half of the total area of the forty-eight contiguous states of the United States. More than half of this immense area lies in Saudi Arabia; there, the annual rainfall is very low, averaging possibly no more than 75 mm. (3 in.) and perhaps less. A fourth of the whole region lies in Egypt, where the rainfall is still lower, averaging probably no more than 25 mm. (one in.) for the whole country. Thus, for the Middle East, as here defined, total annual precipitation may average no more than 10 cm. (4 in.). Some areas do indeed have a great deal more, reaching more than one m. (40 in.) annually in some mountain areas, but the acreage of such comparatively high rainfall areas is very small compared with the great desert areas. In addition to the Saudi Arabian and Egyptian deserts, there are extensive desert areas in Jordan, Israel, Syria, and Iraq.

But even as little as 4 inches average precipitation over such a very large area amounts to roughly 350 million acre-feet total water supply annually falling on the land of the region. A great deal of this water, however, is lost as far as any productive agricultural use is concerned. As we noted in Chapter 1, evaporation is high; it may exceed 6 to 7 feet from a free water surface annually. The rain that does fall in desert areas tends to come in rather heavy downpours; some penetrates the soil where it falls, much runs off, with some penetrating sandy washes or wadis, and the rest quickly evaporating from the surface. The vegetation in such regions is scanty, with little or no grazing value; the wonder is not that the vegetation is so scanty, but that there is any at all. Because of their sheer extent,

the deserts do receive a large total water supply annually, but this water could be collected and used for agriculture only at prohibitive costs. There are a few oases in these deserts, with great value both historically and at present, but the outlook for economic development of agriculture based upon water in the desert areas is very dim indeed.

In smaller areas—but still comprising substantial total acreages—the water which falls does permit an economic agricultural output, even though lack of water seriously limits the magnitude of output. Large acreages provide enough natural plant growth to support a grazing industry; in other areas, nonirrigated crop farming is possible; and in still others, irrigated agriculture can be based on streams or springs or wells that can be developed to take advantage of groundwater fed from infiltration of rain. Some of the natural water supply is inevitably "lost" in the sense that it does not lead to economic agricultural output. Some of the water which falls in agriculturally productive areas comes as flood flows which cannot by used, often cannot be stored economically, and which evaporate rather quickly from the areas where they spread; other stream flows are available in seasons when there is no need for irrigation; evaporation constantly reduces all water supplies; and for other reasons, some of the total water supply cannot be used productively even where water is scarce.

There is something on the general order of 145 million acre-feet measurable average annual water supply in rivers, streams, springs, and sustained groundwater yield in the Middle East, as we define it. Flows in clearly defined streams, which are the more obvious part of this supply and are generally the easiest part to measure, are dominated by inflows to the region (Figure 3-1).

At least half of the water supply from all sources is accounted for by the Nile River. All of its water arises outside the region, primarily in the highlands of Ethiopia, where the Blue Nile has its origin. The White Nile, which arises much farther south in the heart of East Africa, is potentially a much larger river than the Blue Nile, but most of its total water supply evaporates or seeps into the ground in the vast swamps of the Sudd of southern Sudan. There is no addition to the water flow of the Nile after it enters Egypt; on the contrary, evaporation takes its toll. Of the Tigris River, some 40 percent of the total water comes from Turkey and the rest predominantly from the high mountains along the northern border of Iraq and Iran. Of the Euphrates River, nearly all the water arises in Turkey. Thus, over 80 percent of the measurable water supply in rivers, streams, springs, and sustained groundwater yield flows into the region from outside. In this respect, the deserts of the Middle East are similar to those of some other parts of the world, as northern Chile or the American Southwest, where rivers *cross* but do not arise; but differ

---

[1] This chapter is based on Appendix B and the references cited therein. Text references to water management and future utilization are omitted for the sake of an uninterrupted narrative, and the reader is referred to Chaps. 4 and 13 for the pertinent discussion.

from other deserts, as Sahara and Mongolia, where such crossing rivers are absent.

The Nile

The Nile is a huge river; no one who has seen its relatively uncontrolled flood flows in the past can fail to be impressed by this broad, swift brown current of water, sweeping mostly to the sea. Nearly one hundred years of measurement put the average annual flow of the Nile at

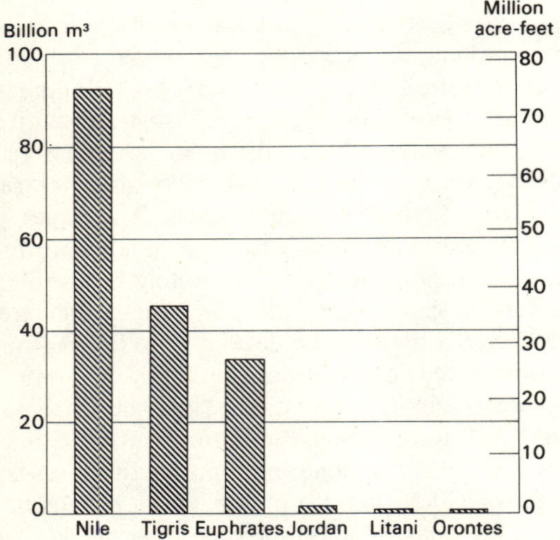

Fig. 3-1. Average Annual Water Supply, Major Middle Eastern Sources (Excluding Groundwater)

Source: Appendix B—Nile, Table B-1, average 1961-65, at Aswan; Tigris, Table B-2; Euphrates at Hit, Iraq, Table B-10. For Jordan, Litani at Khardale, and Orontes, see last section of Appendix B. No years given for average, Tigris, Euphrates, and Orontes; for Litani, average 1953-1967.

Aswan at nearly 75 million acre-feet; it thus compares closely with the Columbia River at Grand Coulee Dam or with the Ohio River at Cincinnati, or with the Mississippi River just after it receives the Missouri. The natural flow has varied considerably from year to year in the past (Figure 3-2). While the natural river will presumably continue to vary in flow from year to year in the future, the High Aswan Dam will create a large reservoir that will permit some evening out of supply over a period of years.

For millennia, the Nile has flooded each year. While every flood produces some damages, and considerable damage if flooding is great, the floods have been the foundation of the kind of agriculture which developed, and have provided great benefits. The system of agriculture for seven thousand years or more was based upon annual flooding; with the completion and proper operation of the High Aswan Dam, flooding will become a thing of the past, and a greatly different resource management program must accordingly be developed and implemented.

There is a strong seasonal pattern to the flows of the Nile, similar in years of high and of low flow (Figure 3-3). Low flows are fairly similar from year to year, with monthly volumes in the general order of 1.7 to 3.5 million acre-feet from January through June and sometimes considerably later. Flood flows begin in the summer, reaching a peak in September, when total volume for the month has varied from as little as 15 million acre-feet or less in low flow years to 23 million acre-feet or more in high-flow years. Most of the variation in total annual flow is attributable to differences in height and duration of the flood periods. From the September peak, the flow gradually diminishes to the winter low. Although there are not great differences in average monthly flows of the natural river through the low-flow season, in some years in the past the natural flow was so reduced by early summer that irrigation was virtually impossible.

In the floods of the past, the river flowed out of its

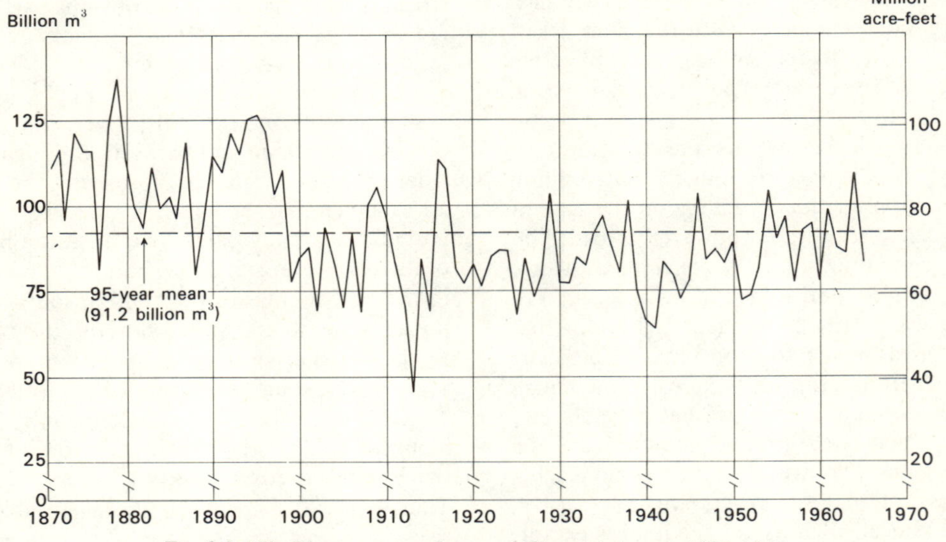

Fig. 3-2. Nile River: measured Annual Flows at Aswan, 1871-1965

Source: Appendix Table B-1

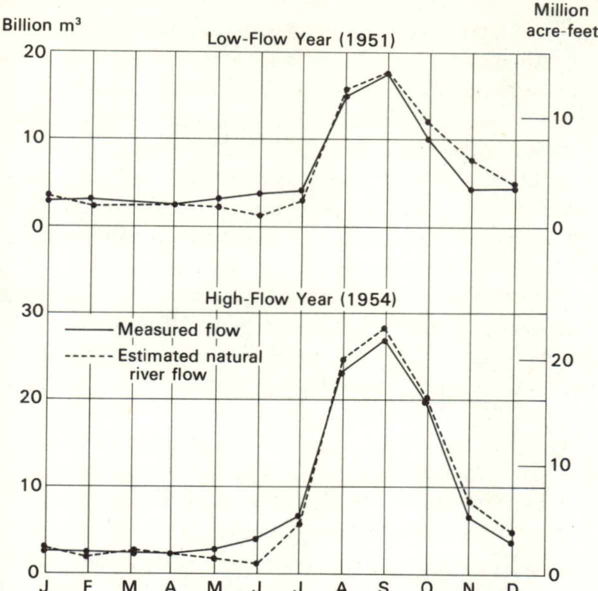

Fig. 3-3. Nile River: Typical Seasonal Patterns of Low- and High-Flow Years at Aswan, Prior to Construction of High Dam

Source: Appendix Table B-1

banks and wetted the soil of the valley to a considerable depth; as the flood diminished, the water receded and gradually drained off the land, or was evaporated from it, thus permitting the planting of crops. Over the past one hundred years, this natural regimen of the river has been gradually modified by man-made vegetation and in the future will be modified greatly by the operations of the High Aswan Dam. The flood waters have always carried a considerable load of sediment, about 2,500 parts per million; most of this sediment has been deposited in the valley, usually close to the banks of the river. These in turn have been built up over the centuries, so that the natural drainage of the valley is not into but away from the river; in this respect, the Nile is similar to many other streams throughout the world. But it does make more difficult the task of agricultural drainage (see Chapters 4 and 13 below).

The fertility value of the sediments deposited annually by the Nile has often been stressed in literature dealing with agriculture in the Nile Valley. The additions to fertility have no doubt been important, but modern irrigation management would stress as vastly more important the flushing action of these floods, which prevented an accumulation of salts harmful to plant growth. This effect becomes very evident when one considers the Mesopotamian Plain, where floods have been neither so regular nor so effective in their flushing action; salt accumulation there has thus reached almost unbelievable proportions and is the greatest single obstacle to greater agricultural output. One thinks of a desert region as one short of water, and in one sense this is true of the vast Egyptian desert; it is less generally realized that deserts may have no soil—not only no soil in the sense of a developed soil profile with all its biological life and organic matter,

but no material suitable for the development of such a soil. Deserts may need far more than water to become productive. In that context—i.e., in relation to acreage of good land upon which to put it—the supply of water furnished by the Nile River appears ample.

Tigris-Euphrates

The Tigris and the Euphrates are separate rivers until near their mouth, where they are joined; but they are similar in many respects, and a considerable part of the Mesopotamian Plain can be irrigated from either, with a direct water interchange possible through the Tharthar and Abu Dhibbis depressions and Habbaniyok Lake. The Euphrates is the smaller of the two; it arises in Turkey, with little net addition of water after leaving that country; it flows through Syria, into Iraq; some of its water might be used in each of the three countries, hence development possiblilities must carefully consider multiple claims.[2] The Tigris also arises in Turkey, but only about 40 percent of the total water supply of this river system comes from that country; development possibilities on this river in Turkey are very limited, and the river only barely touches Syria; a number of rather substantial tributaries enter it in Iraq, all of which arise in the high mountains of northern Iraq and in Iran. The Tigris is thus much more an Iraqi river than is the Euphrates.

The total flow of the Tigris, including its tributaries, averages about 37 million acre-feet and thus compares generally with the Missouri River at Kansas City, or with the Snake River where it joins the Columbia, or with the Ohio River near Parkersburg (after the Muskingum has joined it). The total flow of the Euphrates is smaller, as mentioned, averaging about 28 million acre-feet, thus comparing generally with the Missouri River at the point where it leaves Nebraska, or with the Arkansas River at Little Rock, or with the Susquehanna River, or twice the Colorado at the point where it enters Arizona.

The flow of the main stem and of the chief tributaries of these rivers varies greatly from year to year (Figure 3-4). A considerable similarity in pattern of year-to-year variation is evident from one stream to another; all reflect differences in precipitation, especially in winter snowfall, which in turn arise from year-to-year variations in the major general storms that sweep into the area. Each stream also varies considerably in its seasonal flow, in not too different patterns (Figure 3-5). The low flows come at the same season and at approximately the same volume; the Tigris peaks normally in April, the Euphrates normally in May at a much lower level. Each river, but especially the Tigris, produces destructive floods; much more is known about the floods of the Tigris, and its damages, especially to Baghdad, have been great. However, the greatest floods experienced in modern times have involved a volume of water only about a third of what hydrologists estimate as a

[2] In terms of area, about 45% of the total basin of 440,000 Km² (170,000 sq. miles) lies in Iraq, 40% in Turkey, and 15% in Syria. (Ahmad Soussa, *The Floods of Baghdad in History*, [Baghdad: Al Adib Press, 1965] .)

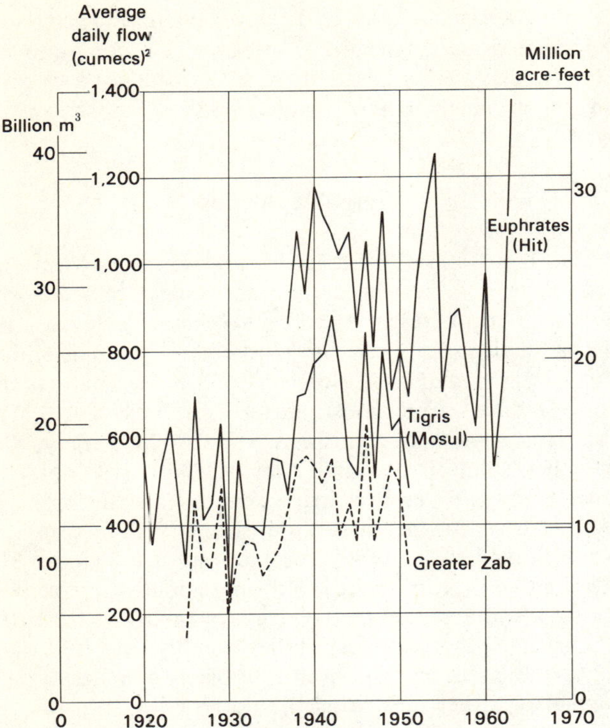

Fig. 3-4. Annual Flow of Euphrates River at Hit, of Upper Tigris River at Mosul, and of Greater Zab River at Eski Kelek, Years of Record, 1920-63[1]

[1] Tigris and Greater Zab: 1920-52 only

[2] Cubic meters/second

Source: Appendix Tables B-2, B-4, B-10

possible flood some day. While the economics of water development depend upon many factors, including the efficiency with which the water is used, the general potential of water development projects for reducing flood hazard, of capturing seasonal peak flows for later seasonal use in irrigation, and for evening out supply from year to year, should be evident from Figures 3-3 and 3-4.

Smaller Rivers and Groundwater

In comparison with the Nile and Tigris-Euphrates systems, all other surface sources of water in the area are very small. The Jordan River has attracted a great deal of attention, in part because of its intimate association with Biblical history, and in recent decades because it has been the subject of bitter international controversy; for it and its tributaries are divided among Israel, Jordan, and Syria.[3] Development and planning of the Jordan River system are fairly complete. The Jordan is a rather small stream with a total supply of about 1¼ million acre-feet for the whole river system, an estimate about which professional judg-

[3] Two reports which review this political controversy are: (1) Kathryn B. Doherty, *Jordan Waters Conflict* (New York: Carnegie Endowment for International Peace, May 1965); and (2) Georgiana G. Stevens, *Jordan River Partition* (Hoover Institution Studies [Stanford, Calif.: Stanford University, 1965]).

ment differs some, but which is a good enough approximation for purposes of this analysis. As such, it compares with small American rivers known to few outside the United States and not to many outside the state in which they flow: the Allagash River in Maine, the Yampa River in Colorado, or the Muskegon River in Michigan.

The Litani, located wholly in Lebanon, and the Orontes, located in Syria and Lebanon, are still smaller, each about a third of the volume of the Jordan system. In addition, there are springs in numerous parts of the region, as at Jericho; near Damascus; as the source of the Yarkon River in Israel; at Ras-el-ain in Syria; the source of the Kabour River; and at many other locations, which have varying amounts of water—all are very small by comparison with the major river.

There are also usable supplies of groundwater in parts of the region, notably on the flanks of the hills and mountains at the eastern end of the Mediterranean, in Israel, Lebanon, Jordan, and northern Syria particularly, where some of the rainfall infiltrates to lower permeable strata from which it can be pumped. These groundwater supplies are permanent, at least in some volume, in the sense that a natural recharge takes place; if withdrawals are limited to recharge—and such limitation may require regulation or some other form of enforcement—water levels and water quality can continue unimpaired.

In addition, large volumes of groundwater are known or suspected to exist in some other areas, such as the Nubian sandstones of the western Egyptian desert and some parts of Saudi Arabia. Much hope commonly rides on these supplies, and imagination tends to elevate them to a position that is not justified on the basis of present knowledge. The recent discovery of groundwater in the Libyan desert that was made in the course of exploratory activity for oil is only the most recent instance. Little is known of the origin of such supplies and the existence and role of recharge. Whether the supplies could support

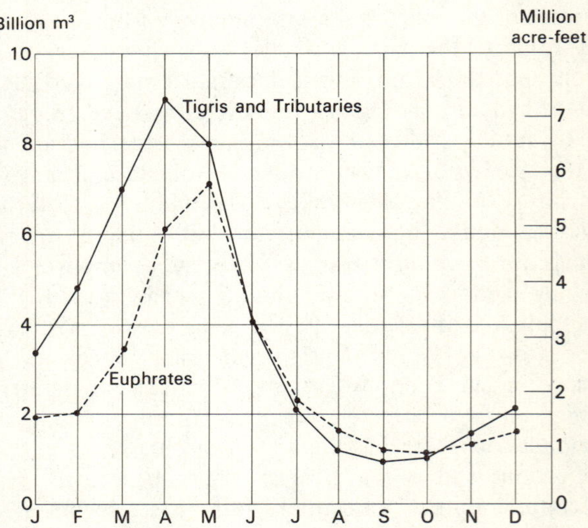

Fig. 3-5. Average Monthly Flow of Tigris and Euphrates Rivers. (For points of measurement and other details, see discussion of the Tigris River in Appendix B.)

sustained agricultural activity of the magnitude to constitute a significant addition to the region's or country's existing plant is a question to which no answers have as yet been found. Perhaps these areas, or some of them, are similar to areas in the United States—of which the High Plains of Texas is the largest and best-known example— where water is being mined from stored supplies to which almost no replacement occurs; and the end of irrigation from this source is clearly in sight. Perhaps they do receive a sufficient recharge to put exhaustion into the far future, though current knowledge does not favor this hypothesis. Frequently too these supplies are of poor quality, because of salinity, and expensive to secure. In any event, sizable funds are being spent, especially in Saudi Arabia, to investigate these supplies and understand their characteristics, so that in time we are likely to receive some answers. On the whole, one would be ill-advised to count these sources as firm long-term supplies and to build large investments for sustained agricultural production around them.

As for groundwater in the Nile Valley, it must be noted that its utilization might at first, by lowering the water table, represent a one-time net addition. But over the long run this would not represent a net additional source of water, though possibly a cheaper one. However, in that event drainage would be a must.

In thus emphasizing the very small volume of the smaller rivers and groundwater sources, one should not underestimate their past importance in the history, especially in the agricultural history, of this region. They have been extremely important in local irrigated agriculture, and around some of them, as at Jericho, there have existed agricultural settlements for thousands of years. These miscellaneous sources of water are still highly important. Their possibilities for greater development are limited, however, since most are already fully used.

PART II

The Present Organization of Agriculture in the Middle East

PART II is a description and analysis of agriculture in the region today, and of how it has developed to its present state. It has been written in the conviction that if one wants to know where he is going, and more particularly if he is to choose where he wants to go, he must understand where he now is and how he got there. It consists of ten chapters, each on a more or less clearly defined aspect of present agriculture, but all interrelated.

Chapter 4 starts by describing the land and water use programs and practices which have evolved, to use the natural resources described in Part I. Chapter 5 then considers the inputs into agriculture, such as machinery, fertilizers, and other chemicals, some of which are produced in the region but much of which are imported. Chapter 6 considers labor as an input into agriculture: how many workers, and their characteristics. Chapter 7 describes the farms, as entrepreneurial or managerial units; these are closely related to labor supply, for on many farms the chief labor source is the farmer and his family, but the focus of the two chapters is different. Then come two somewhat detailed chapters, one on crops, the other on livestock; these, in particular, are based upon quite extensive statistical tables which are included in the Appendix. Chapter 10 translates the previous information, which was largely in such physical terms as acres and numbers of livestock, into value concepts or monetary units, especially gross output, input costs, and value added, and compares these on a per unit of land and per worker basis. Chapter 11 considers present markets for the output of agriculture in the region and discusses the status of marketing capability and organizations. Chapter 12 briefly considers some aspects of rural living, how do the farmers and other rural people live, as a result of their agricultural productive efforts, as well as the infrastructure or the social, institutional, political, and economic framework within which agriculture as a whole and farmers as individuals operate.

The treatment in Part II is thus primarily topical or subject-matter oriented; under each subject, available information is presented for each of the countries in the Middle East. More specifically, tables and charts include columns for each country, wherein information on some aspect of agriculture is presented. The order of countries is uniform in all such tables and charts; beginning with Egypt, each country is taken up in turn, moving clockwise around the region. In a considerable number of cases, information of a comparable kind is also presented for the United States, on the assumption that it would sharpen the understanding of U. S. readers to learn how the magnitudes of relationships under discussion compare with the situation in the United States.

The wide variety of measures of land area, volumes or weight of products, and of other aspects of agriculture that are in common use in the Middle East are a source of confusion and possibly of error. An effort has been made to reduce such measures to internationally commonly accepted units, and to provide conversion factors between the units used and others used in the region or in the United States.

CHAPTER 4

Land and Water Use and Management

IN this chapter we discuss the use that is now made of the land and water in the Middle East.[1] Because this use is conditioned by internal and external factors peculiar to each country, we will consider the countries individually rather than as a region.

A general caution, however, must precede this discussion. Land use or classification statistics are not as meaningful anywhere in the world as one would like them to be, principally because physical and economic criteria are usually mixed up in the assignment to one or the other use category. In the drier parts of the Middle East, however, there is the additional difficulty that land use varies from year to year, according to the highly variable rainfall. Thus one could get land use figures that are true with maximum rainfall, with average rainfall, or under extremely dry conditions. In Table 4-1 (and Figure 4-1 which is based on it) we reproduce the most consistent set of figures we have encountered, namely those presented in FAO's Indicative World Plan. In addition, comments will be found in the land-use discussion for each of the countries, where there is a specific problem or set of problems.

Egypt

Largely because it has a good supply of high quality water for irrigation, Egypt makes better and more complete use of the 3 percent of its land that is arable than any other country of the Middle East. Of the total annual flow of the Nile about 16 percent represents the steady contribution of the White Nile, which has its origins in the lake country of Uganda.[2] The remaining 84 percent comes from the Blue Nile and the Atbara, both of which originate in the highlands of Ethiopia and come mainly as floods between early August and the end of October. This period during which most of the Nile waters are passing toward the Mediterranean Sea does not coincide with the period of water demand by either the summer crops such as cotton, maize, or rice or the winter crops such as wheat, barley, and the fodder crop berseem clover.

About six thousand years ago the people of Egypt began to harness and utilize these violent flood waters by diverting a portion into large basins to a depth of one to two meters. These floodwaters were allowed to remain on the land inside the basin for from two to three months. During this time two things of importance happened. The heavy burden of "silt" in the muddy floodwaters settled on the surface of the soils. At the same time the soil was soaked to great depth by the water. When the level of the river came

down sufficiently in October, the surplus waters of the basins were drained back into the river channel and the soil surface allowed to dry enough for tillage.

A winter crop was then planted on the water-soaked soil. Thus Egypt became one of the most famous granaries of the world because the fertile Alluvial soil produced good crops of grain, especially wheat. The other main winter crop was berseem clover. This valuable forage crop furnished food for the cattle that pulled the plows and furnished the milk for the farm family. As a legume it also left a residue of nitrogen needed by the succeeding wheat crop.

What moisture the wheat required during winter and springtime could be met by the water that soaked into the deep soil during the time it was flooded. By harvest time most of the water in the soil had been lost by transpiration of the plants and by evaporation from the soil surface. No further crops could be planted until the next time of flooding.

One Additional benefit of this system of basin irrigation, perhaps not foreseen in the beginning, was—as noted previously—the removal from the land of soluble salts that are harmful to plants, by the return to the river of the excess of the surface water, as well as that drained through the soil profiles. Thus the soils of the Nile remained productive. The only nutrient element that otherwise would have been in short supply for production of the wheat at prevailing yields—nitrogen—was supplied in large part by the atmospheric nitrogen "fixed" by microorganisms associated with the roots of the clover. To raise summer crops such as maize, rice, and cotton, water could be lifted by hand for small areas near the river. But the bulk of the land could not be watered without there being water in the canals during the spring and summer months.

A barrage to lift the level of water in the canals feeding the Nile Delta was first begun in 1843 and first used in 1863. It failed under the first load because of inadequate design and poor construction.[3] This first barrage was repaired, and it and its successor built in the 1930's have been in use each year since. (A barrage is essentially a dam across a waterway without any storage capacity. Its only purpose is to raise the level of the water so that, as in this case, it would flow into the canals.)

With the advent of the barrage the canals could have water in them all of the time except during a midwinter period set aside for cleaning and repairing canals and drainage ditches. Thus Egypt embarked successfully on the production of cotton, corn, and some rice. Because these summer crops grow during a season of high temperature and have a rather long growing time they require much water.

---

[1] The chapter lays the groundwork for the consideration of land and water development potential taken up in Chap. 13.

[2] H. E. Hurst, *The Nile* (London: Constable, 1952).

[3] W. Laurence Balls, *Egypt of the Egyptians* (New York: Charles Scribner's Sons, 1916).

Table 4-1

Land and Land Use in the Middle East, ca. 1962

(million acres)

Total[a] land area	Total arable	Arable					
		Rainfed			Irrigated		
		Cropped[b]	Fallow	Total	Cropped[b]	Fallow	Total
Egypt 247.1	6.4[c]	–	–	–	6.4	–	6.4
Israel 5.0	1.0	0.6	d	0.6	0.4	d	0.4
Jordan 22.3	2.8	1.3	1.3	2.6	0.1	.04	0.14
Lebanon 2.6	0.7	0.4	0.1	0.5	0.13	.02	0.15
Syria 45.8	16.4	6.9	8.3	15.2	0.9	0.4	1.2
Iraq 110.9	16.7	3.5	4.3	7.9	4.6	4.2	8.8
Total 433.7	44.0	12.7	14.0	26.8	12.4	4.6	17.0

[a]Land area only for Israel and Jordan; total area for others.

[b]Adjusted to exclude multiple cropping.

[c] See pp. 13 to 14 for soil resource discussion. 1968 area about 6.7–6.8.

[d] Negligible

Source: Indicative World Plan, Near East, (Rome FAO, 1966), II.
Note: Parts do not always add to totals due to rounding.

When water is available by gravity and not charged according to amount used, overwatering is sure to occur. The Delta of Egypt was no exception. Raising summer crops meant getting two crops a year from some of the land and in some cases even three crops. But overwatering and the lack of adequate drainage for most of the Delta brought about a condition of high water table and saltiness that has materially reduced crop yields.[4]

More recently, barrages have been put on the Nile upstream from the Delta, and there too water tables and salty soils are now a problem, though not to the degree existing in the Delta.

Not all land is reached through perennial irrigation. In 1950, about one million acres of land in the Nile Valley were still under basin irrigation. However, completion of the Aswan High Dam will eliminate floods, and hence basin flooding will disappear. A conversion program, now almost complete, is bringing water to all of this land on a year-round basis (except of course for the necessary repair and cleaning time). Small pumps are now being used to lift water from the canals to some of the farms.

In order to conserve water and to reduce the hazard of high water tables, the Egyptian government has sought to reduce the overuse of water by the farmer. Since financial charges are ruled out by local ethics, the control is exercised through the size of the field outlet and the possible head on it, by fixing the total flow into the laterals (based on the area served and official irrigation standards); and it is embodied in the nature of the water delivery ditches.

[4] For a good presentation of data, see W. Laurence Balls, *The Yields of a Crop* (London: E. & F. N. Spon Limited, 1953).

Originally, irrigation canals were designed for the water to be about ten inches above ground elevation; but this was conducive to overirrigation. In recently designed canals—such as those on areas converted from basin to perennial irrigation and on newly reclaimed areas—the operating water level is designed to be some twenty inches below the land surface at the point of receipt by the farmer. In addition, partly by design of the irrigation authorities and partly because given laterals have gradually had to serve greater areas, the water level in older laterals too has been lowered, forcing the user to employ a lift device—screw, water wheel, or pump.

There are no official statistics as to which area is supplied by gravity and which area requires lifting. Indeed, the same area may be served in both ways at different times, depending upon the season and the fullness of the canals. The best but informal estimate we have come across is that about one-third of the land in the Valley is served through lifting. Until now manpower appears to have been the prevailing mode, with pumps coming gradually into wider use. The burden that this labor puts on one who is often not too well fed and subject to debilitating health conditions keeps him from lifting excess quantities, or so the theory goes.

For a crop requiring 3 acre-feet of water the farmer must lift about 4,000 tons of water for each acre cultivated. Though the lift of 20 inches is not so great as some lifts in the past have been, it still absorbs an inordinate amount of labor. As pointed out in Chapter 12, it is much to be hoped that a method less reliant on the factor of human fatigue and more efficient and flexible could soon replace this way

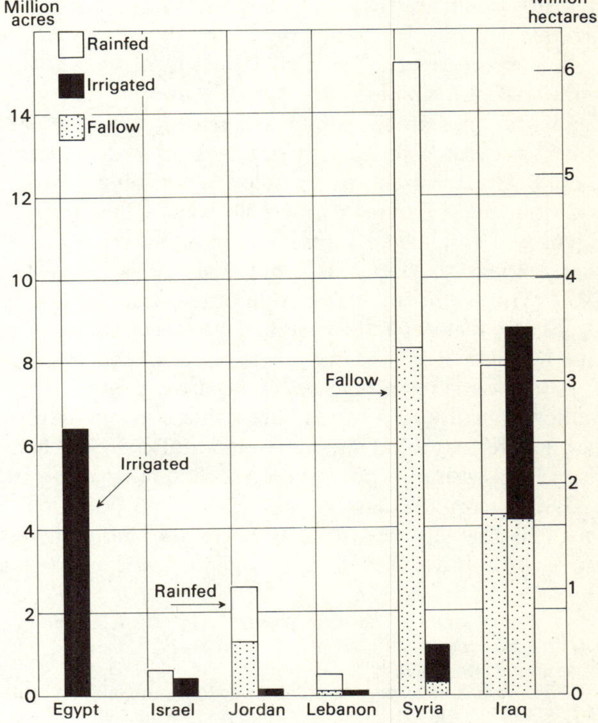

Figure 4-1. Total Land and Land Use in the Middle East, ca. 1962

Source: Table 4-1

of rationing water. This might take the form of a charge for water used.

Land-Water Relationships—The Past

The presently cultivated area of Egypt is between 6.6 and 6.7 million acres, up some 300–400,000 acres from the early 1950's. At that time about 1,010,000 acres were still under basin irrigation, and of these, 625,000 acres depended on flooding only, with only one crop a year. The balance produced an additional grain crop from pumped groundwater. Up to 1967, 728,000 acres have been converted to year-round irrigation, with completion of the conversion scheduled for 1969. It is estimated that the change from basin to perennial irrigation does not materially change the yearly water demand.

According to Kamal and Hashem, agriculture's average gross annual requirement for water, measured at the point where the Nile passes Aswan, is 7,500 cubic meters per feddan or 5.9 acre-feet per acre.[5] This includes an allowance for 727,000 acres of rice. On the 6,525,000 cultivated acres reported for 1959/60 this comes to just over 38 million acre-feet of water, to which must be added 1.2 million for nonirrigation use and flows for navigation of 2.0 million, for a grand total of 41.4 million acre-feet. The available flow at Aswan, prior to the High Dam, was set at 39 million acre-feet, with salvaged and ground flow enough to come up to 40.5 million acre-feet. Thus under prevailing management practices the total available Nile flow was fully committed to the 6.5 million acres under cultivation in 1959/60. Indeed, the agreement is close enough to suggest that the water need was calculated from the amount being diverted.

It is expeced that the High Dam will make available an additional 6.1 million acre-feet of water for irrigation in Egypt, an increase of about 16 percent over past availability. According to current standards of water duties this should irrigate about one million acres of new land. In addition, by the terms of the Nile Waters Treaty of 1959, Egypt will be allowed to use 1.2 million acre-feet of water that belongs to the Sudan, but will not be needed until 1977. This would allow irrigation of another 200,000 acres, at 6.0 acre-feet per acre. The question is, can this much new land be found?

A recent survey suggests a negative reply.[6] Of 14.5 million uncultivated acres that the survey covered in varying degrees of detail, it found only 88,300 acres of Class I soil with high potential for a wide range of crops and 193,800 acres of Class II soil with good potential. This included fine-textured soils, heavy clays, and some loamy sands with heavy subsoils, as well as some coarse sandy

loams. The Class III soil mapped comprised 569,900 acres. However, these Class III soils are not good and cannot be expected to give returns on inputs comparable to those from the soils of Egypt now under cultivation. Many of the Class III soils are shallow to rock; others are covered with windblown sand; some have high water tables without being irrigated; some are coarse-textured, gravelly, and loamy sands that will not use irrigation water efficiently. All of them lie at elevations above the river that range from 30 to 400 feet.

The total area of soils of Classes I, II, and III is about 850,000 acres. This is some 340,000 acres short of the amount of land needed for the 7.3 million acre-feet of water that appears to be available, at least until 1977, without making allowance for the fact that some of the Class III land was very charitably placed. When one considers the apparent overirrigation under present practices and the presently assigned requirements, he reaches the conclusion that Egypt has more water available than she has good land to put it on. "The total area of additional land in Classes I, II, and III, accessible to irrigation from the Nile, is considerably less than the area irrigable by the amount of water expected to be available when the high Dam and Reservoir come into operation."[7] Given the storage capacity of the new High Dam the problem is thus not a shortage of water; rather, it is a shortage of good land on which to live and to grow crops.

Government expenditures between 1960/61 and 1966/67 by the Ministry of Irrigation indicate the priorities in development.[8] Nearly half of the funds were expended for the conversion from basin to perennial irrigation; 17 percent went to reclamation; 13 percent to general and 8 percent to covered tile drainage. Five percent or less went to such tasks as new irrigation works, pump improvement or replacement, development of supplementary water resources, etc. Practically no funds went to research in 1965 and 1966, the only two years for which research is broken out as a separate category of expenditure. These shares, incidentally, apply to total funding of just under $300 million for the seven years, or between $40 and $45 million per year.

As for trends, the most pronounced has been a rise in the reclamation category. In view of the large availability of water from the High Dam completion this is an understandable development.

Water Use and Practices

It has been said that the Egyptian farmer invariably takes all of the water he can get. This happens wherever there is no strong incentive to save on water, however; and the Egyptian farmer is no different from irrigators elsewhere.

Table 4-2 gives estimates from a number of sources of the irrigation (water) requirements for the principal crops.

---

[5] A. Kamal and A. Hashem, *Control and Distribution of Nile Water,* Ministry of Irrigation textbook in Arabic, (Cairo: Government Press, 1967).

[6] *High Dam Soil Survey* (United Arab Republic, General Report, FAO/SF: 16/UAR [Rome: FAO, undated—after 1966]). See also Chap. 2, p. 00. The survey, requested by the Government, was begun in late 1959 and completed with the convening of a panel of international consultants, who endorsed its findings in 1964.

[7] *Ibid.,* p. 10.

[8] Data derived from *Irrigation Developments During 15 Years 1952–67* (Cairo: Directorate of Research and Planning, Ministry of Irrigation, 1967).

Table 4-2

Comparative Irrigation Requirements, Egypt

(m³ per Feddan*)

Crop & Season	1920 (McDonald) Basis		1948–59 Experiments			Estimates by various surveys and persons							
						Cooper[a]		Agriculture[b]		Hurst[c]		General Dev. Div.[d]	
	North	South	North	Center	South	North	South	North	South	North	South	North	South
Winter													
Wheat	1,000	1,400	1,000	1,230	1,690	1,035	1,370	1,140	1,510	1,080	1,430	1,124	1,650
Barley	960	1,460	1,050	1,290	1,780	935	1,430	1,030	1,575	980	1,300	1,124	1,710
Peanuts	960	1,240	790	970	1,340	935	1,221	1,030	1,340	980	1,270	1,124	1,460
Berseem	2,450	3,320	2,220	2,730	3,760	2,390	3,232	2,630	3,560	2,500	3,380	2,870	3,880
Cover crop	1,210	1,210	1,374	1,374	1,374	1,180	1,180	1,300	1,300	1,270	1,270	1,418	1,420
Summer													
Cotton	3,380	4,320	3,250	3,780	5,270	3,400	4,250	3,740	4,640	3,550	4,400	4,080	5,060
Rice	15,000	—	7,750	8,800	—	44,620	—	16,100	—	15,300	—	17,565	—
Corn	2,800	3,100	2,500	2,920	4,100	2,730	3,042	3,000	3,340	2,850	3,170	3,273	3,640
Cane	5,600	7,120	10,170	11,600	17,000	5,460	6,955	6,000	7,660	5,700	7,270	6,540	8,350
Flood													
Rice	9,300	7,600	—	8,200	—	9,100	7,449	10,000	8,200	9,500	7,800	10,910	8,950
Corn	2,480	2,330	2,285	2,720	3,750	2,430	2,279	2,670	2,510	2,540	2,380	2,913	2,740

* 1000 m³ per feddan equals .781 ac. ft. per acre.
[a] Based on survey of Abu Menja pumping area.
[b] Based on Ministry of Agriculture recommendations.
[c] 1900–14 study.
[d] 1946 studies.

Source: Compiled from material published in A. Kamal and A. Hashem, *Control and Distribution of Nile Waters,* Ministry of Irrigation textbook in Arabic (Cairo: Government Press, 1967).

(These are presented in the original versions of thousands of cubic meters per feddan, but may be readily transformed to acre-feet per acre by the formula shown in the table.) From these data giving the range of estimates that various people have made, it is obvious that the water needs of the various crops are not well established. The gross allowance of 6.0 acre-feet/acre of cultivated land cited above (p.00) and the 3.6 acre-feet/acre of cropped land (i.e., counting each acre as many times as it is cropped) obviously present only a very rough reflection of these estimates.

Water is made available to the farmer according to a schedule related to the needs of the crops. The usual summer rotation has been an allowed watering of the land once every eighteen days for crops other than rice, beginning in mid-April. Rice may be watered four out of each eight days. In years of low water availability the canals may have a day or two of general closure in addition to the scheduled periods without water. To take water out of turn during the summer period is a contravention.

In the past, when the flood waters came during August, the schedule was five days of high water, five days of low water, and five days of closure. This lasted until mid-December, when a flush irrigation was given before the winter closure period. The closed period, which helps in lowering the water table and gives an opportunity to make necessary repairs to canals and drainage facilities, lasted from the end of December until early in February. Spring delivery schedules paralleled the ones practiced in the fall.

Land leveling of small areas of soil surrounded by raised borders has been practiced in the Nile Valley and the Delta for a long time. Both dry and wet leveling has been considered necessary to get an even depth of water on the individual area being irrigated. There are no indicators to tell one how precisely these small areas have been made level. Allowing for the limitations of the wooden beam or other instrument pulled by two draft animals, the job has probably been well done. However, what is really needed is precise grading to a constant slope (in contrast to leveling) on larger areas of land so that furrow rather than flood irrigation can be used. This could result in considerable economies of water use and facilitate mechanization, when and where required for other purposes.

Drainage Problems

Most, if not all, observers agree that much of Egypt's good agricultural land is in dire need of better drainage. The high water table prevailing during the growing season greatly reduces crop yields. Increased yields to be expected from installation of field drains have been estimated at from 38 to 75 percent for corn, 40 to 50 percent for cotton, and 20 to 32 percent for wheat.

The need for drainage has undoubtedly existed since the beginning of irrigation agriculture. But for reasons that are implicit in the descriptions given in Chapter 3, the problem became acute when perennial irrigation began to replace flood irrigation in the Delta. It is estimated that the length of open drains (excluding field drains) discharging freely, or by pumping, into the sea, the lakes, or irrigation canals, is presently about 7,500 miles. These provide some amount of drainage for more than one-half of the cultivated area. However, the field area that had been tile drained by the end of the 1964/65 season was only 258,000 acres.

In recognition of the seriousness of the problem a thirty-year plan costing 200 million Egyptian pounds ($460 million at current exchange rates) was drawn up in 1965 for tile draining the whole country. However, by the end of June 1968 only an additional 145,000 acres had been completed instead of the 235,000 scheduled for the first three years. The thirty-year plan has now been reduced to twenty years to accomplish the same objectives. This will require a substantial stepping up of annual expenditures, previously set at about $14 million for the seven years 1965–71. Another serious difficulty lies in the organization responsible for the tile drainage and leaching not having enough engineers—by several fold—to oversee the work on the scale contemplated.

The economics for drainage installation are relatively simple. It is estimated that the replacement of open by closed drains for field drainage and collection, will free about 12 percent of the land drained. Cost of field tile drainage runs about 35 pounds per feddan ($78 per acre), while land costs on the average 500 pounds per feddan ($1100 per acre). The value of the land saved is then almost twice as great as the cost of drainage. Indeed, at this cost relationship the value would be greater in any situation in which drainage saves just over 7 percent, as an acre tiled would then yield more than the $78 of drainage investment.

The above calculation does not, however, include costs of maintenance, and it must be emphasized that no drainage system is of value unless properly maintained. To summarize the first brief discussion of drainage, it would appear that completion and maintenance of the tile drainage system is an absolute necessity if Egypt is to maintain, let alone to improve, the present level of agricultural production.

Crop Rotations

For a long time Egypt had a rotation of wheat and berseem clover. The clover furnished not only forage for the cow or buffalo but also the nitrogen for the soil to make a good crop of wheat. When cotton became a more widely planted crop, there was not enough nitrogen for cotton and wheat, and manufactured fertilizers came into widespread use. Today Egypt is a heavy user of nitrogen fertilizers even

though berseem is still in rotation in nearly all of Egypt. In the meantime, maize, rice, and sugar cane have also become important crops.

Field crops, including beans, lentils, and onions, dominate in the agriculture of Egypt. In 1952 they occupied 98 percent of the cropped area, with only 1 percent each in vegetables and fruits. During the 1960's the relative share of both vegetables and fruits doubled but still left 96 percent in field crops. The main field crops were wheat, maize, millet, barley, rice, cotton, sugar cane, and berseem clover.

Traditionally three seasons for planting have been, and still are, emphasized: winter, summer, and Nili rounds. The winter round starts in October and November with the planting of wheat, berseem, barley, beans, lentils, flax, chick-peas, onions, and vegetables. The summer round starts in February after the shutdown period of the canals and ends in May or June. During this time cotton, rice, millet, sugar cane, sesame, and vegetables are planted. The Nili round has decreased in importance during recent years. It is a late summer round starting in July when maize, millet, rice, and vegetables are grown. At this time the Nile River starts to rise, thus making water available without shortchanging the summer crops already growing. In a year in which the floods started late, the Nili round crops could be delayed until more water was available.

With the completion of the High Dam the original reason for the Nili round crops—i.e., the additional water from Nile river floods—will have largely ceased to exist. The Nili round crops are now down from 21 percent of the cropped area in 1952 to 12 percent. A rethinking of the best possible succession of plantings of the various crops is now in order. For example, rice and to some extent maize have been planted by permission when it was seen that there would be enough water for them. The maize crop could profit by being planted earlier in the year. One of the causes why current yields of maize in Egypt are not good may be late planting.

Berseem clover occupies more acreage than any other crop, though some of this is short-term in winter before cotton. On land that otherwise cannot be utilized during the winter season it seems like an ideal crop. It yields well and Egypt is short on fodder, but in much of Egypt berseem clover has occupied about one-third of the land each year. In an area that is so short on food, alternatives to the large acreages devoted to berseem should be considered.

Grazing Lands

The natural pasture land of Egypt contributes but little to the agricultural production. There are some grazing lands along the coast west of Alexandria. But forage production is sporadic and the control of grazing by the nomadic herds is at best exceedingly difficult. It is unlikely that substantial gains in animal production in this area can be achieved under present conditions.

In the desert areas of Egypt, particularly east of the Nile River, there is some grazing by nomadic herds. But these areas, too, contribute little to the agricultural production.

Israel

In 1967 slightly more than one-fifth of the land area of Israel was cultivated—one million out of a total area of five million acres. Of this, 400,000 acres were irrigated. Essentially all of Israel's potentially arable land is now under cultivation. Perhaps we should qualify this statement somewhat by saying "with present water supplies" (i.e., without taking into account additional water from seawater desalting or weather modification). There is some additional land south of the areas now cultivated that could support irrigated agriculture if more water were available. However, in view of the fact that Israel may even now be somewhat overdrawn on her groundwater supplies and that more water will be required by domestic and industrial needs in the future, one may legitimately judge that Israel's agricultural area is not subject to much expansion in the near future, which of course, is not to say that her agricultural production cannot be increased.

The soils of Israel are for the most part salt free or have less salt in the rooting zone than would be harmful to crops. Salt can become a problem if water with too high a content of soluble salts is used. At times in the past there have been threats of salt water intrusion along the coastal strip, but gradually regulation of pumping of groundwater has taken hold and the problem been brought under control. In general, the fact that over half of Israel's water supply is centrally managed by a country-wide company and the bulk of it is transported in pipes makes it more readily subject to efficient utilization. The soils of Israel are well managed at the present time.

Some small areas have a salt problem, one of them in the Jordan Valley. But the areas are not large, and the problems are local only.

Israel's arable land is well farmed, and per acre production is very high for many crops. It is to a considerable extent the diversity of the crops grown that leads to a high value of output per acre. As against 96 percent of the cropped area in field crop in Egypt, for example, and only 4 percent in fruits and vegetables, Israel has 34 percent in other than field crops (though differences in definition may exaggerate the contrast somewhat). Of this area not in field crops, two-thirds is in orchards and almost one-fourth in vegetables and potatoes. These products yield much higher financial returns per acre than most field crops.

The largest single crop acreage in Israel is wheat, which constitutes about 18 percent of the total cropped area. Another 11 percent is in barley. Altogether grains occupy 32 percent of the planted acreage, though the percentage fluctuates a good deal from year to year. One may question whether an intensively cultivated and highly industrialized country which has limited land and water resources as does Israel should try for self-sufficiency in crops that do not generally bring large returns per acre at world prices (e.g., food and feed grains and sugar). However, grains are largely produced in the drier areas in which, given the pattern of world prices, it is difficult to find better alternatives. Where grains are grown under irrigation, they may form an essential element in the rotation. Moreover, such decisions are not usually based wholly, or even largely, on considerations of economic efficiency, but allow for political contingencies.

Most Israeli farms raise both irrigated and rainfed crops. Indeed, the same land may carry a rainfed winter crop and after its harvest an irrigated summer crop. It is our impression that the irrigated crops are managed with a higher degree of skill and with more nearly optimum inputs than are the rainfed crops.

However, there is no sharp line of distinction. The winter rains are taken into account in irrigated areas, although they do not suffice to raise most of the crops. Citrus, for example, is mostly rainfed in the winter but requires some irrigation throughout the year. The growing season for sugar beets extends over nine months, thus requiring irrigation part of the time. In many of the larger settlements optimization in the use of rainfed and irrigated areas is sought through such techniques as linear programming, with prices, technology, and sometimes even particular social factors considered in the program.

Several types of farming units have evolved in Israel and changes are still taking place. During the earlier days of the increase in Jewish settlement, a type of mixed farming was developed that gave maximum stability and security to the evolving settlement.[9] But this self-sufficient type of farm could not contribute substantially to greater exports or to a reduction of necessary imports for the country. Subsequently, specialized types of farming were developed that were more suited to the available soils and climatic conditions, that could produce for export, or that were in proximity to outlets for raw products, such as centers of population for dairy products. In addition, raw materials and processing facilities were set up together—sugar factories and sugar beets, cotton gins and cotton, etc. Four basic types of such specialized farms will be briefly mentioned.

The Dairy Farm

These farms characteristically have about seven acres of land and an allocation of about two acre feet of irrigation water per acre each crop year. Part of the land, about one acre, is typically in citrus, another two acres in grain or industrial crops for sale, and the balance in fodder or pasture crops for the five or more cows generally maintained on these farms. Some of the land produces more than one crop in a year. Among the fodder crops are alfalfa, sown grasses, clover, corn, sorghum, and fodder or stock beets. Both summer and winter green feed is produced. The tendency for these farms is to use more and more of the total land area for producing fodder crops for the cows and to supplement the roughage with purchased concentrates.

---

[9] Raanan Weitz and Avshalom Rokach, *Agricultural Development, Planning and Implementation* (Dordrecht, Holland: D. Reidel Publishing Co., 1968; also New York: Frederick A. Praeger).

The Citrus Farm

A second type of farm that has been well established produces citrus and field crops. These units embrace typically seven acres, of which about 2.5 acres are in citrus, and the balance, other than the house site, in field crops. The farm unit may also have some livestock, such as chickens. According to Weitz and Rokach, some 1500 of these units have been established in the Central and Southern Coastal Plain. The water allocations are about the same as for the dairy farm.

Thought is being given to modification of this farm unit by increasing the area in citrus to five acres. The balance of the area would be set aside for field crops, primarily peanuts and vegetables for market. The livestock would consist of 200 hens.

The Field Crop Farm

Most of the recent settlements in the southern part of the cultivated portion of Israel have been of the field crop kind. A typical allotment is 9.4 acres of land and 15.5 acre-feet of water in the Negev. The crops include between three-quarters and one acre of fruit or wine grapes in the Negev and citrus on the Coast. The other crops are vegetables, sugar beets, cotton, peanuts, or other industrial crops. Livestock would commonly consist of some 100 chickens, 7 sheep, and 2 goats.

Present plans contemplate an increase in farm unit size to 13.5 acres of cultivated area with a water allotment of 15.5 acre-feet of water. The livestock component would be reduced to 200 chickens. The only other livestock in an area where there is some grazing land would be some sheep for meat and wool. This anticipates a rigid economy with respect to water use.

The Export-Type Farm

The fourth basic type of farm unit in existence and being further developed is one for producing crops for export. These include mainly fresh vegetables, potatoes, melons, and perhaps table grapes. Quality of produce must be high and the transport from farm to consumer swift. The western Negev is an area where this type of farm is especially relevant.

The size of farm is about 13 acres and the water allotment 21 acre-feet per year or 1.64 acre-feet per acre. It is planned that 2.5 acres will be in citrus, with the balance of the farm area in peanuts, export tomatoes, export onions, melons, carrots, and peas for canning.

A range of soil texture in the western Negev area from medium to coarse silt to sands, with some soils having sand over the silts, is suitable for the crops being produced. Citrus, peanuts, and winter potatoes do particularly well.

Grazing Lands

At one time the grazing lands of Israel were badly overstocked, especially by goats. The erosion that was primarily caused by denudation of the vegetation by too many animals brought about a strong reaction, but it now looks as if the pendulum has swung too far in the opposite direction. Controlled grazing could not only produce additional meat, for which there is a ready market, but would reduce fire hazard to tree crops. This is true primarily for the more humid mountainous and hilly areas that can produce much good grazing but that are too rough to be farmed. This undergrazing is in contrast to all of the other countries of the area, which suffer from various degrees of overgrazing.

Israel's southern desert areas are not used to any great extent for animal production even by nomadic peoples. Not only is there very scant vegetation to support animals, but there may also be a security problem.

Jordan

Jordan has a land area of about 22.3 million acres. Total agricultural area is uncertain, as the statistics vary greatly from year to year and even between sources for the same year. Agricultural land for 1965 from one source is 2.5 million acres, while another gives 1.75 million acres for the next year.[10] Variations of this degree must result from differences in definition of categories as well as from the natural yearly change in area. This can be quite substantial, as rainfall is highly variable from one year to the next, and acreage expands and contracts with the amount of rain received. Cropped area, for instance, went down from 1.5 million acres in 1965 to 1.2 million acres in 1966.

Total agricultural area includes cropped area, land left fallow, and some uncultivable land. Most recently, cropped land has been figured at just over two-thirds of the total agricultural area. The use of the 1.2 million cropped acres breaks down as follows: field crops—wheat, barley, lentils, etc.-52 percent; vegetable crops—tomatoes, eggplant, melons, cabbage, etc.-5 percent; fruits—grapes, olives, figs, etc.-10 percent. Fallow land accounts for 24 percent of total area (410,000 acres) and uncultivable area, 8 percent (140,000 acres).[11]

The product mix on the irrigated lands of the Jordan Valley—only 5 percent of the area under crops in 1965—is quite different. Of the 75,000 irrigated acres, over half are in vegetables, one-third in field crops, and the rest in fruit. There are plans to irrigate far more area (actually to double the 1965 acreage), but for the present increases will be less ambitious—the seven year program estimates only 13,000 more acres by 1970. At present only one crop per year is grown on most of the irrigated area, with the net result that about 1.25 crops per year are grown on that land as a whole. It has been estimated that this figure could be raised to 1.5 crops per acre per year, meaning that, on the average, a second crop is raised on every other acre. Thus nothing approaching full use is now being made of the cropping potential. Allocations of water to individual farm units at the present indicate considerable wastage. This is likely to

---

[10] For 1965: *Estimate of the Annual Report of the Ministry of Agriculture for the Year 1964–65* (Arabic), pp. 9–10.

For 1966: *Report on the Results of the Agricultural Sample Survey, 1966* (Amman: undated). pp. 36–39.

[11] *Sample Survey, 1966.*

result in some high water table problems of the kind and consequence discussed earlier.

The dry-farm area of Jordan is in the highlands on both banks of the Jordan River. It comprises about 1,230,000 acres under conditions of rainfall that range from more than 24 inches on the West Bank to only 8 to 20 inches on the East Bank. Since it is estimated that Jordan has at most 6 percent of its total land area—or 1,450,000 acres—with rainfall of 12 inches or more on the average, some of the presently farmed land must be considered marginal for cultivated crops.

The crops of the dry-farm area of 1,230,000 acres are mostly wheat and barley. Other crops grown in considerable acreages are olives, lentils, vetch, tomatoes, grapes and melons, and lesser amounts of other fruits and vegetables. Some tobacco is produced in the West Bank area as a dry land crop that is planted after the last spring rain.

Cereal growing suffers from two big problems. First, yields of wheat and barley, grown in the highlands of the East Bank, are among the lowest in the world. Secondly, there are very wide differences in amount and pattern of rainfall from year to year, and areas planted and harvested vary with the rainfall changes. This is true even in areas of fairly adequate average rainfall.

To illustrate, FAO reports cereal production of 105,000 metric tons in 1963, but 403,000 tons in the subsequent year. Such year-to-year changes are difficult to deal and live with, but are amenable to improvement by better dry-farming practices.

In this connection it should be pointed out that in years of better rainfall the cultivated area is pushed further toward the deserts, where the percentage of years in which a profitable crop can be produced is low. Some of this land could be more profitably used as grazing land provided such grazing could be controlled at a level commensurate with the resources. It cannot be overemphasized, however, that historical, ethnic, and social factors make grazing intensity in Arab countries a very difficult problem.

Jordan has about 145,000 acres of olives nearly all in the dry-farmed areas, for the most part on the rougher land. The production from year to year is even more erratic than is true for grain: in recent years it appears to have ranged more than tenfold within three successive crop years. Despite these fluctuations, olives form an important part of the country's agricultural produce. Better management, including a fertility program, could lead to some stabilization of the annual production.

There are marginal grass lands between the cropped areas and the dry deserts, but these are not very extensive in Jordan and offer only modest potential for grazing. The main potential grazing areas are in the higher rainfall portions of both the East Bank and West Bank, where the land is too rough or too steep for cultivation, but is suitable for good forage production. Here, the key to proper utilization is management. Overgrazing has not been controlled nor have pastures been managed for optimum production. While better use of these areas could materially increase Jordan's meat supply, the obstacles to good management, it must be repeated, are great.

Beyond the fringe of the land that is now dry-farmed and the intermediate grass lands lie the deserts both to the east and the south. These constitute more than 80 percent of Jordan's land area. It is estimated that about 50,000 people live in the desert area, many of them nomadically, though there are a few oases.

The land of Jordan is not nearly so well managed nor the water so well used as in Egypt or Israel. From this one might conclude that there is a greater margin for improvement, and in a purely abstract way, there is. But the extreme vagaries of the annual rainfall amount and its distribution pose formidable problems.

Lebanon

Lebanon is a mountainous country with a narrow coastal plain on the Mediterranean Sea and a valley that is a part of the Great Rift and thus a continuation of the Dead Sea and the Jordan Valley of Israel and Jordan. The total area of Lebanon is 2.6 million acres. Precise figures for the amount of agricultural land are lacking; estimates by different sources at different times range from 740,000 to 865,000 acres. An intermediate value of 806,000 acres of agricultural land was given by the Lebanese Ministry of Agriculture (1963) for 1960.[12] Of this, 773,000 acres were actually in use, with annual crops occupying 589,000 acres and permanent crops, 184,000 acres. Irrigated acreage amounted to 166,000 acres, of which 77 percent were in annual and the balance in permanent crops. Again, however, figures are not firm. A more recent estimate indicates that regularly irrigated areas amount to no more than 150,000 acres.

The agricultural land holdings in Lebanon are small and badly fragmented. A statistical sampling of agricultural holdings in Lebanon in 1961 showed that the average size of the holding was less than six acres. Of these small holdings 88 percent were fragmented—that is, the holding consisted of two or more noncontiguous parcels of land. The average number of noncontiguous parcels making up the fragmented holdings was 5.6. In spite of this condition, which greatly reduces the economic viability of farming, one-half of the workers in Lebanon gain their living in the farming sector.

Crops Grown

The most productive agricultural area of Lebanon is along the 100 miles of Mediterranean coast. The coastal plain proper is quite narrow and even vanishes in places where the mountains touch the sea, but there are less steep secondary associated areas at the foot of the mountains.

The soils and climate are favorable for production of a variety of fruit and vegetables. From sea level to 1200 feet elevation, citrus and banana culture is the most important and most profitable use of the suitable irrigated coastal soils. As much as 80 percent of the banana production

---

[12] Cited in *Agriculture in the Lebanese Economy*, survey prepared for *World Atlas of Agriculture*, reviewed by Adel Cortas and Abdus Sattar (Rural Economic Institute, A.Ec. No. 9, February 1968).

comes from interplantings in the younger citrus groves. When citrus trees approach maturity the banana trees are removed. Apples are the next important fruit crop of the area. These are grown up the mountainside at about 3000 feet elevation where climatic conditions are suitable. Efforts to extend the apple plantings to lower elevations have not been so successful. Apples are second to citrus as Lebanon's most valuable agricultural export. Vegetables are the other important crop in the coastal area. These include potatoes, tomatoes, onions, melons, cucumbers, and beans.

Between the citrus growing area near sea level and the apple zone at higher elevation are intermediate levels where figs, almonds, peaches, plums, and apricots are grown; above the apple producing areas there is only forestry or grazing.

As one moves away from the coast, the land is mostly so steep that rock terraces are necessary to hold the soil in place for the tree plantings. These terraces are very laborious to construct, and the individual holdings here are very small. But it is only by means of the terraces that advantage can be taken of the favorable climatic conditions to grow crops that cannot be produced either at higher or at lower elevations.

Olives are the most extensively grown tree crop in Lebanon, and the olive as a fruit and the oil from it are important items in the Lebanese diet. Olives are grown mostly on terraced nonirrigated land. Production varies much between years, due in part to fundamental physiological characteristics of the tree that dispose it toward uneven years of production, a condition also encountered in other countries where olives are grown. A part of the variation, however, may result from management practices that could be improved. There has been a tendency to replace olive plantings by citrus and garden vegetables, and some are also giving way to houses; but olives still account for about one-third of the country's fruit acreage, and for more nearly one-half in North Lebanon.

The Bekaa Valley, located in a rain shadow of the Lebanon Mountains (and hence not getting so much precipitation as do the west slopes of the Lebanon mountain range or the coastal area), is another distinct farming area. Rainfall diminishes rapidly from about 24 inches in the south to 8 or 9 inches in the north, making agriculture in the north marginal without irrigation; however, irrigation is being expanded both from river sources and from wells.

This valley produces a variety of crops both irrigated and rainfed. It accounts for about one-half of the country's total wheat production and two-thirds of its grapes. The irrigated crops are primarily tree crops, grapes, and vegetables, including potatoes and onions. There is a large area of rough grazing land in the north portion of the Bekaa Valley.

Cropping Patterns

The sequence of crops on a given piece of land is rather variable depending on the preference of the farmer and the anticipated demand for agricultural products. In the annual cropping of irrigated land a winter crop, such as wheat,

barley, or one of the local vetches, is nearly always grown. Usually the land is idle in the summer. The next year a summer crop is grown. This may be potatoes, tomatoes, other vegetables, sugar beets, or corn. Then the land goes back to a winter crop. Some farmers plant a crop of late potatoes or dry beans. On irrigated land, summer crops sometimes follow summer crops.

The dryland sequence of crops is, again, wheat or barley during the winter, then a summer fallow followed by a winter crop of lentils or chickpeas. The practice of having a whole year of fallow is being given up by some farmers who use fertilizers, and that practice has been spreading in the last few years.

In a statistical survey conducted in 1961 it was found that only 30 percent of the farmers used any fertilizer at all. And the use that was made was concentrated in the farms of larger area. Thus a vast majority of the small farms got no commercial fertilizers. Both in the dry-farm and irrigated areas a far more profitable use of fertilizers could be made. It is no wonder that yields of some of Lebanon's crops, particularly wheat and barley, have been so low. However, much progress has been made since 1961 and a higher proportion of the farmers now use fertilizers.

An indication of more recent practices is a sample survey of wheat farmers made in 1966,[13] when 61 percent of the growers reported themselves as users of chemical fertilizer and the area of wheat having received such fertilizer was calculated at 73 percent of all land grown to wheat. Estimates made in connection with FAO's Indicative World Plan suggest that the heaviest users of fertilizers in Lebanon are fruit rather than field crops, with annual applications as heavy as 250 kilograms of nitrogen per hectare. But these appear to be not in the nature of statistical findings but rather informed conjectures.

Lebanon has done remarkably well in turning the mountainsides into terraced farm land, on which oranges and apples for export are produced in large quantities. But field crop production could be greatly improved by better management, including purchased inputs such as fertilizers, insecticides, weed-killers, and improved seeds.

Syria

How much of Syria's total land area of 45.8 million acres is suitable for cultivated crops is not subject to close determination. The main reason is that in the tension zone between adequate and inadequate precipitation the boundary between cultivated land and that suitable only for grazing surges back and forth with the cycles of greater and lesser rainfall.

The land-use picture in Figure 4-2 emerges from official data for the average of 1964—66:

As in the figure, most statistical computations by U. S. and international bodies show "arable" land between 15 and 17 million acres, a classification that is presumably

[13] Abdus Sattar, *Survey of Wheat, Barley, Lentils and Chickpeas, 1966* (Ministry of Agriculture, Republic of Lebanon, A.Ec. No. 3, February 1967).

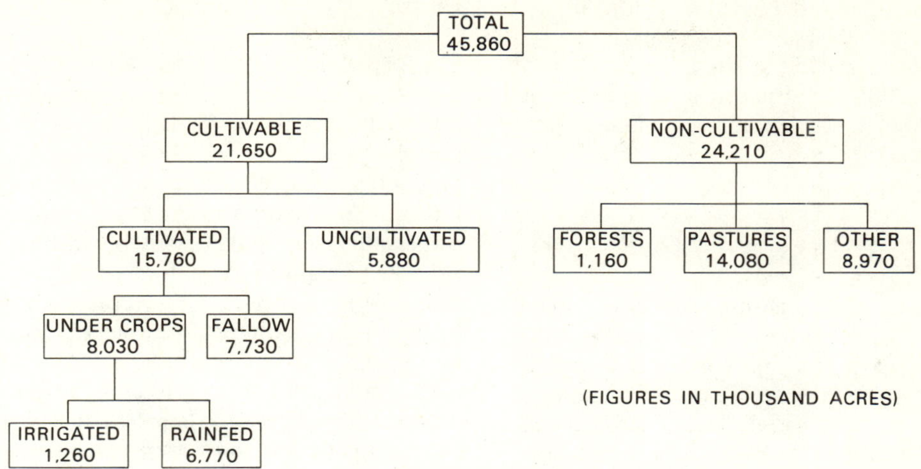

Fig. 4-2. Syrian Land Use, Av. 1964-66

Source: Syrian Arab Republic, Ministry of Planning, *Statistical Abstract 1967* (Damascus), p. 283.

more or less identical with the category termed "cultivated" in the figure and includes large portions of very poor rainfall areas as well as poor soil.

A more useful way of looking at the land is to distinguish roughly between three main areas. First, there is the cultivable area, with an annual precipitation of 24 inches or more, which totals about 4.7 million acres. Another 4.9 million acres have annual rainfall of 16 to 24 inches, and in some years this area is only marginal for rainfed winter crops. Finally, there is the land between the areas now cultivated and the deserts, which could produce crops of winter grain in years of better rainfall. But failures would be frequent and risk great.

In 1967, areas in crops and fallow totaled 15 million acres, of which 8.1 were in crops and 6.9 in fallow. Irrigated land, both cropped and idle, was 1.3 million acres. Of the 15 million acres, as has been shown in the preceding paragraph, 9.6 had precipitation of at least 16 inches. Assuming that the bulk of the irrigated land is in stretches that have less than that amount of rainfall there must, therefore, be about 4 million acres of cropped land and idle crop land in areas that have less than 16 inches mean annual rainfall. It is in these dry areas that, it is generally agreed, extensive winter wheat and barley farming was pushed too far in the 1950's. Statistics suggest, however, that there has been quite a pullback in cropped area: the area in crops in 1958 was 13.8 million acres—5.7 million acres more than in 1967.

Crops

Wheat occupies the largest acreage in Syria, with barley second. Together these two grain crops now (1967) occupy 67 percent of the land, down somewhat from the 73 percent in 1957. Cotton is third with 9 percent, olives fourth with 5 percent and lentils fifth with 3 percent. All dry legumes together occupy 7 percent, vegetables 4 percent, and all industrial crops, mostly cotton, are grown on 10 percent of the land.

Wheat and barley yields are very low: they averaged only 800 pounds per acre in 1967, composed of 13 bushels of wheat and 17 bushels of barley respectively. Yet they are not as low as they have been in other years. In 1955, for example, yields were only one-third of these values. These yields are obtained on rainfed farmland and thus are subject in any one year to all of the vagaries of the erratic rainfall. While all of the variation cannot be eliminated, the range of the swings can be diminished, as will be pointed out in Part III.

Cotton, which began to be an important crop in the 1950's, is produced primarily under irrigation and the yields are quite good, somewhere between those of the United States and those of Egypt. However, the cotton yield is only half that of the high-yielding parts of the United States like California, and of Israel, among countries less remote.

Crop Rotations

Systematic rotations are not rigidly followed in Syria. But a practice of one or more years of fallow is generally observed. A characteristic three-year rotation in the higher rainfall wheat-growing areas is a year of wheat or barley followed by a year of chickpeas or lentils. The third year is fallow—that is, no crop is planted. At this point it should be emphasized that in Syria, as well as in most of the Middle East, a year of fallow means a year of weeds that are grazed by sheep or goats. While this break in the cropping sequence has some value for controlling pathogens and insect pests, it does little if anything to conserve moisture or to restore fertility, an aim more commonly ascribed to fallowing. As we shall have occasion to comment, the fallow year might better be used to grow a purposeful crop—forage, for example, or continuous cereals.

In the drier grain belt the cropping sequence is simply a

year of grain followed by one or two years of fallow, before another year of grain is introduced.

Fertilizers are used in reasonable amounts on cotton and to some extent on vegetables and industrial crops such as tobacco. But the rainfed areas producing grain and dry legumes get no fertilizers.

Soil Limitations

Both rainfed and irrigated agriculture encounter some severe soil limitations. While Syria does have some good deep soils, many are shallow and thus have only a limited volume of soil for the rooting zone. Some are shallow to limestone or gravels and though they cannot store much water, these may not be too bad. But others are shallow over gypseous materials or even have a high proportion of gypsum in the soils themselves. On these soils water from the rainfall season cannot be stored for use in a succeeding crop year, as is done in deep loessial soils, such as are found, for example, in the Great Plains and Pacific Northwest in the United States. Herein lies one of the basic fallacies of the fallow system in Syria.

The underlying gypsum beds of some of the soils proposed for irrigation in Syria pose an even greater problem. Because gypsum is rather soluble in water, the soil can be lowered unevenly by differential solution. Irrigation ditches must be concrete lined and because of very shallow soil depth, irrigation intervals must be very short in the hot weather. Both qualifications spell high costs and need for skilled management.

Grazing Lands and Annual Feed

The grazing lands of Syria are very extensive but nonetheless are overgrazed in dry years. This applies both to the Sierozem soils of the areas between the good rainfed wheat and barley producing farmlands and to the sparsely vegetated deserts. This is a knotty problem, at the root of which lies the fact that, in general, people who have grazing animals do not grow field crops and people who have crops do not keep grazing animals. Though the greatest source of feed for these flocks is in the cultivated areas—grain aftermath and weed fallow—there is no coordination between the feed available and the number of animals that need feed. There is now no significant production of fodder raised specifically for feeding of domestic flocks; and concentrates, such as oilseed cake and milling by-products, are exported though badly needed at home. Even straw is hauled outside Syria for sale in countries such as neighboring Lebanon.

Overstocking of the grasslands and desert pastures is a very serious problem. Far too high a proportion of the feed produced goes into maintenance rations—i.e., only serve to sustain the animals and provide a grossly inadequate production of additional animals for slaughter. The amount of meat provided is very small in relation to the numbers of animal units maintained.

Obstacles to Better Farming

The impediments to better use of the soil and water resources in Syria are numerous. Among the more evident are: lack of reliable rainfall from year to year, lack of knowledge of better farming practices, lack of the means of production (especially capital and credit at reasonable rates), lack of tenure to the land, gross inequalities in land-ownership (many holdings are fragmented and small, others have been far too large for efficient production—though land reform may have reduced the size of some of the large holdings), high cost of transport and lack of market accessibility, and last but not least the historical separation of animal production from field crop farming.

It is relatively easy to conclude that much more could be done with the soil and water resources available, and some directions of change are indicated in Part III.

Iraq

Of Iraq's land area of 110 million acres, rainfed agriculture is practiced on about 9.5 million, all of it in the Northeast. Something of the order of 8 million acres seems a more realistic appraisal of the land suitable for rainfed agriculture. Much of the higher rainfall area is very mountainous, and rainfall is low and uncertain in other locations. Such land is better left in native grass pasturage. There are, of course, some irrigated lands within the area having enough precipitation for winter crops. These are used primarily to grow summer crops.

To ask how much arable land there is in Iraq involves one at once in asking how much water there is for Iraq. For outside the rainfed area the land is not usable for agriculture without a source of irrigation water. Hence the amount of land that can be used for production of crops depends upon the amount of water that can be delivered to Iraqi farmers. This in turn depends, as is discussed elsewhere, on how much of the Euphrates River water is taken out upstream by other countries. Estimates of irrigable acreage range upward from the 7.5 million acres that are now used for irrigated crop production. With the exception of the irrigated land in tree crops—primarily dates—and the rice acreage, nearly all of this is used for crops in alternate years. More acreage in and near the desert could undoubtedly be brought under irrigation if water and financing were available, but use of much land above and far away from the primary sources of water—the Tigris and the Euphrates—would face rapidly rising cost.

Present Water Use

Iraq's current water use from the Euphrates is thought to be about 14.6 million acre-feet out of a total average available flow of just under 24 million. From the Tigris River Iraq now uses about 13.8 million acre-feet on 4.4 million acres of irrigated land. Developments now underway or planned call for new land and intensified land use that would require a total of about 53.5 million acre-feet of water from the Tigris River—a volume that exceeds by 16.5 million acre-feet the average annual discharge.[14] It is obvious that all of these expansions cannot take place. In fact if the land in Iraq that now is sometimes irrigated were cropped every year instead of in alternate years and if more summer crops were grown, the resulting water demand might approach the total available flow in both rivers, leaving no supplies to meet the demand of any new land brought into cultivation. This is certainly true if future use resembles the present lavish use per acre cropped.

Land Available for Irrigated Agriculture

When irrigated agriculture was begun in the Mesopotamian Plain several thousand years ago the soil was very fertile and productive. Today this is for the most part not true. To know how the degradation of the quality of the land came about provides one with some perspective for future action.

All river water contains some dissolved minerals—soluble salts. Those carried by the Tigris and Euphrates are very moderate—200 to 400 parts per million—and thus make for very good irrigation water. But through evaporation of water from the soil surface and from the transpiration of plants, accentuated by the very high summer temperatures, the application of water to the land for growth of crops over thousands of years has left a serious salt residue in the soils, consisting primarily of sodium chloride, calcium sulfate, and calcium carbonate. The harmful accumulation of salts could have been avoided by adequate drainage systems and by providing downward leaching of enough water to carry the excess of soluble salts into the drains. In earlier times perhaps not enough was known about salt accumulations to make the salt removal feasible. But the remedy for the worsening salting-up in the Mesopotamian Plain has been well known for more than forty years. Yet very little effort has been made to remedy the situation.

Present Land Management

Because of the high salt content of the soils and the lack of drains, much of the Mesopotamian Plain cannot now be cropped except by resorting to unusual and wasteful agricultural practices that bring no lasting solution. After the harvest of a crop the soil is allowed to dry out for a whole year. During this time deep cracks develop. Prior to land preparation for the next crop the land is flooded and the salt near the soil surface is washed downward into the soil but not out of it because of the absence of drainage. The soil is then prepared and planted in wheat, or, if too salty for this, in barley which is more salt tolerant. The grain crop may do quite well in the upper portion of the soil where the salt concentration has been reduced, but by harvest time salt has risen back to the soil surface and the cycle is ready to be restarted. Many of the soils south of Baghdad are so salty that no crops at all can be grown. On others, less salty, barley can be grown with very low yields.

That the salts can be washed through the soils into drains has been demonstrated. That when freed of salts these soils can be of good quality there is no doubt. But under the present system of flooding salt into the dried soil without providing drains, high water tables have developed and the water in these water tables is quite salty.

Crops Grown

Wheat and barley have been the principal crops of Iraq from early times, occupying in recent years upward of 80 percent of the crop area. Rice and dates are the next most extensive single crops, with 4 and 3 percent of total acreage respectively, and vegetables occupy perhaps 5 percent of the cropped area. The role of cotton has been diminishing. In 1965, it occupied only 1 percent of the cropped area.

Crop yields are exceedingly low. Estimated wheat yields were 7 bushels per acre in 1959 and 11 bushels in 1965, though in 1968 a crop cutting survey indicated a yield of 17 bushels. Barley yields were a little better than wheat yields in the years listed. Such yields provide much of the wheat and barley for domestic market but only because of the very large acreages devoted to those cereals. In terms of land, water, and not least, the human resources used in its production, this food is, however, very expensive.

No breakdown between wheat and barley produced under irrigation and under rainfed agriculture is available. Since the total acreages in each are comparable and since the overall yields are so low, it would seem without purpose to try to make the comparison. Factors other than rainfall certainly are limiting yields on the irrigated lands, and to a considerable extent on the rainfed lands.

Inputs for Production

Fertilizer use in Iraq is insignificant, and little use is made of weed killers and insecticides. Some fertilizer is used on rice, cotton, and vegetables but none on wheat and barley. It is not surprising, therefore, that land that has been producing for thousands of years without a program of fertility maintenance has now very low yields.

Grazing Lands

The high rainfall area of Iraq is in the mountainous northeast and along the western front of these mountains. To the west and south lie the deserts, including a substantial portion of the floodplain of the Tigris and Euphrates Rivers. Thus most of Iraq's irrigated agriculture is in the desert and near desert areas. However the transition zone between the higher rainfall areas near the mountains and the deserts is a large area of land in which rainfall is intermediate. These stretches—totaling some 20 million acres— were originally and to some extent still are grassed "steppe" country and possess appreciable grazing potential, varying from very good to quite meager.

In general the grazing lands are not in holdings, making control of intensity of stocking impossible under present use patterns. Also, the grazing is decidedly seasonal. This creates a need for systematic production of forage to supplement the pasturage. But since there is neither cooperation nor coordination between the migrant animal owners and the people who cultivate land, there is no provision for producing and storing the forage needed to supplement the grazing.

The failure to use the grazing lands of Iraq in accord with their potential compares in seriousness with the gap between utilization of the croplands and their potential, though, of course, the croplands have far greater potential in terms of food production per unit of land. Coordinated use of the potential of the grasslands and the cultivated lands could do much to move Iraq's agricultural output into new size brackets.

Soil and Rainfall Limitations

As far as rainfall is concerned, some of the steppe soils of Iraq have enough, on the average, to qualify as good rainfed crop land; but soil limitations make them suitable only for grazing. Many of these soils are shallow and have a high content of gypsum; many are underlain by beds of gypseous rocks. These soils simply do not store enough water during the winter rainy season to produce a reasonable crop. At the same time, they are so shallow that dry-farm methods of dry fallow carrying of moisture from one year to the next are not feasible. Under good pasture management such soils can produce more profitably in native grasses.

Another limiting factor in these areas that lie between the good rainfed crop land and the deserts is the extreme range in year-to-year precipitation. Not only is the uncertainty of rainfall in any one year very great, but sometimes there are a series of years with low and erratic rainfall that can leave the animal population drastically reduced. These deficiencies and irregularities make proper pasture management and planned fodder production even more essential if good use of the potential of these lands is to be realized.

In summary, it is impossible not to conclude that the present use of Iraq's soil and water resources is highly unsatisfactory, and that for this reason the country has the greatest unrealized agricultural production potential of any of the countries of the area under review.

CHAPTER 5

Agricultural Inputs

Obviously a great deal more goes into the final bushel of wheat, ear of corn, or orange than the seed, sun, soil, and water which nature provides. Labor, machines, the chemicals present, all enter in different quantities, in different relationships to one another according to the stage and kind of agriculture of a region.

In the Middle East these relationships are changing as the region passes from traditional to modern agriculture. This change involves better and more productive use of the individual, the machines, the chemicals. While it raises many social, educational, and financial problems, the problems of no change would be even greater. And at the same time this change defies the making of valid generalizations about the present state of the agricultural arts.

Many persons not well acquainted with the region think of the Middle East as an area with a generally primitive agriculture. They have heard of, and perhaps seen women harvesting a meager barley crop by pulling the plants up by the roots, as the authors of this book saw in May 1968. By pulling up the barley, rather than cutting it, every inch of straw that might provide livestock fodder was saved; and by pulling by hand, a crop of possibly no more than four bushels per acre became economical to harvest—since the opportunity cost of labor was presumably not far above zero—when harvesting by machines would not have been. A woman will still carry burdens on her head while the man rides the overburdened small donkey.

But along the same roads one can also see modern tractors and other farm machines at work; commercial fertilizers are applied to some land for some crops, sometimes at quite heavy rates; a wide variety of chemicals are used to help control insects, plant disease, and weeds—not generally, perhaps not enough, or not properly, for best results, but nevertheless in significant volume. The Middle East thus presents the picture of agricultural practices in transition.

In the previous chapters we have reviewed the natural endowment and its use in Middle Eastern agriculture. We shall now examine the present use of other inputs—in this chapter the mechanical and chemical, in the next the human. We are leaving to Part III the discussion of how these relationships can and must change for a more productive agriculture.

Farm Machinery

Farming in the Middle East is only lightly mechanized. There are, it is true, a variety of farm machines in all the countries, but their use is neither widespread nor intensive. Large areas still rely on the simple farm methods and implements that have been used for decades and, more often, centuries. Figure 5-1 gives a rough idea of the level of mechanization of each country, related to the United States by reducing the number of tractors to a common area basis, in this case the arable land plus land under permanent crops, as it is reported to the FAO. (This comparison of countries has its faults, as it takes no account of different tractor sizes, cropping patterns, like double cropping, landholding arrangements, etc.) Out of this comparison Israel emerges about as mechanized as the United States, though probably less so if one were to put the comparison on a horsepower basis.

In Lebanon, next to Israel the country with the largest number of tractors per acre, mechanized farming has been stimulated in part by rising costs of labor. While the very poor areas, and the terraced ones where specialized machinery would be needed, have not shared in this trend, mechanization has been gaining rapidly.

This is a relatively recent development of the past decade or so. The findings of the 1961 Census of Agriculture reveal little mechanization. Nearly three-fourths of all farms then relied upon animal power only, only 1.5 percent had mechanical power only, about 6.5 percent had both animal and mechanical power, and about 12 percent had no power source of their own. As might be expected, tractors were found more commonly on large than on small farms, but many large ones had only animal power and a few small ones had only mechanical power.[1]

Egypt had thousands of tractors, where the other Middle Eastern countries had only very few; but by the same token, relative growth has been more moderate in the recent past, and a substantial amount of work that machines could do more efficiently is still done by manpower or draught animals. But Egypt, limited severely in the amount of new land it can bring under the plow, faces the problem of how to mechanize its agriculture without displacing more farm labor.

Specific information, beyond mere numbers, tends to be episodic, but there is little doubt that the other countries—Jordan, Syria, and Iraq—have been slow to adopt modern, mechanized farming.

Although we lack comprehensive data on the use of tractors, a number of reports have briefly mentioned their manner of use and problems associated with them. One source describes how farms in Iraq prepare land for winter cereals by plowing with horses and oxen; and how the soil is not turned, but simply stirred, and to a depth of only three or four inches.[2] Another observer, writing in 1967,

---

[1] *Census of Agriculture 1961* (Beirut: Lebanon Ministry of Agriculture, March 1966).

[2] A. P. G. Poyck, *Farm Studies in Iraq* (Wageningen, Netherlands: Laboratory of Agricultural Economics of the Tropics, Agricultural University, *1962*).

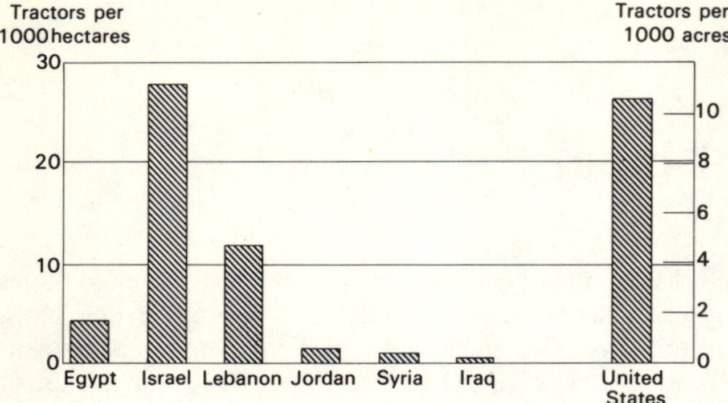

Fig. 5-1. Number of Tractors in Relation to Area of Arable Land and Land in Permanent Crops, by Countries, ca. 1964

Source: Table 5.1 and *FAO Production Yearbook, 1967.*

Table 5-1

Tractors* in Use, by Countries, Years of Record 1948–66

Year	Egypt	Israel	Lebanon	Jordan	Syria	Iraq
1966		11,430		2,068	7,424[d]	
1965		10,470		1,772	7,675[d]	
1964		10,190	3,556	1,462	7,274[d]	
1963		9,300		1,169	6,698[d]	
1962	10,994	8,505		1,081	5,591[d]	
1961	10,994[a]	7,485	580	890	4,314[d]	2,404
1960	10,994[a]	7,075		883	4,754[d]	2,404
1959	10,994[a]	6,349	580	807	3,772[d]	2,404
1958	10,994[a]	5,307	440	664	3,406[d]	2,188
1957	12,086[a]	4,700	224	566	2,792[d]	2,188
1956	10,750[a]	4,500	195	439	2,074[d]	2,096
1955	10,750[a]	4,990	160		1,786[d]	1,715
1954	10,355[a]	3,535	135		1,454[d]	469
1953	8,850[a]	3,386	112		1,155[d]	
1952		3,119	134	175	977[d]	439
1951			126	84	947	662
1950	b	2,300[c]	121	84	642	617
1949		3,800[d]		88		
1948	5,400[c]	2,146[d]		74		

[a] Includes garden tractors.

[b] Donald C. Mead, *Growth and Structural Change in the Egyptian Economy* (Homewood, Ill.: Richard D. Irwin, 1967), p. 334, quotes figure of 9,972 tractors of all sizes in 1950.

[c] Unofficial estimate.

[d] Tractors for all purposes.

* Tractors in agriculture only and excluding garden tractors, except as noted.

Source: FAO Production Yearbook.

stresses the inadequacy of tractor and other farm machinery to meet needs, and stresses also the high costs of its use by farmers.[3] An earlier source, the FAO Mediterranean Development Project report for Iraq, states that

most of the farm tractors in that country are used in the northern rainfed zone; and describes the generally unsatisfactory soil preparation in irrigated areas, where farmers are severly limited in their ability to prepare the land by reason of their simple implements and their limited animal power.[4] That this may still be true, one would conclude from similar observations made as recently as 1965.[5]

The special problems of operating tractors, even small two-wheeled ones, in the small, terraced, and often rocky fields of Lebanon are described in another report.[6] And a recent FAO report on Jordan states that tractors, mainly wheeled tractors of 40 horsepower and over, are largely confined to the grain fields of the north, where their usefulness is limited in that they tend to bring marginal land into use rather than contribute to better cultivation of areas already being exploited.[7]

There is little hard information on domestic manufacture of modern farm equipment. Some tractors are produced in Egypt, but the extent to which these are assembly operations and the share of imported components is not readily available information. Imported tractors come from various countries, including the United States; and several makes are usually imported even from the same country. The result is a diversity of tractor models, sizes, and types. Even in the absence of data, analogous conditions elsewhere lead one to assume that tractor costs are high because of the supply situation.

Maintenance throughout the Middle East is poor. A serious shortage of trained men to service the equipment exists almost everywhere, and this bad situation is aggravated by the difficulty of getting spare parts for such a great variety of tractors. As a result, far too often they lie idle for lack of necessary repairs.

Even less information is available about farm machines other than tractors. Grain production in the rainfed areas, especially those where rainfall is erratic and hence yield highly variable, is mostly carried out with combine harvesters, grain drills, and other equipment similar to that used in similar grain-growing areas in the United States, Canada, Australia, and Russia. Where pump irrigation is practiced, as in Iraq and Syria, the pumps are also imported.

A great deal of the machinery used is relatively simple, drawn either by animal power or used by hand. Such simple machines and tools are generally made locally; many utilize age-old designs and traditional materials.

Farm mechanization in general, and substitution of mechanical for animal power in particular, is often looked upon as a means of labor saving—of making human labor more productive. In countries such as Egypt, where there exists a very large agricultural labor force that could not be

[3] Sayed Abdul Kadir Hasso, "The Directorate General of Farm Machinery and Agricultural Implements: Its Policies, Problems, and Proposals for Future Action" (paper presented at National Seminar on Agrarian Reform in Iraq, Baghdad, April 2–13, 1967.

[4] FAO Mediterranean Development Project, *Country Report* (Rome: FAO, 1959).

[5] H. Charles Treakle, *The Agricultural Economy of Iraq* (ERS-Foreign 125, Economic Research Service, USDA [Washington, D.C., August 1965]).

[6] Farm Mechanization—The Introduction of Specialized Machinery for Terraced Areas (Beirut: Institute of Rural Economics, undated).

[7] FAO Mediterranean Development Project, *Jordan Country Report,* (Rome: FAO, 1967).

quickly absorbed in urban employments if it were displaced from agriculture, it may be argued that agricultural mechanization is unwise for the nation, even if it should seem profitable to the farmer. But farm mechanization does far more than save labor. Primarily and above all, as far as the Middle East is concerned, it would permit a type of soil and crop management in terms of timeliness of operations—so especially important when multiple cropping is practiced—and of thoroughness of each phase of the cycle from plowing to harvesting—which animal power does not provide and which is essential to greatly increased yields and output.

While mechanization would accomplish relatively little unless accompanied by better irrigation and drainage, greater fertilizer use, better crop varieties, better control of weeds and crop diseases, and by other components of a technologically advanced agriculture, conversely, most if not all of these components call for mechanization to be effective. The resulting dilemma between the fruits to be reaped from more efficient production and the price to be paid in terms of declining employment opportunities and perhaps sharpening social conflict is well illustrated by FAO's strong advocacy of mechanization in its production-oriented Indicative World Plan and its warning, for the less-developed countries generally, of the impending issues that will derive from shortages of jobs.[8]

Substitution of mechanical for animal power requires purchases of machines from outside the country, at least until domestic manufactures can be developed. As this development requires foreign exchange, mechanization is sometimes opposed on this ground. However, this argument may rest upon an incomplete examination of the facts. A country such as Egypt, which depends partly on imports to feed its population, is incurring foreign-exchange costs for its domestic animal power on farms; if draught cattle were reduced, land could be freed to grow crops for human consumption. Thus the foreign-exchange argument is seen to lack substance, once it is put in a broader context. The matter of optimum farm mechanization at each stage of development for each country is a rather complex one that deserves more attention than given here. But it is important, above all, to view it comprehensively in all its implications.

Fertilizers[9]

Most soils of the Middle East are reasonably fertile, but most of them respond to fertilizers. As is true in all desert regions, its soils are low in nitrogen; the system of agriculture followed for centuries had not led to much addition of plant material to the soil nor to the development of humus, and the hot climate has tended to burn out that which was added or developed. There are exceptions to this generalization; in places where drainage was impaired, as in the Huleh marshes of Israel or in the Ghab depression of Syria, humus

did accumulate, as it did also in rather unusual farming situations, such as in the date groves which line the banks of the Shatt Al Arab in Iraq, where the soil surface is heavily shaded. The cultivated soils in the Nile Valley of Egypt are relatively high in organic matter, simply because

Table 5-2

Consumption of Commerical Nitrogenous Fertilizers, by Countries, Years of Record 1946—66

(Hundreds of Metric Tons of N)

Year	Egypt	Israel	Lebanon	Jordan	Syria	Iraq	Total Reported
1966	2,500	247	115	20	97	48	3,027
1965	2,852	239	100	18	129	26	3,364
1964	2,606	229	110	14	121	23	3,103
1963	2,271	215	100	10	86	20	2,702
1962	1,961	192	59	6	150		2,368
1961	1,919	213	79	4	140	10	2,365
1960	1,766	188	69	2	108		2,133
1959	1,057	173	57	4	44		1,335
1958	1,771	157	62	4	36		2,030
1957	1,573	148	56		36		1,813
1956	1,233	116	56		35		1,440
1955	1,152	126	59	2	46	1	1,386
1954	1,121	96	32	2	46	1	1,298
1953	1,112	113	33		20		1,278
1952	1,039	81	33	4	16		1,173
1951	1,056	43	16		16		1,131
1950	1,010	78	15		6		1,109
1949	930	38	14		5		987
1948	760	28	11		5		804
1947	649		7		3		659
1946			5				5

Source: FAO Production Yearbook.

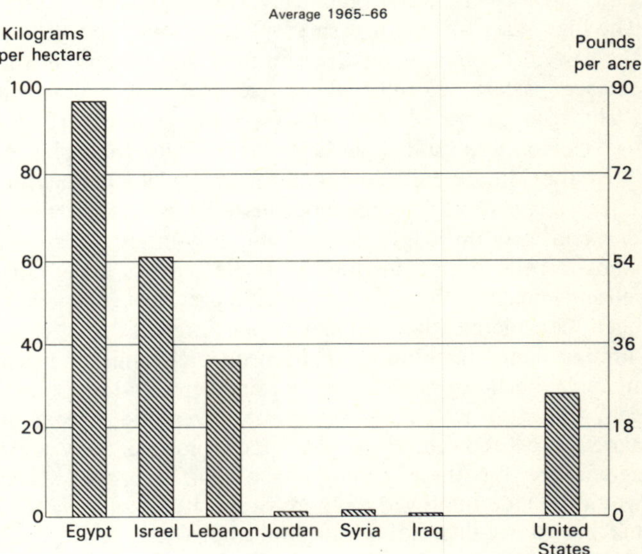

Fig. 5-2. Nitrogen (N) Fertilizer Use per Unit of Arable Land and Land in Permanent Crops, by Countries

Source: Table 5-2 and *FAO Production Yearbook.*

---

[8] See, for example, speech by A. H. Boerma, Director-General of FAO, at Notre Dame University, October 24, 1969. For an extended recent treatment, see B. F. Johnston and J. Cownie, "The Seed Fertilizer Revolution and Labor Force Absorption," *American Economic Review,* September 1969.

[9] See Appendix Table C15-4.

the mud which the Nile brought from the highlands of Ethiopia, before the High Dam was built, contains a good deal of organic matter, aiding soil texture as well as fertility. But a great deal of the present or potential cropland of the Middle East is seriously lacking in nitrogen.

All of the countries of the region report the use of nitrogenous fertilizers (Table 5-2). The amounts used per country differ greatly, but an upward trend is apparent in all. Since World War II, Egypt has increased its use of nitrogenous fertilizers fourfold, Israel by eightfold, Lebanon fivefold, and Syria no less than twenty times. But these rapid increases start from very small bases, such as 3,000 tons in Syria in 1947.

These data are more meaningful if reduced to some common unit of area. Figure 5-2, which does this for the two most recent years, uses the same unit as for tractors—area of arable land plus land under permanent crops—and it suffers from the same limitations. On this basis, Egypt emerges as the heaviest user of nitrogenous fertilizers, with an average rate of almost 100 pounds N per acre. Both Israel and Lebanon report relatively high use of such fertilizers, at rates well above those prevailing in the United States, for example, where a considerable acreage of idle cropland is included in the base and nitrogenous fertilizers are rarely used on the extensive semiarid wheat-growing lands. The other countries in the region lag far behind—indeed, for practical purposes may be said not to use such fertilizers at all.

Information on the crops and types of land on which the nitrogenous fertilizers are used is both scanty and hard to evaluate as to reliability. It is only for Egypt that detailed statistics of fertilizer use by crops are available. One such compilation was attempted in the course of work on FAO's Indicative World Plan and was kindly made available to the authors. It is presented in Table 5-3 in terms of per-acre application as well as shares of the total. As can be seen, these are fairly high rates of nitrogen use, less high in the case of phosphates, and practically nonexistent for potash. The high rates for major cash crops stand out especially in the case of sugarcane, onions, and fruit. But because of the acreage pattern, cotton, wheat, rice, and onions between them account for 75 percent of all nitrogenous fertilizer use. Cotton also ranks high in its share of superphosphates.

In the Middle East in general it seems likely that fertilizer, and especially nitrogenous material, is used primarily or exclusively on irrigated lands, and only rarely on rainfed lands—especially not on those of the latter with lower rainfall and cereals as the only crop. In the past, it has generally been considered that available moisture was the chief limiting factor for crop production on such lands, and that there was little or nothing to be gained from adding nitrogen; this may well have been true, given the system of agriculture followed. However, a technologically advanced agriculture for these grain-growing rainfed areas (which would include improved methods of moisture conservation, much better control of weeds, and improved seeds) would require application of nitrogenous fertilizers on these lands as well.

All of the countries report use of phosphate fertilizers

Table 5-3

Fertilizer Use* by Crops, Egypt, 1961/63

Crop	Total use				Use per unit of land[a]			
	Nitrogen		Phosphate		Kg. per ha.		Kg. per feddan	
	Th. M.T. N	Percent of total	Th. M.T. P_2O_5	Percent of total	N	P_2O_5	N	P_2O_5
Wheat	40	18	–	–	70	–	27	–
Barley	2	1	–	–	40	–	16	–
Berseem	–	–	12	34	–	12	–	5
Beans	1	b	3.5	10	–	25	–	10
Onions	2.4	1	0.8	2	110	35	43	14
Cotton	49.3	22	7.4	21	80	12	31	5
Corn	52	24	–	–	75	–	29	–
Sorghum	15	7	–	–	75	–	29	–
Rice	24	11	2	6	60	5	23	2
Sugarcane	6	3	–	–	110	–	43	–
Vegetables	18	8	3	9	75	12	29	5
Fruits	7	3	1.5	4	110	25	43	10
All other	4	2	5	14	25	30	10	12
Total	220	100	35	100	–	–	–	–

* Potash used only for potatoes and fruits, averaging 1,500 tons in 1961-63.

[a] Here derived from total use and acreage figures, but in the underlying calculations presumably rates were assumed first and total use derived by multiplication with acreage figure. This accounts for relative uniformity of rate and round figures in Kg/ha. column.

b Less than 0.5 percent.

(Table 5-4). Again, considerable differences between countries are evident. The trend in use is upward and, in total, at about the same rate as the upward trend in use of nitrogenous fertilizers—about a fourfold increase since the end of World War II.

The quantity of phosphate fertilizer per unit of land area (Figure 5-3) varies greatly from country to country, and in a pattern that differs from the variations in use of nitrogenous fertilizer. For phosphate, it is Israel and Lebanon who use the most, relative to their area, with Egypt high but at a somewhat lower level; the per acre use of phosphate in these countries is of the same general order of magnitude as that in the United States. The other countries in the region use very little. As with nitrogenous fertilizer, we lack information on soil types and crops to which the phosphate fertilizer is applied; however, it is probably used primarily on irrigated lands.

The amount of potash fertilizers used also varies greatly among the countries of the region (Table 5-5). The available data are somewhat irregular; where data are missing it is probable, but not certain, that use was relatively small. Unlike nitrogenous and phosphate fertilizers, potash fertilizer shows no clear upward trend in use. Use in relation to land area also varies greatly among countries and, again, in a somewhat different pattern than either of the other two types of fertilizer (Figure 5-4). Potash use is high in Israel and Lebanon, but definitely lower than in the United

Table 5-4
Consumption of Commercial Phosphate Fertilizers by Countries, Years of Record 1947−65
(Hundreds of Metric Tons of P_2O_5)

Year	Egypt	Israel	Lebanon	Jordan	Syria	Iraq	Total Reported
1966	550	96	60	8	24	18	756
1965	547	119	36	8	44	8	762
1964	431	114	117	13	43	4	722
1963	480	102	74	11	44	2	713
1962	416	117	39	15		7	584
1961	484	127	46	7		7	671
1960	360	121	42	12	81		616
1959	250	144	20	5	14		433
1958	277	144	25	2	11		459
1957	275	146	69	4	8		502
1956	740	154	26		10		930
1955	632	152	41	4	9		838
1954	150	132	22	1	9		314
1953	150	98	19		6		273
1952	165	75	8				248
1951	173	99					182
1950	150	65					215
1949	140	74					214
1948	135	25	5				165
1947	76						76

Source: FAO Production Yearbook.

Table 5-5
Consumption of Commercial Potash Fertilizers, by Countries, Years of Record 1946−66
(Hundreds of Metric Tons of K_2O)

Year	Egypt	Israel	Lebanon	Jordan	Syria	Iraq	Total Reported
1966	10	46	25	30	4	2	117
1965	10	44	11	25	2	1	93
1964	9	29	23	20	2	3	86
1963	10	25	19	8	18	2	82
1962	13	23	18	4	10	2	70
1961		23	30	8		1	62
1960	32	22	28	10	5		97
1959	21	21	17	6	1		66
1958	23	20	34		4		81
1957	15	20	69		5		109
1956	15	38	19				72
1955		16	14		4		34
1954	5		16		4		25
1953	5	40	16		2		63
1952	1	30	16		1		48
1951	15		15				30
1950							
1949							
1948		7	11				18
1947		20	8				28
1946			5				5

Source: FAO Production Yearbook.

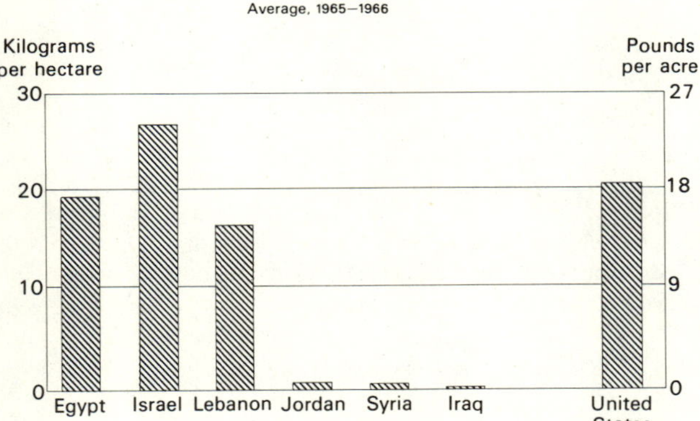

Fig. 5-3. Phosphate (P_2O_5) Fertilizer Use per Unit of Arable Land and Land in Permanent Crops, by Countries

Source: Table 5-3 and *FAO Production Yearbook, 1967.*

States; Jordan is next, but far behind Israel and Lebanon; Egypt has some use, and the other countries very little.

These rough data on fertilizers may be briefly summarized: Iraq uses each kind of fertilizer least of all countries reporting, and at a very low rate of application; Syria and Jordan have a low total fertilizer use. In contrast, both Israel and Lebanon use all three kinds of fertilizer at relatively high rates, which in total are higher than in the United States. Egypt shows the highest rate of use of nitrogenous fertilizer, a fairly high rate of use of phos-

Fig. 5-4. Potash (K_2O_5) Fertilizer Use per Unit of Arable Land and Land in Permanent Crops, by Countries

Source: Table 5-4 and *FAO Production Yearbook, 1967.*

phates, but a very low rate of use of potash; its total of all fertilizers combined is the highest in the region, and roughly double the rate of the United States.

It is hazardous to draw any general conclusion from these data. Reasons for differences in rate of use of any of the three nutrients may be in type of soil, kind of crops, presence or absence of drainage, availability of funds, etc. As has been learned, application of fertilizer may have little value or even be counterproductive when associated inputs are missing. Excessive vegetative growth may cause lodging, etc. Nevertheless, one is probably correct in saying that such dangers are far removed from countries like Syria, Iraq, and Jordan—all of which could probably apply large additional amounts. Some speculations on this will be found in Chapter 17.

Those unfamiliar with the area will perhaps be surprised that Egyptian agriculture consumes about twice as much of all nutrients per acre as does the United States. But it must be kept in mind that all of Egyptian agriculture is irrigation agriculture, in which the combination of water and fertilizer offers much better possibilities than that in dryland farming. In addition, the discipline of a large and traditional cash crop, cotton—which is sold in foreign markets and is thus subject to external price competition—is an important factor in pushing toward higher yields, such as cannot be achieved without fertilizers. And finally, the data presented take no account of multiple cropping—or intensity. Thus, in Egypt any given amount expressed in terms of acres should logically be related to cropped acres, not to physical acres. In that case, Egyptian rates of use would be lower by perhaps one-third.

The relationship between production and consumption of fertilizers differs greatly among countries as well as among kinds (Table 5-6). Only Israel, among the countries of this region, produces potash fertilizer; this is derived from the evaporation of the saline brines of the Dead Sea. Having access to the Dead Sea, Jordan, too, could produce potash and has for some time had plans to do so. Prior to the partition of Palestine in 1948 a potash plant did exist at the north end of the Dead Sea, but it was dismantled. Israel has a large export surplus of potash fertilizer, which moves in international trade; although the other countries of the region must import their potash, it may be assumed that none of the Israel potash moves to them.

Phosphate fertilizer is produced in Egypt, Israel, Jordan, and Lebanon in considerable volume; the latter three countries produce export surpluses, while Egypt needs moderate imports. The other countries of the region must import all or nearly all the phosphate fertilizer they use. Israel produces the nitrogenous fertilizer it uses, with no significant import or export. Egypt produces vastly more nitrogenous fertilizer than does Israel, but only about half what she consumes; and Lebanon too now produces some of the nitrogenous material she consumes. The other countries in the region except Jordan have plans to start producing nitrogenous fertilizer in the near future.

One cannot terminate this brief description without calling attention to the fact that in fairly close proximity to

Table 5-6

Annual Production and Consumption of Commercial Fertilizers, by Countries, Average 1962/63-1966/67

(Hundreds of Metric Tons)

Country	Production	Consumption
Potash (K_2O)		
Egypt	none	10
Israel	2,147	33
Lebanon	none	19
Jordan	none	17
Syria	none	7
Iraq	none	2
Re	2,147	98
Phosphate (P_2O_5)		
Egypt	362	485
Israel	127	110
Lebanon	98	65
Jordan	212[a]	11
Syria	none	41
Iraq	none	8
Region	799	720
Nitrogen (N)		
Egypt	1,387	2,438
Israel	241	224
Lebanon	70[b]	97
Jordan	none	14
Syria	none	117
Iraq	none	25
Region	1,698	2,915

[a] Jordan, Department of Statistics and Jordan Phosphate Co. Estimated on the basis of an average of 18% of P_2O_5 from a reported output of 1,180,000 tons of rock from phosphate rock.
[b] 1966-67 only.

Source: FAO Production Yearbooks, 1966 and 1967.

the countries here covered there are the oil—and gas—rich areas of the Middle East. These offer enormous possibilities for production of nitrogenous fertilizer, and some activity in this direction has already occurred. In the early 1970's, Kuwait alone is expected to possess a nitrogen equivalent capacity for ammonia production of 800,000 tons.[10] Thus, in the long run, the countries reviewed should be able to profit from this regional factor; and Iraq, of course, being an oil producer herself, would not even need to await developments elsewhere. A plan to erect a plant producing some 50,000 tons of nitrogen equivalent has been reported, production to commence by the end of the 1960's.

Contrary to what one might have expected, it does not

[10] *Chemical Week,* April 5, 1969 (McGraw-Hill, New York).

Table 5-7

Fertilizer Prices Paid by Farmers, by Countries

(Prices in U.S. $ per 100 Kilograms,
Average 1964/65-1966/67 –or Fewer Years
in This Period if Data Unavailable)

	United States	Egypt	Israel	Lebanon	Jordan	Syria	Iraq
Nitrogenous fertilizers (N)[a]							
Ammonium sulphate	28.0	30.2	29.1	25.1	26.4	35.8	34.7
Ammonium nitrate	25.0	30.4	30.6	26.2	23.9	29.7	31.5
Sodium nitrate	41.4	37.0		45.1		51.1	
Phosphate fertilizers (P_2O_5)[b]							
Superphosphates, low	22.8	18.4	20.7	14.7	22.7	25.5	
Superphosphates, high	19.5			17.0	21.9		25.2
Potash fertilizers (K_2O)[c]							
Potassium sulphate		11.2	13.4	13.8	14.1	19.0	
Muriate of potash, over 45%	10.8		10.0	10.4	9.9		

[a] Price per 100 kilogram of plant nutrient: United States, prices paid by farmers at varying points of delivery, some cost-sharing by government, which is not considered a subsidy; Egypt, official retail prices at nearest delivery centers, no subsidy; Israel, price at factory gate, subsidy to manufacturer; Lebanon, no subsidy; Jordan, price at retail store except sodium nitrate which is f.o.b. port, no subsidies; Syria, prices charged by local agents, 5% reduction to co-operatives; Iraq, prices at retail store.

[b] Prices paid by farmers per 100 kilograms of plant nutrient, average bagged and bulk; prior notes about prices paid by farmers apply here also; "low" is less than 25%, "high" is 25% of more.

[c] Prices paid by farmers per 100 kilograms of plant nutrient, average bagged and bulk; prior notes about prices paid by farmers apply here also.

Source: FAO Production Yearbook, 1967.

appear from publicly available statistics that fertilizer prices paid by farmers in Middle Eastern countries are uniformly higher than in the United States (Table 5-7). Some are and some are not. Reported prices for Iraq and Syria are generally higher than for the United States, suggesting perhaps that low fertilizer use in these countries is in part a consequence of the high prices. Fertilizer prices in Israel run below those in the United States; in Lebanon, some prices are higher and some lower. Prices in Egypt are higher than in the United States for nitrogenous fertilizers but lower for phosphates. Without more detailed knowledge and study— including the distorting effect of currency conversion rates over time—than these data provide or than we are able to undertake, one cannot enter into a comprehensive discussion of the reasons for variations in fertilizer prices in these countries. Moreover, the significant factor is the relationship between fertilizer prices and the prices of crops in the production of which the applied nutrients are employed.

Lack of such data deprives one of a major tool of analysis, and one can only hope that such data will become available in the course of future research. Finally, it would seem that even if fertilizer prices at the farm gate are no higher than in the United States, they are then grossly higher than the cost of locally based inputs—above all, human labor— compared with the corresponding relationship in the United States. If 100 kilograms of ammonium sulphate cost $28 in the United States, equivalent to say 10 manhours of labor, the same price in Egypt, would more likely be the equivalent of 190 man-hours of labor. Thus, the decision to fertilize, when weighed in terms of other inputs—as well as income—is not easily reached.

Such studies on crop responses to fertilizers as have been made suggest a highly favorable response, not only in physical but also in monetary terms. In this connection, the response to fertilizers will depend to a major extent upon the general level of agricultural technology. If more nitrogenous fertilizer is applied to a dryland field of wheat, where available water has not been conserved, where the field is weedy, and where an old and unresponsive variety is planted, then the fertilizer response may be small or even negative—it may lead to more weeds or to more straw, but not to more grain.

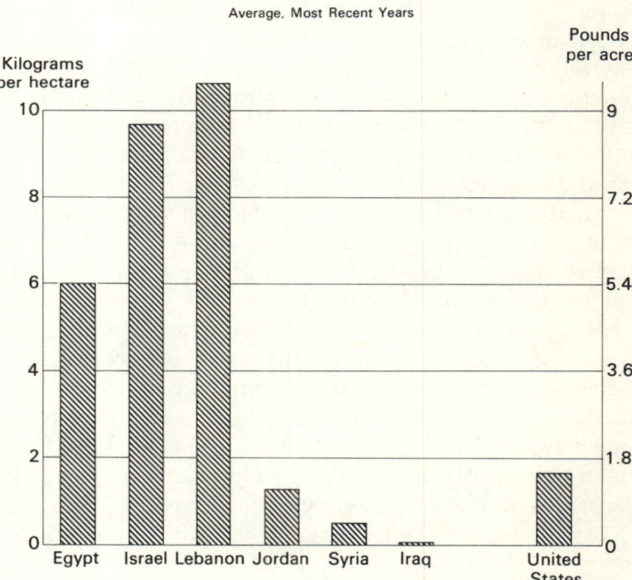

Fig. 5-5. Pesticides and Other Agricultural Chemicals,* per Unit of Arable Land and Land in Permanent Crops, by Countries
* Excluding fertilizer.

Source: FAO Production Yearbook, 1967
Note: These values are greatly influenced by the inclusion of mineral oils and sulfur for some countries. These two substances comprise 26% and 55% of the total of Egypt and Israel respectively. Mineral oil was 19% of the total in Lebanon and sulfur was 89% of the total in Jordan. For the United States mineral oil comprised less than 0.1% of the total reported while sulfur was 31% of the total. Thus, the total amounts of material used per hectare is not a good measure of the effectiveness of pest control in the various countries. They do indicate that the countries are aware of the problems and that some action is taken.

Pesticides and Other Chemicals
(Excluding Fertilizers)

Modern agriculture requires a large volume and diverse kinds of chemicals to control insects, plant diseases, and weeds, and to be used for other purposes. Insects, diseases, and weeds reduce crop yields, often in ways which farmers do not fully comprehend, and their control has been a problem. It is possible that the adverse influence of these factors is now greater than it once was; surely, they all spread from country to country, wherever the environment is hospitable, and commonly with greater rapidity than do improved crops and better farming practices. Improvements in crop varieties, greater use of fertilizers, irrigation, and other measures may be relatively useless unless a reasonable degree of control can be exercised over insects, diseases, and weeds—all of which benefit from the same growth-promoting elements that are applied to enhance agricultural production. And the war is never fully won; each pest or disease shows remarkable ability to adapt and to continue in the face of control measures.

We have added into a single total the volume of all chemicals used in agriculture upon which FAO gets reports from member countries.[11] Because of the extreme diversity of the chemicals included in this total, not much importance can be attached to the specific quantity; at best it may serve as a rough index of the use of modern chemicals in agriculture in the countries concerned. But even this is of only approximate value, since these quantities include sulfur and mineral oils, a fact that renders the data even less useful, as pointed out in the footnote to Figure 5-5. Lebanon and Israel lead in the use of such materials and an examination of the detailed data on which Figure 5-5 rests shows that they reported the use of almost every one of the many chemicals listed above. Egypt, too, reported a high usage at somewhat more than half that of Lebanon, but Egypt has about three times the usage per acre of such chemicals in the United States. Use in Jordan is lower, and in Iraq and Syria far lower still. These comparisons are all for an average of three recent years; Appendix Table C6-5 shows that the use of such materials has increased greatly in Egypt in recent years, by nearly tenfold in twelve years.

Readily available information does not permit a judg-

---

[11] DDT and related compounds; benzene hexachloride and lindane; aldrin; dieldrin; chlordane; toxaphene; organo phosphorus compounds; arsenicals; mineral oils and dinitro compounds; botanical insecticides; fungicides—sulphur and compounds; fungicides—copper and compounds; fungicides—mercury and compounds; fungicides—dithiocarbamates; fungicides—captan and others; fumigants; herbicides; rodenticides; and other pesticides and animal dips.

ment as to the use of such materials—that is, whether use has been fully adequate, excessive, or inadequate. Information available does suggest that some materials were in short supply in Egypt in some years.

Intraregional Comparisons

The information provided in this chapter is that which was readily available and which includes the major inputs other than labor and land; it is anything but complete, but permits some general conclusions:

(1) Iraq and Syria operate at a relatively low level of input into agriculture; their use of fertilizers and of chemicals is extremely low and their use of mechanical power low;

(2) Israel and Lebanon show up with high rates of inputs; Israel exceeds all other countries with chemicals and tractors, ranking high too in use of fertilizers; Lebanon is also high on most input factors;

(3) Egypt, too, has a high level of inputs for most fertilizers and for chemicals—although much less so for tractors;

(4) Jordan holds an intermediate position on most indexes—higher by far than Iraq and Syria, but lower by far than Lebanon and Israel.

Possibilities of Regional Cooperation

Potentialities for future economic development of agriculture in the region are considered in Part III. But at this point we may point out, very briefly, that great possibilities exist for regional cooperation in the *supply of inputs* to agriculture. The manufacture of some farm machinery, of chemicals needed for agriculture, and of fertilizers could be done on a regional basis; plants in one or more countries could supply the entire region. This cooperation would have the added advantage that implements appropriate for the area could be engineered and that servicing, commonly a weak link, could be organized on a more efficient scale and pattern. It is probably unrealistic for the near future to talk about the possibility of trade between Israel and the Arab countries, but trade might be developed within the latter. A larger volume of output than any country could hope to market domestically would surely lead to production economies. To the extent that regional sources of supply can be developed further, imports from outside the region can be reduced; to the extent that prices of inputs can be lowered, farmers will use more, with consequent increases in agricultural output and a new round of demand for the inputs.

CHAPTER 6

Agricultural Manpower

AGRICULTURE represents both a way of living and a means of earning a living. It has always been difficult, therefore, to distinguish between people who live on the land—i.e, rural people—and those who are gainfully occupied in agricultural pursuits. The difficulty exists mainly on family farms, and is especially pronounced in the case of women and children. This is true even in the developed countries, where distinctions between urban and rural environments are sharper and where statistical recording has become sufficiently sophisticated to distinguish between "living" and "working." It is doubly true in the less-advanced countries and it beclouds intelligent discussion of labor supply and demand. Yet, an appraisal of both the size and the quality of available farm labor is an indispensable element in the evaluation of agriculture's potential.

How necessary attention to this factor is may be illustrated by recent studies of Egyptian farm labor.[1] The predominant view until recently was that Egyptian farm labor was in vast surplus supply, that its marginal productivity was at or near zero, and that up to half the agricultural population could be removed without adverse effect on production. New statistical findings—as well as studies that look not at agriculture as a whole, but analyze it regionally, by tenure, by size of farm, and by sex and age of laborer—have led to substantially different conclusions. They are, to anticipate what appears in greater detail below, that permanent unemployment is probably of a low order of magnitude; that it differs greatly by sex, age, and region; that seasonal underemployment is substantial, but not nearly as large as has been judged in the past; and that there are a great many rigidities that impede the free flow of agricultural labor. Thus surplus and scarcity often coexist. Obviously, it matters greatly for the prospects and methods of achieving agricultural growth whether there is a reasonably fully employed labor force, an easily movable large labor surplus, an efficiently functioning labor market, etc. The closer one can come to correctly judging these conditions, the less he is likely to be led astray when it comes to considering the odds on expansion.

Agricultural Population

How much then is known about the agricultural population in each of the countries of concern? Table 6-1 presents the aggregate data as of the mid-1960's. Except for Israel and Jordan, the population that derives its livelihood directly from agriculture exceeds 50 percent. It is lowest in Israel, where the percentage is approaching 10 percent. In Jordan about one-third of the population may be classified

as agricultural. While the definitions and actual statistical efforts that underlie these measurements no doubt greatly differ from country to country, the order of magnitude is what matters; and one could probably obtain general agreement to the statement that of the 50 million people now residing in the region, about half may be called agricultural.

It is noteworthy that, by and large, the part of the population called "agricultural" is about the same as that called "rural," though again this is probably due as much to lack of clear definitions as to the preponderance of true agricultural pursuits in the nonurban areas of the Middle East. Where income is very low, it will not support much of a population that provides services; or conversely, a poor agricultural environment does not favor the growth of service functions in rural areas. In any event, published statistics provide no insight into the degree to which the rural population divides its working time between agricultural pursuits proper, functions associated with agriculture, other gainful occupation, housework, and no work at all.

The figures of Table 6-1 do not permit any unequivocal conclusions. For example, it would be correct to say that the low share of agriculture in the total labor force of Israel denotes an economy that is not primarily an agricultural one. Indeed its percentage of only twelve is low enough to be found elsewhere only in highly developed countries. But it would also be true—and this is equally applicable to neighboring Jordan—that the natural resource endowment simply does not furnish the basis for a larger agricultural economy. As the FAO report on Jordan puts it: "The relatively small share of the labor force in agriculture . . . is a reflection of the poverty of the natural resources of Jordan rather than an indication of advanced industrial and service sectors and hence of a highly general level of productivity and output."[2] In the case of Israel, there is, as well, the additional factor of relatively high yield and high productivity, which in combination with resource limitations make for a small—and in recent years diminishing—share of agriculture in total employment.

Comparison of rural and agriculture population affords a rough indication of the degree to which nonagricultural pursuits are continuing to be located in a nonurban environment. With the exception of Israel and Jordan, the differences are small,[3] suggesting either that people who live in villages or other types of rural communities do in fact engage principally in agriculture or that they do so indicate to the census-taker. In Israel, the Government's deliberate policy of locating industrial operations within or next door

---

[1] Especially in the work of such investigators as Bent Hansen, Robert Mabro, and A. H. Mohiedin. See footnote[5] below.

[2] FAO Mediterranean Development Project, *Jordan Country Report,* (Rome: FAO 1967), p. 6.

[3] The sizeable difference in Iraq is of doubtful significance because the two sets of figures are from different sources.

Table 6-1
Population and Labor Force, 1965*
(Thousands and %)

	Egypt	Israel	Lebanon	Jordan	Syria	Iraq	Total
Total Population[a]	29,600	2,563	2,405	1,976	5,345	8,180	50,069
Rural population[b]	16,940	472	1,560	1,045	3,207	4,614	27,838
Rural as percent of total population	57	18	65	53	60	56	55
Agricultural[a] population	16,347	310	1,323	651	2,940	3,859	25,530
Agricultural as percent of total population	55	12	55	33	55	46	51
Number gainfully[c] occupied in agriculture	4,905	110	360	165	670	1,020	7,230
Gainfully occupied in agriculture as percent of total gainfully occupied	55	12	55	33	50	50	50

*Data are shown for 1965 since this is the most recent year for which most of the data are available for most countries.

[a] Source is *FAO Production Yearbook, 1967,* which carries newly revised population data for 1965. These revisions are important for Lebanon, the total population of which had been raised by the UN from 2,050,000 to the figure shown, and from 1,127,000 to the figure shown for the agricultural population.

[b] Source is FAO's *Indicative World Plan for the Near East,* released in 1966—except for Israel, which is taken from that country's *Statistical Abstract* for 1967, and for Iraq, the source of which is *Iraq Statistical Abstract, 1966.* The figure for Lebanon is the authors' rough up-scaling of the estimate given in FAO-IWP, which was based on the lower total population figure; the implicit assumption that the rural population bears the same relationship to the revised as it did to the original total population base may well be fallacious, but there are no indications on how to arrive at a better estimate.

[c] Source is *Yearbook of International Labor Office.*

to agricultural settlements accounts for the substantial number of rural people that are not engaged in agriculture.

It is not clear whether the large gap in the case of Jordan is due to the inclusion in the "rural" estimate of refugee settlements, or whether the large number of unemployed in rural areas results in their omission from the agricultural population estimate. Moreover, the gap would be even larger if higher estimates of rural population that have been made[4] were accepted as more nearly reflecting the true facts. In its 1967 report on Jordan FAO had this to say: "Of the 'registered refugees' in Jordan 653,732 were not self-supporting and were fully dependent on UNRWA rations and educational and health services. This figure represents 32 percent of the total Jordanian population in mid-1966. About 226,000 of these refugees live in special camps scattered around the settled part of the country." Under those conditions the division between rural and urban populations, tenuous under most circumstances, must be considered as practically meaningless. We shall return to the peculiarities of the Jordanian manpower picture later.

Manpower for Agricultural Production in Egypt

Accepting the fact that manpower statistics are beset by uncertainties and that in most of the area one is reduced to generalities such as "about half the population is engaged in agriculture," what then can we learn about the most interesting question—i.e., to what extent does available manpower exert either a drag or a stimulus on raising the level of agricultural output?

To supply even elements of an answer, one must make a sharp distinction between labor supply at the current stage of organization and technology and the labor input that might be needed in a reorganized agriculture. Perhaps the most celebrated instance of a debate that has tended to gloss over this distinction is that concerning the magnitude of disguised unemployment in Egypt, briefly alluded to above. In the past, most attempts to measure the degree of underemployment have calculated the number of man-hours needed to produce the country's agricultural output, based on accepted norms of a specified number of man-hours per unit of area. The result was then compared with available man-hours—arrived at by assuming certain annual working hours for the agricultural population—and the gap was taken to indicate the degree of underemployment, or, put differently, of surplus manpower. On the basis of such calculations, it was often suggested that unemployment in Egyptian agriculture ran no less than 30 and perhaps as high as 50 percent.

Even if some of this gap could be discounted as seasonal, there was enough left to indicate large-scale unemployment. Recent investigations have thrown strong doubts on that judgment.[5] It has been pointed out that, first of all, a clear distinction must be made between employment of men, women, and children. Employment opportunities for these groups vary widely, and the differences tend to get lost when consideration is limited to aggregates. These difficulties are reflected in the way female and child labor are

---

[4] See, for example, the estimate of 1.5 million in *The Agriculture of West Asia* (ERS-Foreign 143, Economic Research Service, USDA [Washington, D. C., February 1966]).

[5] See especially: (1) Bent Hansen *Development and Economic Policy in the UAR—Egypt* (Amsterdam: North-Holland Publishing Co., 1965). (2) Robert S. Mabro, of the School of Oriental and African Studies, University of London, in unpublished manuscript; this in turn draws on some useful analysis performed by A. H. Mohiedin, in "Agricultural Investment and Employment in Egypt since 1935" (unpublished Ph.D. thesis, University of London, 1966). (3) Donald Mead, *Growth and Structural Change in the Egyptian Economy* (Homewood, Ill.: Richard D. Irwin, 1967).

reported in successive census statistics and in hard-to-reconcile differences between the censuses of population and agriculture. To what extent a farm woman who does household chores for part of the year and pitches in on farm work when necessary is or is not counted in the labor force has been the subject of much guessing. Similarly, the large role played at certain times of the crop year by children—especially in cotton—raises a general problem of both statistics and basic measurement of labor force and unemployment. As Bent Hansen has put it: "Children between six and fifteen who are employed between one-third and one-half their time are certainly *overemployed* by any reasonable, modern, social standard."[6] Yet there are others who, leaving social objectives aside, would consider these children underemployed. Hansen goes on to speculate that, if much or all of child labor were taken over by women, these in turn might become "overemployed" in a social sense, with the result that eventually there might be a substantial call for additional male labor.

The reason for mentioning this controversy is to draw attention to the importance of the framework within which one contemplates and analyzes statistics.

As to the statistics, perhaps the most illuminating set has been derived from a sample survey that was conducted jointly by the National Planning Institute of the UAR and the International Labor Office in 1964—65. Table 6-2 presents an overall summary of this data, but one that surely does not do justice to a very interesting statistical document.

Apart from differences due to timing, etc., it seems obvious that the census considered many persons as not in the labor force who indeed were in the labor force. Because of the painstaking detail with which labor status was ascertained in the sample survey, one is inclined to prefer its findings to those of the census. The problem of highly fluctuating female labor participation alone is a major factor in putting less faith in the census figures. Mabro suggests that (1) farmers are generally reluctant to go on record as having their wives and daughters officially in the labor market; and (2) it is often difficult to distinguish farm chores from household chores. Because of differences in definition, Mabro contends, half the housewives omitted by the census were included in the sample survey. Similar undercounting prevails for children. The sample survey turned up interesting facts in this connection. In terms of a "standard" amount of hours to be expected (2,468 hours per year from males, and 794 hours per year from women), the only age group that worked "full" hours among women were girls aged 12 to 15!

The younger group, 6 to 11, worked 27 percent, and the next oldest, 16 to 19 years, 69 percent of "full." Among men, the 100 percent level was reached among those aged 30 to 34, and 40 to 44, with a slight drop for the 35 to 39 and 45 to 49 age-group. By and large men aged 25 through 49 worked full time. Table 6-3 shows labor inputs, by age and sex.

---

[6] Bent Hansen, "Employment and Wages in Rural Egypt," *American Economic Review,* LIX, No. 3 (June, 1969).

Table 6-2
Manpower in Rural Egypt

	According to 1960 Population Census		According to 1964-65 Sample Survey[a]	
	(in %)			
Total Rural Population	100		100	
In Labor Force	31		46	
Employed		31		42
Unemployed		0		4
Not in Labor Force	44		24	
Housewives		25		13
In school		19		11
Unable to Work	25		30	
Children under 6		19		19
Disabled, old age, etc.		6		11[b]

[a] The Sample Survey takes into account mainly small growers and landless laborers. Part of its usefulness derives from the use of size of household and size of holding as variables, and the fact that the statistics are based on actual interviews with workers regarding hours spent in productive labor. Also, age and sex characteristics do much to make the survey a most illuminating document.

[b] Includes school-age children not attending school.

Source: Unpublished report by Robert Mabro, School of Oriental and African Studies, London University, citing a sample survey conducted jointly by the National Planning Institute of the UAR and the International Labor Office.

Table 6-3
Index of Labor Input in Egyptian Agriculture, by Sex and Age, 1964/65

Age	Male	Female
	(2,468 hours per year = 100)	(794 hours per year = 100)
6—11	14	27
12—15	67	100
16—19	83	69
20—24	92	42
25—29	96	67
30—34	100	85
35—39	96	90
40—44	100	79
45—49	93	84
50—54	89	68
55—59	77	29
60—64	81	31
65—69	73	21
70 +	58	3

Source: Same as Table 6-2.

The tabulation seems to support the hypothesis advanced in more recent research that if not all, then much of what had earlier been designated as unemployment, may in fact be merely a reflection of the lower participation rates of different age and sex groups, and that adult male labor is, in fact, substantially employed in Egyptian agriculture as now constituted.

But there are other considerations in judging manpower availability. One of primary importance is health, stretching from nutrition to the effect of specific diseases. This obviously makes it difficult to identify or even describe true "underemployment." Schistosomiasis, the disease that is said to reduce productivity by 25 to 50 percent, affects a large part of the population.[7] It is usually not directly fatal but debilitating and so far has been difficult to control, since short of eradication of the disease-carrying snail from irrigation waters, the danger of reinfection is ever-present. Whether most recent advances in the drug field will be as efficacious as is hoped by some, remains to be seen. In the meantime, the extension of perennial irrigation, permitted by completion of the High Dam, poses a new threat of further spreading the water-borne disease.

The widespread prevalence of diseases rests on poor sanitation, poor housing (overcrowding, lack of ventilation), poor handling of manure and other materials that provide breeding grounds for flies, etc. Thus, the remedy lies not along the lines of control that have led to such dramatic reduction in the incidence of malaria, cholera, etc. How one might correct labor force statistics for such adverse health factors we do not pretend to know, but one must surely judge cautiously in appraising employment levels.

The question of seasonality of labor use deserves a few more comments. Again, we take Egypt as an illustration. May–June and September are the months of peak demand, one for harvesting wheat and cultivating cotton, the other for harvesting cotton. The seasonal trough falls in January. Attempts to measure unemployment at that time suggest 8–9 percent, but this is without taking account of part-time employment.

There have been suggestions that the degree of seasonality has commonly been overstated, largely because estimates have reflected field-crop work only and thus have accounted only for a portion of farm work. This is borne out by the fact that sample surveys turn up a much smoother seasonal employment graph than do those based on cropping patterns. Mabro suggests that field work accounts for no more than half of men's work and much less in the case of women. Hansen arrives at a similar judgment. Other agricultural work consists of caring for livestock and general maintenance, especially of drains and irrigation works (dredging, weeding, remodeling of banks and sides, etc.).

----

[7] Mead (*ibid.,* pp. 28–29), cites a figure of 60%, but it is not clear whether this rate applies to the rural or the total population, though, since the disease is typically a rural one, the former is more likely. In any event, incidence is very high. "As surveys are extended into rural populations in infested areas, it becomes increasingly clear," states a summary report of its work in this field by the Rockefeller Foundation (*Rockefeller Foundation Quarterly,* No. 2 [1969], p. 7), that "any individual old enough to walk to the village watering spot will become infested." The report adds that of Egypt's rural population, 100% of those aged over two are likely to carry the disease often referred to as bilharzia.

Farm Size and Labor in Egypt

There is yet another dimension in which the agricultural universe is far from homogeneous, and that is in the matter of size of farms. The share of very small family farms has been rapidly increasing since 1950. Land reform has tended to aggravate an existing situation based largely on the Moslem Inheritance Law that makes for fractionation of holdings.[8]

As a result the labor force is concentrated in small farms (less than five acres), on which are found nearly three out of every four members of the agricultural labor force and which account for less than 40 percent of acreage but nearly 85 percent of all holdings. The labor force density in such farms is 2.2 per farm and 1.5 per acre, in contrast to large farms where it goes to 7.7 and 0.1 respectively, as shown in Table 6-4.[9]

Not only does this structure tie down a large portion of the permanent labor force—and even many of the landless rural residents are "attached" to give locations by some kind of kinship or traditional ties rather than constituting a large mobile labor force—but there is evidence from the previously cited sample survey that a significant portion of small holdings resort to the hiring of labor, though most of this is of a temporary or seasonal character. The sample survey turned up the following information.[10]

Size of Holdings (In Acres)	Percent of Holdings Employing Outside Labor
0.5–2	24
2–5	36
5–10	13
10 and more	85

Another tabulation sheds additional light on the need for outside labor in small farms. It showed that nearly 25 percent in a sample of holdings less than ten acres large had a family labor density of less than one-half unit per acre (where a unit is equivalent to one man, and age, sex, and other differences are weighted into the composite). Such holdings must be assumed to require hired labor at some times of the year. Though by and large the farm labor surplus is found in small farms, especially among men, and labor scarcity if found on large farms, any simple association of small farms with family labor and large farms with hired labor turns out to be not borne out by the facts. The picture is much more complex. A surplus of women and children is much less marked on small farms and may turn into a shortage at harvest times on medium-sized farms, while on large farms there is a marked surplus of female and child labor in the off-season, between December and March. One concludes, tentatively, that at least in Egyptian agri-

----

[8] There is also a question whether apparently large farms, prior to land reform, were not frequently aggregates of smaller units, from an operating point of view.

[9] Robert Mabro *op. cit.* (above, n. 5), citing A. H. Mohiedin.

[10] According to figures cited by Robert Mabro, *op. cit.* (above, n. 5).

Table 6-4
Egypt: Labor Force Density on Farms, 1961

Size group (feddan)	Density of labor	
	Per feddan	Per farm
Less than 1	3.2	1.5
1 to less than 2	1.5	1.9
2 to less than 3	1.0	2.2
3 to less than 4	0.8	2.5
4 to less than 5	0.6	2.8
Less than 5	1.1	2.0
5 to less than 20	0.4	3.4
20 and over	0.1	7.7
All farms	0.6	2.3

Source: Same as Table 6-2.

culture, surplus labor and labor scarcities coexist, made possible not only by seasonal swings but also by lack of mobility. Both factors, it should be noted, have important implications for mechanization. A highly mobile fleet of tractors and other machines could shave the seasonal peak as well as supply labor-short areas, with potential large-scale freeing of rural labor.

If the above statistics and their interpretation suggest that Egyptian agriculture presently does not have the large agricultural labor surplus that used to be taken for granted, it is the continuing emigration to urban areas that has drained it off. Table 6-5 shows the pattern between 1947 and 1960. Governorates are here assumed roughly to represent the urban and provinces the rural population.

The tabulation indicates the shifts toward urban areas, the substantial increase in the number of those carried as "not employed," the very large increase in the number of those employed in services—both in urban and rural areas—and the very modest increase in agricultural employment. All of which suggests that rural surplus labor, instead of accumulating on the farms, has clogged other portions of the economy, including prominently the ill-defined "services" category, the nooks and crannies of which tend to be the refuge of the near-unemployed and the near-unproductive. In any event, agricultural growth in Egypt has clearly not been rapid enough to absorb the natural increase of the agricultural population, even at the prevailing labor utilization pattern described above that seems highly wasteful of labor.

Two developments working in opposite directions, will be important in the future: the elimination of child labor and portions of female labor from employment, which, if pursued radically, would create employment opportunities for men and perhaps actual shortages during seasonal peaks; and mechanization, which will tend to offset emerging shortages. But it is equally clear that any throttling imposed on the rural-urban population flow would be bound to create a severe rural surplus labor problem, regardless of one's judgment as to whether one does or does not exist now.

Table 6-5
Egypt, Population Shifts, 1947–60
(Thousands)

Category	Change Between 1947 and 1960				
	Population in 1947	Natural Increase	Actual Increase	Calculated Migration	Population in 1960
1. *Governorates*					
Population	3,544	1,316	2,151	835	5,695
Adult Males:					
Total	1,113	304	540	236	1,653
Employed	943		469		1,412
In agriculture	27		38		65
Not employed	171		71		242
In services	694		307		1,001
2. *Provinces*					
Population	15,478	5,747	4,912	−835	20,391
Adult Males:					
Total	4,648	1,268	1,032	−236	5,680
Employed	4,303		879		5,182
In agriculture	3,112		383		3,495
Not employed	345		153		498
In services	899		418		1,317

Source: Donald Mead, *Growth and Structural Change in the Egyptian Economy* (Homewood, Ill.: Richard D. Irwin, 1967).

Labor in Agriculture in Other Countries

There are several reasons why much has been said about the labor force of the United Arab Republic. For one, there exists a good deal of published material; secondly, the available material is sufficiently detailed yet inconclusive to have aroused the curiosity of scholars; and thirdly, some of the easy generalizations about agricultural labor in the developing countries of the Middle East are best tested in this context.

When we look at the other Arab countries, both volume and quality of information diminish rapidly, while the situation in Israel is well documented but, because basic concepts are different, difficult to compare with the other countries.

Syria

According to latest available official figures about 3.4 out of Syria's 5.4 million people—or 63 percent—live in rural areas. Of these, no more than two million are estimated to constitute the farm population. FAO judges that about 670,000 of these are gainfully employed in agriculture; other sources puts the figure as high as one million. No doubt, the definition as to what constitutes gainful employment is even more difficult here than in the United Arab Republic, since a larger part of the agricultural population is nomadic. Assuming six members per family, the two million cited above as constituting the farm population would represent some 330,000 families. In that event, 670,000 actively engaged in agriculture might be just about right on a narrow definition and low on a more generous one—i.e., drawing a wider circle of what constitutes employment.

A similar conclusion is reached when one considers that the 1960 census showed only 22.3 percent of the Syrian population, or just over one million people, in the total labor force. Even if 60 percent were engaged in agriculture, the result would only be 600,000 thus employed; but the percentage is probably lower. In one of his very careful studies of the Ghab Project, N. S. Randhawa[11] puts it at 51.1 percent, but adds that in the less-developed countries the size of the agricultural labor force is often underestimated. We would add "on a narrow definition," and add further that it is often overestimated, using a broad definition.

It is worth noting in this context that in planning the Ghab project provision was made for family labor to supply between 80 and 90 percent of all manpower requirements and to provide between 160 and 200 man-days of work per year per equivalent male worker for field crops alone (i.e., disregarding livestock chores, etc.). There are no statistics to support this, but one can assume that these are much higher intensities than prevail in agriculture at large.

One characteristic revealed by recent statistics is the small number of wage earners in agriculture. Whereas in the United Arab Republic the landless wage earner represents around one-third of total agricultural manpower, in Syria the percentage is reported as more nearly 12 percent.[12] Owner-operators and their families account for the remainder, the former for about 40 percent, the latter for 48 percent.

Iraq

Another country in which there is a wide range of estimates of rural and farm population, as well as of the share of the agricultural labor force is Iraq. Of the estimated 8.3 million people living in Iraq in 1965, according to the 1965 census, those living in rural areas are estimated at 4.6 million, or somewhat over half.

The population directly involved with farming, however, is smaller by a substantial margin, indicating the presence of substantial small-scale craft or service functions in Iraq's rural areas. There are no census figures for 1965 that show the size of the farming population, but it was hardly higher than 3.9 million (Table 6-1) and more likely near 3.7 million persons, up half a million from the 3.2 million indicated in the 1957 census. That would put it nearly a million below the rural population estimate for 1965. On the basis of these figures the farming population would have grown at an annual rate of about 1.8 percent per year between 1957 and 1965. If in these same nine years the total population

lation has risen from 6.3 to 7.2 million, as the census data state—or 1.7 percent per year—then farming population would have grown somewhat less rapidly than the population as a whole. Given the quality of the data, one might say that the two populations were moving at the same rate. This does not prove a lack of a migration to urban areas, a condition that would be in substantial contrast to what has happened in the United Arab Republic; for higher natural growth plus migration to urban areas could also be compatible with equality of total and farm population growth rates. But it is possible that in fact migration has not been large, perhaps because there are enough employment opportunities in farming to absorb population growth, or because a growing volume of rural unemployment is held back on the countryside owing to a lack of governmental measures to accomodate growing numbers in the cities.

There are no direct data to support one hypothesis rather than the other, though observers report a substantial flow of labor from the countryside to urban areas. Some circumstantial evidence, however, may be noted. Of the 3.2 million persons estimated to have been deriving their livelihood from agriculture in 1957, only about one quarter of a million were landowners or tenants, totaling with their families perhaps 1.2 million. The residual 2 million would be landless farm laborers and their families. A substantial number of these must have been unemployed, partially or totally; for statistics for 1959 put the agricultural labor force at 2 million, with actual employment in agriculture at 80 percent of this. Since 70 percent of the labor force is estimated to have consisted of landowners, landholders, and their families, and since many of those can be assumed to work only sporadically, unemployment among the farming population would seem to be very high. FAO, in its Indicative World Plan, puts the number of gainfully occupied in agriculture at only slightly above one million in 1965. If this were so, then the number of those of the farming population who are not gainfully occupied must be very large indeed, and this would be compatible with a low volume of out-migration. Another element that strengthens this conclusion is that reportedly as much as 75 percent of the agricultural labor force of the country is unemployed at seasonal troughs. Since we shall see later that under proper resource management Iraq offers the greatest opportunities in the region for agricultural growth, the suggestion of a large labor surplus is of considerable interest—not only because it could provide needed manpower once the land becomes fit again to be tilled, but also because Iraq has at times been looked upon as a possible recipient of surplus population from other countries of the region. One would doubt that a country having a rural surplus of its own could reasonably be counted upon to perform such a role.

Before leaving the subject of Iraq labor, it is worth citing a comment made by the Beirut UN office in a recent publication. Discussing the second five-year economic plan, the article refers to the fact that in 1957, about 45 percent of the population was less than 15 years of age and " . . . that this concentration of young people will make itself felt precisely during the Plan period, with the result that apparent unemployment will surge from an estimated 76,000

---

[11] U.N. Special Fund *Ghab Development Project* (Land Use and Production Economies, Series No. 4 [Damascus: Syria Arab Republic, Ministry of Agriculture, July 1965]).

[12] International Labor Office, *Statistical Yearbook 1966* (Geneva).

in the base year of the Plan (1964) to 400,000 in 1970." And the report goes on to say that " . . . a considerable improvement of manpower statistics should be achieved before a national evaluation of employment prospects can be attempted."[13] Both comments, it is believed, are worth considering in conjunction with the preceding discussion.

Jordan

As has been observed early in this chapter, the labor force of Jordan forms an uncommonly small portion of the total population. In 1967 it accounted for only 23 percent. Part of the explanation of this phenomenon is the country's age structure, above all the heavy component (46 percent) of those aged less than 15. As a consequence the median age in Jordan is about 17 years. If labor force is related only to those aged 15 and up, the economically active population rises to 42 percent (though this figure is undoubtedly too high, since it can be assumed to include in the numerator an unknown number of persons less than 15 years old). A second factor explaining the relatively small labor force is the small number of gainfully occupied women—fewer than 30,000 in 1967. Out of the total population of about two million this is small indeed, though reportedly women who are seasonally employed in agriculture are omitted.

Of the gainfully occupied, about one-third is estimated to be employed in agriculture—the largest single group, though not nearly as large as in the other countries, except Israel. As has been pointed out, limitations on the development of agriculture rather than, as in the case of Israel, strongly developing secondary and tertiary employment sectors, account for the relatively small employment in farming.

Of those engaged in agriculture, a little over half worked as owners or holders of land, and, if family workers are included, nearly three-quarters, leaving less than 30 percent as wage earners.

The Jordan River divided the population fairly evenly, with a slightly higher share living in the East Bank portion of the country, (though some shift towards the East must have resulted from refugees crossing the Jordan after the 1967 war). But given the much smaller area, density on the West Bank was nearly fifteen times that of the East Bank (this, of course, conveys a misleading impression, unless one notes at once that much of the land east of the Jordan is not inhabitable).

Population and labor force statistics in Jordan are further beclouded by the refugee situation. According to UNRWA, there lived in Jordan in mid-1966 over 700,000 people classified as refugees from Palestine, and of these over 90 percent, equivalent to one-third of Jordan's total population, were "fully dependent on UNRWA rations and educational and health services. ...About 226,000 of these refugees live in special camps scattered around the settled part of the country."[14] Without entering into the discussion of the number of "refugees proper" and their degree of dependence on UNRWA, it is obvious that under such conditions, comparisons of labor force and employment statistics with those of other countries are not very meaningful.

Two other characteristics of Jordan's labor force need be mentioned. One is the low degree of participation in the labor force not only of women, but also of men. A special survey made in 1960 found that a large portion—22 percent—of adult males were not merely unemployed but were not seeking work.[15] Again, the refugee situation may have a depressing effect, since refugee families are entitled to assistance if their income falls below that of an unskilled laborer. But whatever the reason, labor force participation is small and unemployment is thus inadequately measured if cast only in terms of those who are seeking but not finding work. This was at a level of 7 percent in 1961. Additional unemployment probably resides in such factors as seasonality and short hours. Nor has working abroad offered any solution for those engaged in agriculture. While other trades and professions had between 8 and 20 percent of their total numbers working abroad in 1961, the corresponding figure for agriculture was only one percent. Those who put total unemployment at 12 to 15 percent of the labor force may hit the right level.

Unfortunately, no recent data are available to judge whether there has been an upward trend in the labor force or in employment, both in and out of agriculture. But the statistics leave one with the general impression of a very large manpower reservoir that, while perhaps not currently high in quality, would be available to become a very significant addition to Jordan's labor pool, if employment opportunities should arise.

Lebanon

Unhappily, the statistical picture is even less illuminating in neighboring Lebanon. Neither population census nor social security statistics are available, but episodic estimates and conjectures take their place. Discrepancies between successive estimates are such as to make them at best marginally useful. The latest population census was taken in 1932, and food rationing registration of 1944, as well as estimation of civil service personnel in 1953, "and a number of other guesses," are employed to derive population growth estimates.[16]

Most estimates agree that about half of all active manpower is employed in agriculture, but the actual figure may vary between 250,000 and over 400,000, depending on the widely varying guesses as to the size of that manpower pool. Once more, who is and is not employed in agriculture is not a judgment easily rendered by those engaged in manpower and labor matters in developing countries.

A tabulation that appeared in the 1961 Lebanese census

---

[13]U.N. Economic and Social Office in Beirut, *Studies on Selected Development Problems in Various Countries in the Middle East,* 1968 (New York), p. 7.

[14]FAO Mediterranean Development Project, *Jordan Country Report* (Rome, 1967), p. 4.

[15]*Ibid,* p. 5.

[16]*Summary* (FAO Near East Commission on Agricultural Statitics, Fourth Session, Baghdad, September 10–17, 1968 [15]), p. 48. See also Table 6–1, n. b, above.

of agriculture and was published in 1966 throws little light on the matter. The census reports some 127,000 farm families as having farm holdings (half of which families, incidentally, do not derive their main livelihood from agriculture) and totaling with family and dependents nearly 800,000 people. It follows that some 40 percent of the population were members of families with farm holdings, but the division corresponds neither to an urban-rural nor to a farm-nonfarm split.

Statistics presented in early 1968 by the Rural Economic Institute[17] give perhaps the best indication of the population's economic structure. Of a total population of "just over two million" (the official estimate of 2,267,000 as of January 1, 1965, is thought to include many Lebanese who have in recent years left the country, but note footnote 13 above), the rural population is put at 1.3 million, a figure that probably includes a good many people living in cities but reported in villages. The total manpower pool is estimated at 1.2 million, but only 850,000 are in the active labor force—600,000 permanently and 250,000 seasonally. Finally, of the active permanent labor force, 44 percent, or 265,000 are believed to be engaged in agriculture: 60 percent as employers or self-employed, and 40 percent as wage earners.

Official unemployment figures do not exist, but it is likely that the large number of emigrants has kept unemployment relatively small.

Israel

The population of Israel is of the same order or magnitude as that of Lebanon or Jordan: some 2.5 against 2.4 or 2.0 million. But Israel's rural population is only 18 percent of the total, compared with 65 and 53 percent in Lebanon and Jordan, respectively; and those employed in agriculture constitute only 12 percent of all gainfully occupied compared with 55 and 33 percent.

These differences strongly influence per-worker return, as described in Chapter 10. But they do not stop there. Age composition of the Israel population does not show the heavy proportion of the "less-than-15" group of Jordan, though with just over 34 percent it is still heavy in that bracket and reflects the gaps in the older age groups resulting from their decimation in Europe between 1933 and 1945; for of the 1965 Jewish population of Israel, nearly one out of two came to the country as an immigrant. Only 650,000 were there in 1948; 563,000 were born since; and 1,087,000 arrived from other countries. Even in the period of lowest immigration, one out of every six additions to the population was an immigrant.

With this large influx from abroad, the annual rate of population growth from 1948 to 1965 has been 7.4 percent among Jews as against 4 percent among non-Jews. And density has tripled since 1948, from 111 persons per square

mile to 332 in 1965. But at that it is less than one-sixth the density that prevailed in 1965 in the United Arab Republic. In fact, density in Egypt was higher, and substantially so as far back as 1882, and, if we had data, undoubtedly beyond.

Of those gainfully employed on the land in Israel, 60 percent are self-employed, the balancing 40 percent, hired labor. But since total agricultural employment is small, farm labor amounts to only 15 percent of all gainfully occupied, again in much contrast to the countries discussed above. Actually, the total number of persons who are engaged in physical farming is even smaller, for the 110,000 gainfully occupied in agriculture, about 15 percent are engaged in administration, crafts, transportation, and other services. The basic production labor is furnished by 37,000 farmers, statistically recorded as "skilled," supported by 27,000 semiskilled and 30,000 unskilled workers. A survey made in 1964 throws some further light on this structure. It shows that 25 percent of those engaged in agriculture had no schooling—a group perhaps roughly similar in scope to the "unskilled" category; 12 percent had one to four years; 34 percent, five to eight years; and the remaining 29 per-

Table 6-6

Hours of Work in Agriculture, Israel, 1962–65

Number of hours worked per week	Percent of Persons in Designated Group[a]			
	1965	1964	1963	1962
1-14	4.4	3.9	3.9	2.4
15-44	38.8	40.2	38.7	31.1
45-49	35.8	31.5	33.2	48.2
50+	16.0	17.6	17.7	11.1
Not known & absent	5.0	6.7	6.5	0.7

[a] Year-to-year differences are most likely due to differences in timing of this survey.

Source: Statistical Abstract of Israel, 1966 and earlier issues.

Table 6-7

Land and People, 1965

	Population[a]		Arable land			Irrigated land		
	Total	Rural	Total	Per capita[b]		Total	Per capita	
				Total	Rural		Total	Rural
Iraq ...	8.2	4.6	18.5	2.2	4.0	9.08	1.11	1.97
Israel ..	2.6	0.5	1.0	0.4	2.0	0.38	0.15	0.76
Jordan .	2.0	1.0	2.8	1.4	2.8	0.15	.08	.15
Lebanon	2.4	1.6	0.7	0.3	0.4	0.15	0.06	0.09
Syria ...	5.3	3.2	15.1	2.8	4.7	1.29	0.24	0.40
UAR ...	29.6	16.9	6.9	0.2	0.4	c	—	—

[a] Table 6-1.
[b] *FAO Production Yearbook.* Includes land under permanent crops.
[c] Same as arable.

---

[17] *Agriculture in the Lebanese Economy,* survey prepared for the *World Atlas of Agriculture* (Beirut, Rural Economic Institute, February 1968).

cent, nine years and more. These figures suggest, though they do no more than this, that years of schooling and position in the hierarchy of skills were significantly correlated.

There are no figures on unemployment within agricultural labor, but the shrinking percentage that agricultural represents of total employment indicates that there has been a continuing shift from agriculture to other types of occupation: in 1955, 1960, and 1965, agriculture (and forestry) accounted for 17.9, 17.2, and 12.7 percent of total employment, respectively. Not everybody works a full week in agriculture, of course. Table 6-6 shows distribution of hours worked for a few recent years and suggests that there is a good deal of part-time work, but much of Israel agriculture is so patterned as to not only permit but to encourage combinations of agricultural and other work.

Moreover, much of Israeli agriculture is planned to provide a balanced labor input so that productivity increases are probably the principal way of obtaining added production from the agricultural labor force as it now exists.

Summary and Conclusions

From this broad survey of agricultural manpower in the countries under review, perhaps the most clearly emerging generalization is that, Israel apart, we know exceedingly little about the size and makeup of agricultural labor. Some interesting attempts have been made in the Egyptian case to disentangle a picture confused by lack of confidence in the nature of the data, but the controversy over the size and existence of disguised unemployment is far from settled. In the other countries one is confined to obsolete data and vague conjectures. For whatever they are worth, they suggest varying, but substantial degrees of agricultural unemployment, relieved so far by a continuing drain towards urban areas. One would not err in stating flatly that in none of the countries, Israel included, is agriculture in a position to absorb the sector's own natural population increase. However, this conclusion requires two qualifications, pointing in opposite directions. Firstly, if child labor were eliminated, there would no doubt arise additional employment opportunities for adults. How extensive can only be guessed in the case of Egypt.[18] The same may be true to a lesser degree for female labor, though the picture here is more complex; that is, how extensive female labor participation really is may not be apparent in official statistics

since for cultural and other reasons statistical surveys may heavily underreport this item. Secondly, the above conclusion is true for agriculture as presently organized, often with a most inefficient and wasteful use of human labor. Any turn toward efficiency would be bound to further diminish employment opportunities. If economic development produced a strong demand for labor in the cities, agriculture could quickly be restructured to release large numbers of workers.

Here again, a qualification is called for, and that is the possibility of large-scale land reclamation, as in Iraq. In that event, new employment opportunities would arise. More generally, the fact that population density, in terms of arable land, is so much lower in Iraq and Syria than it is in the United Arab Republic, Lebanon, and Israel (Jordan is an intermediate case) has given rise to considering Iraq and Syria as countries suffering from a manpower shortage (see Table 6-7). That the picture is very different between these two groups of countries is undeniable. What is equally undeniable is that agriculture as presently organized and production resources as presently available have provided the agricultural population with neither sufficient employment nor adequate income to keep it from draining off to the towns and cities—and this despite the persistence of human labor in many farming operations long mechanized in some other countries. As recently as 1966 an official survey of wheat production in some parts of Iraq showed that on over 40 percent of the acreage grown to wheat, harvesting employs neither machines nor animals.[19]

Migration estimates for Iraq furnish the following statistics of rural-urban movements:[20]

1947–57 330,000
1958–Present .... 45,000 per year

One cannot, therefore, speak of any "labor shortage" in Iraq's agriculture unless one speaks of one that is related to a future, fuller utilization of the potentially cultivable land area instead of circumstances as they have prevailed up to the present. Similar comments could be made for Syria, but the data base here is even slimmer. The issue is further discussed in Chapter 16.

---

[18] See above, pp. 49 and 51 and Hansen, *op. cit.* (above, n. 5).

[19] W. Rasaputrum and F. Al-Hilli, "Survey of Costs of Production of Wheat in Five Liwas of Iraq for 1966" (mimeo., Central Bureau of Statistics, Baghdad).

[20] Sayed Hassal El-Rawi, "Land and Utilized Mechanical Power" (paper presented at National Seminar on Agrarian Reform in Iraq, Baghdad, April 2–13, 1967).

CHAPTER 7

Farms

IN the foregoing chapters the natural resources of the Middle East, the soil and water management practices that have developed over the centuries, and some of the major inputs have been considered. The present chapter explores how these inputs are combined into productive enterprises that in the aggregate are responsible for the outcome of the productive process—in terms of specific crop acreages, yields, and production, of livestock numbers and production, of gross agricultural output, and the like—all taken up in subsequent chapters.

General Nature of Farms

The equivalent of the business firm in manufacturing is less easily found and defined in agriculture. It is relatively unequivocal in U. S. practice, though, and therefore it is perhaps helpful to comment on that practice as an example. In the United States, "farm" is a term meaning an economic enterprise in agriculture, at the core of which lies decision-making—what to produce, how, with what inputs, where and how to market, how to use credit and capital, how to manage labor, and all the rest of it. Although the definition of a farm has changed somewhat from one agricultural census to another (the census is a major source of information about farms), it has been reasonably similar over the decades. For one thing, it is clearly understood that all land under the farmer's control, by whomever it might be owned, should be included in his farm. Secondly, the emphasis in both professional and popular literature in the United States, is upon the farm as an operating unit or economic decision-making unit, which may differ greatly from conditions of ownership. To cite one example, nearly one-fifth of all U. S. farmers both own some land which they farm and also rent some land, under various arrangements, which they also farm. Usually called part-owners, they could equally well be called part-tenants; because they average much larger than full owner-operated farms, the part-owners operate nearly half of all land in the United States. Still other farmers are outright tenants, owning no land, but renting under one or another of numerous rental arrangements, all the land they farm. By and large, far less is known about farm land ownership in the United States than about farm-operating units—beyond such broad, general knowledge that the rented land is owned by widows of former farmers, by retired farmers, by small businessmen, and by other relatively small owners, and that the great concentrations of large landholdings which are typical in some other parts of the world do not exist in the United States.

These U. S. concepts of farm operation and land ownership have frequently been applied—or rather misapplied—in other parts of the world, especially in the matter of confusion between farm operating units and land ownership units. In an attempt to assemble data on numbers of farms in various countries, the Food and Agriculture Organization has used the "concept of agricultural holding, that is, consisting of all land which is used wholly or partly for agricultural production and that is operated by one person—the holder—alone or with the assistance of others, without regard to title, size or location (livestock kept for agricultural purposes without agricultural land is also considered as constituting a holding)." This definition is similar to the agricultural census ones in use for the United States, except that the latter also include minimum acreage and value of output criteria.

In supplying requested statistics in accord with its definition, FAO reports as follows: "A number of countries, however, somewhat deviated from this definition by restricting the enumeration to those holdings which conformed to certain additional criteria and which fell above certain lower limits as to size of holding, or size of operation, or both."[1] Of the Middle Eastern countries here considered, only Iraq and Lebanon submitted data on agricultural holdings to FAO (the numbers they reported are those used in this study). One is tempted to conclude that these concepts of agricultural holdings, or farms as operating businesses, do not well fit conditions in the other Middle Eastern countries, nor indeed in many other countries around the world. However, the matter of farms as business decision-making units is so important that we propose to consider them as such, using the best data we can find, in spite of the fact that these concepts may not exactly describe the kind of agricultural organization which exists in some Middle Eastern countries.

In Egypt, some fellaheen own land in more or less the Western sense of the term "ownership"; others are buying land from the State, on long-term contracts; still others lease or rent land from private or public owners. Before the major land reforms of the past twenty years, a great many farmers were seasonal tenants; they grew crops on such land as they owned, but rented additional land for a particular seasonal crop. A very large number of persons at that time owned very small tracts of land, not suitable for operation as separate farms, and these were typically rented to larger—but still often quite small—landowners who operated the combined holding as a farm. Under these circumstances, it was often difficult to say just which were separate farms or how large each was. The land reforms and the changes they have made in landownership and operation are considered later in this chapter.

In Israel, different forms of landownership and operation prevail. Almost all the land is the property of the State or one of the Jewish agencies. It is typically rented to farmers

---

[1] *FAO Production Yearbook, 1967,* p. 666.

or groups of farmers on a nominal rent basis, with hereditable leases. Moshav farms—known as "moshavim" in the plural—have been established in many settlements; each farmer has his own tract or tracts of land, leased as noted above; and in addition, the village sometimes has acreages of land operated by the village cooperative, from which the individual may buy the forage or other crops. These villages are characterized by a high degree of cooperative action, in owning and using farm machinery, in buying farm supplies, in getting and using credit, in marketing output, and in other ways; but decisions are made by the individual farmer—subject, of course, to many conditioning influences from other sources. Another major type of farming enterprise is the kibbutz—or kibbutzim in the plural. These are collectives; the whole farm is operated as one unit, with members working at tasks assigned to them; everyone shares equally in the income, which is mostly in kind (food, shelter, etc.); there is a communal dining room, the children live in a children's house while the parents have small apartments, and there are other distinctive features of kibbutz life, apart from strictly agricultural practices. From some points of view, the kibbutz is but a single farm, since there is a single decision-making process; from other points of view, it is a collection of many farmers.

In the other countries of the Middle East, there are varied forms of agricultural operation, and varied forms of tenure; but the concept of a farm as an economic operating unit can be applied reasonably well at least to most crop farms. In each country there are also some nomadic livestock producers who normally own no land, although they may have more or less recognized grazing rights to particular areas. Usually each family owns its own livestock, but the livestock graze together.

From what little has been set out above, one may surmise that the usual statistics on numbers of agricultural holdings omit or understate the numers of nomadic livestock producers (but these numbers are generally small in relation to numbers of crop-producing farms); and that they somewhat overstate the number of actually operating small-crop farms, since many persons may own some land which they do not customarily operate as a separate farm, and these may inflate the statistics to some degree.

It would be desirable to present statistics on numbers of farms in each country, with meaningful classifications by size and other characteristics, not only for the sake of such information but also as a basis for calculations of average figures per farm. Indeed, in an earlier draft of this report we did just that; using such data as we could find and making such adjustments as we could for differences in concepts of farms, we presented such statistics. In gross approximation, we arrived at about 3.2 million farms in Egypt, 400,000 in Syria, 250,000 in Iraq, 125,000 in Lebanon, 100,000 in Jordan, and 70,000 in Israel. The reactions of various reviewers convinced us that the confusion introduced by the attempt to compare seriously unlike data and concepts was greater than the enlightment provided by their use, and hence we have omitted any statistical data relating to numbers of "farms" and to averages based upon such numbers.

Land Tenure

Land tenure involves the relationships among men over the use of land. More particularly, it involves the relationships between society as a whole, operating through some form of organized government; the "owner" of the land or the one who has either legal claim or possessory claim to the land; and the actual land user or cultivator—who often but by no means always is the same person as the owner. The rights and obligations of these various groups differ greatly from one society or cultural group to another, but everywhere they greatly influence the ways and the efficiency with which farmland is actually used. Land tenure in the Middle East is extremely complex and includes many forms of tenure and land-use relationships not familiar to Western land economists. A complete discussion of land tenure in all the countries of our region is beyond our capabilities and in any case is not necessary for our purposes. However, a relatively brief review of the situation is necessary.

As one looks at land tenure in the Middle East as a whole, particularly in a longer historical perspective, a few facts or relationships stand out rather sharply:

(1) The matter of titles or ownership of land has been confused and insecure throughout nearly all of the region's long history; repeatedly, peasants or users of land have lost control over the land they cultivated. Larger and more powerful landowners, military leaders, or political figures have been able to take advantage of various situations to acquire land for themselves, and to extract from the land cultivator substantial payments for use of the land. Control over, and titles to, land have repeatedly been reshuffled after wars or changes in ruling dynasties.

(2) Ownership and use of land in the Middle East has always had a religious and a moral aspect which has been largely or wholly lacking in the West. A great deal of land tenure in Arab countries rests on Muslim beliefs and law, as well as upon governmental laws and economic forces.

The countries considered in this study were all part of the old Ottoman Empire. Wickwar traces the development during the Empire of cadastral surveys, to demark and identify land ownership boundaries, and of land-title registration, to record who owned the land.[2] The primary objective of the Turkish government in undertaking this registration was originally to obtain more revenue from taxes based on land; but a more secure land title also often enabled the land user to achieve a greater economic output, thereby benefiting the Empire far more, even though more indirectly. Various forms of tenure were recognized and recorded, each with its distinctive rights. Wickwar also shows how the actual administration of such laws often operated to favor the large landowners or others with some measure of economic strength, at the expense of the poor peasants.

There is a particularly long and involved history of land tenure in Egypt, about which perhaps more has been writ-

---

[2] W. Hardy Wickwar, *Modernization of Administration in the Near East* (Beirut: Khayats, 1962), pp. 59–82.

ten than for the other countries. Baer traces landownership from 1800 to 1950.[3] In 1800, land was, de facto, owned by the central government; it farmed out the collection of taxes on this land to various local dignitaries or leaders, who in turn extracted from the peasants as much as they could. Various local leaders had possession of significantly large areas of land, which they used almost as if they owned them. Throughout the nineteenth century, large estates were formed and sometimes broken up as ruling dynasties changed. Rulers of Egypt, notably Muhammad Ali and his heirs, built up quite large estates during the nineteenth century; so did others, including various creditors who foreclosed on land mortgaged by smaller landholders, or used their hold to continue collecting interest and establish bondage in fact, if not in word. The distinction between State lands for whose administration the government was responsible, and private property of the ruler was not observed by these rulers of the nineteenth century. In more modern times, King Farouk similarly built up large private estates, with little regard for society's rights in government lands. Thus, throughout the nineteenth and early twentieth centuries, the fellaheen of Egypt, who actually cultivated the soil, frequently experienced changes in masters but rarely did they have any substantial degree of freedom and security in use and possession of land.

One distinctive feature of Arab countries has been the development of waqf land, which is land dedicated in perpetuity to some religious or charitable purpose, to specifically identified individuals, or is, as Baer defines it, "endowed property not subject to normal transactions, the income being assigned by the founder." Many landowners, either out of piety or public conscience, or in fear of losing control of their land in any case, so dedicated it and thus effectively removed it from the market. "Conversion into waqf safeguarded property against sale for the payment of debt or for other purposes; nor could it be seized for the debts incurred by beneficiaries. It could not be converted if it was already mortgaged, or merely as a device to prevent its falling into the hands of creditors."[4] Such dedications of land as waqf have generally been respected, even by new rulers or conquerors. In fact, so deeply has the waqf designation been respected that it has been difficult politically, in many instances, to change the ownership and use of such lands even when the original designation had long since served its purpose. Waqf lands, while benefiting from not being fragmented by successive heirs, have often been badly administered, and often primarily for the profit of the administrators rather than for the benefit of the original designee. In the land reforms of recent years, discussed below, waqf land—which at its peak amounted to perhaps 10 percent of all agricultural land in Egypt—has generally been treated more or less as State land, and subdivided to those needing land or more land.

By 1950, landownership in Egypt had reached a situation which was highly unsatisfactory from many points of view.[5] On the one hand, there were a great many very small land ownership units—many smaller than an acre. A substantial proportion of these very small ownership units were too small to provide even a bare subsistence existence for a peasant, and they were rented to other landholders—who may themselves have had but a very few acres. Land in Egypt at this time and for decades had represented the chief form of economic security; the peasant or laborer who could manage to accumulate a little savings put it into land, or the man who inherited a small piece of land clung to its ownership even when he could not operate it. Poor as was the condition of these extremely small landowners, they were somewhat better off than the completely landless rural workers. On the other hand, there were a very few very large landholdings, many running into the thousands of acres. Some of these were held by foreign owners, others by ruling families or by economically entrenched families. The government and the politics of Egypt were dominated by these large landholders. Rental terms for their land were generally onerous, and rents had risen over the decades until the annual rental of land typically involved a far larger sum than the cost of the labor required to produce a crop.

It was the social injustice and the economic inefficiency of this landownership situation which led the military leaders of the coup against Farouk to initiate a sweeping land reform in 1952. Moreover, they sensed that, if their revolution was to persist, they had both to destroy the power of the entrenched landholding classes and win the popular support of the rural masses; land reform accomplished both objectives. The first land reform of 1952 established a limit of 200 feddans (about 208 acres) for one owner, except that minor children could be granted land up to an additional 100 feddans; lands in excess of this limit could be sold by the owner, for a short time; thereafter they were taken over by the government. Some large estates were confiscated. All of these, plus some State lands and some waqf land, were distributed to small landowners or to landless rural workers. There is some question over the speed at which these reform moves took place, but those finding it too slow must, of course, contend with the counterargument that a minimum of production efficiency had to be preserved, especially where there was indivisible equipment on the larger farms.

The lands taken over were to be paid for, over a period of years, at a price set in relation to the previously assessed value for taxation purposes; in general, the price set was far below that which the land would have brought in an open market prior to land reform. Small purchasers were to pay for the land at the price paid by the government, plus an overhead charge of 15 percent. The second land reform, in 1961, lowered the limit for one owner to 100 feddans. Prices paid to owners were lowered, too, at least in terms of interest rates paid on deferred payments; and the same

---

[3] Gabriel Baer, *A History of Landownership in Modern Egypt, 1800–1950* (London: Oxford University Press, 1962).

[4] *Ibid.*, p. 166.

[5] In addition to Baer (above; n. 3) see: Gabriel S. Saab, *The Egyptian Agrarian Reform, 1952–1962* (London: Oxford University Press, 1967), and Bent Hansen and Girgis A. Marzouk, *Development and Economic Policy in the UAR (Egypt)* (Amsterdam: North-Holland Publishing Co., 1965).

reductions were made to purchasers of these lands. At the same time, limits were put on the rents that could be charged by private landowners or were charged for State lands rented to farmers. The Egyptian land reform, like that of almost every country in the world where land reform has been reasonably successful, has involved a degree of expropriation of land values—indeed, in its absence it would have been extremely difficult, if not impossible, for purchasers to pay for the lands they received. The open market price of land, before land reform, capitalized on the subsistence wages of the workers; and it was precisely to improve their lot that land reform was undertaken.

Official statistics suggest that land reform has substantially changed the distribution of land ownership in Egypt (Table 7-1 and Appendix Table C7-1). The number of small ownerships has increased by about 15%, while their combined area has risen almost 75%. Ownership beyond the prescribed size has vanished, according to these data; the 0.2% of the largest landowners who owned nearly 20% of the land in 1952 have either sold out or now have much smaller acreages. The intermediate size group—which includes a wide range in sizes—has also increased about 15% in numbers and nearly 25% in acreage. Relatively little of the land sold by the State went to these intermediate owners; their increase reflects largely private sales and the residue of the larger ownerships after reform had been carried out.

The economic consequences of land reform in Egypt have undoubtedly been significant—resources have been shifted from one very small group of large landowners to a much larger group of small owners, rents charged for land have decreased, wages or imputed returns to labor have in-

Table 7-1
Number of Land Holdings and Total Land Area in Egypt, by Three Size Groups, Various Dates

	1952[a]	1952[b]	1961[c]	1965[d]
Number of holdings (1,000):				
less than 2 hectares	2,642	2,841	2,919	3,033
2–42 hectares	154	162	180	178
over 42 hectares	5	5	0	0
Total	2,801	3,008	3,101	3,211
Total area (1,000 hectares)				
less than 2 hectares	891	1,168	1,332	1,551
2–42 hectares	945	1,013	1,223	1,163
over 42 hectares	678	333	0	0
Total	2,514	2,514	2,555	2,714

[a] Prior to promulgation of Agrarian Reform Law.

[b] After Agrarian Reform Law set maximum ownership of 200 feddans (84 hectares).

[c] After new Agrarian Reform Law, which set maximum ownership of 100 feddans (42 hectares).

[d] Excludes State-owned lands and Agrarian Reform lands not yet distributed.

Source: See Appendix Table C7-1.

creased—even if one assumes that land rent control has not been fully effective. In carrying out a land reform in Egypt, choices necessarily had to be made between considerations of economic efficiency and considerations of welfare of rural people. The numbers of rural persons without land or with too little land for even a minimum subsistence unit were simply too large, in relation to the land area available, for everyone to share in the available land and at the same time provide even the smallest practical farm unit. Units of maximum economic efficiency would have been very much larger than the very small farms. By and large, the choice in land distribution was for welfare, not for efficiency—that is, while a minimum unit was provided, as many such units were created as possible. The formula for deciding minimum farm units was rather complicated, taking into account size of family and productivity of land, but generally units of less than five acres, and often of less than two acres, were established. "The size of the holdings allotted," Saab states, "was calculated to give each beneficiary and his family an annual income exceeding bare subsistence expenses by 10 percent . . ."[6]

Far more significant than the direct economic effect of the land reform, however, were the political consequences and the subsequent economic changes. By breaking the power of the entrenched large landowners, for most of whom agriculture was merely one of their many investments, the way was opened for major economic reforms and development. Industrialization and urban employment generally are being developed faster since the land reforms. In the long run this may have a great impact upon agriculture by creating urban employment, domestic markets, and domestic input supplies.

One important feature of the Egyptian land reform was the development of agricultural cooperatives in the villages where land reform was carried out. Some agricultural cooperatives had existed in Egypt for many years, but many more were established as a result of land reform. While at first they were promoted in the villages where land reform was a major factor, the government shortly moved to promote agricultural cooperatives in all parts of the country. The total number of agricultural cooperatives in Egypt has increased almost threefold since land reform began; the agricultural cooperatives now serve as virtually the only source of agricultural credit and agricultural supplies and as the only channel for marketing cotton. While the cooperative is figuratively under the management of its directors, who are local farmers, each is supervised by a governmental employee; there is some reason to believe that the latter in fact dominates the cooperative, so that its actions may not always be those which its directors would have chosen. Clearly, the cooperatives have fast become the primary channel by which the government seeks to deal with farmers, all the way down to production decisions on each field; some degree of leadership and even of direction by the government employee may well contribute considerably to the competence of the cooperative.

The situation in Israel is considerably different. When the

___
[6] *Ibid.*, p. 37.

Zionists began to migrate to Palestine before World War I, then still under Turkish rule, land for agricultural settlements could be obtained only through purchase. Such land was often hard to find, expensive, and its purchase was often accompanied by long delays. In order that the land so purchased would not be lost to Jewish settlers, through sale to others, title to the land was acquired by and remained in the Jewish National Fund—which also put up the necessary purchase price. Land was leased to individual settlers in the moshavim and to groups for kibbutzim under long-term leases, with nominal rents, and providing for inheritance but not for sale of the leases. After the establishment of Israel in 1948, control over nearly all the land within the country was assumed by the new national government. Rubner estimated the land ownership situation in 1957 as follows:[7]

Government and government agencies	76.5%
Jewish National Fund	18.0
Private Jewish landlords	3.0
Private Arab landlords	2.5

The same practice of long-term, hereditable leases at nominal rents was continued, and still continues. Rents which were low, or nominal, when first established have tended to remain unchanged or only slowly changing in monetary terms, in spite of continued and relatively rapid inflation, so that today the land is almost rentless. In addition, water prices to farmers are heavily subsidized, directly and indirectly. Rubner has well pointed out the serious distortions in resource use rising out of artificial prices, at far less than productivity values would suggest. However, as he comments, lack of irrigation water of amounts which would otherwise be demanded has forced a type of resource rationing or allocation upon Israeli farmers, in a manner similar to that which land rent would do in a more nearly competitive market.

The farms established in Israel by this process have varied only moderately in size, depending in part upon soil productivity and market possibilities. The typical farm size has been 28 dunums (7 acres) for the moshavim and a similar size per family for the kibbutzim, of which part would normally be irrigated and part unirrigated. The great variations in farm size which characterize most countries of the world thus do not exist in Israel. A recent development, however, is tending toward a greater range in farm size. The size and type of farms originally established was designed to provide productive employment for one man and some family help for a total of about 400 man-days per farm annually. With increased labor efficiency, many farms are unable to provide so much productive employment, even with greater intensity of farm operations; at the same time, employment opportunities in the cities and towns are drawing many of the farm youth. The result has been that many small farms have been subleased by their former operators, to other farmers in the village, whose operations have thereby been increased in area.

The situation is still different in Lebanon and in Jordan. In those two countries, most land is operated by its owners, and farms are generally rather small. The agricultural census in Jordan in 1965 showed that 65% of the agricultural lands were operated by their owners, that another 20% was in farms partly owned and partly rented, and that only about 12% was in farms wholly rented (with a small area in miscellaneous tenures). In Lebanon, about 82% of the land is farmed by its owners and only 18% is rented. The typical rental arrangement in each country is sharecropping, although some cash renting is found in each. In Jordan, a land reform or land distribution program was put into operation when the East Ghor irrigation project—in the Jordan Valley—was developed in the early 1960's. Before that land distribution took place, a relatively few owners with 125 or more acres each possessed about 10,000 acres or nearly a fourth of the land and a somewhat larger but still rather small number had 25 to 125 acres each for a larger total area (30% of the total; see Appendix Table C7-4). Because of a complete discrepancy in size classes, exact comparisons with the postdistribution situation are not possible; yet after land reforms, an extremely small number of owners with more than 50 acres each had only 3% of the land, and an equally small group with 32 to 50 acres each had only 5% of the land. Thus, a substantial redistribution of land ownership took place in this one area, as irrigation was extended to it. It might be added that the economic capability of many of the new smaller units under irrigation was probably as high as, or higher than, that of the much larger units on unirrigated land from which they were created.

Both Syria and Iraq undertook land reforms, each closely modeled after the Egyptian land reform, but neither has pushed it to the end originally intended when it was begun.

In February 1958, Syria entered into formal union with Egypt, to form the United Arab Republic. Almost immediately, an Egyptian-type land reform was initiated.[8] Data are lacking to describe the land ownership situation precisely in the pre-reform period, but it has been estimated that fewer than 3,000 individual owners held over 40 percent of the farm land of Syria, their holdings averaging over 12,500 acres each.[9] These large holdings had been assembled in large part by the foreclosures of creditors on small farmers, whose indebtedness had been growing and who had experienced particularly low incomes in a series of climatically adverse years. These creditor-landlords were largely urban merchants. In addition, there were large feudal estates dating back to Ottoman days, when favored citizens were granted large estates as a reward for cooperation with the Sublime Porte. Finally, tribal chiefs had taken advantage of the land-registration program to register tribal lands in their individual names.

Like its Egyptian model, the Syrian reform of 1958 limited holdings to 200 acres of irrigated land, or to 750 acres of unirrigated land; areas in excess of these limits were to be appropriated by the State and distributed to small

---

[7] Alex Rubner, "The 'Price-Less' Land of Israel," *Land Economics*, XXXIV, (November 1958).

[8] Kenneth H. Parsons, "Land Reform in the United Arab Republic," *Land Economics*, XXXV, No. 4 (November 1959).

[9] George S. Medawar, unpublished data.

farmers in lots not exceeding 20 acres irrigated or 75 acres unirrigated land. The expropriation price was set at not more than 10 times annual rental value; State bonds were paid to landowners, carrying a 1½ percent interest return and payable over 40 years; prices charged to purchasers were the same, plus 10 percent for administration. As in Egypt, provision was made for the promotion of agricultural cooperatives.

This land reform program aroused intense political opposition in Syria. "The application of most of these laws to Syria, where the landlord class was still powerful and the native bourgeoisie traditionally much more vigorous; where state intervention was a much more recent and restricted phenomenon; and where a remarkable rate of growth between 1945 and 1957 had been succeeded by a depression—partly owing to poor rainfall—following the union with Egypt in 1958 had dramatic consequences. Discontent, which had been smoldering for some time, flared into the September revolt which dissolved the union."[10]

Nevertheless, the land reform in Syria went forward, but very slowly. It had been estimated that about 4 million acres of private land, mostly rainfed rather than irrigated, would be subject to expropriation, and about 900,000 acres of State land, or nearly 5 million acres in all, would be available for distribution. By the end of 1966, or eight years after the program began, only about three-fourths of the private land had been sequestrated; and only about 25 percent of the total expropriated area had actually been distributed to new owners. The farms actually established have been about half the size provided for in the land reform law; as in Egypt, policy has faced a choice between economic efficiency, or larger farms, and welfare of rural people, or smaller farms; and, as also in Egypt, practice has swung more toward welfare than toward efficiency. The State lands not distributed to new owners have been rented to farmers. Although the land reform called for the promotion of agricultural cooperatives, in Syria they have not reached anything like the proportions attained in Egypt; by the end of 1965 less than 10 percent of all farmers in Syria belonged to cooperatives, and they are not the exclusive channels for agricultural supplies that they are in Egypt.

Although modeled after the Egyptian land reform, that of Iraq has gone still differently. Statistics of distribution of landholdings before the land reform was initiated in 1958 are shown in Table 7-2. The situation is complicated greatly by the diverse land tenures. Privately held land can be seen to have been highly concentrated in a relatively small number of ownerships, with very large numbers of small owners; in addition, there were various kinds of more or less public lands—Miri Sirf, or land purely owned by the State; waqf land, previously discussed; and unsettled land, whose ownership is unclear or clouded, but may be regarded as State land. By June 1968, 2,464 landowners were considered subject to the Agrarian Reform Law, and 2,409 had been served with sequestration decrees; the total area of land involved by all these owners, including government lands held by them but not cultivated during three consecutive years, was about 5 million acres. Including the various forms of State land, a total of more than 15 million acres was estimated as ultimately available for distribution to small farmers; however, by June 1968 only about 10 percent of this had actually been distributed. A considerable part of the remainder was leased annually to farmers. Although cooperatives were to be pushed, only about a fifth of all farmers belonged to one in 1968.

The Iraq land reform has been severely criticized for other reasons than its slowness. Construction of irrigation works has lagged somewhat; construction of drainage (an absolute requisite, as we have shown in other chapters) has lagged even more seriously; and many of the drains built have not been maintained in effective operation. The new farmers who have been granted lands have not been provided with adequate credit, technical advice, or supervision.

The Iraq land reform established a ceiling of 620 acres irrigated land or 1,235 acres unirrigated land—much higher than in either Egypt or Syria, reflecting the generally much lower pressure of population on land in Iraq. Land was to be distributed in lots of at least 18½ acres, but not exceeding 37 acres irrigated land.

Agricultural output of the three main crops of wheat, barley, and rice dropped by 25 percent in the first three years of agrarian reform. In large part, this was due to the peculiarities of the Iraq crop rotation, discussed in Chapter 4. The necessity of fallowing irrigated lands, so that they could dry out, and then cropping them only one year before salt accumulations forced another year of fallow, was described there. This practice was difficult enough when fields and farms were large; when they were smaller, it was difficult to dry out one field when adjacent ones were being irrigated.

Table 7-2

Iraq: Land Tenure Before 1958 Revolution

Area per holding (in dunums[a])	Landowners		Land area	
	Number	Percent of total	Thousand dunums[a]	Percent of total
Less than 4	57,958	34.5	73	0.3
4–30	56,725	33.7	697	3.0
30–100	30,119	17.9	1,677	7.2
100–1,000	20,126	11.9	5,025	21.5
1,000–10,000	3,143	1.838	9,090	39.0
10,000–50,000	251	0.15	4,554	19.5
50,000–100,000	19	0.01	1,334	5.8
100,000 and more	5	0.002	877	3.7
Total	168,346	100.0	23,327	100.0

[a] One Iraqi dunum = 0.25 hectares = 0.61 acres.

Source: Abdul Jalil El-Hadithy and Ahmed El-Dujaili, "Problems of Implementation of Agrarian Reform in Iraq," in *Land Policy in the Near East,* compiled by Mohamad Riad El-Ghonemy (Rome: FAO, 1967).

---

[10] Charles Issawi, *Egypt in Revolution—An Economic Analysis* (London: Oxford University Press, 1963), p. 61.

Size of Holdings

The only information we have on size of holdings in the Middle East is a certain amount of data on number of farms classified according to their total land area. But acreage of all land, without regard to its productivity, or even according to whether or not it is irrigated in most cases, and without regard to the many other inputs into farming, is a poor measure of economic size of farm. Moreover, for reasons discussed previously, one cannot always be certain that the available data relate to farms as operating units rather than to landownership units. With the limitations in mind, the data on percentage of farms by size are nevertheless interesting (Table 7-3).

The overwhelming impression given is that farms of the Middle East are predominantly small in area. The percentage of all farms with fewer than ten acres of land of all kinds is 97% for Egypt, 86% for Lebanon, 65% for Jordan, 70% for the irrigated areas of Syria and 34% for the unirrigated areas, and perhaps 50% for Iraq. Since Egypt has no unirrigated crop land, these data clearly relate to irrigated land (except to the degree a little waste or overflowed land may be included); for Syria, irrigated and unirrigated lands are shown separately; for the other countries, the data intermingle irrigated and unirrigated land, and presumably even noncrop land to the extent that this is found on farms, indiscriminately.

Yet there are some very large farms in each country except in Egypt where the land reform led to the subdivision of all farms larger than about 100 acres. Although very large farms may be few in number, they may contain a very large proportion of all farmland. Unfortunately, we do not have data for some countries. In Lebanon, less than 1% of all farms with more than 124 acres including 15% of all farmland; in Iraq, much larger holdings with 1,240 acres or more, comprising slightly less than 1% of all farms, contained about 55% of all farmland. However, lacking information on land quality by size of landholding we do not know to what extent, if at all, the larger holdings were predominantly lower-yielding dryland or grazing land. Thus comparative analysis on a size basis is beyond our means at this point.

Types of Farms

In the United States, it has long been customary to classify farms according to type. The dominant classification is the one used by the census of agriculture; it is based primarily upon the value of products sold. Farms are classed as cotton, tobacco, dairy, poultry, or something else, depending upon the relative income from the specified products. While farms are grouped into about a dozen major types, there is obviously a continuum in sources of income, from the farm which has every dollar of its income

Table 7-3

Percentage of Landholdings, by Size Group, by Country[a]

Hectares	Acre equivalent	Egypt[b]	Lebanon[c]	Jordan[d]	Syria[e] wholly irrigated	wholly unirrig.	Iraq[f]
Under ½	Under 1.24		35		8.9	6.2	
½ to 1.0	1.24 to 2.47	94.5	19	36.4	10.3	6.2	29
1.0 to 2.0	2.47 to 4.94		19	10.9	19.8	8.0	
2.0 to 3.0	4.94 to 7.41	2.4	8	10.0	17.8	7.1	
3.0 to 4.0	7.41 to 9.88		5	8.1	13.3	6.6	28
4.0 to 5.0	9.88 to 12.4		5	5.8	7.7	6.5	
5.0 to 10.0	12.4 to 24.7	1.9	6	15.2	14.5	17.6	12
10.0 to 20.0	24.7 to 49.4	0.9	2	8.6	5.9	19.3	12
20.0 to 50.0	49.4 to 124	0.3	1	4.0	1.3	17.3	12
50 to 100	124 to 247	-		0.7	0.1	3.8	4
100 plus	247 plus	-		0.3	0.4	2.4	3
Total		100.0	100	100.0	100.0	100.0	100

[a] No comparable data for Israel available.

[b] Landholdings in 1965; class intervals do not exactly coincide with those shown; see Appendix Table C7-1.

[c] Number of agricultural holdings in 1961; see Appendix Table C7-2.

[d] Number of holdings 1965; see Appendix Table C7-3.

[e] Number of farmers 1962; omits farms partly irrigated and partly unirrigated; see Appendix Table C7-6.

[f] Number of cultivated units and holdings, 1958/59; see Appendix Table C7-7.

from a particular source to one which has several sources of income, though one is dominant. The census of agriculture presents a wide array of data about farms of each type—numbers of farms, acreages, distribution by size, amounts of income, tenure, expenditures for a major range of inputs, and the like. All of this information is extremely valuable in understanding agriculture in a large country, such as the United States, where natural and economic conditions vary greatly from one area to another.

There is nothing remotely comparable to this for any Middle Eastern country. Various studies in some countries have discussed farms of different types, although rigorous definitions of types have usually not been made; but the national statistics on agriculture do not include an analysis of farms by type. Accordingly, it is impossible to present any statistics or any reasonably comprehensive analysis.

In general, in the Middle East as a whole, grazing of livestock is a separate type of farm: livestock graziers do not raise crops, crop farmers do not graze livestock on native pastures. The division between settled farmer and nomadic grazier, which we shall explore a little further in Chapter 9, is of ancient standing. In some parts of the region, notably in some parts of Syria and Iraq, there are specialized grain farms, primarily on lands with marginal or only slightly supramarginal rainfall; they utilize mechanical power and combine harvesters, and often are relatively large. This type of grain farming is found in many parts of the world—the Great Plains and elsewhere in the United States and Canada, in Australia, in Argentina, and in parts of the Soviet Union—and has many similarities wherever it is found.

In Egypt, most farms grow some cotton, some wheat, some other grains, and perhaps some vegetables and may possess one or more head of cattle (most frequently a buffalo). If the U. S. definitions of farm type were applied to Egypt, probably nearly all these farms would qualify as cotton farms; while cotton might occupy a third or less of their crop acreage, the value of the cotton crop is likely to overshadow all other sources of income. Apparently only a few farms in Egypt specialize in the production of vegetables and fruits, although the relative importance of these crops may vary from farm to farm, or village to village, depending in part on market opportunity. Rice is grown in some parts of the Nile Valley, and probably some farms would be classed as rice farms, if source of income were the criterion.

In Israel, there are a considerable number of specialized citrus growing farms or groves, many of which produce relatively little else. In addition, some citrus is grown on farms where only a minor part of the total acreage is in this fruit although it may constitute a major part of the total income. The typical Israeli farm, as it developed prior to 1948 and as it continued in large measure after that date, was a diversified one; some vegetables and some fruits were grown for sale, as well as for farm-family consumption; some dairy cows were kept, utilizing various feeds grown on the farm; and often a poultry enterprise was included. Even a decade or more ago, however, there were varying degrees of farm specialization within this general pattern; and as

time has gone on, farms have tended to become somewhat more specialized. In the Judean Hills, farms have tended to specialize in deciduous fruits, for example. Grapes are grown in still other somewhat specialized areas. Farmers in many of the newer agricultural settlements have emphasized vegetable growing more, partly because it is a labor-intensive type of enterprise, where they can utilize their often ample labor supply productively; and many older farms have reduced their vegetable production or gone out of it entirely. There has been some tendency to specialize in dairy production on some farms, although dairy cows are still found widely; and specialized poultry farms have grown up, to produce broilers or eggs or both. But in spite of this trend toward farm specialization, farms in Israel tend to be rather diversified; and diversity characterizes the agriculture of each part of Israel.

In Lebanon, there is some tendency—though less marked than in Israel—in the direction of diversification. Many farms grow various fruits, including olives and grapes; they may also grow some vegetables; and many are likely to grow some grain as well. Similarly, in the irrigated areas of Syria and Jordan, and in the higher rainfall, rainfed areas of each, there is some tendency in that direction, but an even less marked one. Many farms are primarily grain and olive farms; grain is grown in winter, and the deep-rooted olive trees are able to produce, even though the summers are rainless. Irrigation in these countries has not merely enabled the production of higher yields, but has permitted the development of new crops, or the expansion of crops previously known but not much grown because of moisture limitations. Throughout the region, there are farms which probably would be classified as subsistence level under the U. S. classification scheme; more than half of their total output is consumed on the farm, by the farm family, and yet they are basically farmers, working at off-farm work little or not at all.

Managerial Decisions

The processes by which a farmer reaches decisions on production and marketing of farm commodities, and the factors which influence, if not govern his decisions, are extremely important. Knowing very little about this aspect of Middle Eastern agriculture, we must presume that Middle Eastern farmers or farmers anywhere make those choices which they think best, in terms of their long-run welfare, and taking due account of risks of losses—within the limitations of knowledge and of personal resources of labor and capital, and within the whole institutional structure in which they find themselves (capital supply, credit availability, technical information, farm supplies or inputs, transportation facilities, market facilities, and the like). Many farmers in the Middle East are illiterate or nearly so; but, as has often been said, they can count even if they can't read.

Government plays a large role in farmers' decisions in almost every country of the world; but a convincing case may be made that the role of government is greater in the Middle East than in many other countries. Much of the decision-making—what to produce, how much and what kinds of inputs to use, how and where to market, and the like—has been shifted from the operating farmer to the government bureaucrat. This does not necessarily mean that the resulting decisions are less efficient, in an economic sense, than they would be if the farmer had wider scope for action. It may be argued, in these countries, that a government expert with his information and professional expertise is able to make an economically rational decision better than is a farmer with his more limited knowledge. The reverse might also be argued; we know of no research in the Middle East which really tests either argument.

In Egypt, there are legal limitations as to the maximum and minimum acreages that may be grown of certain crops, notably wheat and cotton. The availability of irrigation water—not only whether any water is available for each farmer—is determined by the Ministry of Irrigation and further controlled by the extremely cumbersome and indirect method of lowering water level below the land surface (see Chapter 4).

The agricultural cooperatives have become essentially the only source of seeds, fertilizers, chemicals for insect and disease control, credit generally, and other basic farm inputs. Although the co-ops are theoretically directed by the farmers, each has a government-representative adviser; and considerable doubt exists as to the degree to which such advisers actually dominate the co-ops, so that they become, in effect, instruments of the government. Marketing of cotton is solely through the co-ops, and they are playing an increasing role in the marketing of other crops. In times past, the government has had an active role in marketing cotton abroad or in storing it as a means of influencing the price; since Egyptian cotton is long staple, not directly competitive with the shorter staple cottons, some have argued that a monopsonistic policy by the Egyptian government might increase total returns to its cotton producers. For some years, a substantial share of the Egyptian cotton production has been committed to the servicing of loans from the U.S.S.R. In all of these ways, the decision making of the Egyptian fellaheen is strongly influenced or sharply controlled.

In Israel, there has always been intensive central planning for agriculture, by the Ministry of Agriculture for the established and older settlements and by the Jewish Agency for the newer settlements. The amount and the terms of agricultural credit have been determined by decision of these agencies, and this had been a powerful weapon in influencing, if not determining, decisions by farmers. Irrigation has been developed by a government agency, which also allocates water to settlements.

Marketing quotas have been established for some commodities, notably vegetables, and allocated among settlements by one or both of the planning agencies. Imports have been controlled, or at least strongly influenced by the government, for farm machinery and other inputs. Subsidies have been paid for several agricultural products. Citrus marketing has been influenced by government planning. In these and a great many other ways, the Israeli moshav farmer or the agricultural kibbutz is strongly influenced in its production and marketing decisions by the two agencies listed.

Much less seems to be known about the factors affecting farmer decision-making in the other countries of the Middle East. In Iraq, any major extension of irrigation to presently irrigated lands and to potentially irrigable lands will depend upon government construction of major facilities. Much more important, productive use of much land for which water is now available, and still more for any additional land that may be irrigated, depends on the government construction and maintenance of adequate drains. All evidence indicates that the drainage program has been handled very badly in Iraq to date. Some drains have been constructed, but often not maintained, and soon become ineffective. Provision of irrigation water and drainage will not be very productive unless companion measures are instituted to provide technical services, transportation, credit, necessary inputs such as improved seeds and fertilizers, marketing facilities, and the like. While there is growing recognition in Iraq of the needs for much better governmental programs, yet competent managerial and technical manpower has not been available even to spend the funds appropriated for such programs, let alone doing an adequate job. This is a matter to which we shall return in Part III, where we consider future possibilities.

In Syria, one interesting form of agricultural decision-making has arisen and been quite successful. Much of the growth in cotton production in that country, which we consider in Chapter 8, has been on lands irrigated by pumping from the Euphrates River. The capital and the entrepreneurship for this irrigation development, and for cotton production, has come primarily from urban merchants; the farmers have typically been tenants, operating under conditions closely controlled by these merchant landlords. While it may be argued that the landlords have prospered while the tenants have not, yet this type of entrepreneurship has defintely produced a rapidly expanding efficient cotton production. This is one of the few examples in this region of private urban entrepreneurship reaching into agricultural areas.

In all the countries of this region, as in most other countries around the world, the provision of technical advice and information has been in the hands of various public agencies. It is quite probable that this technical information has been provided with varying degrees of competence and skill.

CHAPTER 8

Crops[1]

HAVING so far dealt with the resource base, inputs—including labor—and organization, we consider in this chapter the physical aspects of crop production in the Middle East—acreages, yields, and total production. Several factors stand out in a general view of crop acreages in the Middle East (Table 8-1). Firstly, there are very large differences in the size of the agricultural establishment. Egypt has by far the largest acreage of crops harvested, with about one-third of the regional total; Syria and Iraq have almost equal acreages, and these two, with Egypt, account for about 85 percent of the total; Israel, Lebanon, and Jordan between them have a smaller crop area than any other country. Secondly, the crop area of the region is dominated by cereals, with about two-thirds of the total area devoted to grains of various kinds. Wheat is the dominant cereal, with a third of the total crop acreage and with over half of the cereal acreage.

This ranking holds only for physical data. On a value basis, the relationship among the crops is very different. As we shall show in Chapter 10, industrial crops such as cotton, fruits, and vegetables have far more per acre value than the cereals, Moreover, on a tonnage production basis the relationships are different, as the data in the appendix tables show; fruits and vegetables produce far more weight output per unit area than do the grains.

The relationship between the area of crops harvested and the area of arable land (including land under permanent crops) is an important characteristic of agriculture, as was noted in Chapter 7. The area of crops harvested may be larger than the area of arable land if two or more crops are harvested annually from the same tract; in the Middle East, this is generally possible only if the land is irrigated. The area of crops harvested may be less than the area of arable land if some land is fallowed each year, or if some is idle, or if there are crop failures. A great deal of the rainfed area of the Middle East is on an alternate crop-fallow system; so is some of the irrigated land in Iraq, as was noted in Chapter 4.

The relation between crop acreage harvested and acreage of arable land differs sharply between the individual countries (Table 8-2 and Figure 8-1). Egypt gets 1.6 crops per year, on the average, per acre. A good deal of the irrigated land in that country bears two or three crops annually; on the other hand, some crops, such as sugar cane, occupy the land throughout such a long growing season that no second crop can be harvested during a twelve-month period. O'Brien has shown how the ratio of crop acreage harvested to cultivated area has risen from 1.0 in the first half of the nineteenth century to 1.2 toward the

end of the century, to 1.45 by the time of World War I, to 1.55 by the latter 1920's and 1930's, to about 1.6 by the postwar period.[2] The rise in this ratio is closely related to the increase in acreage of cotton grown, for cotton is a summer crop, supplementing the age-old winter crops of grain.

In Israel, the ratio of crops acreage harvested to arable land is slightly below 1.0; this ratio is higher on irrigated land, some of which produces more than one crop annually, but the ratio is much lower on the rainfed land, much of which is on an alternate crop-fallow system. In all the other countries, the ratio is well below unity and is much nearer one-half; the acreage of fallow and failure each year far outweighs the acreage from which two or more crops are obtained. As noted above, in Iraq even the irrigated lands are on a crop-fallow rotation.

For the sake of perspective, one might mention that the ratio of crop acreage harvested to arable land is well below unity also in the United States. Due to a number of factors, including the country's relatively generous inheritance of good crop land, double-cropping is relatively uncommon in the United States. On the other hand, a great deal of arable land is fallowed, or is in conservation reserve, or is seeded to pasture annually—all of which reduces the harvested-arable ration. From this quick resume it is obvious that one promising method for raising output is to boost the intensity of land, first to, and then above, unity.

Wheat

The Middle East is the ancient home of wheat. Here wild wheats were domesticated and developed as a cultivated crop. From here, wheat spread around the world, finding in many areas an excellent if not better environment than in its original home. All Middle Eastern wheat is winter wheat; irrigated in the Nile Valley and some other areas; rainfed on extensive areas of Iraq, Syria, and down into Lebanon, Jordan, and Israel where rainfall will permit a crop. The cool winter climate is also more congenial to wheat than is the hot summer climate of this region.

The acreage of wheat grown annually has expanded since World War II in Syria and Iraq, as cultivation has spread onto lands with lower rainfall (Appendix Table C8-1). This type of wheat production is usually mechanized, and is similar to wheat production in climatically similar regions of USSR, the United States, Canada, and Australia. There is some evidence to indicate that this postwar expansion of

---

[1] See Appendix Tables C8-1 to C8-44 for details of acreages, yields, and output—by crops, by countries, and by years.

[2] Patrick O'Brien, *The Revolution in Egypt's Economic System, from Private Enterprise to Socialism, 1952-1965,* (London: Oxford University Press, 1966), p. 5.

Table 8-1

Crop Acreages, by Countries, Average 1960-64

(1,000 Acres)

	Egypt	Israel	Lebanon	Jordan	Syria	Iraq	Total reported
Cereals:							
Wheat	1,443	144	162	648	3,616	3,726	9,739
Barley	135	161	34	218	1,859	2,761	5,168
Maize	1,795	4	15	f	18	8	1,840
Rice	830	––	0	0	1	222	1,053
Other[a]	571	71[d]	31	144	600	187	1,604
Subtotal	4,774	380	242	1,010	6,094	6,904	19,404
Industrial crops:							
Cotton	1,816	36	0	0	663	83	2,598
Sugarcane	125	––	e	0	e	c	125
Sugar beets	c	14	3	0	12	c	29
Tobacco	c	6	12	12[g]	23	33	86
Other	c	29	10	0	21	c	60
Subtotal	1,941	85	25	12	719	116	2,898
Vegetables:							
Tomatoes	165	8	6	49	37	54	319
Cucumbers	40	7	7	52[h]	f	25	131
Potatoes	70	12	15	4	8	f	109
Onions & garlic	67	5	4	8	13	13	110
Other	224	25	28	29	184	208	698
Subtotal	566	57	60	142	242	300	1,367
Fruits:							
Citrus	113*	101	22	4	3	f	243
Olives	3*	29	68	139	304	f	543
Grapes	25*	29	60	48	171	f	333
Other[b]	54*	44	65	39	134	277	613
Subtotal	195*	203	215	230	612	277	1,732
Forage crops:							
Berseem	2,532	0	0	0	0	0	2,532
Other	c	150	––	––	f	f	150
Subtotal	2,532	150	––	––	––	––	2,682
All other	662	74	––	––	––	––	736
Total harvested crop area	10,669	949	543	1,394	7,667	7,597	28,819

* If data unavailable for one of these years, another year
  substituted or average based on fewer years.
[a] Includes legumes used as grains—such as lentils and chick peas.
[b] Includes almonds.
[c] If any, included in "all other."
[d] Including pulses.
[e] Less than 0.5.
[f] Not reported separately.
[g] *FAO Production Yearbook,* average 1961-65.
[h] Includes watermelons and melons.

Sources: Egypt: Appendix Tables C8-10, 13, 15.
Israel: Appendix Tables C8-16, 18, 19.
Lebanon: Appendix Table C8-22.
Jordan: Appendix Table C8-29.
Syria: Appendix Table C8-34.
Iraq: Appendix Table C8-42.

wheat acreage has been in part upon submarginal lands, where annual rainfall is too low and too variable for sustained economic wheat production. Elsewhere in the region, wheat acreage has been rather stable. In Egypt this grows out of the rigid crop rotation described in Chapter 4. In Lebanon and Jordan, wheat has long occupied most of the rainfed lands capable of growing economical yields.

Outside of Egypt, wheat yields per unit of cropped area have been highly variable from year to year (Appendix Table C8-2 and Figure 8-2). In Egypt, yields seem to have

Table 8-2

Total Acreage of Harvested Crops, Average 1960-64, in Relation to Arable Land (plus Land Under Permanent Crops), ca. 1964, by Countries

	Egypt	Israel	Lebanon	Jordan	Syria	Iraq	United States
1. Total acreage of harvested crops, average 1960-64 (1,000 acres)	10,669	949	543	1,394	7,667	7 597	296,946
2. Arable land plus land under permanent crops, 1964 (1,000 acres)	6,603	1,016	731	2,817[a]	16,442	18,528	458,034
3. Ratio, 1 to 2	1.616	.934	.743	.49	.47	.41	.648

[a] This is the highest of any estimates that have come to the authors' attention (see especially Chapter 4, p. 32). It is carried here in order to have the same source for all countries. Probably the ratio shown above is substantially higher, but there is no firm basis for determining it.

Source: FAO Production Yearbook, 1966.

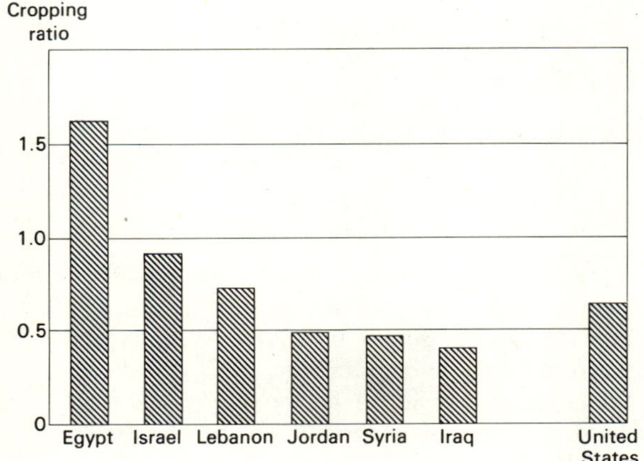

Fig. 8-1. Cropping Ratio,* by Countries, ca. 1964
* Total acreage of crops harvested, average 1960-64, divided by area of arable land plus land under permanent crops, ca. 1964.

Source: Table 8-2.

risen somewhat slowly but significantly up to 1940; they fell sharply during World War II. Fertilizer supply was restricted during that war, and other factors may have caused difficulties in grain production. By the mid-1950's wheat yield in Egypt had regained the prewar level, and since then seems to have risen steadily and steeply (this conclusion rests, as do all other analyses in this book, upon official statistics, and if these have tended to be somewhat on the optimistic side, then the conclusions are similarly affected.) Wheat yields in Egypt have always been above those in the United States as a whole; the latter are held in check by the extensive areas of relatively low yielding rainfed areas. Wheat yields in Egypt are not especially high, however, when one considers that all Egyptian wheat is *irrigated* wheat; in the United States, on the Columbia Basin irrigation project, where winter wheat is grown, yields in recent years have been about 85 bushels per acre, or about double recent Egyptian yields.

In Iraq, wheat yields are low and highly variable; they mostly fall in the 5 to 12 bushels per acre range. Here, most wheat is grown on rainfed land, in spite of extensive areas of irrigated land, in large part because the latter is too saline for wheat to grow. Wheat yields in Iraq show no upward trend for the years for which data are available. As we shall point out later, wheat yields on these lands could be increased greatly. Wheat yields in Syria are generally similar, in both level and variability, to those in Iraq.

For nonirrigated wheat, the year-to-year variation in yields is primarily a matter of moisture supply, and this in turn is closely related to total precipitation. The relationship between wheat yield and annual rainfall in Jordan provides a typical example (Figure 8-3). Not only does more rain mean higher wheat yield, but the increase in yield is more than proportionate to the increase in rain—an increase of 10 points in the rain index leads to an increase of about 17 points in the wheat yield index. Even this probably understates the true relationship, for in low rainfall years some land is not harvested because yields are too low to repay even the harvesting cost and thus does not enter the yield statistics, while in higher rainfall years such land would be harvested. This relationship is derived from data for natural rain, but the same results could obviously be obtained by better conservation of the moisture which does fall, such as much more efficient fallowing as a means of increasing available moisture and thus of increasing wheat yield.

Total wheat production is greatly affected by yields per unit of area, hence is rather stable from year to year in Egypt and is highly variable from year to year in other countries (appendix Table C8-3). In Syria and Iraq, for example, total output in the better years may be fully three times total output in poorer years. In high output years, an export surplus of wheat may be produced; in low output years, imports are necessary to provide for domestic consumption. The economic possibilities of storage to even out the domestic supply has for years been high on the agenda of foreign assistance agencies and remains so to this day.[3]

---

[3] Almost uniformly, the early World Bank missions focused heavily on the need for additional storage capacity.

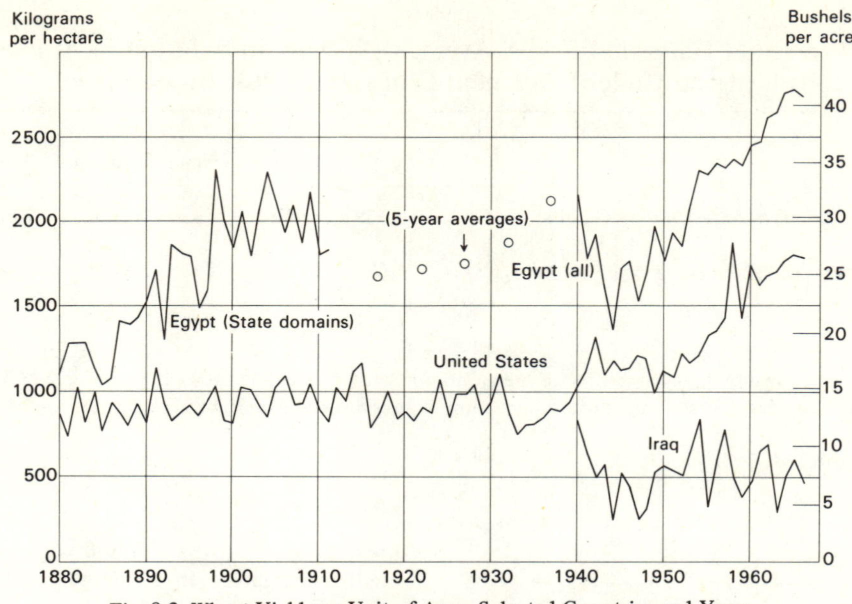

Fig. 8-2. Wheat Yield per Unit of Area, Selected Countries and Years

Source: Egypt and Iraq: Appendix Table C8-2; United States: *Agricultural Statistics* (U.S. Dept. of Agriculture)

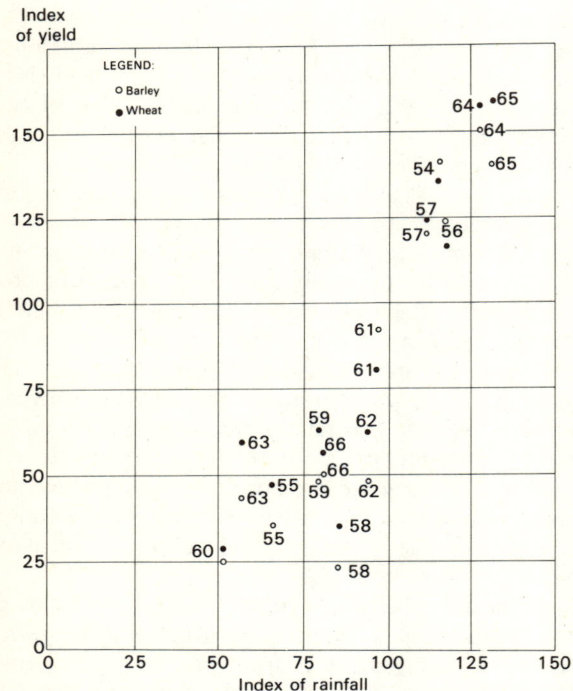

Fig. 8-3. Indexes of Annual Rainfall, Wheat Yields, and Barley Yields, Jordan, 1954-66. (1954-56 = 100)

Source: FAO *Jordan Country Report, 1967,* p. 52.

NOTE: Points show for year indicated, yield and associated rainfall, in terms of 1954-56 average.

The costs and difficulties associated with sporadic export movements as well as the costs of moving in the necessary supply in years of low output, could justify a considerable storage cost to smooth out domestic supply.

Barley

Barley is another ancient crop of the Middle East. It is generally similar to wheat in its moisture and climatic requirements, and is grown in winter, on irrigated land in Egypt and in Iraq, and on rainfed lands elsewhere. One big difference from wheat, as far as the Middle East is concerned, is that barley is more tolerant of salts in the soil than is wheat, for which reason it is extensively grown on the very salty irrigated lands of Iraq, for example. Others are its earlier maturing and the fact that it is less demanding of spring rainfall, and thus more dependable in areas of marginal moisture supply.

The acreage of barley grown in Egypt is only about one-fifteenth that of wheat (Appendix Table C8-4). In contrast, the acreage of the two crops is about equal in Israel. Much less barley than wheat is grown in Lebanon and in Jordan. The same is true in Syria and in Iraq; here the relative differences are not as great, although in absolute terms they are considerable.

Despite some advantages just cited, barley yield per unit of crop area is low and erratic in the rainfed areas, for the same general reason that wheat yield is low and erratic in the same areas—lack of moisture (Appendix Table C8-5 and Figure 8-4). Barley yield in the rainfed areas shows no upward trend since the war, and is far below average barley yield in the United States, where it is raised on unirrigated land but in regions where rainfall is relatively ample.

In Egypt, barley yields have a marked long upward trend. Data for yields to World War I which apply only to State land may not be representative of all crop lands in Egypt—in the case of wheat yield, on Figure 8-2, such lands seem to have yielded higher than all other crop lands, while for barley, as shown in Figure 8-4, they yielded much less. But,

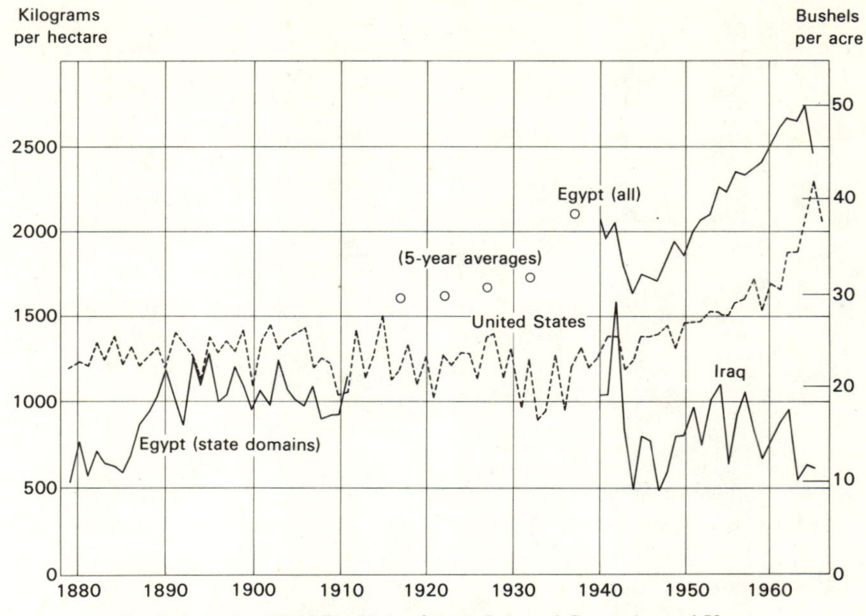

Fig. 8-4. Barley Yield Per Unit of Area Selected Countries and Years

Source: Appendix Table C8-5.

for all crop lands in Egypt since World War I, a notable upward trend in yield is evident, from about 30 bushels per acre to nearly 50 bushels today. In spite of this marked upward trend in barley yields in Egypt, the acreage of barley there has rather steadily declined over the same period. Apparently, the comparative advantage of growing barley, as contrasted with wheat, has declined. Or it may be that the total market for barley is more or less constant, so that acreage has declined about as yield has increased.

Total barley production in the rainfed areas is highly variable from year to year (Appendix Table C8-6). Output has fallen by three-fourths, from one year to the next, as in Jordan and Syria from 1954 to 1955 or in these same two countries from 1957 to 1958; or it has about doubled from one year to the next. Such variability in feed grain production is not conducive to stable livestock feeding operations.

Maize (Corn)

Maize is an important crop in the Middle East only in Egypt, where it fits into the rotation as a late summer crop. Its acreage is thus fairly stable, from year to year (Appendix Table C8-10), ranging from 1.5 to 2.0 million acres in most years.

No upward trend in maize yield per unit of cropped area is evident until 1960 (Appendix Table C8-11). Assuming that the data are accurate, yields since 1960 have risen considerably; it remains to be seen how far this represents a trend and is associated with the introduction of hybrid corn at that time. However, even in spite of this marked improvement in yields in recent years, Egypt's yields are still low by American standards, especially for corn under irrigation where moisture is not a major limiting factor. The highest reported Egyptian yield was reached in 1966—about

60 bushels per acre. In the United States, corn harvested for grain yielded on the average about 25 bushels per acre until World War II; but, influenced by hybrid varieties and other improvements, yields have risen greatly in the last three decades. In 1966, the U.S. average yield was 72 bushels, that for Iowa was 89, and each has continued upward. As a very rough judgment, it would appear that a considerable increase in maize yield is possible in Egypt.

Total production of maize in Egypt has been fairly stable (Appendix Table C8-12), ranging in recent years from 70 to 80 million bushels.

Rice

Three-fourths of all the rice grown in the Middle East is grown in Egypt and nearly all the rest is in Iraq. The area has risen considerably in each country since the war (Appendix Tables C8-10 and C8-42). Yields have risen more or less steadily in Egypt (Appendix Table C8-11). Rice yields have been highly variable in Iraq, with no clear trend evident (Appendix Table C8-43). There are encouraging reports of major improvements in rice yields in Iraq in 1967 and 1968, partly based on special fertilizer allocations, but it is too soon to judge whether these mark a sustained upward trend from the very low levels of about 1,200 pounds per acre in the preceding years. Rice yields in Egypt are higher. In the early sixties only Italy, Spain, and Australia exceeded them; and U. S. yields were about 1,000 pounds lower per acre (always in terms of rough rice). In 1966—68, however, the situation was reversed, and U.S. yields have shot ahead of Egyptian yields by 4-500 pounds; and Japan now too has forged ahead of Egypt. With the spread of new varieties the world picture will undoubtedly undergo changes, but it is too soon to evaluate recent statistics.

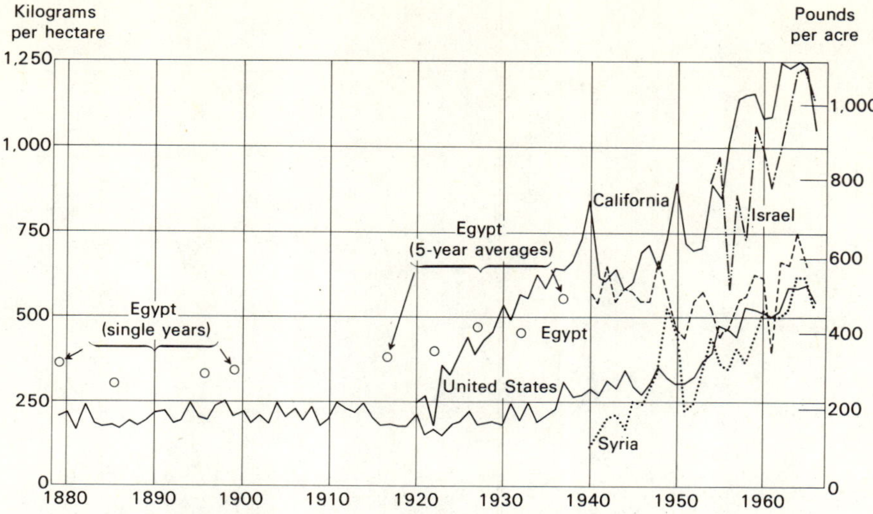

Fig. 8-5. Yield of Cotton (Lint), per Unit of Area, Selected Countries and Years.

Source: Appendix Table C8-8.

Cotton

Cotton has been grown commercially in Egypt for well over 100 years, and for shorter periods of time in other countries of the region. It was grown commercially in Egypt in the first half of the nineteenth century; but its real expansion came during the American Civil War, when cotton from the South was largely cut off from its European markets, and as a result cotton prices rose sharply. Cotton is a summer crop in the Middle East, as it is almost everywhere that it is grown; and in the Middle East it is almost always an irrigated crop. In order that it could be grown under irrigation in the Nile Valley, it was necessary that dams be constructed to divert the low summer flows of the river into some of the canals and channels that previously had been used only for flood waters, and then more often to carry the water off the land than to bring water to it. As a result of keeping these canals full throughout the summer, as well as actually spreading water on the land, the Nile Valley, especially in its lower (northern) parts suffered a serious rise in the level of the ground water. There is good reason to think that as a result, yields, especially of cotton, remained below the level they might otherwise have achieved.

Cotton acreage in Egypt rose during the second half of the nineteenth century and continued to rise up until World War I, since which time no sustained increase in acreage has taken place (Appendix Table C8-7). The acreage in recent decades has rarely dropped below 1½ million acres or run over 2 million acres. During World War II, acreage fell off sharply; fertilizers, insecticides, and other production materials were in short supply; and shipping was not available to get the cotton to market. Cotton piled up in storage, and was purchased by the government as a means of helping the farmers. After the war, cotton acreage increased again so that by 1948 or 1949 it had reached the prewar level. Even in 1950-52, when there was no governmental restraint on cotton growing and when cotton prices were abnormally high (due in large part to the Korean war), cotton acreage hardly exceeded 2 million acres. The stability of cotton acreage in Egypt is due in large part to its peculiar place in the crop rotation, as discussed in Chapter 4.

Cotton is a much newer crop in Israel, having been introduced only since 1950, and yet being grown on a relatively small scale. While it has been grown much longer in Syria, the real expansion of cotton in this country is a postwar phenomenon. Nearly all of Syria's cotton is irrigated, mostly by pumping from the Euphrates River. As noted in Chapter 7, much of this cotton development was initiated, financed, and directed by urban merchants who acquired land, installed pumps, and developed cotton production by tenant farmers. Some cotton is grown in Iraq but the trend in acreage there seems to have been downward since World War II.

There seems to be a general belief that cotton yields in Egypt are exceptionally high, as compared with other countries. While there was some merit in this view once, today cotton yields in Egypt are not high by international standards (Figure 8-5 and Appendix Table C8-8). The available data indicate a long slow rise in cotton yields in Egypt from the earliest data for which records are available until the outbreak of World War II. Even in this period, it was only the introduction or development of improved varieties, the greater use of fertilizer, and the greater use of insecticides which prevented a down trend in yield,[4] caused by the waterlogging of the soils. This in turn was a by-product of the summer irrigation in large areas: the water table was less than two meters, and in some areas no more than one meter (40 inches) below the ground surface. From 1940 to 1960, there is no clear upward trend in cotton yields in

---

[4]W. Laurence Balls, *The Yields of a Crop* (London: E. & F. N. Spon Limited, 1953).

Egypt; since the later 1950's or 1960 the reported figures show that cotton yield has advanced considerably. Since 1960, cotton yield has averaged slightly above a bale (500 pounds) per acre.

The trend in cotton yields in the United States has been quite different; no appreciable increase in yields is evident from 1880 to 1930, as cotton spread westward into drier areas and as the cotton boll weevil reduced yields; but a major rise in cotton yield began about 1930, resulting in a yield more than twice as high at present. Part of this increase has been due to the shifting of cotton acreage still further west, onto the irrigated lands of California and the Southwest; part has been due to a greater use of fertilizers; and part to other factors. For the last decade, U. S. cotton yields have been about as high as Egyptian cotton yields, though U. S. average yields are much depressed by extensive areas of rainfed cotton, some in areas where yields are rather low. A comparison with California, where, as in Egypt, all of the crop is irrigated, is instructive. Here yields are about double those for the United States as a whole and for Egypt. Moreover, the really significant fact about the California cotton yields are the way they have risen over the past forty years, compared with those in Egypt.

Syria, too, has shown a rapid rise in cotton yields in the postwar years, until today its yields are about comparable with those of the United States and of Egypt. These rapidly rising yields in Syria have been experienced at a time when cotton acreage was also expanding rapidly—in fact, high yields may well have provided much of the economic stimulus for increasing cotton acreage. A still more interesting case is that of Israel; although cotton is a new crop here, as noted previously, yields have always been well above those in Egypt and have trended upward steeply though somewhat irregularly. Today they are nearly double those in Egypt, and are almost as high as those in California. The point to note is that climatic and soil conditions in the Middle East do permit high cotton yields, when production inputs are also high.

Total cotton production in Egypt has averaged more than 1.8 million bales in recent years; in Syria, it has averaged about 650,000 bales; and in other countries, much less. Egyptian cotton is a long staple, of a different type than the shorter staple cotton grown elsewhere in this region and in most parts of the world; but Egypt does not have a monopoly of this type of cotton. The special marketing problems of Egyptian cotton are discussed in Chapters 11 and 16.

Vegetables

A large variety of crops are included under this general heading; for the Middle East, "vegetables" include potatoes, onions, and melons—which is not always the case in U. S. data and discussions. The total acreage varies considerably among the countries of the region, as we saw in Table 8-1; moreover, as that same table shows, vegetables as a percentage of all crop area varies from 3 or 4 percent in Syria and Iraq, to about 11 percent in Lebanon and Jordan. Some differences in makeup of the vegetable total are also shown in this table. Rather noteworthy is the small role played by potatoes, as compared with its role in the northern countries, such as those of northern Europe, for example.

With year-to-year fluctuations the trend in vegetable acreage is up for the countries of this region (Table 8-3). Total acreage of all vegetables increased by 50 percent or more during the decade from the late 1950's to the present. The various vegetables all have some seasonal characteristics, due in part to their climatic requirements. It is not always clear whether the official statistics report all vegetables grown, including those which the farmer consumes in his own family, or whether they are primarily marketed vegetables.

Partly because of these seasonal characteristics and for other reasons also, one must be careful about drawing too positive conclusions about vegetable yields, yet the available data are interesting (Table 8-4). Yields for each crop seem to vary considerably between countries, and mostly are lower than in the United States. Potatoes seem to yield about as high in Egypt and Israel as in the United States, but in other countries yields are only half as high. Tomato yields everywhere are lower than in the United States, being less than a fourth as high in Iraq; a similar situation seems to exist for onions. However, cucumbers and cabbage seem to yield as well as in the United States, although variations among countries are great.

Table 8-3

Acreage of All Vegetables (Including Potatoes and Onions), by Countries, Years of Record 1950-67

(1,000 Acres)

Year	Egypt[a]	Israel[b]	Lebanon[c]	Jordan[d]	Syria[e]	Iraq[f]	Total Reported
1967			75		302		377
1966	680		78	115	176		1,049
1965	647	60	75	170	226	398	1,573
1964	630	57	81	171	258	373	1,570
1963	605	58	65	141	361	314	1,544
1962	551	56	57	151	249	289	1,353
1961	531	53	50	143	199	274	1,256
1960	511	54	47	106	146	250	1,114
1959	494	52	52		179	166	942
1958			51		170		221
1957		55	45				100
1956		55	42				97
1955		53					53
1954		56					56
1953		55					55
1952		46					46
1951		37					37
1950		32					32

[a] See Appendix Table C8-13.
[b] See Appendix Table C8-19.
[c] See Appendix Table C8-22.
[d] See Appendix Table C8-29.
[e] See Appendix Table C8-35.
[f] See Appendix Table C8-42.

Table 8-4

Yield of Major Vegetables, by Countries, 1965

(Metric Tons per Hectare*)

	Egypt[a]	Israel[b]	Lebanon[c]	Jordan[d]	Syria[e]	Iraq[f]	United States
Potatoes	19.25	21.8	12.0	9.6	11.1		23.5
Tomatoes	15.25		10.6	9.2	8.2	7.0	30.3
Cabbage	25.04			14.8[g]			20.0
Cucumbers	12.47		5.5	7.0		6.6	9.7
Onions			11.5	6.0[h]	7.7	8.4	32.4
Watermelons			10.3	7.4			11.7

* To obtain short tons per acre, multiply by 0.446.

[a] See Appendix Tables C8-14–1964.

[b] See Appendix Tables C8-19, and 21.

[c] See Appendix Table C8-23.

[d] See Appendix Table C8-30.

[e] See Appendix Table C8-36.

[f] See Appendix Tables C8-42, and 44.

[g] Including cauliflower.

[h] Including garlic.

The available statistics seem to suggest a somewhat smaller variety of vegetables in most countries than in the United States (see appropriate tables, Appendix C-8). The relatively small size and generally similar climatic conditions in most parts of each country would suggest that this might be true. One might also expect somewhat poorer seasonal availability of vegetables than in a large country like the United States; and the probability of a seasonal discrepancy between available supplies and effective demand is probably greater. The general lack of vegetable-processing industry greatly reduces the possibilities for balancing seasonal production with seasonal demand. Much hope is pinned on the role of vegetables (and fruit) in the years to come in the Middle East, but in this chapter we limit description to what exists now.

Fruits

The available information about fruit production in the Middle East is somewhat variable from country to country, in part because of certain difficulties which seem to plague such data in all countries, and in part because of the great variety of specific crops found under this general label. The appendix material carries what information is readily available, but differs in physical factors covered, number of years, varieties included, etc. No summary text table is, therefore, presented.

Olives and grapes are ancient crops in this ancient land, being frequently mentioned in the Bible; together, they include today about 40 percent of the total fruit area of the region. Each is a deep-rooted crop and thus capable of drawing soil moisture from deep levels to sustain itself and produce a crop during the long rainless summers which characterize this region. Many of these trees and vines are found in the hill country, often on terraced land, with many very small terraces—a situation not well suited to most other crops. Many of the olive trees are quite old and yield poorly. Since these are each long-lived plants, the acreage—though not the yield— is stable from year to year; there has been a rather steady increase in their area in Syria during the past two decades, and grape area has been expanded considerably in Israel in the same period; elsewhere, the acreage of each has been rather stable.

Citrus—primarily oranges—is relatively more of a newcomer to this region, at least on a large scale; it now accounts for about 10 percent of the total fruit acreage. Its acreage has expanded greatly in Israel in the past two decades, and it is an important earner of foreign exchange for that country. The oranges grown there are of very high quality, yields are high, and the fruit has been marketed to good advantage in Western Europe, being among the first oranges in the season to reach these countries.

Dates are another ancient crop in this part of the world. They are found chiefly in Iraq and Saudi Arabia, and are the chief fruit for those two countries. Indeed, Iraq is the largest date exporting country in the world.

A wide variety of fruits are found in the hill country of Israel, Lebanon, Jordan, and Syria, although the total acreage involved is not large. There has been a recent major expansion of such fruits into the hills of Israel, accompanied by a major improvement in their quality. And Lebanese apples reach into most Arab countries and beyond.

As with vegetables, so with fruits there is relatively little processing for use outside of the fresh season. Dates are dried, of course, some wine is made, and some other processing takes place; but not on anything like the scale which characterizes fruit production in the United States.

Yields of citrus in Israel, Lebanon, and Jordan compare well with U. S. standards (Table 8-5). Elsewhere, citrus

Table 8-5

Yield of Citrus, Olives, and Grapes, by Countries*, ca. 1965

(Kilograms per Hectare)

	Egypt[a]	Israel[b]	Lebanon[c]	Jordan[d]	Syria[e]	United States[f]
Citrus	9,378[g]	19,511	19,000	19,748	3,000[g]	22,576
Olives	10,043	909	1,900	621	560	
Grapes	9,232	7,273	4,600	4,000	2,940	16,515

* No information available for Iraq.

[a] See Appendix Table C8-15.

[b] See Appendix Tables C8-18 and 21.

[c] See Appendix Table C8-23.

[d] See Appendix Table C8-30.

[e] See Appendix Table C8-36.

[f] Data for citrus from *Agricultural Statistics 1967,* USDA; data for grapes from *FAO Production Yearbook, 1966.*

[g] Oranges only.

yields are lower, ranging downward to marginal in many areas. Olive yields are high in Egypt, where, like all other crops, they are irrigated. Elsewhere, olives are almost never irrigated; and consequently yields are very low, the fruits are very small, and harvesting becomes expensive in terms of amount of labor required. But these small olives are also highly flavored, and much liked by many people who know them. Grape yields in all countries are low by U. S. standards, and especially so in the hill country. However—and matters of this sort make comparison difficult—it is also true that hill-country grapes for wine in the United States yield much lower than the larger acreages of more common grapes for eating fresh, making raisins, or for more common wines. Few raisins are made in the Middle Eastern countries considered in this report; most of the grapes are eaten fresh, although, as noted, some are used for wine.

Summary

The Middle East raises some high-value crops, such as cotton and most of its fruits and vegetables; but its total crop acreage is dominated by large areas of cereals—crops which are usually relatively low-valued per acre, especially so in the Middle East because of the low yields obtained.

With exceptions noted above, the Middle East is, by and large, a region of rather low crop yields. Yields of cereals are especially low; yield of cotton is moderately good but not outstanding; fruit and vegetable yields are only fair. Natural resources would allow yields of nearly all crops to be *much higher* than they are now; in Part III, we return to this subject, and suggest some of the ways in which crop yields could be substantially improved.

CHAPTER 9

Livestock[1]

Afew outstanding characteristics of livestock production in the Middle East may be noted at the outset:

(1) We know much less about livestock numbers, production, management, and marketing than we do about the comparable aspects of crop production. There is a serious paucity of data about livestock; there is also reason to suspect that available data on livestock are less accurate than data on crops—to some extent, this probably reflects the nature of the livestock production, as described in this chapter. In Volume II of FAO's Indicative World Plan for the Near East this is said about livestock data: "None of the countries have a complete set of livestock industry statistics. A number of the countries possess incomplete estimates for various sectors; in some countries different sources yield conflicting data, especially in the production sector, where estimates sometimes vary by 300 percent or more."[2]

(2) Crop production dominates the agriculture of the Middle East, with livestock production decidedly in second place. Livestock produces less than one-fourth of total agricultural output in Syria, about one-third in Egypt, Jordan, Lebanon, and slightly more than two-fifths in Israel and Iraq; in contrast, in the United States, livestock accounts for about 56 percent of total agricultural income. More data on this are found in Chapter 10.

(3) There is a substantial divorcement between much of the livestock production and any form of crop production. Much of the livestock obtains its feed from grazing on native range lands, with little use of feed from harvested land or at least from land operated by the man who owns the livestock. On most farms, the livestock which is kept is distinctly subsidiary to the crop production; there is rarely that close integration of cropping and livestock programs which typifies the Corn Belt farm of the United States, for example.

(4) The nomadic grazing of livestock is regional and international, rather than primarily national. The flocks and the herds follow the grass (or other forage), and these movements are in turn highly influenced by comparative rainfall in the different areas. From time immemorial, herds and flocks have moved over considerable distances, as feed availability dictated; modern nationalism has imposed considerable boundaries to such livestock movement, and most countries are making an effort to persuade their Bedouin to take up permanent residences. But such livestock still move about a good deal, a condition that contributes to the unreliability of the data. Whenever feasible, owners market their livestock output where market conditions are best. Conse-

quently, livestock output in each country each year is not necessarily an accurate reflection of the production of the livestock claimed by that country. Tempered by these cautions, the following may be said about the livestock economy of the countries under review.

Livestock Numbers

There are about four million head of cattle in the Middle East, of which about 40 percent each are in Egypt and Iraq. Except for Iraq, a good share of such cattle as are found are located on farms; and in Egypt, all of them are on farms, since the Egyptian deserts provide no feed adequate for cattle. Elsewhere, the native grazing is less well suited to cattle than to sheep, goats, and camels. Most of the cattle are native breeds; but some herds of dairy cattle of Western breeds (chiefly Holstein-Friesian) are found in all countries, and in Israel these are dominant. On irrigated or other crop farms, the cattle often perform a number of functions; they may be used as work animals, they produce some milk, and they as well as their offspring provide meat. In Egypt, and to a lesser extent in Iraq, the cow is perhaps no more important than the buffalo, which performs to some extent similar functions. There are about as many buffalo as cattle in Egypt, but in Iraq their number is only about one-sixth as great—and elsewhere they are relatively unimportant or altogether absent.

The number of cattle in the region as a whole and in most countries has trended modestly upward since World War II, according to the best data available. The increase in Egypt has been modest, roughly 20 percent; in Israel, a very much larger increase has taken place, as agricultural colonization and development have proceeded, so that present numbers are six times or more those of twenty years ago; Lebanon, too, reports a major relative increase—about fivefold over the same period—but absolute numbers there, as well as in Israel, are not large; Syria reports a modest upward trend somewhat greater than that of Egypt; Jordan, in contrast, shows no clear upward trend; for Iraq, the data are erratic, but it too may lack an upward trend.

There are about 20 million head of sheep in the region, of which about half are in Iraq and about one-fourth in Syria. These are mostly or wholly range sheep and so are most of those in Jordan and Lebanon; in Israel and in Egypt, on the other hand, the sheep are kept primarily on irrigated farms. Sheep in the Middle East are an important source of milk; as elsewhere in the world, they also produce wool and meat. While most sheep in the Middle East are of breeds native to this region, they differ greatly in their milk-, wool-, and meat-producing capabilities.

Trends over time must be viewed with caution. In Egypt,

---

[1] For additional information, see Appendix Tables C9-1 to 18.
[2] *Indicative World Plan, Agricultural Development 1965–85, Near East* (Rome: FAO, 1966), II, xxvi.

sheep numbers show no consistent trend for thirty years or more, although numbers have fluctuated somewhat during that period; as with cattle, numbers have experienced a large relative rise in Israel and Lebanon but actual numbers are small; data in the earlier years seem suspiciously variable in Jordan (a fivefold increase between 1948 and 1954, which about exceeds the reproductive capacity of the sheep) but seem to have trended upward in the past decade; Syria shows a 75 percent increase over the past twenty years; Iraq also shows a major increase but not quite as marked as in Syria.

Since it is extremely doubtful that natural sources of feed have increased in this period—indeed, due to the plowing up of the better grazing lands to convert them into marginal grain lands, the opposite is more probable—one cannot help but wonder how much of the reported increase in sheep numbers is real and how much is a "statistical phenomenon."

In U. S. range management practice, when estimating feed demand, it is customary to convert sheep to cattle equivalent on the basis of five sheep per head of cattle. The proper conversion ratio depends in part upon whether cattle and sheep of all ages, including the very young, are counted or whether only the mature animals are counted— and also partly upon the season of the year, since the cow or ewe nursing her young makes a heavier demand for feed than does the animal on a maintenance basis only. On the basis of five sheep per head of cattle, the feed or forage demands of cattle and sheep in the Middle East are about equal. However, as we have noted, a large share of the cattle are on farms and most of the sheep are not, so the direct competition between them for forage is not great.

There are over seven million goats in the Middle East, of which more than half are in Iraq and Saudi Arabia, with comparatively large numbers in Syria, and smaller numbers in the other countries. As with sheep, goats are an important source of milk in the Middle East; in addition, they produce mohair and meat. Some of the goats—as are the sheep—are kept on farms, particularly in Egypt and Israel; but many are herded on to natural pastures or ranges. There has been little or no consistent upward trend in goat numbers in any country except Israel, nor in the region as a whole.

The goat has long had a very bad, and generally undeserved, reputation for destructiveness in grazing on natural ranges. It is true that the goat will eat many types of forage that neither cattle nor sheep will eat, particularly leaves and small stems of shrubs. The goat is also physically more active than sheep, literally climbing small trees in some cases to eat leaves that sheep cannot reach. But, if the goat has eaten many forage plants literally into the ground, the fault lies with the owner and manager who has held too many goats too long on the same tract of land. Properly managed, the goat is no more destructive than the sheep or the cow, and is a valuable animal for grazing some types of forage most productively.

Nearly a million camels are reported from the countries of the region, with about one-third each in Iraq and in Saudi Arabia, and most of the rest in Egypt. No consistent trends in numbers are apparent in most countries and these statistics may be particularly suspect. There is some reason to believe that the camel is becoming less useful in the region, as mechanized forms of transport gradually take over some of the functions once performed by this animal. The camel is also a source of milk and of meat in this region.

Meat Production

In considering data on meat production, it is necessary to distinguish between meat produced by slaughtering of animals raised in the country—"meat from indigenous animals"—and meat from animals slaughtered in the country but raised in and imported from other countries. The data presented in Table 9-1 are for meat from indigenous animals only; if any meat of animals had been exported (most of these countries are meat importers) it would also be included. Such data are thus a measure of meat production—not a measure of meat availability for consumption, which is considered in Chapter 12. As a further complication, data on meat production are highly variable in meaning and in accuracy. We have used the data published by FAO; some are total slaughter, some are commercial, some are inspected. The reported numbers may be somewhat too low, to the extent that farm or uninspected slaughter is excluded; however, the amount of such slaughter in these countries is probably not great. The usual data on meat production also exclude weight of edible offal, such as liver; these are shown separately, and included in the lower part of Table 9-1. In this region, as in the United States, "meat" frequently excludes chicken and other poultry, in spite of the obvious interchangeability of such food with the traditional red meats. The available data on the latter are also included in Table 9-1.

Meat production from cattle, buffalo, sheep, goats, and pigs in the countries reviewed totals about 400,000 metric tons annually; this is supplemented by nearly 200,000 metric tons of meat from poultry and from miscellaneous sources (chiefly edible offal and camels). Pork production is negligible; both Muslims and Jews refrain from eating pork on religious grounds. Of the meat from animals, roughly 60% is from cattle and buffalo, and roughly 40% from sheep and goats; or, if the comparison is based upon total meat production, roughly 40% is from cattle and buffalo, and 30% each from sheep and goats, and from chicken. For perspective it is interesting to compare this with the different makeup of total meat production in the United States: although beef and veal in the United States account for about 46% of the total, or somewhat more than in the Middle East, turkey production is three times that of lamb, pork is 29%, and all kinds of poultry about 25%. These differences in kinds of meat produced reflect, in part, differences in kinds of feed readily available to produce meat animals and birds; and, in part, differences in food consumption habits and demands of consumers, which in turn reflect differences in meat availability over long periods of time.

Of particular interest is the rapidly changing role that

Table 9-1

Production of Meat from Indigenous Animals and Poultry, by Countries, 1948–66

(1,000 Metric Tons for Animals;
100 Metric Tons for Poultry)

Beef, Veal, and Buffalo

Year	Egypt	Israel	Lebanon	Jordan	Syria	Iraq	Total reported
1966	185	15	2	2	9	43	256
1965	182	16	3	2	9	42	254
1964	178	19	2	3	9	40	251
1963	175	16	3	2	9	39	244
1962	171	15	3	2	9	35	235
1961	166	13		1	9	37	
1952–56 av.	164	2	2	4	8	29	209
1948–52 av.	130	1	2	4	5	21	163

Mutton, Lamb, and Goat

Year	Egypt	Israel	Lebanon	Jordan	Syria	Iraq	Total reported
1966	38	3	4	7	52	48	152
1965	36	3	3	7	51	47	147
1964	35	3	3	8	50	45	144
1963	34	3	3	6	50	47	143
1962	32	2	3	6	50	47	140
1961		2		6	50	47	
1952–56 av.	37	1	1	6	40	33	118
1948–52 av.	36		1	5	43	25	110

Pork

Year	Egypt	Israel	Lebanon	Jordan	Syria	Iraq	Total reported
1966	1	3	a	a	a	a	4
1965	1	3	a	a	a	a	4
1964	1	3	a	a	a	a	4
1963	1	4	a	a	a	a	5
1962	1	4	a	a	a	a	5
1961	2	4	a	a	a	a	6
1952–56 av.	1	2	a	a	a	a	3
1948–52 av.	1	1	a	a	a	a	2

Total Animals

Year	Egypt	Israel	Lebanon	Jordan	Syria	Iraq	Total reported
1966	224	21	6	9	61	91	412
1965	219	22	6	9	60	89	405
1964	214	25	5	11	59	85	399
1963	210	23	6	8	59	86	392

Total Animals – Continued

Year	Egypt	Israel	Lebanon	Jordan	Syria	Iraq	Total reported
1962	204	21	6	8	59	82	380
1961		19		7	59	84	
1952–56 av.	202	5	3	10	48	62	330
1948–52 av.	167	2	3	9	48	48	277

Poultry (100 metric tons)

Year	Egypt	Israel	Lebanon	Jordan	Syria	Iraq	Total reported
1966	700	819	184	50	28	29	1,810
1965	650	740	150	48	27	26	1,641
1964	600	745	125	46	28	25	1,569
1963	600	670	106	38	28	25	1,467
1962	529	664	102	20	25	24	1,344
1961	510	546	63	10	21	23	1,173
1952–56 av.	408	126	16	5	22	19	596
1948–52 av.	240	63		3	17	14	

Other Sources of Domestic Meat (100 metric tons)[b]

Year	Egypt	Israel	Lebanon	Jordan	Syria	Iraq	Total reported
1966	150	28		30	20	36	
1965	150	30		29	20	35	
1964	150	32		28	17	33	
1963	150	30		27	19	33	
1962	260	27		26	22	33	
1961	259	24		25	23	48	
1952–56 av.	247	7		8	14	15	
1948–52 av.	163	4		6	22	17	

Total, All Meat (1,000 metric tons)

Year	Egypt	Israel	Lebanon	Jordan	Syria	Iraq	Total reported
1966	309	106	24	17	66	97	619
1965	299	99	21	17	65	95	596
1964	289	103	18	18	64	91	583
1963	285	93	17	14	64	92	565
1962	283	90	16	13	64	88	554
1961		76		10	63	91	
1952–56 av.	267	19	10	11	52	65	424
1948–52 av.	207	9		10	52	51	

[a] None reported.
[b] Primarily camels and edible offal from all animals.

Source: FAO Production Yearbooks, 1966 and 1967—except poultry, Lebanon, which data is from Khaled M. Abed and Abdus Sattar, *Production and Supplies of Agricultural Products in Lebanon, 1954-1966* (A. Ec. No. 5, Ministry of Agriculture, Republic of Lebanon, May 1967).

chickens play in providing meat for the Middle East. Over three-quarters of the total meat production in Israel and just about that much in Lebanon is from chickens. In these two countries, the role of chicken has changed dramatically in the past decade; output in the latest year shown in Table 9-1 (and it is probably higher now) is more than ten times the output in the earliest period shown. For the most part, this represents intensive broiler production, similar in technique to that used in the United States. For the Middle East the chicken is the intensive converter of grain to animal protein, which the pig has traditionally been in the United States and in northern Europe.

Poultry production accounts for between 20–25 percent of total meat production in Egypt, and recently somewhat more than that in Jordan; some of this may be commercial broiler production, but much is from small flocks on farms. The trend in output has been upward in Egypt also, though at a slower rate. In the other countries of the region, poultry is an unimportant source of meat and shows only a modest rise.

To understand livestock production of the Middle East, it is helpful to express meat production, as presented in Table 9-1, as output per unit of livestock numbers—the latter as shown in Appendix Tables C9-1 to 3, supplemented by FAO data on numbers of buffalo, which are not shown here. The resulting figures—to which U. S. figures are added for comparison—are the outcome of birth and death rates among the livestock, age at which animals are slaughtered and other marketing practices, feed availability as reflected in weights per animal slaughtered, and other factors. The available data for 1965, are as follows:

Meat Production (Carcass Weight)
Per Head of Livestock Reported
in Inventory in 1965—in Kilograms

	Beef, veal, and buffalo	*Mutton, lamb, and goat*
Egypt	56	13
Israel	76	9
Lebanon	29	6
Jordan	27	4
Syria	17	8
Iraq	25	4
United States	82	10

Source: Table 9-1, Appendix Tables C9-1 to 3, and *FAO Production Yearbook, 1967.*

The generally low output of meat per head of livestock in the inventory is clearly evident. For the most part, this represents the cumulative effect of a number of factors— chief among them relatively low birth rates, relatively high death losses, disease, inadequate feed for maximum gains, slaughter of animals without intensive fattening. Some data on carcass weights of slaughtered animals are published by FAO, but such data are largely meaningless without further information on the age of the animals concerned. A low weight per animal slaughtered might be due to the slaughter of relatively young animals, or it might be because the

animals were small and thin for their age. The matter of animal nutrition is discussed below.

Other Livestock Products

The livestock products here considered are milk, eggs, and wool; though to the Egyptian fellaheen manure would possibly be a more important product than these.

In the Middle East, the cow is only one of several producers of milk (Table 9-2). This is in sharp contrast to the United States, where cows are so nearly the only producer of milk that Department of Agriculture statistics do not bother to specify that the "milk" they report is cow's milk, nor do they include any data on milk from other sources— which would presumably be totally insignificant in amount. For the Middle East as a whole—always as defined in this study—cow's milk is somewhat more than a third of the total supply. But it varies greatly: in Israel, it is 90 percent, in Saudi Arabia it is not much over 10 percent, and in the other countries it is at various intermediate percentages. For the region as a whole, the buffalo produces almost three-fourths as much milk as does the cow; but production is limited to Egypt (where the buffalo produces twice as much milk as the cow) and to Iraq. No milk from buffalo is reported from any other country of the region. For the whole region, milk from sheep and goats represents about one-quarter of all milk production, with goats contributing about half as much as sheep. As with other sources, this relationship varies considerably among the countries of the region.

Milk is an excellent food for humans, as well as for the young of each animal species; dietitians in the West have placed great importance on the consumption of milk as a part of an adequate and well-rounded diet. But there are serious problems to the production and consumption of milk in a relatively hot region, with generally small farmers, and usually poor marketing facilities. Milk is also a good food for many strains of bacteria, including some quite harmful to the human. Unless milk can be produced, transported, and marketed under conditions which keep such bacteria out of the milk, kill them by pasteurization or other means, or inhibit their growth by refrigeration or in other ways, or do all of these things, then the milk as it actually reaches the consumer may not only not be beneficial but can be harmful. A high per capita consumption of fluid milk is a Western custom and reflects Western standards of dietary adequacy; and this in turn arises from the fact that milk production is economically more efficient in cooler regions and its production and transport is easier where farms are larger. It may well be that the nutritional needs of people in the Middle East, as in other near-tropical countries, can better be met in other ways than by increased milk production, though the use of fermented products of milk lessens the problem. This is a subject to which we shall return in Part III.

The yield of milk per animal is an import measure of livestock production. It can be calculated on either of two bases, each of which has usefulness: (1) as total milk output per head of all animals, of all sexes and ages; and (2) as milk

Table 9-2
Milk Production, by Kind, by Countries, 1948–66
(1,000 Metric Tons)

Year	Egypt	Israel	Lebanon	Jordan	Syria	Iraq	Total reported
Milk from Cows							
1966	408	360	48	23	104	182	1,125
1965	402	333	48	20	101	182	1,086
1964	379	315	40	17	92	194	1,037
1963	374	307	49	12	88	194	1,024
1962	408	326	44	10	86	194	1,068
1961	391	292		12		186	
1952–56 av.	301	149	40	3	106	190	789
1948–52 av.	362	97	35	4	96	186	780
Milk from Buffalo							
1966	754	0	0	0	0	27	781
1965	744	0	0	0	0	27	771
1964	749	0	0	0	0	30	779
1963	739	0	0	0	0	30	769
1962	785	0	0	0	0	30	815
1961	761	0	0	0	0		
1952–56 av.	638	0	0	0	0	32	670
1948–52 av.	722	0	0	0	0	29	751
Milk from Goats							
1966	8	28	15	29	30	44	154
1965	8	28	15	26	30	44	151
1964	8	25	15	22	29	62	161
1963	8	25	19	13	21	62	148
1962	8	25	23	11	27	62	156
1961	8	26	10	13	19		
1952–56 av.	7	22	10	15	59	67	180
1948–52 av.	10	11	10	21	48	74	174
Milk from Sheep							
1966	12	18	12	34	125	166	367
1965	11	18	12	33	117	166	357
1964	11	20	12	26	105	142	316
1963	10	20	11	17	99	142	299
1962	10	19	11	13	88	142	283
1961	10	19		14			
1952–56 av.	7	8	10	9	92	145	271
1948–52 av.	9	5	10	12	69	150	255
Total Milk							
1966	1,182	406	75	86	259	419	2,427
1965	1,165	379	75	79	248	419	2,347
1964	1,147	360	67	65	226	428	2,293
1963	1,131	352	79	42	208	428	2,240
1962	1,211	370	78	34	201	428	2,322
1961	1,170	337		39			
1952–56 av.	953	179	70	27	213	434	1,876
1948–52 av.	1,103	113	65	37	247	439	2,004

Source: FAO Production Yearbooks, 1966 and *1967.*

output per milking animal. The first reflects livestock management practices— herd makeup, marketing practices, and others; the second shows what the actual producing animals do. Data for each, as far as they are available, are provided in Table 9-3. By either measure, milk output in Israel per head is higher than in the United States. According to the first measure, this reflects largely the fact that cattle in Israel and nearly all of the milking type (Holstein-Friesian, predominantly, and rather highly bred), whereas in the United States many are primarily beef cattle; according to the second, Israel has developed a very intensive type of dairying—fully as advanced as that in any part of the United States—whereas in the United States a good many cows of rather lower yields are milked, especially in the areas producing manufacturing milk.

The extreme gap between the two measures in Lebanon and Syria is interesting. On the one hand dairying in these countries—while not quite as intensive as in Israel—is in

Table 9-3
Milk Yield per Head of All Animals and per Milking Animal, by Kind of Livestock, by Country, 1965

	Milk yield (kg. per year)	
Country	Total milk divided by total number of animals, all ages and sexes	Per milking animal
I. Cattle		
Egypt	247	674
Israel	1,649	4,902[a]
Lebanon	457	2,150
Jordan	274	850
Syria	193	2,100
Iraq	125	
United States	517	3,767
II. Buffalo		
Egypt	452	
Iraq	120	
III. Goats		
Egypt	10.1	
Israel	184.2	
Lebanon	33.9	
Jordan	34.2	
Syria	36.1	
Iraq	23.8	
IV. Sheep		
Egypt	5.6	
Israel	92.3	
Lebanon	133.3	
Jordan	33.4	
Syria	21.6	
Iraq	15.0	

Calculated from Appendix Tables C9-1 to 3 and from Table 9-2.
[a] From crossbred cows only.

many respects roughly similar as to types of cattle, methods of feeding, methods of marketing milk, and the like; and those cows that are milked are high producers. On the other hand, only a minor proportion of all cattle are milked, so that milk output per head of all cattle is quite low. In Egypt, both measures are rather low; perhaps this reflects the fact that many of the cattle are also used for draft and other work purposes.

The great differences from one country to another in milk per head of goats and of sheep, of all ages, doubtless reflects in large part the kinds of sheep kept, their methods of management, and the degree to which the sheep are regarded as a source of milk. No data are available on output per milking animal. While it cannot be demonstrated, it is altogether possible that better breeding and better feed could greatly raise milk output per head.

The importance of egg production varies greatly among the countries of the region (Table 9-4). By and large, the same countries that are heavy producers of chicken meat also produce many eggs. Israel and Egypt lead in egg production, each with about 40 percent of the regional total. Nearly all of the eggs in Israel are produced under very intensive management, in rather large flocks—using individual cages for the hens, feeding them intensively, and the like. Since Israel is an exporter of eggs, these figures do not reflect egg consumption per capita. Egypt's large egg production reflects primarily its large number of farms, rather than intensive production. Production in the other countries is much less, but every country reports a marked upward trend in production since 1948, with Iraq showing the least relative increase, at about 40 percent; Egypt reports nearly a trebling, and other countries an even greater increase. The region as a whole shows a threefold increase.

Wool is a relatively minor product of the region, although it may be an important source of income to some livestock operators. Iraq produces about half of the regional total, and Syria a large share of the remainder. As with other animal production, output of wool per head is rather low, using U. S. production as a standard of comparison; but the statistics are probably not reliable enough to draw very useful conclusions.

Output of Wool per Head of Sheep, 1965

	Kilograms	Pounds
Egypt	1.4	3.1
Israel	2.1	4.6
Lebanon	5.6	12.3
Jordan	2.8	6.2
Syria	2.4	5.3
Iraq	1.2	2.6
United States	4.4	9.7

Source: Table 9-4 and Appendix Table C9-2.

Livestock Grazing and Land Use

On a very large part of the world's land surface, domestic livestock are produced by grazing upon native forage plants. The United States has an extensive range livestock industry,

Table 9-4

Egg and Wool Production, by Countries, 1948—65

Year	Egypt	Israel	Lebanon	Jordan	Syria	Iraq	Total reported
Hen Eggs (Millions)							
1966	1,244	1,223	240	110	222	212	3,251
1965	1,092	1,296	220	110	306	212	3,236
1964	1,095	1,279	204	107	293	208	3,186
1963	988	1,113	120	93	259	204	2,774
1962	915	1,273	99	89	226	200	2,802
1961	875	1,290	70	84	147	196	2,662
1952—56 av.	575	433	34	49	185	171	1,447
1948—52 av.	474	300	30	24	118	149	1,095
Wool (100 Metric Tons, Greasy Basis)							
1966	29	4	5		112	127	277
1965	28	4	11	28	132	127	330
1964	32	4		22	111	127	296
1963	40	4		9	94	127	274
1962	35	4		13	80	132	264
1961	32	3		15	63	127	240
1952—56 av.	34	2			87	151	274
1948—52 av.	20				76	135	231

Source: FAO Production Yearbooks, 1966 and 1967.

primarily in the western half of the country; and grazing is the largest single land use. Extensive areas in Siberia, Mongolia, Africa, Canada, Australia, and parts of South America are also dominantly range livestock producing regions. These areas have a number of common characteristics. One is that native or natural forage plants are used, although sometimes forage plants have been introduced from other countries. Secondly, the lands are generally not cultivated, largely because they are too dry, the soils are too poor or too rocky, or the growing seasons are too short. Furthermore, forage production per acre is typically low, highly variable from year to year, and often has a marked seasonal pattern of growth and availability for livestock. The dominant factor affecting forage production each year is the amount of precipitation, which tends to be highly variable in these areas.

Herding of domestic livestock to harvest the low yield of roughages from native ranges is the only economical way to convert these plants into food. To do so, flocks and herds must be moved from one area to another—as the available feed supply dictates—and this in turn depends upon the ups and downs of precipitation. Sometimes these livestock movements follow well-developed seasonal patterns, rather regular from year to year—especially when the movement is to higher mountain areas in summer and to foothills and deserts for winter; but sometimes the livestock movements are more irregular and less predictable. Throughout the world, in recent decades, plowing for grain production has encroached upon the better natural grazing lands. The tractor and other machines have made it possible to grow grain—sometimes with only indifferent financial success—on land that could not be used for this purpose when animal draft power had to be used.

In these and in other respects, the livestock grazing industry of the Middle East is similar to that in other parts of the world. Or rather, it was probably in the Middle East that nomadic grazing of domestic livestock began, long ago. In its essentials this form of agriculture has remained unchanged from its ancient beginnings.

In most of the grazing regions of the world, the owner of the livestock used land without firm tenure; he does not have an exclusive use, but shares that use with other livestock owners. Even when the law states that he should have exclusive use, enforcement against trespassers on widely flung lands which are not continuously occupied presents problems. The first man on the land with his livestock gets the grass; frequently, there is a race and a struggle for available grass. From a land-use point of view, the significant factor here is that under these circumstances no livestock operator has any interest in either managing or conserving forage. One major element in modern range management is the leaving of a substantial proportion— often as much as half—of the forage ungrazed, both to protect the vigor of the growing plant and to permit the production of some seed for replenishment of stands. In many situations, it can be shown by careful experiments that, over the long run, taking only half of the annual plant growth provides more forage than taking all of it year after year. But no one has an incentive to observe this practice if he knows that it will merely lead to someone else's taking advantage of his forbearance.

Overgrazing of range on grazing lands is thus an inevitable consequence of insecure land tenure. But firm land tenure is a necessary, not a sufficient, condition for maximum forage production from range land; for some livestock producers will overgraze their lands out of ignorance, failing to understand that by grazing off all the forage of the land each year, the actual forage production declines relative to the potential production from such lands, and in the end everyone loses. There is general agreement that overgrazing on dry lands of the Middle East is severe, in relation to that found in other regions of the world. Every range management technical advisory group to the Middle East—and there have been several—has emphasized that the numbers of livestock grazed are far greater than the number that would permit maximum forage production.[3]

But overgrazing also means that the animals are poorly fed. Most of such feed as they do obtain must be used to maintain them alive and in at least tolerable flesh; relatively

little energy is left for growth and for reproduction. Experiments have shown quite conclusively that a given amount of feed, if used by the proper number of animals, will produce a greater total output of livestock products than if it is grazed by a larger number of animals. Again, every range management survey or study of the Middle East has emphasized this fact.

An especially uneconomical land and livestock management practice in the Middle East is that of allowing weeds to grow on fallow land in the grain-fallow rotation, and then allowing grazing by range livestock. The primary function of fallow in a modern grain-growing program is the accumulation of moisture; moisture from one year is conserved, to supplement the moisture available in the year of grain growing. As was noted in Chapter 8, modest additions to available moisture supply result in much greater than proportionate additions to grain yield. But this function of fallow is completely lost if weeds grow on the land; they then take the moisture that the grain will need. Moreover, the weeds are a poor livestock feed—poor in both volume and quality. Both more feed for livestock and more grain would be produced in this way, and at lower total costs, if some of the lower-yielding land were retired from grain production and planted to permanent grasses, and on the better lands a strict grain-fallow rotation would be practiced with no weeds robbing the fallow of its moisture—or if a crop of grain or feed were grown each year.

Although we return to the subject in more detail in Part III, a few suggestions—none of them novel—for improvement of livestock output from grazing lands are noted at this point. Any single measure has limited value without the others. To be effective the package of measures for improving livestock grazing must include—to some degree—all of the following:

(1) A reduction of livestock numbers on each area to match the carrying capacity of the land; and this in turn requires some form of firm land tenure control and an informed and competent degree of land management, whether by licenses, leases, or other devices.

(2) Improvements on the range, such as reseeding of the best land, grazing rotation to permit seed production and reproduction of the plant species, development of livestock water to spread grazing use more evenly over the land, water spreading where topography is favorable, and others.

(3) Improved breeding of livestock, by systematic use of improved sires.

(4) Control of disease among range animals.

(5) Marketing to the best possible advantage of the products of the range livestock production. In that connection, in the longer run, a transition from animal to mechanical draft power would lead to the kind of livestock enterprise that focuses on production for its own sake and encourages appropriate practices.

---

[3] See, for example, Harold F. Heady, *Report to the Government of Saudi Arabia on Grazing Resources and Problems* (Rome: FAO, 1963); J. P. H. van der Veen, *Range Management and Fodder Development, Report to the Government of Syria* (Rome: FAO, 1967); Kamal M. Ibrahim, *Summary Report and Bibliography, The Pasture, Range and Fodder Crop Situation–The Near East* (Rome: FAO, 1967); Michael Baumer and O. Milton Hackett, "The Development of Natural Resources in Jordan," in *Nature and Resources*

(UNESCO), Bulletin of the International Hydrological Decade, I, No. 3 (September 1965); W. J. van Liere, *Classification and Rational Utilization of Soils* (Rome: FAO, 1965); Marvin Klemme, *Pasture Development and Range Management* (Rome: FAO, 1965).

CHAPTER 10

Inputs, Output, and Value Added

IN Chapters 1 to 3 of Part I, and in Chapters 4 to 9 of Part II, agriculture has been dealt with primarily in physical terms—areas of land, volumes of water, tons of fertilizer, numbers of tractors, numbers of workers, numbers of farms, tonnage of crop and livestock production, and the like. All of these matters are basic to an understanding of agriculture in the Middle East today and of its possibilities for the future. But they are also, to a degree, a prelude to the payoff: cost of inputs, value of output, and value added. These form the subject of this chapter.

As elsewhere, farmers in the Middle East produce both for their own use and for sale to nonfarm customers. Many farmers in this region consume a substantial proportion of their output, yet even these farmers have some surplus to sell, and it is the amount of income they can derive from such sales that often determines their welfare. Very few, if any, farmers in this region practice truly subsistence farming, in the sense of consuming virtually all that they produce and of producing virtually all that they consume. Most of them live in a money economy: it is money from sale of

Table 10-1

Gross Value of Agricultural Production (Excluding Intermediate Products*), by Major Sources, by Countries, ca. 1965

(Million U. S. $, at Official Exchange Rates)

	Egypt[a]	Israel[b]	Lebanon[c]	Jordan[d]	Syria[e]	Iraq[f]	Total reported
I. Crops:							
Cereals		12	6	28	91		
Cotton[g]	} 1,180	18	–		} 117	} 149	} 1,642
Other field crops		16	24	1			
Vegetables[h]	189	45	14	25	29	83	385
Citrus	} 76	79	} 47	} 16	} 51	31	} 352
Other fruits		52				31	
Others	15	–	–	–	–	–	15
Subtotal	1,460	222	91	70	288	263	2,394
II. Livestock:							
Milk	129	41	11	} 10	60	82	328
Eggs	25	37	9		7	8	91
Poultry Meat	80	45	9	7			
Animals: cattle	205	29			} 10	} 98	} 568
sheep & goats		6	} 8	} 12			
other	} 55	4					
Other		13	3	8	5	34	63
Subtotal	494	175	39	37	82	222	1,049
III. Miscellaneous	–	3	–		5	13	21
Total	1,954	400	130	107	375	498	3,464

* As far as the data permit, this utilizes the U. S. concept of "realized" farm income. It includes all products sold or consumed by the farm family but it excludes value of feed and grain crops produced and fed on the same farm to livestock (although feeds sold by one farmer and purchased by another are included). It excludes also the value of animal manure; it includes, to the extent data permit, increases in value of animals on farms, but excludes changes in value of land, buildings, farm machinery, etc.

[a] See Appendix Table C10-1.

[b] See Appendix Table C10-11.

[c] See Appendix Table C10-13.

[d] See Appendix Table C10-14. Cereals include legumes.

[e] Based upon data obtained from Ministry of Planning, "Real Estimates of National Income of Syria for 1964," (Mimeo No. 175, Damascus, February 1968). Data are for 1964.

[f] See Appendix Table C10-17.

[g] Including cottonseed.

[h] Including potatoes.

Table 10-2

Index of Agricultural Output, by Countries, Years of Record, 1872–1966
(1953 = 100)

Year	Egypt[a]	Israel[b]	Lebanon[c]	Jordan[d]	Syria[e]	Iraq[f]	United States[g]
1966		357.7	184.4		115.1	103.8	121.5
1965	130.2	335.9	180.8		158.1	103.8	123.7
1964	129.7	326.6	171.8		159.3	95.6	120.4
1963	126.8	292.9			147.7	88.2	120.4
1962	126.8	290.0			151.2	100.0	116.1
1961	108.0	267.8			100.0		115.1
1960	124.1	239.3	123.9		79.1		114.0
1959	120.9	230.7			89.5		110.8
1958	120.0	199.6			88.4		109.7
1957	115.1	171.7			124.4		102.2
1956	110.5	155.6			116.3		104.3
1955	106.7	130.8	100.0		88.4		103.2
1954	110.4	120.0			112.8		100.0
1953	100.0	100.0			100.0		100.0
1952	108.5	92.4			89.5		98.9
1951	97.5	75.1					95.9
1950	101.1	71.4					92.5
1949	106.0	54.5					93.5
1948	108.2						94.6
1947	90.7						87.1
1946	88.2						90.3
1945	89.4						87.1
1944	80.9						89.2
1943	76.4						86.0
1942	86.5						88.2
1941	97.1						78.5
1940	107.8						75.3
1939	108.5						73.1
1938	102.3						72.0
1937	113.5						74.2
1936	106.7						59.1
1935	103.4						65.6
1934	91.9						54.8
1933	98.0						62.6
1932	87.2						68.0
1931	86.3						70.1
1930	96.2						64.0
1929	104.4						65.3
1928	96.8						66.7
1927	88.2						64.0
1926	92.3						64.0
1925	93.0						62.6
1924	87.7						60.6
1923	83.9						60.6
1922	81.0						59.9
1921	68.9						54.6
1920	77.8						62.0
1919	73.4						57.3
1918	74.9						57.9
1917	84.8						57.9
1916	71.7						53.9
1915	76.9						59.3
1914	78.2						57.9
1913	89.3						52.5
1912	80.7						58.6
1911							53.2
1910							53.2
1909							53.8
1908							55.7
1907	77.3						53.8
1906							56.8
1905							53.0
1904							53.8
1903							51.5
1902	73.3						50.8
1901							50.0
1900							50.4
1899							50.0
1898							50.8
1897	66.7						48.1
1896							45.4
1895							41.7
1894							40.5
1893							37.5
1892							37.1
1891							47.0
1890							39.4
1889							39.8
1888							37.9
1887							36.4
1886							36.7
1885							36.7
1884							37.5
1883							34.5
1882							34.8
1881							31.1
1880							33.3
1879							32.2
1878							31.1
1877							29.9
1876							26.5
1875	35.9						26.1
1874							24.2
1873							24.6
1872							25.0

Footnotes to Table 10-2, see facing page.

their surplus production in which they are primarily interested, and it is money which is needed to buy a wide variety of producer and consumption goods.

But monetary comparisons of agricultural inputs, output, and value added always present problems, and these are particularly acute in the Middle East. First, comparisons over time, within each country, may be confused by a persistent upward drift of prices in general. This is a problem in each country of this region. It is particularly acute in Israel, where the general price level has increased by nearly twenty times in the past twenty years. Secondly, comparisons between countries for a given year, and even more so for a series of years, are difficult because of the exchange rate problem. We have used official exchange rates, to convert local currencies to U. S. dollars as the most widely used common measure of value, familiar to most readers of this book; but official exchange rates may be "nominal" in that they fail accurately to reflect real exchange rates or the basic economic conditions in the different countries. However, any attempt in this context to construct more "realistic" exchange rates would probably raise more problems than it solves. As a way of adjusting for price-level changes over time we have, wherever possible, made comparisons in terms of constant prices; but this adjustment too may be insufficient, since price relationships existing during the base period may have shifted over time. In a number of situations, however, differences among some countries are so extreme that no reasonably conceivable adjustment in exchange rates would affect them.

Gross Value of Agricultural Production

In 1965, the Middle East had a gross value of agricultural production of about $3½ billion (Table 10-1). This was about 8 percent of the comparable figure for the United States in the same year. Egypt had more than half of the total; Lebanon and Jordan less then 5 percent each. More than two-thirds of the total was from crops, the balance from livestock. Although available data are somewhat incomplete, it appears that cotton was the most important single product, accounting for possibly as much as a fourth of the total, reflecting above all the great importance of cotton in Egypt, but also its rising production elsewhere.

The trend in agricultural output varies between countries, but for the region as a whole is generally upward (Table 10-2). There are no historical data for Jordan, and the record is short and uncertain in direction for Iraq. Syria's output has fluctuated sharply, but recent years show much higher levels than do the earlier years. The trend has been more steadily upward in Egypt, and has been steeply and regularly upward in Israel.

The Egyptian situation deserves a closer look (Figure 10-1). Fortunately, it has been possible for various scholars, as well as for Egyptian official sources, to construct an agricultural production series that goes back, though not on an annual basis, for several decades—even though not all of

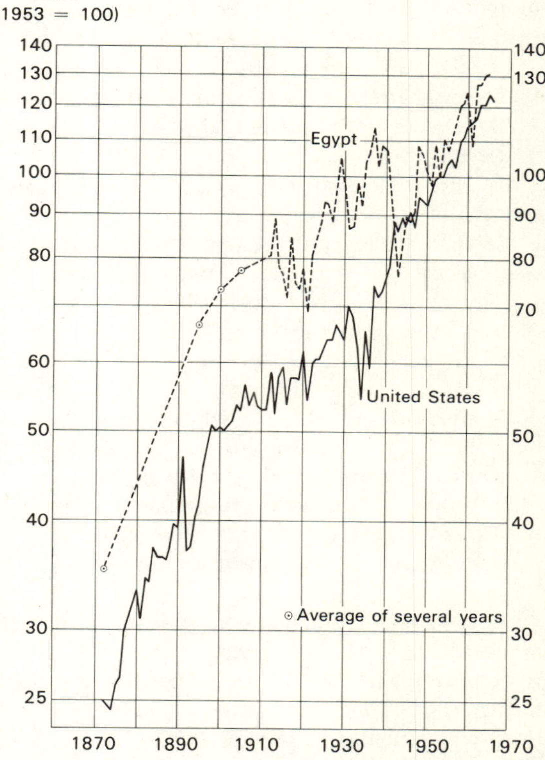

Fig. 10-1. Indexes of Agricultural Output, Egypt and United States, 1872-1967.
Source: Table 10-2.

---

[a] Data for 1956 to 1965 taken from Appendix Table C10-2, calculated to a basis of 1953 = 100.0; data for 1913 to 1955 inclusive taken from Donald C. Mead, *Growth and Structural Change in the Egyptian Economy* (Homewood, Ill.: Richard D. Irwin, 1967), p. 320; data for earlier years taken from Patrick O'Brien, "The Long-Term Growth of Agricultural Production in Egypt: 1921-1962," in *Political and Social Change in Modern Egypt*, ed. by P. M. Holt (London: Oxford University Press, 1968)—data for 1895 to 1910-14 from p. 188, for 1872-78 from p. 185, all calculated to 1953 base as nearly as possible.

[b] Data from Appendix Table C10-12; those data on agricultural year, Sept.-Oct., are entered here in year in which the greater part of the period lies; data are in constant prices, but include intermediate products in all years, including the base year; all calculated to base 1952/53 = 100.0.

[c] Data taken from Appendix Table C10-13; value of gross agricultural production at constant prices; calculated on base 1954/56 = 100.0, since no data available for 1953.

[d] No data available; Appendix Table C10-14 shows value of agricultural output but in terms of current prices.

[e] Data taken from Appendix Table C10-15; recalculated on basis of 1953 = 100.0

[f] Data taken from Appendix Table C10-17; calculated on basis of 1962 = 100.0, since no data available for 1953.

[g] Data for 1934 to 1966 taken from *Agricultural Statistics 1967*, p. 544; data for 1910 to 1933 inclusive taken from *Agricultural Statistics 1952*, p. 661, in each case data for "Farm Output"; data for 1870 to 1909 inclusive taken from Neal Potter and Francis T. Christy, Jr., *Trends in Natural Resource Commodities* (Baltimore: Johns Hopkins Press, 1962), p. 148, for all data, calculated to base of 1953 = 100.0.

the data in this series may fully reflect the real situation, and the concept of output may not be the same for all years. Again, a comparison with U. S. experience lends some perspective to the data. (Obviously, comparisons with other countries could be made and might, for certain purposes, be more appropriate and instructive. The choice of the United States, here as elsewhere, reflects the authors' more intimate acquaintance with its history and statistics and the anticipated makeup of the audience.) From it one concludes that up to 1910 total agricultural output rose about as rapidly in Egypt as it did in the United States, though for different reasons. In Egypt, additional land was brought under irrigation in these decades; cotton became a major summer crop complementing the historic winter cereal crops; and the ratio of crops harvested to land under cultivation—i.e., the cropping intensity—rose. In the United States, the area of farmland rose sharply in these decades, as the whole country was settled and new lands were brought under the plow.

During the two decades or so between World Wars I and II, the output of Egyptian agriculture again rose about as rapidly as did that of the United States. Since World War II Egyptian agriculture has not been as much above the pre-war level as agriculture has been in the United States; if it had advanced at the same pace as did the United States, Egyptian agricultural output would today be about 50 percent above its current level. As it was, agriculture in Egypt was hit hard by both world wars, which created serious supply and marketing problems, growing out of shipping difficulties. Output dropped in the first war, but it plummeted in the second—and each time it has taken some years to recover to postwar levels. Indeed, it is highly doubtful if in either case it recovered to a position on the prewar trend line.

World War II brought the Egyptian agricultural output index down to that of the United States; in a sense, this is but a statistical result of using 1953 as the common base. But in another and more serious sense, it represents a real loss in Egyptian agricultural progress. Since the end of World War II, the agricultural output indexes of the two countries have moved closely parallel; though output in the United States has varied less from year to year, and the upward trend has been steadier.

The comparisons contained in Figure 10-2 are *not* evidence that Egyptian agriculture has moved upward in the past as fast as was reasonably possible, nor do they imply anything about future trends. Nevertheless, it is significant that Egyptian agriculture, laboring under some rather serious handicaps, has been able to make the substantial increases in output which this chart shows.

A more detailed presentation of the situation since World War II, for which data are available for Syria and Israel, as well as Egypt and the United States, is made in Figure 10-2. Again, one may note the general similarity between trends in Egypt and the United States; output in Syria, on the other hand, has been erratic, going down considerably and for several years, then bounding rapidly upward, to end the period about 50 percent above the beginning. Increases in output have been dramatic in Israel.

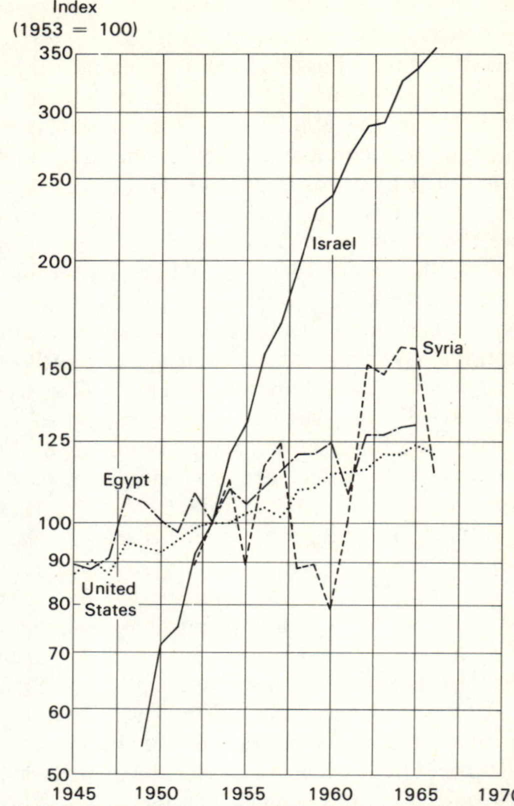

Fig. 10-2. Indexes of Agricultural Output, Four Countries, 1947-67. *Source:* Table 10-2.

Output at the end of the less than twenty years of record was six times that at the beginning; each year has been higher than the preceding one, by a rate which approximates 15 percent per year for the whole period. Many new agricultural settlements were established after 1948; heavy immigration offered rapidly expanding markets; irrigation water was developed; agriculture has been intensified and received major assistance at the highest levels of government; and, of course, output started from a very small base. There are some slight signs of a slowing down, beginning with 1959, but the upward rate in Israel is still far higher than in any other country of this region.

The experience in Israel well demonstrates that agricultural output is not merely the result of areas of land or numbers of livestock multiplied by output per unit of land or of livestock; it is also a matter of the "mix" of output. Some tonnages "weigh" more heavily than others, depending on their unit value. True, per acre yields of some important crops have trended upward in Israel, as have per unit outputs of some kinds of livestock. But it has been the development of the intensive fruit, vegetable, poultry, and dairy enterprises—in contrast with the cereals and staple crops that cannot be expanded as fast and as far and are economically less productive—which has been primarily responsible for the steep upward trend in output of Israel's agriculture. These in turn have depended largely upon the growth of per capita income generally within the country, enabling consumers to buy these foods.

Inputs

Data on agricultural inputs are poorer than data on agricultural outputs; both because they are less complete and available for shorter periods of time, and because there is considerable doubt that the same range of items is included for the different countries. Here, as elsewhere in this book, we have used official statistics, including their definitions. The concept of agricultural output, noted in footnote* to Table 10-1, is the same one used in U. S. agricultural statistics. There may be doubt that all of the data conform to it; in particular, it may be that the values of some livestock feeds have been included when they should not be, and in spite of our efforts to exclude them; even if true, this does not greatly affect the totals for each country. The data on inputs should include only production materials and services purchased from outside the agricultural sector, but should include all such goods and services as well. Thus, value of livestock feed and of manure produced on the farms and consumed there should not be

Table 10-3

Cost of Agricultural Inputs, by Countries, ca. 1965

(Million U.S. $)

Item	Egypt[a]	Israel[b]	Lebanon[c]	Jordan[d]	Syria[e]	Iraq[f]	Total reported
Seeds[g]	59	1		2.9	17	2	81.9
Fertilizer[h]	218	9		1.1	6	4	238.1
Pesticides[i]	13	6		0.3	1	1	21.3
Livestock feeds[j]	152	61		5.3	27	17	262.3
Fuel, oil, & lubricants[k]	31	6		*	5	4	46.0
Machinery[l]	4	11		1.5	9	4	29.5
Depreciation[m]	7	32		*	*	*	39.0
Irrigation water[n]	0	16		1.0	1	*	18.0
Packaging materials[o]	*	18		*	2	*	20
Transport[p]	*	14		*	*	*	14
Govt. services[q]	*	5		*	*	*	5
Interest[r]	*	*		*	1	*	2
Land tax	*	*		*	*	2	2
Miscellaneous[s]	5	6		0.9	1	29	41.9
Total	489	185		12.9	71	63	821

* Not listed separately.

[a] Appendix Table C10-3; 1965, but at 1959 prices; at 1965 prices and exchange rate, total is 540 instead of 489.

[b] Appendix Table C10-10a, 1964/5.

[c] No data available.

[d] Appendix Table C10-14, 1965.

[e] Based on data from Ministry of Planning, Syria, 1964.

[f] Appendix Table C10-18, 1965.

[g] In Israel, includes seed cleaning; in Jordan, includes seedlings.

[h] Although this item should include only fertilizer purchased from outside of the agricultural sector, there is reason to suspect it includes farm-produced manures in some countries; in Egypt, for instance, nearly half is "natural fertilizers."

[i] In Egypt, listed as "insecticides"; in Israel, as "plant and animal protection materials."

[j] Although this item should include only feeds purchased from outside of the agricultural sector, there is reason to suspect that in some countries it includes the value of feeds produced within agriculture; in Egypt, listed as "fodder"; in Israel, "feedingstuffs," apparently all purchased outside agriculture; in Jordan, "animal feed"; in Syria, "feedstuff"; in Iraq, "animal feed". In none of these three latter countries is there any indication of the extent, if any, to which the feeds come from within the agricultural sector.

[k] In Israel, includes electricity also.

[l] In Egypt, "maintenance," which we assumed was for machinery; in Israel, "spare parts, repairs, and tools"; in Jordan, "machinery expenses"; in Syria, "maintenance of machinery"; and in Iraq, "machinery."

[m] In Egypt and in Israel, "depreciation," but not clear on what investment.

[n] In Jordan, "irrigation expenses."

[o] In Syria, "sacking and packing."

[p] Average distance over which goods are transported multiplied by unit cost per/ton-kilometer.

[q] Indirect taxes paid to central and local government.

[r] In Syria, "interest (banks)."

[s] In addition to items listed as "miscellaneous" or "miscellaneous services," this includes: in Egypt, "eggs for hatcheries"; in Jordan, "imported chicks"; and in Iraq, "waste" is $6 million and "other requirements of animal production" is $22 million.

included; we suspect that it is, in some countries, and this, of course, mars comparability. Moreover, it appears that some items of expense are not included in some countries. In spite of these probable shortcomings, the data offer some interesting comparisons, though one should refrain from pushing them very far.

On the basis of the available data, livestock feeds are the largest single item of expense in agricultural production in the Middle East (Table 10-3). However, this regional total is dominated by Egypt, and may largely reflect the value of farm-raised, farm-fed livestock feeds, probably including an imputed value for the clover and other fodder feed to cattle; it is also large in Israel, and there it is rather clearly the imported feeds that are utilized by the extensive poultry enterprises and, to a lesser extent, the dairy enterprise. The expenditure for fertilizer is next largest in size; it may include the value of farm-produced manure in Egypt, but, in any event, fertilizer from outside the agricultural sector is a large item of cost in several countries. The cost of seeds is also large, but again the extent to which these are farm produced is not clear.

On the other hand, reported expenditures for some items seem too low. Only Israel and Syria explicitly mention packaging materials, for instance; the costs for machinery seem extremely low in some countries. Only Syria shows any cost for interest, or for capital borrowed outside the agricultural economy. Only Israel shows any costs for transport—possibly other countries measure farm output at the farm, but even then there would be transport costs for other inputs, which might be included in cost of such inputs. Only Israel shows a specific charge against agriculture for government services, but these may be included in other items in other countries. Other items could be mentioned, in which one cannot help but wonder if comparable definitions and data coverage have been used.

Useful data on the trend in agricultural inputs are available only for Egypt and Israel (Table 10-4). In Egypt, inputs vary from year to year without any consistent trend. In Israel, inputs have trended sharply upward, increasing by about three times in the ten years for which data are available; the greatly increased output, noted above, required a more or less proportionate increase in inputs. For comparison, the U. S. experience is shown on Table 10-4. Total inputs into agriculture have remained constant, with the sharp drop in labor input offsetting the increases in some others. The dramatic increase has been in fertilizer and lime, with more modest but still substantial increases in (1) feed, seed, and livestock purchases, and (2) miscellaneous. Inputs of machinery (not shown in Table 10-4) have increased but little during these years.

Value Added

Since value added by agriculture is the difference between the value of output and the cost of the inputs—i.e. purchased goods and services—it inherits the errors in both. But it is also the most valuable comparison, because it shows how much the agricultural producer has left to spend outside of agriculture. In the Middle East, the ratio of the cost of inputs to the value of the output varies greatly from country to country. Assuming that data limitations do not render comparisons meaningless, the ratio is low in all the Arab countries, ranging from 12 percent in Jordan to 27 percent in Egypt. This phenomenon may be looked at in two ways. In one sense, a low ratio is a measure of high efficiency; the output is high in relation to the input. But in another, and we think more important sense, such a low ratio is a measure of inefficiency; inputs are basically only land and human labor—and with low inputs, output per worker, and (except in Egypt and Lebanon) per acre of cropland, and, qualified by the uncertainties attaching to the meaning, per farm (not shown in the tabular comparison), is low.

The comparisons among the countries are highlighted in Figure 10-3 and shown in detail in Table 10-5. Value added

Table 10-4
Measures of Quantity of Agricultural Inputs— Egypt, Israel, and United States, 1948–66

| | Egypt[a] | | Israel[b] | United States index, 1957-59 = 100[c] | | | |
| | | | | Total input | Fertilizer & lime | Feed, seed, & livestock purchases | Miscellaneous |
Year	Value	Index					
1966				105	185	130	128
1965	129.2	102					
1964	128.5	102		103	155	123	120
1963	119.4	94		102	141	124	115
1962	121.4	96	185	101	124	121	113
1961	118.0	93	174	101	116	123	109
1960	115.4	91	176	101	110	109	106
1959	115.3	91	154	102	109	106	105
1958	128.1	101	141	99	97	101	100
1957	135.7	107	124	99	94	93	95
1956	124.0	98	114	101	91	91	98
1955	126.6	100	100	102	90	86	94
1954	125.2	99	91	102	88	82	91
1953	97.9	77	79	103	83	80	91
1952	95.8	76	67	103	80	81	88
1951				104	73	80	88
1950				101	68	72	85
1949				101	61	69	82
1948				100	57	72	74

[a] Appendix Table C10-6; "gross value of the requirements of agricultural production" at 1954 prices; index, 1955 = 100.0. Value is million Egyptian pounds.

[b] Israel, Central Bureau of Statistics, *National Income Originating in Israel's Agriculture, 1952-1963* (Jerusalem: 1964); see Table 16, pp. 44-45, "Quantity index of the purchased agricultural inputs . . . ," 1955 = 100.

[c] *Agricultural Statistics 1967,* p. 543, "Total farm input: Index numbers of farm input by major subgroups . . ."; original includes additional grouping, here omitted for lack of space.

Table 10-5

Value of Agricultural Output, Inputs, and Value Added in Agriculture, by Countries, ca. 1964/65

	Egypt	Israel	Lebanon	Jordan	Syria	Iraq	Total reported
I. *National totals:*[a]							
Value of output	1,954	423	131	108	389	492	3,497
Value of input	540	185	33	13	71	65	907
Value added in agric.	1,415	238	98	95	318	427	2,591
II. *Percentages of total output:*							
Value of output	100	100	100	100	100	100	100
Value of input	28	44	25	12	18	13	23
Value added in agric.	72	56	75	88	82	87	77
III. *Size of the agricultural plant:*							
Arable land[b]	6,603	1,016	731	2,817	16,442	18,528	46,137
Crops harvested[b]	10,669	949	543	1,394	7,667	7,597	28,819
Labor force[c]	4,905	110	360	165	670	1,020	7,230
IV. *Per acre of arable land:*							
Value of output	296	416	179	38	24	27	76
Value of input	82	182	45	5	4	4	20
Value added in agric.	214	234	134	34	19	23	56
V. *Per acre of crops harvested:*							
Value of output	183	446	241	78	51	65	122
Value of input	51	195	61	9	9	9	32
Value added in agric.	133	251	180	69	42	56	90
VI. *Per worker:*							
Value of output	399	3,845	364	655	581	482	484
Value of input	110	1,682	92	79	106	64	125
Value added in agric.	288	2,164	272	576	475	419	358

[a] Data differ in some instances from those in Appendix Tables C10-1 and C10-13 because of source and definition. Data for Egypt: Appendix Table C10-1—natural fertilizer omitted from value of inputs and output; Israel: Appendix Table C10-9a—data for 1964/65; Lebanon: Appendix Table C10-13—in absence of input data, value was assumed to be 25% of output; Jordan: Appendix Table C10-14; Syria: based on "Real Estimates of National Income for Syria for 1964," Ministry of Planning (Mimeo. No. 175, Damascus, February 1968)—data are for 1964; Iraq: Appendix Tables C10-17 and C10-18—1965 data at 1962 prices. All figures in million U.S. $ at official exchange rates.

[b] Table 8-2 (thousand acres).

[c] Table 6-1.

per acre of arable land is slightly higher in Israel than in Egypt, relatively high in Lebanon, and very low in the other countries. When the comparison shifts to a basis of crop acreage harvested, Israel is much higher, Lebanon rises to second place, Egypt is moderately high, and the lag of the other countries becomes less marked. The differences between these two types of comparison are due to the differences in cropping patterns, discussed in previous chapters. Finally, on a per worker in agriculture basis, Israel is again extremely high by regional standards; but the differences among the Arab countries are relatively less, although Jordan is double Egypt and Lebanon. In view of the possible divergences in definitions, discussed previously, not too much reliance should be placed in relatively small differences among countries.

While the cost of purchased inputs in relation to value of output presently varies considerably among the countries of the region, none of them show any marked trend in this ratio (Table 10-6; nor, therefore, any different trend in

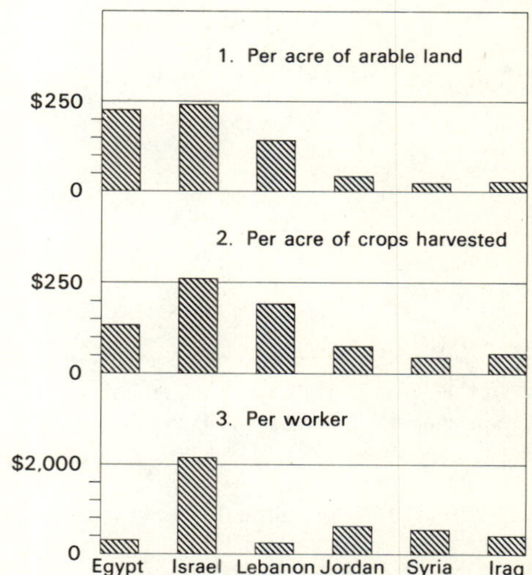

Fig. 10-3. Value Added in Agriculture, on Various Bases of Comparison, 1965.

Source: Table 10-5.

value added from value of outputs, as in Table 10-7), except possibly in Jordan, where there may be a downward trend. In this regard, all the countries differ sharply from the United States, where there has been a marked upward trend in cost of inputs in relation to value of output—from about 50 percent at the end of the war to nearly 70 percent today. At the root of this phenomenon is the fact that the American farmer is becoming more and more of a specialist, who processes materials and services purchased from outside of agriculture and adds some value in the process; but that value added now does not amount to more than a third of the value of output. The degree to which the farmer relies on purchased inputs for his production in the Middle Eastern countries is still quite small and little changing, with the exception of Israel, which requires large imports of feedstuff to support its livestock economy.

Some further light on the relationship between inputs and output is shed in Figure 10-4. In general, if workers in

Table 10-6

Cost of Inputs into Agriculture, as a Percentage of the Value of Agricultural Output, Selected Countries and Years, 1947−66

Year	Egypt[a]	Israel[b]	Jordan[c]	Iraq[d]	United States[e]
1966		47	15	11	67
1965	24	44	12	13	69
1964	24	42	10	12	69
1963	23	43	13	12	70
1962	23	46	13	12	69
1961	27	44	15		68
1960	23	48	20		69
1959	23	44	27		70
1958	26	42			67
1957	29	43			68
1956	27				65
1955	29	44			66
1954	29				64
1953	24				61
1952	22	39			62
1951					60
1950					60
1949					57
1948					54
1947					50

  [a] Appendix Table C10-2, gross value of agricultural production and value of requirements, in constant (1954) prices.

  [b] Appendix Table C10-9a, value of output and value of input, in current prices.

  [c] Appendix Table C10-14, agricultural income and expenditure in current prices.

  [d] Appendix Table C10-17, value of agricultural output and of requirements, in current and in constant (1962) prices.

  [e] *Agricultural Statistics 1967*, p. 570, realized gross income from farming, and production expenses.

Table 10-7

Trends in Value Added by Agriculture, Selected Countries and Years, 1952−66

	Egypt[a]		Israel[b]		Syria[c]		Iraq[d]	
Year	Value	Index	Original index	New index	Value	Index	Value	Index
1966							155.6	105
1965	405	101					152.5	103
1964	405	101			1,055.6	83	142.1	96
1963	402	100			1,037.8	82	131.3	88
1962	400	100	310.5	100	1,265.5	100	148.6	100
1961	326	82	285.7	92	843.7	67		
1960	394	98	249.7	80	676.9	53		
1959	381	95	250.0	81	765.2	60		
1958	365	91	203.3	65	745.6	59		
1957	337	84	169.1	55	1,118.3	88		
1956	330	82	159.8	51	981.0	78		
1955	317	79	129.7	42				
1954	309	77	128.7	41				
1953	313	78	101.1	33				
1952	332	83	100.0	32				

  [a] Appendix Table C10-2; value added in constant (1954) prices; converted to index, 1962 = 100. Value added in million Egyptian pounds.

  [b] Table 5, pp. 12-13, *National Income Originating in Israel's Agriculture, 1952-1963*, (Special Series No. 165 [Jerusalem: Central Bureau of Statistics, 1964]); original index, 1952 = 100; recalculated to 1962 = 100.

  [c] Appendix Table C10-16; national income originating in agriculture, constant (1963) prices; converted to index, 1962 = 100. Income in million Syrian pounds.

  [d] Appendix Table C10-17; value added by agricultural production, in constant (1962) prices; converted to index, 1962 = 100.

agriculture are to produce valuable outputs, they must have the tools and materials, or else the value added by their labor will inevitably be low. As the intensity of agricultural operations rise, the cost of inputs rises; but the value added, or the net difference between output and purchased input, rises vastly faster. In the systems of agriculture of the Arab countries, inputs cost mostly in the range of 10 to 30 percent of value of output, but value added per worker does not for the most part exceed $500. There is not a high correlation of these two factors within the group of Arab countries—i.e., a higher ratio of input to output does not seem to be associated with higher value added per agricultural worker. On the other hand, Israel has a much higher ratio of costs and a much higher value added per worker; and the United States is still higher on each count.

Added inputs might, of course, be largely wasted; hence merely increasing the cost of inputs does not in itself guarantee higher incomes—but only low incomes are possible with low inputs. A high rate of inputs makes net returns sensitive to changes in the ratio of input and output prices. If the ratio is unfavorable, value added drops

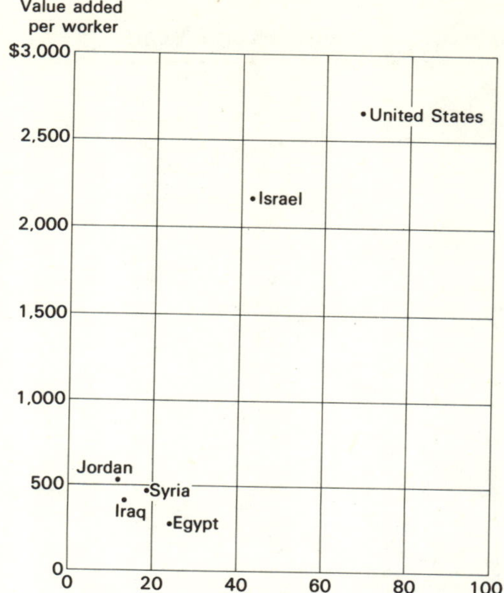

Fig. 10-4. Relationship Between Importance in Agriculture of Non-farm Inputs and Value Added per Worker, by Countries, (circa 1965).

Source: Tables 10-5 and 10-6—except for United States, which data is taken from *Agricultural Statistics.*

sharply; but, when it is favorable, value added rises steeply. Incidentally, Figure 10-4 suggests that relatively modest revisions in any of the data would not greatly affect the basic relationships.

One more point emerges from Figure 10-4, and that pertains to Israel, which is of course in the Middle East, physically and geographically: since its climate, soil, and water are the same as those in neighboring Arab countries, the results which are possible with its natural resources are also possible with the similar natural resources in the neighboring countries. But not only, as is a commonplace observation, is Israel not part of the political and social structure as the Arab Middle East; as Figure 10-4 suggests, it belongs also to a different economic world. Although natural conditions are very similar, its results from agriculture are of a different kind than the results from agriculture in the Arab countries.

Agricultural Income in the National Economy

Unfortunately, data on the contribution of agriculture to the national income are rather incomplete (Table 10-8). In Egypt, there is some evidence of a downward trend in the share which agriculture contributes to national income, in the sense that the latest years are much lower than the earlier years; but for most of the intermediate years, no clear trend is evident. In Israel, because of the rapid and sustained rise in agricultural production for most of its existence, agriculture's share of the national income has fallen only in the last few years, and then not strikingly. The same is true for Lebanon, based on national product rather than national income data. The available data for Syria and Iraq are inconclusive.

These relationships are in marked contrast to the United States. Here, over the past twenty years or so, agriculture's share of national income has fallen by two-thirds.

A downward trend in the relation of agricultural to national income, at a time when agricultural output is rising, is possible only because the nonagricultural sectors of the economy are growing even faster. Because of this, a falling percentage of agriculture in the national income is always viewed as a healthy and desirable economic development.

Table 10-8
Share of National Income Contributed by Agriculture, * by Countries, 1945–66
(%)

Year	Egypt[a]	Israel[b]	Lebanon[c]	Jordan[c]	Syria[d]	Iraq[e]	United States[f]
1966						20	2.7
1965		9				20	2.5
1964		10	18	25	32	20	2.5
1963	26	11	18	19	33	20	2.6
1962	25	11			37	22	2.7
1961	28	12	18	16	32		2.9
1960	30	12			28		2.8
1959	31	12		18	32		2.8
1958	31	14	17		31		3.5
1957	31	13			41		2.9
1956	31	12			38		3.4
1955	31	12					3.4
1954	32	13					4.0
1953	33	12	19				4.5
1952	33	12					4.8
1951	28						5.3
1950	30		20				5.4
1949	34						6.3
1948	36						7.2
1947	40						8.7
1946	42						8.4
1945	44						7.1

* Includes forestry and fishing in some countries.

[a] Donald C. Mead, *Growth and Structural Change in the Egyptian Economy* (Homewood, Ill.: Richard D. Irwin, 1967), pp. 44-45; for later years, figures entered in year of latter half of split year (i.e., 1963 for 1962-63; for later years, used gross national product estimate I).

[b] Central Bureau of Statistics, *Statistical Abstract of Israel 1966,* pp. 176-177.

[c] *Share of Agriculture in Gross Domestic Product, at Factor Cost,* Near East Commission on Agricultural Statistics, Fourth Session Baghdad, September 10-17, 1968, Item 5: "Economic Accounts for Agriculture."

[d] Appendix Table C10-16.

[e] Appendix Table C10-19.

[f] *Agricultural Statistics 1967,* pp. 508-9; realized net income of farm operators divided by national income.

Table 10-9
Major Components of Cost of Producing Crops, Selected Countries and Years

Country	Year	Kind of data[a]	Costs in monetary terms[b]				Percentage of cost		
			Labor	Land & water	All other	Total	Labor	Land & water	All other
Egypt	1964	A—cotton	17.68	24.29	15.71	57.68	31	42	30
		A—sugarcane	16.37	31.81	26.91	75.09	22	42	45
		A—rice	11.38	9.70	13.98	35.06	33	28	39
		A—millet	7.08	13.99	8.03	29.10	24	48	47
		A—sesame	5.75	10.14	5.86	21.76	26	47	38
		A—peanut	9.16	11.90	10.06	31.12	29	38	37
		A—maize	6.07	8.79	11.76	26.62	23	33	45
		A—wheat	4.39	14.56	11.94	30.89	14	47	41
		A—barley	3.44	11.07	7.64	22.15	16	57	37
		A—broad beans	3.76	13.44	8.71	25.91	14	53	36
		A—lentil	3.57	13.95	8.59	26.11	14	53	33
		A—onions	8.48	15.83	22.22	46.53	18	34	53
		A—chick peas	3.94	14.08	9.27	27.29	14	52	34
Jordan	1965	B	11.60	12.75	4.61	28.96	40	44	16
Syria	1966	A—wheat	76.7	31.5	36.6	144.8	53	22	25
	1966	A—barley	65.1	30.5	27.5	123.1	53	25	22
	1965/66	A—cotton	511.0	342.0	475.5	1,328.5	38	26	36

[a] A = crop cost of production data; B = national input and output data; Egypt: Appendix Table C10-8; Jordan and Syria: from an unpublished manuscript by George S. Medawar, with data on Syria based on Sing-Min Yeh, "A Brief Study of Production, Consumption, Price, and Marketing of Wheat, Barley, and Cotton" (Damascus, July 1967). In Syria, "fixed costs" of rent, management, and interest have been taken as equivalent to rent.

[b] In Egypt, Egyptian pounds per feddan (1.038 acres); in Jordan, millions of Jordanian dinars; in Syria, Syrian pounds per hectare.

Cost of Producing Crops

It would be highly informative to know how, country by country, the value added by agriculture is divided among labor, land and water, and capital; but such data are largely lacking for the Middle East. Some suggestion of what this division might look like can be had from the available data on the costs of crop production, although even these data are available only for a few countries (Table 10-9). In Egypt, for all crops except rice, rent of land (there are no charges for water) is more—in most cases, several times more—than the value of the labor used to produce the crop. This is probably evidence of the large supplies and hence relatively low returns, of agricultural labor; and of the scarcity of land—which, as we have noted earlier—finds expression in the small size of the average farm. In Syria, in contrast, the cost of labor used for major crops is about double the cost of land and water. This might reflect differences in the relationship between agricultural workers and land availability, but the data are not nearly good enough to come to more definite conclusions.

CHAPTER 11

Marketing and Trade

THE term "marketing" covers a great many subjects: the division of agricultural production between consumption on the farm where produced and outside it; the division between domestic consumption and exports; the division between fresh and processed use; the organization of getting the product from producer to consumer—i.e., the pattern of distribution, including wholesaling and retailing functions; the state of the art of marketing in a narrower—and more modern sense—i.e., quality control, organization of markets, preservation, storage, advertising, price policy, etc.

In the context of this study we are primarily interested in two of these aspects: (1) in what proportions do the various products go to the principal categories of consumers—i.e., domestic vs. foreign, and direct vs. processed? and (2) how well are the countries equipped to handle an increasing volume of products, possibly of greater variety and, with rising incomes, reaching new layers of the population? In this chapter we discuss the current situation in both of these respects as a background for some speculation in Chapter 18 as to the marketing and trade potential in the next two to three decades. The data at our disposal are not always for the most recent years or years, largely because some of the rather complex statistical compilations needed were made by FAO using 1960–62 or 1961–63 as the base. We have found it convenient to utilize these data, persuaded that the basic characteristics have not changed sufficiently in the intervening years to render our conclusions invalid.[1]

Any detailed discussion of the complex of internal marketing organizations that have in some countries operated for many years, and are of more recent origin in other countries, has been omitted. These organizations include grain and cotton marketing boards, fruit marketing boards, and others. They have among their functions that of bringing farm products from the area of production to the nonproducing consumer, of setting prices, of controlling export prices and quantities, of providing storage capacity, etc. Since the amount of detail needed to adequately deal with the subject goes far beyond the scope of this study, only the more relevant aspects of marketing—those mentioned above—will be dealt with.

Disposition of Agricultural Production

It would be intellectually satisfying as well as statistically neat to begin by showing what amount of production is retained and consumed on the farm and what is turned over to the market, either for consumption elsewhere in the country or for export. But such disposal statistics have not come to our attention. Indeed, the literature does not even yield regrets over their failure to exist. That the deficiency is not due to the lack of diligence is largely borne out by an interesting discussion in a recent FAO publication,[2] which for a number of countries shows estimates of imputed values of subsistence production as percentage of total value of agricultural production. Alas, though, outside Western Europe and North America, which one would expect to be represented, only Africa shows up with a sizeable number of countries (eleven) that have the appropriate statistics; and the United Arab Republic is not among them. Only four Asian countries are shown, and none of them is in the Middle East. Some specific crop data for India round out this meager list (they show that between 60 and 80 percent of the various grains never becomes a "marketable surplus").

In the apparent absence of similar data on the countries of interest in this study, what are we to assume regarding the marketable surplus?[3] The nearest we can come to some

[2] K. C. Abercrombie, "Subsistence Production and Economic Development," *Monthly Bulletin of Agricultural Economics and Statistics* (FAO), XIV, No. 5 (May 1965).

[3] Pertinent information is not always where one looks for it. In its 1967 edition, a regular FAO publication, *National Grain Policies*, carries this comment in a footnote to a tabulation showing grain delivered to the Wheat Bureau of Lebanon:

> The small total amount delivered to the Wheat Bureau by growers, despite the support price, may be explained by the small size of farms, resulting in half the production remaining on the farm for self-supply.

The statement serves to remind one that size distribution of farms is undoubtedly a significant factor in the disposition pattern.

Similarly, the *Jordan Country Report* of FAO, released in mid-1967, carries some statistics for the East Jordan Valley of percentages of specific crops sold in that area in 1961, as follows:

Crop	Percent Sold
Bananas	99.5
Citrus fruits	94.7
Other fruits	98.0
Olives	79.3
Tomatoes	95.7
Other major vegetables	95.1
Wheat and barley	16.3
Yellow maize	72.9
White corn	34.2
Sesame	92.6
Local tobacco	100.0

But the report also cautions that (1) "the conditions in the Valley are not representative of the country as a whole"; and (2) repeats a warning from the source of the data that ". . . it is not

[1] *Indicative World Plan, Near East* (Rome: FAO, 1966), II.

appraisal is to offer a set of tables (Appendix Tables C11-1 to 7) that show commodity balances for important crops or groups of crops, as well as livestock products, for each of the countries, in terms of origin and disposition of total supply. This tells us what proportion of supply is derived, respectively, from domestic production and from imports, and what proportion of supply is channeled into domestic consumption (food and nonfood uses considered separately)—and into export.

To illustrate, in the period for which the data were compiled, about half of Egypt's wheat supply was imported, thus implying a well-developed marketing setup. Similarly, about one-quarter of all rice was exported, leading one to assume again the existence of marketing channels.

But such data are poor substitutes for what one would like to know; for it would still be possible that even when imports loom large as a share of total supply, most local production is consumed where grown, while imports satisfy the demand of urbanized areas. What is more nearly likely, however, is that the farming population feeds both itself and a large part of the nonfarm population. This leaves open the question as to what percentage of the farming population participates in this feeding of the nonfarmers.

The data serve better to illuminate the question of domestic vs. foreign origin and the importance of export markets than as an indicator of the "marketable surplus." Under that aspect one might construct a presentation somewhat as in Figure 11-1. Here both import dependencies and export achievements stand out clearly.

In relative terms, Egypt is the least import-dependent. About half its wheat is imported, and some 10—20 percent of its sugar, oils, and fish. But given the size of the population, the absolute amounts are large despite the relatively low shares of the imported portion. At the other end of the scale are Jordan and Lebanon. In Lebanon, for example, fruits and vegetables are the only categories in which the imported share is less than 20 percent; and Jordan, in addition, imports no eggs. In recent years, Lebanese agricultural imports have been two to three times agricultural exports. The remaining countries hold an intermediate position: Syria imports only small portions of its supply of coarse grains, pulses, vegetables, fruits, meat, eggs, milk, and oils and, of course, practically no cotton. It does import all its rice and substantial portions of starchy roots, sugar, most of its fish, and significant amounts of wheat. Iraq has a substantial import dependency in rice, starchy roots, sugar, oil, and cotton. Finally, Israel imports much wheat, almost all of its rice and major portion of its coarse grains and oils.

In sum, the statistics suggest that even at prevailing levels

---

certain whether 'non-marketable' production has been included in the total yield in every case, or how such production has been disposed of." In other words, all percentages might be too high (which the very high figures for fruits and vegetables seem to make likely). Moreover, irrigation introduced into the valley after 1961 raised production and undoubtedly sales, especially of wheat, barley, and corn.

In any event, all this suggests that there are probably more scattered data available than have as yet been brought together.

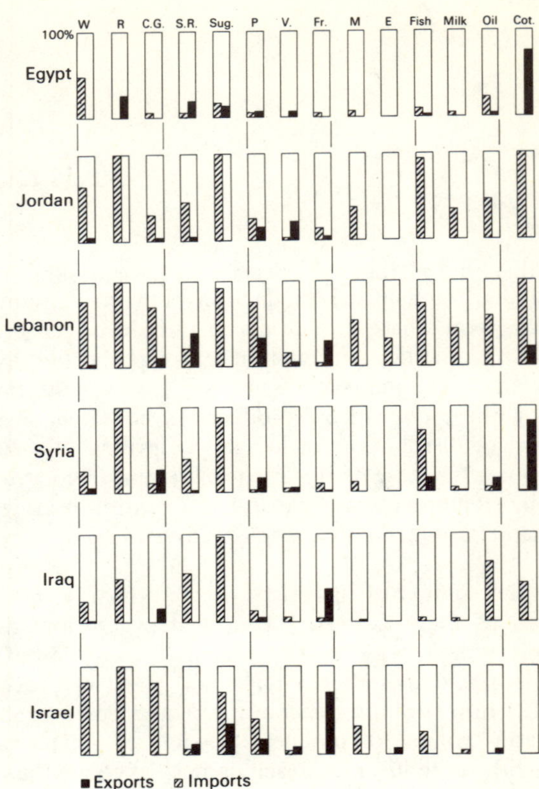

Fig. 11-1. Tonnage of Imports and Exports as Percent of Supply.

W   — Wheat
R   — Rice (paddy)[a]
CG   — Coarse Grains
SR   — Starchy Roots
Sug — Sugar (raw equivalent)[b]
P   — Pulses
V   — Vegetables (incl. melons)
Fr   — Fruit (excl. melons)
M   — Meat (excl. offal)
E   — Eggs
Fish — Fish and products (fish equivalent)[c]
Milk — Milk and products (liquid equivalent)[c]
Oil — Oilseed and vegetable oils (oil equivalent)[c]
Cot — Cotton and products (raw cotton equivalent)[d]
   [a]Israel: milled.
   [b]Israel: refined.
   [c]Israel: Aggregate tonnage.
   [d]Israel: Not available.

Source: Appendix Tables C11-1 through 7.

and structure of per capita consumption, the region as a whole is, by and large, deficient in grains, sugar, and oils, while some parts additionally depend on imports of potatoes, meat, fish, and pulses.

Turning to marketing abroad, we find a good deal of specialization. As examples, Egyptian and Syrian cotton are outstanding and so is Israel citrus fruit. Dates from Iraq give that country a large export quota in the fruit category. Egypt also exports a substantial share of its rice, onions, and potatoes. Jordan exports some one-fifth of its vegetable production; and Lebanon, potatoes, pulses, and prominently a variety of fruit—especially apples—and vegetables so that over 50 percent of her total exports have in recent years consisted of fruits and vegetables. Syria exports coarse grain in substantial amounts—about one-third of its production—as well as some pulses, fish, and oils; indeed it the only country that is not a substantial importer of oils. Iraq, apart from the above-mentioned data, registers some export (between 10 and 15 percent) of food and feed grains; but still is a net importer of wheat and a heavy one of rice.

Upon closer examination of foreign trade, it is of some interest to discover the extent to which the countries involved trade with each other (excluding Israel, with which there is no trade by any of the other countries). Table 11-1 shows the role of agricultural exports as a whole, for each of the countries. Not surprisingly, the role of exports of all kinds in the economic life of the country, as represented by the size of the Gross Domestic Product, is highest in the two petroleum-producing countries—Saudi Arabia (not shown in the tables in this chapter) and Iraq. By contrast, agricultural exports play a very small part, whether as shares of total GDP, of its agricultural portion only, or as a share of total exports. In Egypt, Jordan, and Lebanon, between two-thirds and three-fourths of all exports consist of primary agricultural products, and in Syria the percentage is even higher; while exports as a whole represent only a modest percentage of the GDP, from less than 5 percent in Jordan to not quite 20 percent in Syria. But in each of these countries, with the exception of Jordan, agricultural exports are high—amounting to about 40 percent of the agricultural gross domestic product; this reflects high-priced agricultural specialty exports on the one hand and low-priced staple crops for domestic consumption on the other. Jordan is in a separate class. As has been pointed

out elsewhere, resource limitations have kept its agricultural plant small, and within it there is not much that lends itself to large-volume exportation. Nor is there much evidence of nonagricultural exports; tourism, a major invisible income item, is not part of merchandise trade. As a result agricultural exports, such as they are, form the bulk of total exports; but both total and agricultural exports are but a small item in the country's total output.

Israel's position is intermediary, but on the whole agricultural exports are small when compared to either total exports or total agricultural product; and they are becoming smaller as market limitations keep Israel's mainstay in foreign markets—citrus—from growing as rapidly as are nonagricultural exports. The role played by citrus in Israel's total agricultural exports is shown in Table 11-2 and 11-3. Although noncitrus exports have shown a tendency to increase in recent years, largely through rising exports of cotton and poultry products, citrus still occupies a dominant position.

Intraregional Trade

Within this pattern of trade in agricultural commodities, how does intraregional trade stack up? For the purpose we have used compilations made within the context of FAO's Indicative World Plan, despite the fact that the area is both narrower (it excludes Israel, a deficiency which is immaterial when we look at trade) and wider (it includes a number of other countries). Fortunately, sufficient detail is provided to permit the data to be made useful for our purpose.

Tables 11-4 and 11-5 bring together the pertinent material. The differences in importance of intraregional trade as part of total trade are striking. Again, the absence

Table 11-1

Role of Exports, 1961—63

(In Million U. S. $)

	Exports			Gross domestic product at factor cost			Agricultural exports as % of agricultural GDP $(2) \div (5) = (7)$
	Total (1)	Primary agricultural commodities (2)	(2) as % of (1) (3)	Total (4)	Agri. only (5)	Exports as % of GDP $(1) \div (4) = (6)$	
Egypt	428	309	72	3,669	1,001	12	39
Israel[a]	351	88	25	1,996	220	18	40
Jordan	14	9	62	323	66	4	14
Lebanon	52	33	64	655	93	8	35
Syria	151	128	85	837	329	18	39
Iraq	700	35	5	1,800	336	39	10

[a] Figures for 1963 only. Net instead of Gross Product. Use of GDP would substantially reduce figure in col. (7).

Source: Other than Israel: *Indicative World Plan, Near East* (Rome: FAO, 1966).
Israel: *Statistical Abstract of Israel, 1967.*
United Nations, Statistical Yearbook, 1967.
Note: Years may not in all cases be the same, and export figures, for this and other reasons, not coincide exactly with those shown in Table 11-4.

Table 11-2
Israel Agricultural Exports
(In Million U. S. $)

	Total agricultural exports	Citrus exports	Citrus as % of total
1950	17.0	16.8	99
1955	34.2	31.6	92
1960	63.1	46.6	74
1965	86.5	71.2	82

Source: Israel Economic Development, Past Progress and Plan for the Future, Final Draft (Prime Minister's Office, Economic Planning Authority, Jerusalem, March 1968).

Table 11-3
Israel: Exports as Percent of Output

	1949	1954	1959	1964	1967
Total agriculture	19.7	18.3	17.2	14.2	19.8
Citrus	67.8	73.1	80.2	64.6	82.6
All Other	0.0	1.5	4.4	3.6	5.1

Source: Israel Economic Development, op. cit. (above, Table 11-2).
Note: Figures underlying this special computation are in 1966 prices. Exports are calculated by applying 1966 exchange rate to export earnings in dollars. Agricultural output includes intermediate products. For these reasons the data are not comparable to those based on net national product at factor cost in Table 11-1 above.

or presence of petroleum is the most crucial factor. Thus, only 5% of Iraq's and 18% of Saudi Arabia's exports (not shown in the Tables) stay in the Near East. Because of worldwide use of its cotton, only 7% of Egypt's exports remain in the region. And, of course, these figures would be smaller still if adjusted to the narrower geographic definition used in this study: not quite 2% for Iraq, not quite 3% for Saudi Arabia, and 3% for Egypt. When one draws the circle even smaller and compares agricultural rather than all intraregional exports with total exports, they amount to very little indeed in the three countries named.

By contrast, intraregional trade looms large in the case of Jordan—69%—and even more so in the case of Lebanon, of whose exports fully 90% remain in the Middle East, broadly defined (and over 50% when defined as in this study). Syria occupies an intermediate position, with 27% of exports intraregional, most of them falling within the countries here considered.

Lebanon not only is the principal intraregional exporter, but is also the leading intraregional importer, receiving nearly 40% of all such trade. Similarly, Iraq not only ranks last as shipper but also as importer of trade originating in the region, receiving less than 10% thereof. Again, if we were to restrict this analysis to agricultural commodities only, all the percentages would be substantially reduced.

Intraregional trade is influenced by the existence of the Arab Common Market, based upon the "Agreement to Facilitate and Organize Trade between Arab Countries," under which import duties have been abolished for agricultural products. But statistics of agricultural trade in the years preceding the agreement do not show a basically different pattern. Where intraregional trade is strong—true

Table 11-4
Intraregional Trade, 1961–63
(In Million U. S. $)

From	Total exports (1)	Exports of selected agricultural commodities[a] — Total (2)	In intraregional trade (3)	Intraregional trade — Total (4)[b]	As % of exports to all countries (5)	Iraq (6)	Jordan (7)	Lebanon (8)	Syria (9)	UAR (10)	Others in region (11)
UAR	472	57	9	31	7	0.8	2.6	6.4	1.7	–	19.8[c]
Jordan	17	7	5	9	69	1.4	–	2.2	2.4	–	3.4[d]
Lebanon	52	18	9	46	90	2.4	6.2	–	6.4	0.6	30.6[e]
Syria	155	37	16	41	27	5.0	7.3	21.9	–	0.8	5.9
Iraq	711	28	4	36	5	–	0.6	8.6	1.2	0.2	25.3[f]

[a] Cereals, pulses, oilseeds, fruit, vegetables. For breakdown see Table 11-5.
[b] Sum of amounts shown in Columns (6)-(11).
[c] Of which 60% to Sudan.
[d] Of which 56% to Kuwait and 40% to Saudi Arabia.
[e] Of which 46% to Kuwait.
[f] Of which 50% to Federation of South Arabia.

Source: Indicative World Plan, Near East (Rome: FAO, 1966), II.

Table 11-5
Selected Agricultural Commodities in Intraregional Exports, 1961–63
(In Million U. S. $)

From	Cereals	Pulses	Vegetables	Fruits	Oilseeds	Total
Egypt	4.4	1.1	2.1	1.1	–	8.7
Jordan	–	0.6	3.7	0.5	0.4	5.2
Lebanon	–	0.4	1.0	7.9	–	9.3
Syria	7.4	2.2	1.4	1.5	3.1	15.6
Iraq	1.8	–	0.1	1.4	0.5	3.8

Source: Indicative World Plan, Near East, (Rome: FAO, 1966), II.

especially for Lebanon and Jordan—the type of products involved, the personal and commercial ties with neighboring countries, and the competitive strength of other producers, especially in the Mediterranean region, almost dictate the flow of trade, with commercial policy only a secondary consideration. By the same token, other regional groupings, such as the European Common Market, have practiced an exclusionary policy that has further strengthened the role of inter-Arab trade.

Special Situations

As we have noted, there are some agricultural commodities in which one or the other country occupies a special position. For the most part these can be readily recognized by the role that exports play in their disposition. They are cotton and, to a much smaller degree, rice in Egypt; vegetables in Jordan; tubers, fruit of all kind, and pulses in Lebanon; cotton and coarse grains in Syria; and fruit in Iraq and Israel.

For greater detail one has to go beyond the tabular material here presented. One then finds that Egypt also ranks high as an exporter of onions, vying with the Netherlands for first place as the world's leading onion exporter; that in recent years exports of dates have constituted up to half of all of Iraq's agricultural exports, but that on and off—depending upon the size of the crop—the country has exported substantial amounts of barley, though over the years the importance of this export has diminished; that nearly half of Lebanon's fruit exports has consisted of apples; and that nearly 80% of Israel's agricultural exports has consisted of citrus fruit, with cotton and poultry products now accounting for most of the balance.

Thus, for most of the region's countries, exports consist of one or two highly specialized products, a condition that is important to keep in mind when one comes to speculate about broadening of foreign markets in the years to come. In this connection it is useful to note how insignificant are the exports of some of the items that tend to attract attention in such speculations—i.e., in the main fruits and vegetables, and especially those that can be grown at times of the year when indigenous producers in Western Europe and even traditional exporters to such markets cannot compete.

Since it is sometimes contended that Egypt should slowly become a specialized, large-scale vegetable grower and exporter, let us look at the Egyptian situation as an example. The total area devoted to vegetable raising more than doubled between the mid-1950's and the mid-1960's. By 1965, tomatoes alone were raised on some 200,000 acres, twice the mid-1950's average. But almost all of the crop—about 1.2 million tons in 1965—is marketed in the country.[4] Practically none is processed, and exports are insignificant, even though year-around exports could easily be sustained. The year 1960 was a high point in exports, with 4,200 tons—or one-half of one percent of total production. By 1965, exports had shrunk to 238 tons. Assuming a yield of 15 tons per hectare—or 6 tons per acre—the 1965 exports if grown under average conditions would represent the yield from 40 acres. To make tomatoes into a significant Egyptian export item would surely demand a major marketing effort, no matter how favorable the natural circumstances of cultivation.

Another product often mentioned is fresh grapes. In Egypt, not quite 25,000 acres are currently—and have over a long time been—devoted to vineyards, yielding an average of 100,000 tons of grapes annually. Of these a maximum of 150 tons gets exported, equivalent to production from 30 acres of average yield condition. Again, any increase in land devoted to grapes with the idea of establishing a flourishing and significant grape-export industry would appear to represent a major undertaking.

The difficulties are less obvious in Lebanon, where large-scale fruit and vegetable exports have been traditional for many years. However, as we have seen above, most of these exports are directed toward Middle Eastern countries, many of which would presumably attempt—and some have plans to that very effect—not only to become self-sufficient in fruit and vegetable production but to become in- and out-of-season exporters in their own right. Similar considerations apply to Jordan, currently an important exporter of tomatoes and some other fresh vegetables (mainly to other countries in the region).

The situation in Israel is the reverse. Exports, limited by and large to citrus, are directed to none of the countries in the region, but overwhelmingly to Western Europe. Here, of course, they encounter the full competition of European producers and their closely allied non-European associates (such as North African producers). Thus, for Israel it is not a matter of establishing a marketing organization, but of maintaining and if possible raising her relative share in well-established markets.

Marketing Know-How

The last few paragraphs have led up logically from consideration of past and current patterns of disposal to what one might call "ability to market." This consists, in simple

---

[4] The resulting weekly consumption of nearly two pounds per person seems quite high, suggesting that a good deal might be wasted or there may be more processing than meets the eye.

terms, of getting the right items to the right customers at the right time in good condition and at prices that are remunerative to the seller and attractive to the buyer. Specifically, the considerations involved are cost of production, proximity to markets and availability of appropriate means of transport, and meeting of consumer preferences—a consideration which embraces selection of varieties to suit tastes, grading, packaging, and various technical problems of preserving and storing to overcome differences in timing between production and consumption. Basically, such a system starts on the farm and ends at the receiving dock in the country of destination.

In recognition of the need to improve all of these elements, there was set up in 1964 in Amman an Agricultural Marketing Center, under the auspices and with the principal help of the UN Development Program and aided by FAO. It engages in research, training, demonstration, and advisory services; and it is through a number of studies carried out and published by the Center that those not resident in the area are beginning to learn and appreciate the status of marketing know-how and the difficulties to be overcome.

The "market" is not, of course, an unknown phenomenon in the region. On the contrary, the Middle Eastern "bazaar" is a time-honored institution, in fact as well as in fiction. But the bazaar is a long way from the organized wholesale, or retail market; indeed, its charm for the foreigner often lies precisely in his inability to perceive anything resembling organization.

Two distinctions must be drawn in this discussion of marketing: one between domestic and export-import marketing and the other between staple crops and the balance of production—especially dairy products, meat, and fruits and vegetables or what in this country would generally fall under the category of "produce."

In the absence of comprehensive studies—and excepting Israel, whose public-sector oriented economy has given rise to a highly developed central marketing and distribution system—one gains the impression that domestic marketing of produce is on the whole not highly organized, but continues to consist of fragmented, small-scale operations, largely guided by day-to-day experience. Advisory and research reports coming out of the Jordan Marketing Center and from advisers generally start out by describing and lamenting the absence of regulated, organized markets, of market news (price information, etc.), failure to select and grade, poor hygienic conditions, lack of storage space entailing high percentage of loss in perishable commodities, and so on. The net result is that the grower suffers substantial losses, since his product must be sold as it matures; is subject to deterioration from various sources; fails to benefit from grade differentiation that would lead to sale of some products at premium prices; and fails to benefit equally from processing facilities which, if present in adequate capacity, would provide an outlet for low-quality products and would upgrade the balance of the crop.

To some extent, this is a chicken-and-the-egg proposition. Which comes first: higher incomes of consumers or higher prices for better quality agricultural products? That they need to—and probably do—develop together is an obvious comment; but it is also true that avoidance of

waste in marketing represents a true increase in productivity and as such would enable producers to bear some of the cost of waste-avoidance. As a result, prices could be kept from rising in proportion to such added costs and everybody would gain in the process. Thus, better marketing can legitimately be seen as an independent variable that can be independently pursued.

The difficulties in the path of improved marketing are graphically illustrated in a recent study dealing with the export of Jordanian tomatoes to Syria.[5] It was conducted principally to discover whether income from such shipments, which are customary, could be increased if the product were properly conditioned, sized, and graded, rather than disposed of without such differentiation. "As far as is known," the resulting report states, "this was the first occasion on which conditioned, graded, and sized tomatoes were offered for sale on the Aleppo market." (p.1) Then, having commented on the difficulty of locating a cooperating supplier, the report states: "Even with careful briefing, however, the degree of control that it was possible to exercise over the quality and type of produce coming into the factory left much to be desired. The problem of procurement is emphasized here because it is a basic and recurring problem." (p. 2)

But finding a source was only the initial difficulty. "For the first three shipments transport delays caused tomatoes to be sold in Aleppo on the fifth day after transit. On three of the 11 scheduled days the plant [the packing plant] did not pack for Aleppo. On two of the days the pick-up...was missed because of faulty communication between the parties involved, and on the third occasion the plant generating equipment broke down." (p. 4)

Communication difficulties are illustrated in this passage: "Physically the plant is isolated and telephone communication with it is extremely tenuous. To date, no telephones have been installed in the plant, and calls have to be relayed via the Wadi Yabis [the location of the plant] cooperative office. However, the cooperative offices are closed during the afternoon and evening, which are the normal times of plant operation. In any case, calls between Amman [project offices] and Wadi Yabis most frequently take 3—5 hours although the period may be longer." (p. 5)

The report goes on to discuss such questions as differences between degree of product maturity as required by the plant's machinery (not past pink stage) and that preferred by the customers (firm red); the fact that the tomatoes designated as "third grade" were of such poor quality as to prevent their sale altogether "even on local markets": that the boxes tended to be packed beyond preset capacity, resulting in "extra damage and bruising in boxes whose dimensions are in any case not suitable for soft products such as tomatoes, for which purpose the boxes indeed were not designed"; that supply to the Aleppo market regularly exceeded the demand, even though Jordan is in a virtual monopoly position in the winter months on Syrian markets (these are estimates, since

---

[5] John G. Clarke, assisted by C. O. Emmerich and I. Lanham, *A Report on Tomato Trial Shipments to Aleppo in March 1967* (Research Report No. 3, Agric. Marketing Center [Amman, Jordan: April 1967]).

no statistics are collected in the Aleppo Market on either quantity or price of products sold), etc., etc.

The most encouraging result was that tomatoes of graded quality did in fact sell at a premium of over 10 percent over unconditioned shipments; but because a general condition of excess supplies depressed the market generally, the operation suffered a heavy financial loss, with the mere cost of the purchase of tomatoes before conditioning at the plant exceeding by over 10 percent the total receipts.

We have quoted at some length from this experiment because it illustrates in concrete terms the difference between working out a theoretical "exportable surplus" and the innumerable down-to-earth obstacles, some predictable and some not, that stand in the way of fulfillment—many of them integral components of general economic, social, and institutional development. Though much can, of course, be done marginally and gradually to improve its efficiency, marketing is hardly a function or aspect that can be made to emerge at will and in isolation. We shall have more to say about the promise of new exports that exploit to the fullest the specific locational and climatic advantages of the region in Chapter 18.

It is worth noting in this connection that these difficulties continue to hamper also Israel, which enjoys a long-standing experience in agricultural exports. In the recent plan for economic development,[6] there appear these comments:

> Other items, such as fruit, vegetables, melons, flowers and seeds, have a considerable export potential, since Israel enjoys relative advantages in their production. However, their export expanded very slowly and reached only about $6 million in 1967. There were many reasons for this: a lack of knowledge about foreign markets; lack of know-how in the transportation of perishables over long distances; the attitude to export as a solution for incidental surpluses; the fact that varieties and types available here did not meet the customers' specific demands; problems of organizing exports and the preparation and packing of the produce. Recently the Government has begun to encourage such exports through incentives and fairly extensive direct activities. It is to be hoped that this will have positive results in coming years.

What has been described above in an effort to highlight the broad front on which progress in marketing must be made does not in anywhere near equal measure apply to the marketing of staple crops, primarily grains and cotton. In those fields, because either exports or imports play an important role and because, grains being a basic part of the diet, access to them by nonproducing consumers is a primary condition of civil peace, governments have for various lengths of time operated marketing boards, with differing types and functions, to purchase and resell the grower's product. Therefore, the basic problem here tends to be different. Usually it is one of price policy; and as in many other developing countries the tendency has been to buy at

the lowest possible price in order to keep the cost of living from rising.

To illustrate, a thorough FAO study delves deeply into Syria's marketing of wheat and barley, in which the government, through the Bureau of Cereals, has had a hand since 1951, when it began buying and selling. The study finds that "...in the last 9 years the official purchase price (for wheat) usually fell below the wholesale price by 4–24%." (p. 12)[7] Since 1965 the government has been the exclusive buyer and seller of wheat, barley, and cotton for export, and thus the price-setting policies of the appropriate agencies (in the case of grains, the General Agency for Cereals and Flour; in the case of cotton, the Cotton Marketing Organization) have assumed growing importance. In 1965, according to the FAO study, the government purchased over 25 percent of the country's wheat crop, most of which is allocated not to exports but to nationalized flour mills; but the price at which the wheat was purchased was fully 15 percent below the wholesale price. Assuming the validity of the statistics, one is forced to judge that government price policy is a crucial factor in the degree to which production of a given crop is encouraged; and measures to improve and increase production at the farm level are doomed to failure if the grower finds himself unable to recover his cost or much above it.

The degree to which government intervenes in the supply and distribution of staples varies. Thus Iraq has for a long time had a Grain Board, which has general supervisory function over private merchants and is directly responsible for market research, licenses to buy and sell, and, most importantly, directly builds and operates storage facilities. The latter gives the government additional means of establishing and enforcing standards.[8] Substantial progress has been made in the availability of grain storage. When the International Bank conducted its survey in 1951 it reported that modern storage facilities were limited to 5,000 tons in Baghdad. By mid-1964, according to the USDA source just cited, facilities had grown to 65,000 tons in Basra; and there was an equal amount in existence elsewhere in the country, which jointly would cover some 15 percent of an average annual crop. The total substantially exceed what the International Bank, in 1951, thought the country might need. We have not, however, encountered any description or analysis of the usage of these facilities and their role in stabilization of prices or improvement of quality. A more recent FAO study,[9] discussing financial aspects of marketing boards, comments that "Where...a board is unable to secure full use of the facilities it has provided, as, up to the present, [in the case of] the Grain Board in Iraq, its operating deficits and capital costs must be met by government subsidy." One would thus assume that for reasons not stated, but perhaps associated with either fear of taxation

---

[6] *Israel Economic Development: Past Progress and Plan for the Future,"* Final Draft (Prime Minister's Office, Economic Planning Authority, Jerusalem, March 1968), p. 336.

[7] Sing-Min Yeh, *A Brief Study of Production-Consumption Price and Marketing of Wheat, Barley, and Cotton in Syria* (Damascus: United Nations Expanded Technical Assistance Program, July 1967), p. 12.

[8] H. Charles Treakle, *The Agricultural Economy of Iraq* (Washington, D. C.: Economic Research Service, U. S. Department of Agriculture, August 1965).

[9] J. C. Abbott and H. C. Creupelandt, *Agricultural Marketing Boards, Their Establishment and Operation* (Rome: FAO, 1966).

or with inefficient quality control—both mentioned as possible obstacles to storage utilization by the International Bank mission in 1951—the facilities now exceed the demand made on them by Iraq's grain producers.

Jordan has operated a government purchasing program since 1954, designed to even out both supply and price fluctuations; and for that purpose to provide for storage and preservation, grading, cleaning, transportation, seed improvement, collection of information on many aspects of grain production and marketing, setting of local prices, making loans against grain as collateral, etc.

The above-cited FAO report calls the Jordan Grain Office typical of Near Eastern marketing boards generally. Since then, however, developments in Egypt seem to have gone a considerable distance beyond—in that purchasing, selling, as well as milling have either formally or factually been nationalized. At the same time, producer prices have been fixed at levels above those of world prices.

As was said at the beginning of this chapter, the issue of marketing boards is a large one. But it is well described and systematically covered in available literature and a fuller analysis here would take us far afield. Perhaps the principal question one might want to raise is one of the appropriate division between private and governmental functions. One's impression tends to be that strong private bodies might accomplish a good many things in the marketing field better and at less cost than do government agencies. The difficulties are (1) that the impression is hard to support factually, as opposed to episodic reporting; and (2) that it is precisely the absence of such private bodies in most of the countries (Lebanon and less so, but for reasons specific to it, Israel, are probably the primary exceptions) that is an earmark of their low stage of economic maturity and has made governmental bodies the focus of development.

Summary

There has been no intention in this study to provide a detailed description of marketing in all or even most of the countries of all or most agricultural products in all or most of its aspects, but rather to point at those of its principal features that bear upon the future unfolding of agricultural production:

(1) The great bulk of agricultural output is consumed domestically. Exports consist as a rule of no more than one or two special items, accompanied by very small quantities of a larger variety of products. How much of the domestically-consumed portion undergoes marketing is not known.

(2) Agricultural exports form varying portions of total merchandise exports. The principal distinction is between the oil producing countries, for whom agricultural commodities form only a minute share of total exports, and those countries like Lebanon and Jordan that produce little else that is exportable. Israel is at an intermediate stage when agricultural exports are substantial but are slowly being outdistanced by other goods.

(3) Because agricultural exports tend to run to specialties, any expansion to new lines is likely to require a major investment in a marketing effort that stretches all the way from raising the appropriate crop to finding the best customer. Current conditions suggest that the task is not an easy one, and that it is difficult to advance at a pace faster than economic, social, and technological development generally. Moreover, costly improvements at the production end with consequent substantial increases in output, can easily be stultified by the absence or deficiencies of reasonably efficient channels and methods of marketing.

(4) With few exceptions (Egyptian cotton, Lebanese pulses, Syrian barley, Iraqi dates, Israel citrus), agricultural exports tend to remain within the area, and their volume—the specialized items apart—is exceeding small. Neither Western nor Eastern Europe are at this time major outlets, nor are Asian countries to the East.

(5) All countries have set up a more or less complex machinery for intervening in the marketing of staple crops. This tendency is not likely to abate. Ideological reasons and the need for assured access at reasonable prices of the growing urban population might even widen their scope and number.

CHAPTER 12

Rural Communities and Agricultural Services

OUR concern in this book, up to the present point, has been primarily with natural resources, their use, and their management—and with the financial aspects of resource use. Since one of the purposes of stepping up agricultural production is to raise the level of living of farmers, we now consider the way in which rural people in the Middle East presently live (using a wide variety of indicators, from specific to broadly descriptive statistics) and how well they are served by what is often called the "infrastructure," or the sum total of services and institutions that agriculture must rely on for its efficient functioning.

Rural Living

Unfortunately, there is a serious lack of direct, comprehensive, inclusive, fully comparable data, for comparisons between countries and for comparisons over time in the same country. Many countries that have reasonably accurate data on crop acreages, yields, and production have almost no data about actual living conditions among the rural people. This is true of many developed countries; much more so of the developing countries of the Middle East. In this chapter, we review such statistical data as exist and draw such inferences as we can about living conditions among farm people.

Per Capita Incomes

Value-added data contained in Chapter 10 comes close to personal income data, but is not exactly the same. There we dealt mostly in national totals, with some comparisons per acre and per worker. Here the focus is on income per person. But in drawing intercountry comparisons we are, as in other parts of this book, limited by official exchange rates; these may well overstate the real incomes in some countries, including countries where average incomes are very low. And to the extent that exchange-rate adjustments lag behind inflationary price increases, as has been true at times in Israel, for example, they will lead to overstatements. Moreover, monetary income may be misleadingly low, especially in very low-income countries where farmers produce a great deal of their own food supply. But there is little choice, and when all is said and done, monetary income data provide the best measure devised as yet for comparisons between countries, population groups within a country, and time periods.

Per capita incomes in most of the Middle East are low (Table 12-1). Except in Israel, they averaged less than $350 in each country in 1965, and greatly less in some of the more populous countries. The comparable figure for the United States in the same year was nearly $3,000. Even when the latter figure is adjusted downward for such bars to direct comparability as the fact that some kinds of

services—for instance, laundering—are included as income when done commercially but not when self-performed, the gulf between the United States and the Middle East remains wide.

Although the data are incomplete and somewhat fragmentary, per capita incomes of farm people are much below those of nonfarm people—roughly by half. The same general situation exists in the United States; indeed this major discrepancy between farm and nonfarm per capita incomes is not a characteristic of low-income countries, but is found in nearly all high-income countries as well. It is not confined to countries whose agriculture seems stagnant, but extends also to countries in the midst of rapid agricultural revolutions; while its causes are complex, some factors have in the past been the high rate of population increase among farm people, which has contributed to continuing migration away from or excessive subdivision of farms, and the relatively inelastic demand for agricultural commodities. In some countries it probably also relates to the uneven distribution of "political muscle."

The comparison in Table 12-1 is between farm people and the whole population, which includes farm as well as nonfarm population. If data on per capita incomes were available for the nonfarm population alone, the gulf would be still wider. In Egypt, for instance, where the agricultural population is roughly half of the total, it is easily calculated that the nonfarm group must be receiving average incomes three times as high as are the farm people, in order for the average of all incomes to be double that of farm people.

When per capita incomes are very low (under $200), then farm people simply cannot live well by modern standards. Such amounts will not provide enough food, whether farm-raised or purchased, and will still leave much over for other necessities.

Food Supply per Capita

The available data on food supply per capita in the Middle East, unhappily, relate to the total population, not to the farm population. The latter have much lower incomes than nonfarm people (by half, more or less, as we noted); but this fact does not necessarily mean that they will have less food or poorer food or both, than nonfarm people. Farm people in this region raise a considerable part of the food they eat (its imputed value is included in their income figure). This food tends to be much cheaper than food purchased in the cities by urban populations, and is cheaper also relative to costs of nonfood items. What the extent of this influence is and whether it is more than offset by the very low incomes of the farm people is a matter of conjecture; local observers seem to feel strongly that farm people are better fed than those in urban areas. Others have their

Table 12-1

Per Capita Income, by Countries, National Averages and Farm (or Rural) Population, ca. 1965*

(U. S. $, at Official Exchange Rates)

Country	Source[a]	Year	Per capita income	
			National average	Farm population
Egypt	UN '67	1965	161	
	our estimate[b]	1965	135	63
Israel	UN '67	1965	1,076	
Lebanon	UN '67	1965	338	
	FAO '59[c]	1958	245	101
Jordan	UN '67	1965	192	
	FAO '67[d]	1961	(733)[i]	(414)[i]
Syria	UN '67	1963	163	
	MP '68[e]	1964	183	160
Iraq	UN '67	1965	217	
	Iraq '68[f]	1966	225	
	Iraq[f]	1965		123
	FAO '59[g]	1956	135	105
United States	UN '67	1965	2,910	
	USDA '67[h]	1965	2,424	1,545

* Note carefully the definitions; some figures are comparable or useful only for some purposes, not for others.

[a] UN '67 means *Statistical Yearbook 1967,* United Nations; in all cases, national income data used.

[b] Based upon data in Bent Hansen and Girgis A. Marzouk, *Development and Economic Policy in the UAR (Egypt)* (Amsterdam: North-Holland Publishing Co., 1965), p. 34, which show 57% of the economically active population engaged in agriculture; this percentage applied to UN Yearbook 1967 population estimate for Egypt of 30,147,000; Donald Mead, *Growth and Structural Change in the Egyptian Economy* (Homewood, Ill.: Richard D. Irwin, 1967), p. 53, in a footnote cites but does not endorse an estimate of national income, by sectors, in 1964-65; this process, at best, yields a general indication of the gap in per capita income between agricultural and other people in Egypt at the present.

[c] FAO Mediterranean Development Project, *Lebanon–Country Report* (1959); on p. II–6, it is estimated that 40% of population rely primarily on agriculture for their livelihood; on p. II–3, that total population of the country is probably about 1.8 million; on p. II–13, Table II–6, some data are given on net national product by sectors; this process, at best, produces an estimate of the disparity between per capita incomes for rural and other people.

[d] FAO Mediterranean Development Project, *Jordan–Country Report* (1967), p. 19; data are for value added per worker, hence are *not* comparable with other figures in this table, but may show the disparity between incomes of agricultural and other people.

[e] Ministry of Planning, Syria (Memo. No. 175, February 1968); income per capita in agricultural sector.

[f] Unpublished estimates by Central Bureau of Statistics, Iraq, summer 1968; income per capita of "agricultural population."

[g] FAO Mediterranean Development Project, *Iraq–Country Report* (1959), p. 13; income per capita in "non–oil sector."

[h] *Agricultural Statistics 1967,* p. 573; figures are for "per capita disposable personal income," which is not fully comparable with other data in this table, but they do show disparity in income between farm and other people.

[i] Note particularly definition of this item in footnote[d] above.

doubts, and the absence of data leaves matters up in the air. Moreover, we are dealing here with averages, and it is well possible, indeed likely, that rural *poor* are better fed than urban *poor;* but whether the rural *averages* exceed or even equal the urban *averages* is not a question that can be judged on the basis of impressions or logic.

The consumption per capita of the more expensive foods is responsive to differences in average per capita income, be these between countries or between different points of time. (Table 12-2 and Figure 12-1). This responsiveness is particularly marked for meat and eggs, but it applies also to milk. By and large, increased consumption of these foods is likely to mean an improvement in the quality of the diet. The consumption of sugar too is somewhat responsive to increases in per capita income; though in this case it is probable that the higher income countries consume more sugar than nutritional considerations alone would counsel. In contrast, the per capita consumption of wheat declines greatly as per capita incomes rise, although not quite so regularly as the opposite is true for the consumption of meat. In the United States, to take a similar example, potatoes have often been regarded as a food whose consumption declined as income rose—a poor man's food; historically, per capita consumption of potatoes has declined in the United States as average per capita incomes have risen. Comparing the countries included in Table 12-2, however, one would conclude that per capita potato consumption and incomes are positively correlated. However, suitability of potatoes as a crop adapted to climatic conditions may well be the more important factor in its spread.

Some of the figures in Table 12-2 raise questions and call for comment. Why, for instance, should per capita consumption of vegetables in Jordan be more than double the figure in any other country, including the United States? These vegetables are, in the main, tomatoes, watermelons, and other members of the cucumber family, which loom large in terms of weight, but in terms of calories account for only 10 percent or so of daily intake. Thus, what elsewhere, especially in temperate zones, might be—and usually is taken to be—an indicator of a high standard of nutrition (i.e., high per capita vegetable consumption), is here simply the reflection of a grossly unbalanced diet. If the data in this table are reasonably correct, the average consumer in this region is rather well supplied with vegetables and fruits; however, this finding needs careful qualification as just suggested.

Total caloric content of the average diet in Egypt, according to these data, is extremely high; likewise, it seems very low in Iraq, leading one to wonder how accurate all these figures may be. Indeed, FAO in the Indicative World Plan departed from the official figures for Egypt and used a total caloric content, based upon a reduced cereal consumption, of 2,390 calories per day (as compared with the official calorie count of 2940 per day).[1] The total

---

[1] One of the reasons for the difference might be an underevaluation of the volume of consumed cereals, especially corn, fed to livestock, and in more recent years, poultry. Also, waste allowances are reported as very small. In any event, FAO comments that ". . . the very high level of cereal intake shown on the official

Table 12-2

Net Food Supply per Capita, by Countries, ca. 1963–65

(Grams per Day)

Kind of food	Egypt	Israel	Lebanon	Jordan	Syria	Iraq	United States
Cereals	599	285	340	386	438	355	183
Potatoes & starches	36	99	54	35	25	15	132
Sugars & sweets	48	105	72	74	44	81	131
Pulses, nuts, seeds	32	28	74	28	39	15	21
Vegetables	279	310	284	626	169	156	274
Fruits	228	411	486	268	396	196	218
Meat	35	133	87	28	30	55	276
Eggs	3	61	10	7	4	3	50
Fish	15	18	6	2	–	2	13
Milk	124	391	268	115	108	207	673
Fats & Oils (fat content)	16	48	31	36	26	10	58
Total calories per day[a]	2,940	2,830	2,630	2,540	2,330	2,100	3,150
	(2,390)	–	(2,420)	(2,230)	(2,350)	(2,140)	
Protein per day:							
Total (grams)	85.1	86.6	77.6	66.8	69.3	60.7	93.2
Animal (grams)	12.5	39.7	25.3	10.5	9.3	16.8	62.3

[a] Note figures in parentheses taken from FAO's *Indicative World Plan,* which declined to accept national data as accurate. The U. S. Department of Agriculture, in it most recent published attempt to analyze the world food situation (*The World Food Budget 1970,* Foreign Agro. Econ. Report No. 19, [Washington: Economic Research Service, USDA, October 1964]) uses the following caloric intakes per head per day: Egypt, 2,300; Israel, 2,840; Lebanon, 2,480; Jordan, 2,200; Syria, 2,300; and Iraq, 2,210.
Source: FAO Production Yearbook, 1967, pp. 420–433. Supplies at retail level, available for consumption.

protein in the average diet also differs considerably among countries, and the share from animal sources even more so.

Other Statistics on Living Conditions

Some additional data are available on other aspects of living conditions in the Middle East (Table 12-3). Generally speaking, and despite a rising trend in most of these areas, rural people in this region have poor health facilities, live in small and crowded houses, have poor facilities in those houses, have a low participation in education—especially among females—and have limited newspaper and cinema participation. Most of these data are for the entire population, and the foregoing description mostly applies to city as well as to farm people; but where scattered data are available on the situation on farms, it is much worse. For instance, 39% or urban dwellings in Jordan have electricity, but not long ago only 1% of farm dwellings had it—a percentage that might by now have risen perhaps to 5 or 10% (in Syria, the figures are 88% and 10%, respectively). Although available data suggest a modest degree of schooling for girls in Arab countries, it is highly probable that nearly all of this is in towns and that farm girls typically get no schooling.

Again, the number of these various facilities and the amount of participation in their use is rather responsive to per capita income, though the paucity of data only conveys a hint of this, as well as of the variability of facilities among the countries. Israel, as one might expect, lies intermediately, for most factors, between the Arab countries and the United States. Figures for the latter are supplied, as elsewhere in this study, to serve the American reader at least as a guide to interpretation.

In view of the scarcity of statistical indicators of living conditions in the rural Middle East one is tempted to inform himself through the more impressionistic writings of visitors, travelers, and other observers. One such study, for example, is *The Egyptian Peasant* by Henry Habib Ayrout,[2] another is a more scholarly work by Charles W. Churchill, a sociologist at the American University of Beirut.[3] The main trouble with these and similar works is that the observations are usually at least fifteen to twenty years old.

Perhaps all one can and should say is that conditions of living are what they can be at annual per capita incomes that range from barely $100 to two or three times that level. Including as they do the imputed value of self-supplied food, these are not amounts that permit consumption of goods and services above the barest minimum, considering especially that these are averages and thus much of the rural population must be expected to exist below them.

____

food balance sheet has provisionally been lowered toward the USDA estimate." (Indicative World Plan, Near East,Vol. II, p. xviii.)

[2] Boston: Beacon Press, 1963.
[3] Charles W. Churchill, "Village Life of the Central Beqa' Valley of Lebanon," *Middle East Economic Papers* (Beirut, Lebanon: Economic Research Institute, American University of Beirut, 1959).

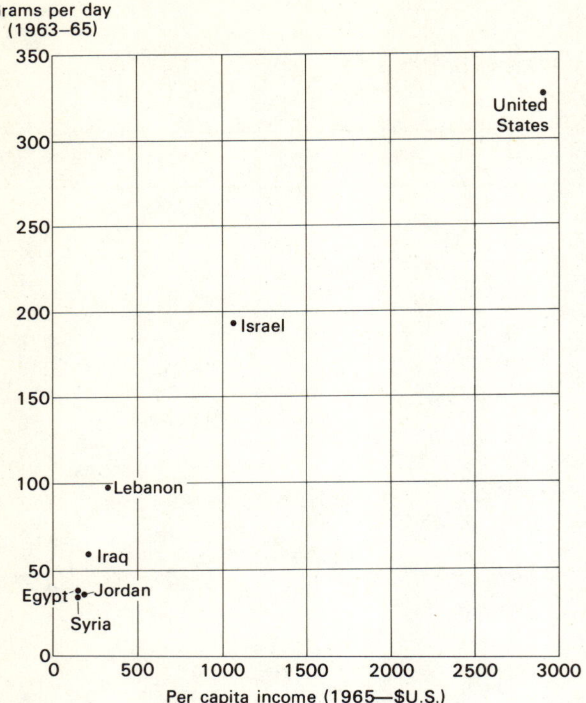

Fig. 12-1. Daily Consumption of Meat and Eggs in Relation to Average per Capita Income, by Countries, ca. 1965.

Source: Tables 12-1 and 2.

Services to Agriculture (Infrastructure)

Infrastructure means the whole body of social arrangements and of physical means within which the social and economic life of the nation and the individual takes place. Among social arrangements, the general educational system is a good example; among physical means, the transportation system is another one. The infrastructure of a nation is critical to its social and economic life. Beyond the capacity of the individual to supply for himself, it must be the product of the government or some other large groupings of individuals.

Some parts of the infrastructure condition the course of the economy in general. Among these are the quality of government, the level of health and education, the network of transportation and communications, the quality and comprehensiveness of basic data on the economy, and others.

About each of these elements a good many things could be said. Some of the comments might be rooted in general reasoning about the nature of developing countries, especially those countries here under review. Other comments might be based on observation by individual scholars and by organizations, both national and international. And some might reflect available statistics and other quantitative indicators. A careful review of the data and the literature that have come to the authors' attention, however, has convinced them that each of these sources is afflicted with such deficiencies—from the specific point of view of this book—that made it the better part of

wisdom to refrain from an extended exposition of those sources.

A major deficiency is that the most comprehensive and most objective literature is badly out of date, going back to the early and middle 1950's. This is true above all of the otherwise excellent mission reports of the International Bank. But it also holds for many sources cited in books that have been published more recently but contain data by now one to two decades old. Residents of the countries involved can legitimately object to a description published in 1970 that harkens back fifteen years or more and can merely end on a note of assumption that things have probably improved since.

Another reason for our abstaining is that hard data and facts are extremely scarce, especially any that would reflect a continuing development. Others are so indirect that conclusions are at best hazardous. Others still are of the case or episodic type, perhaps suggesting a condition in a given place at a given time, but not conducive to broader-based conclusions.

For such reasons, several general topics, despite their demonstrable relevance to agriculture, have not been here included. These are health and education (touched upon in Chapter 6), transportation, and communications.

Another of these topics, efficiency and stability of government, is dealt with in the context of obstacles to growth in Chapter 19.

There are, however, strands of the infrastructure that—as we have pointed out on several occasions—strongly shape agriculture's role and efficiency. For a sustained period agriculture cannot move far ahead of the rest of the economy or society (though it can indeed remain far behind, where programs of rapid industrialization are relentlessly pursued). In the subsequent pages we attempt some description and evaluation of those parts of the economy's services and institutions that are of special relevance to agriculture; but such attention as is afforded each of them is general and brief, lest the study become a respository of miscellaneous information of the various countries' economy.

Infrastructure to the farmer comprise (by this or any other name) such goods and services as technical information, agricultural credit, general supplies or inputs, water supply and irrigation systems, markets, transportation, and other elements that form the economy and social structure beyond his farm. In fact, from the viewpoint of the national society and economy, what we call agricultural infrasturucture may be considered as part of agriculture broadly defined, and not as infrastructure—though what is internal to agriculture as a sector, viewed from outside of it, becomes external to the individual farmer. A number of related but more or less separately identifiable aspects of the agricultural infrastructure can be distinguished.

Agricultural Research, Education, and Extension

Most countries maintain a structure of publicly supported agricultural research, formal agricultural education in

Table 12-3
Selected Measures of Living Conditions, by Countries, ca. 1965

Item	a	Egypt	Israel	Lebanon	Jordan	Syria	Iraq	United States
Health:								
persons per physician		2,380	410	1,390	4,040	5,110	4,760	670
Housing:								
rooms per dwelling	T	3.6	2.1	4.3	2.3	2.5		4.9
	U					2.9		4.8
	R					2.4		5.1
persons per room	T	1.6	1.8			2.3		0.7
	U					2.1		0.6
	R					2.5		0.7
% of all dwellings with inside water	T		90		21			93
	U		93		49			99
	R		74		2			79
gas	T		84			12		52
	U		88			28		66
	R		64			3		14
electricity	T	38	90	91	17	38	17	
	U		94		39	88		
	R		70		1	10		
bath	T		88		9		10	88
	U		91		18			96
	R		57		2			69
flush toilet	T		81		10		34	90
	U		87		23			98
	R		53		1		3	70
Education:								
% of pop. enrolled		15.4	26.6	18.7	20.1	17.4	15.3	28.6
% females, students		36	48		38	29	28	
expenditure per student	$	49	112	38	28		63	522
Newspaper circulation per 1,000			143		8	11	12	312
Cinema—annual attend. per person		2	20	14	3	4	1	12

[a] T = total; U = urban; R = rural.

Source: Statistical Yearbook 1967, United Nations.

schools (primary, secondary, and advanced), and adult or extension education; and in many countries there are private efforts in the same or similar directions. The system in many countries is quite similar to the USDA-land grant college system of the United States; indeed, some have been closely modeled on it. Others resemble the land-grant college system, but owe their origins to other stimuli. Such a research-education-extension structure, when adequately organized and competently managed, is highly productive: money spent for these activities will produce income far in excess of its costs. But such productivity often becomes apparent only over a considerable period of time. Extension must operate with farmers as they are, and it may take a new generation to adopt available new methods; research often takes some years, even at the best, to produce really valuable new results; etc.

Of the countries here studied, each has a publicly supported agricultural research, education, and extension organization or structure. A mere listing of experiment stations, projects, schools, etc., which would be feasible gives little indication of their value or competence. On the other hand, it has simply been beyond our capacity to undertake a careful appraisal of the agricultural research-education-extension structure of each country in the time available for this study. Instead, we present a number of conclusions that can be drawn from the literature on this subject.[4]

[4] The remainder of this section draws upon (1) FAO Mediterranean Development Project reports for Lebanon, Jordan, and Iraq; (2) International Bank for Reconstruction and Development, *The Economic Development of Jordan* (1957), *The Economic Development of Syria,* (1955), and *The Economic*

Everyone recognizes that much good work has been done and that there are many sincere and dedicated men working in this general field; but the following weaknesses have been noted for most countries and by most observers:

(1) All of the programs are on a wholly inadequate scale. Funds available for research are pitifully small in relation to the farm areas or numbers of farms; the number of students in schools is but a fraction of the farm boys and girls that should be receiving training; the number of villages and farmers that the extension people can work with is but a fraction of the total, and so on. For example, in its Indicative World Plan (Vol. I, p. 165), FAO estimates that in much of the Middle East, except in Egypt (and Israel, which is not covered), there may be up to 8,000 farm families per full-time extension worker, compared with what FAO believes is a reasonable ratio of 1 per 1000 families on nonirrigated and 1 per 500 on irrigated farmlands. In Egypt, the 1:500 ratio is judged by FAO to have been nearly achieved.

(2) Every country, with the probable exception of Israel—which some years ago began to export its expertise to other parts of the world—has far too few adequately trained men. In many instances, present staff is not adequately trained to carry out the tasks which have been assigned to them; were an attempt made to provide a fully adequate scale of research, teaching, and extension, the deficiency in trained men would be most severe. The body of trained men is growing and presumably will continue to grow; more and more men are being sent out of each country for training abroad, and domestic educational capacity is rising; but the rate of increase is far too low in relation to the need.

(3) There has been too high a rate of turnover of trained men in the various activities as established so that such men have had insufficient opportunity to carry out the kinds of work of which they were capable and to grow professionally in their jobs. In many instances, professionally trained and competent researchers or educators have been converted into administrators; while understandable—and perhaps productive, in a national sense—yet this robs these specialized agricultural activities of their most promising men.

(4) In agricultural research, there have been too many small research stations, too many specialized research projects, and too often a tendency to repeat what has been studied elsewhere—so that the research efforts have not been as productive in meeting the needs of the farmers of the country as they might have been. A different organization of the available research talent might have been more productive. We suspect, from the statements made in various reports, that the different research stations have tended to work too much in isolation, even within each country, and much more so if a regional or international view is taken. Moreover, there is too little contact between researcher and producer.

(5) A related complaint is that research projects tend to be tailored to the talent and preferences of the individual researcher rather than to objective needs. And since the majority of researchers prefer to stay in the capital or at least the large cities, the research is usually not field-oriented. This often merges with the problems created by a top-heaviness of highly and typically foreign-trained people and a great scarcity of technicians at intermediate levels.

(6) In agricultural extension, the professional workers have been severely limited by a lack of really useful research results to take to farmers; by the lack of necessary farm inputs—such as better seeds and fertilizers—to make their teachings fully effective; and by unresponsiveness of farmers in the face of this situation, aggravated by illiteracy and resistance to change.

(7) Financial limitations often keep needed auxiliary equipment or facilities from being furnished in sufficient volume. Transportation is one item frequently mentioned. So are foreign books and journals.

If agriculture in the Middle East is to become more productive in the decades ahead, then the weaknesses above must be corrected and the research-teaching-extension range of activities must be both broadened in scope and improved in quality.

Agricultural Credit

Borrowing in agriculture is more common among wealthier than among poor farmers, and more common in high-income countries than in low ones. This is in large part because of the costliness or unavailability of credit for low-income farmers or in low-income countries; oftentimes, although these farmers desperately need more capital, they do not have the security for a loan or the prospects for its repayment.

Agricultural credit is often divided into three classes, dependent upon the term for which the loan is made: (1) long-term, from perhaps 10 to 30 years, especially for the purchase of land, and for the making of permanent improvements on the land; (2) medium-term, mostly 3 to 5 years, to finance less costly improvements, to purchase major farm machines, to buy livestock, and the like; and (3) short-term, primarily for one year or less, often to buy seed, fertilizer, livestock feed, and other current production inputs which must be bought at a given time of the year but whose returns do not come for a few months.

Every country has governmental or quasigovernmental credit programs for agriculture, with more or less complete coverage of the three major types. Long-term credit to buy land has been available primarily in connection with land reform;[5] in Egypt, for example, peasants obtaining land

<hr>

Development of Iraq (1952) (Baltimore: Johns Hopkins Press). (3) H. Charles Treakle, *The Agricultural Economy of Iraq* (ERS-Foreign 125, Economic Research Service, USDA [Washington, D. C. August 1965]); (4) Felix J. Brucher, *Agricultural Services and Institutions in the Syrian Arab Republic* (Damascus, January 1965); and (5) Michael Baumer and O. Milton Hackett, "The Development of Natural Resources in Jordan," *Nature and Resources* (UNESCO), I, No. 3 (September 1965).

<hr>

[5] Under Syria's land reform, the peasant now obtains the land without payment.

under the land reform program were originally to repay the cost of the land (which was the price paid by the government to the former owners plus 15% for expenses) over a 30-year period, with 3% interest; later, this was extended to 40 years and the interest rate lowered to 1½%. In Israel, in contrast, there is no credit of this type, since land is not sold to farmers but is made available to them on long-term leases. The other countries are, generally, intermediate between these two.

Each of the countries also has various governmental or quasigovernmental credit facilities of short-term and medium-term duration; these typically operate through cooperatives, either special credit cooperatives or general purpose cooperatives. By far the greater part of these types of credit, however, is provided from private sources—merchants, moneylenders, and banks. In Syria, for instance, the increase in cotton production in the last fifteen years or so has been financed, as well as often directed, by city merchants. The moneylender is an indispensable part of the economic and social structure of rural areas, in most countries of this region.

Without attempting to trace the details of these various credit sources and their operations, some generalizations can be made, some or all of which appear in virtually every report written about agriculture in this region, which mentions credit.[6] Not all of them will be true for all countries, but they characterize the region as a whole.

(1) Most farmers lack capital, especially to undertake new methods of farming; if credit were generally available on favorable terms, many of them could use a great deal, even though many have farms too small or lack managerial capacity to make the best use of credit.

(2) Although numerous governmental and quasigovernmental farm credit programs have been established, and although private sources are available in most countries, there is not enough agricultural credit available, on any terms, to meet the needs of intensive farming.

(3) A large share of what credit is available goes to the larger farmers, with the smaller farmers getting little or none. This was long true in Egypt, during the early part of this century; but it is still a general complaint in the region. This circumstance is not unexpected, since the larger farmers probably can make a more productive use of credit and are better risks.

(4) Such credit as is available is too costly; this is especially true for credit from private sources. No systematic data are available, but statements are frequently made that interest at a rate of 50 percent annually, and higher, is not uncommon on private loans. Any program for increased agricultural production will require cheaper credit.

---

[6] See, for example, Garbriel S. Saab, *The Egyptian Agrarian Reform 1952–1962* (London: Oxford University Press, 1967), Chaps, VI and X; Charles Issawi, *Egypt in Revolution* (London: Oxford University Press, 1963), pp. 259–67; the economic development studies of the World Bank on Jordan, Iraq, and Syria (previously cited; see n. 4, Reference 2), especially pp. 88–95 on Syria, 259–60 on Iraq, 131–142 on Jordan; FAO Mediterranean Development Project *Jordan Country Report* (Rome: FAO, 1967), pp. 98–194.

(5) Credit from the governmental and quasigovernmental sources is usually too slow in being made available, and is too much entangled in red tape to have maximum utility for farmers. Much of this credit, especially in Egypt, is in kind—seeds, fertilizer, chemicals for insect control, and the like—rather than in cash. In several countries, extension of credit to low-income farmers has been accompanied by a considerable degree of supervision from government; this is understandable, given the low level of managerial competence of many farmers, and may well be productive, but too often this has been a delaying factor also.

Agricultural Marketing System[7]

A "marketing system" includes accepted weights and measures, usually supervised by a government agency; often a system of grading commodities, either publicly or privately managed, and often some supervision over sanitation, especially for meat and milk; some means of collecting and pooling small lots from farmers into larger lots for sale to city merchants or others; transportation from rural area to market; storage, always specialized as to product, from season of surplus to season of demand (and, desirably, in both producing and consuming areas, and perhaps enroute also); processing facilities, at least for many products; a means of organizing the buying and selling process itself, in specialized exchanges or elsewhere; and other related processes. Although the small farmers of the Middle East tend to consume a substantial part of what they produce—through far less so than the typical Indian farmer—yet they are producing some commodities for cash sale; hence a marketing system is critical for them.

There is great variation among countries and commodities as far as the marketing system is concerned. For some commodities in some countries, as described in Chapter 11 above, the market is specialized and highly organized; this has more often been true for export than for domestic crops. For instance, marketing of the cotton crop in Egypt has long been highly organized, either privately or by government in recent years. Grades and classes have been established, there is inspection for grading, an organized market provides a means for buyers and sellers agreeing on price, credit is available for stored cotton, the government has exercised coniserable control over sale of the product abroad, etc. More recently, the government has organized markets for other commodities. A less elaborate but still considerable control over marketing has been exerted in Iraq for such products as grains, dates, tobacco.

Cooperative marketing associations play differing roles in the countries of the region. In Israel, nearly all agricultural produce is handled by Tnuva, the agricultural marketing cooperative. In Egypt, the agricultural cooperatives are playing a larger role in marketing in recent years, with all the cotton and a substantial part of the fruits and vegetables sold being marketed by them. In the other countries, the role of cooperatives is smaller.

---

[7] The observations made in the following paragraph supplement the more specific and quantitative description in Chap. 11.

A special problem exists for wheat, and to a lesser extent for other grains, in Lebanon, Jordan, Syria, and Iraq, where yields from rainfed lands are so highly variable from year to year. In a year when yields are high, current output may exceed current consumption; the surplus must either be sold abroad, which often presents serious problems because exports are irregular in timing and perhaps not of high quality, or it must be stored. When wheat yields are low, either necessary supplies must be taken out of storage—if they exist there—or wheat must be imported. The costs of transport of imports, or of exports if these arise, can be considerable; some means of storing surpluses from year to year is essential, possibly with some government subsidy. Jordan, in particular, has faced this problem. If grain yields from rainfed lands should be increased significantly in the future, even to a lesser degree than we think is practical, then this matter of occasional surpluses might become more serious.

Attention is called to Chapter 11 for comments on marketing. In addition, a number of reports dealing with agriculture in the Middle East have made the following general points about agricultural marketing:

(1) The agricultural marketing system in most, if not all, countries is inadequate for its task: its operations are too costly, it is too inefficient, there is often too wide a spread between the prices which the farmer gets and prices which the consumer pays.

(2) Many farmers lack good market outlets for what they now produce. They would be even more handicapped could they produce more. Any attempt to increase agricultural output, especially in some areas, will be limited by the kind of agricultural marketing system that can be established.

(3) The marketing system fails adequately to convey the consumers' desires for agricultural commodities to the farmer.

Agricultural Supply Structure

As we noted in Chapter 10, the average Arab farmer in the Middle East buys relatively few inputs to produce his output; his low output per man reflects this. The average Israel farmer both buys more and produces more; and the average U. S. farmer still more. A considerable range of supplies is needed for modern agriculture. To be useful such supplies must be available in volumes to meet the needs of all farmers; available when needed; and, perhaps as important as all other factors, at reasonable prices—i.e., at prices that when put into the user's cost calculations leave him the prospect of a profit.

In the Middle East, all too often these conditions have not been met. With exceptions, domestic production is limited to older types of farm tools and machines or to assembly operations of imported parts; farm manures have traditionally been the major form of fertilizer; various traditional methods of insect, disease, and weed control have existed, often largely or totally ineffective; and some other domestic inputs have been available. But improved farming requires both more and different inputs. Some of these, like farm machines, can be, and have been, imported.

But this is costly: the types imported are not always fully suitable to conditions in the country; the problem of getting and keeping an adequate supply of spare parts requires modern management and credit facilities; and the costs may raise serious foreign exchange problems. Moreover, a machine without adequate training facilities is bound to render unsatisfactory service, and much complaining can be heard on that score. Sometimes the needed supplies have not been imported in time; the yield of Egyptian cotton dropped a third in 1961, largely because insufficient crop dusting materials had been provided. If any substantial increase in agriculture in the Middle East is to occur, then the supply system for inputs must be greatly improved.

One of the most promising lines of improving the supply situation would be, as noted elsewhere, regional cooperation in the production of various agricultural inputs. Cooperation between Israel and Arab countries in this sphere, as in others, is probably unrealistic to consider for some time to come; but cooperation among Arab countries would seem possible. Machines, fertilizers, and other chemicals (Jordan now produces phosphate in substantial amounts; Iraq is a petroleum producer and thus potentially a producer of most chemical inputs), as well as other inputs could be developed or tested in research organizations, and produced at a few locations in economically efficient plants. At first, it might be uneconomic to produce tractors totally in some country, as for instance Egypt; but a cooperative venture might be undertaken with some economically more advanced country, where some critical parts might be produced, while other parts could be produced in the Middle Eastern country and the whole assembled there.[8] Or a chemical factory in some favorably located country (Lebanon, for instance) might produce some of the more common and readily manufacturable farm chemicals in cooperation with a firm in some advanced country, from which could be imported other necessary chemicals. In short, the possibilities of regional cooperation and specialization seem considerable. Timely availability of supplies adapted to the region would result, and foreign exchange saved. Moreover, new technology, wherever applied, has a habit of stimulating technological advance elsewhere.

Institutions for Irrigation and Drainage Management

Under some circumstances, irrigation is possible by efforts of the individual landowner or farmer; he may be able to pump water from streams or wells, or in some instances he may be able to divert small streams by gravity, using simple ditches or canals. Availability of groundwater in India and

---

[8] According to FAO's *Indicative World Plan*, some countries have moved ahead and greater scale would assist them: "In Syria and Jordan there are already a number of examples of implements and machines being made locally by enterprising small workshops, which are better adapted to local requirements than anything which can be imported, e.g. special stationary threshers, cultivators adapted for manual seeding down a tube, disc harrow ploughs all made locally except for the discs, and harvest and general farm transport trailers."

Pakistan, for example, combined with the prospects of high yields from new varieties has in recent years stimulated a great deal of small tubewell development. Some present irrigation in Syria and Iraq is based upon pumping from wells or rivers in which the water level lies below the surface of the land. Most irrigation in the Middle East, however, is from canals which divert water from a major river—principally the Nile, Euphrates, and Tigris—and from their tributaries. Such canals are relatively major works, with capacity to irrigate considerable areas of land; hence they have uniformly been public works, constructed and managed by the government, with the water used by individual farmers.

In modern times, the central government of each country, or the state, thus has responsibility for irrigation and drainage, and for the building and management of dams, canals, and drains. These arrangements then are a part of the agricultural infrastructure of supreme concern to the farmer. The specifics vary from country to country, and so does the scope of government activity.

In Israel, for example, the government, directly or through a governmental body, also undertakes most of the construction and management of wells as a source of irrigation water; these have been integrated with surface sources of water, as indeed they should be in any competent water management plan in any region.

With the exception of privately owned wells and pumps, irrigation water is provided to the farmer at times and in volumes decided not by him but by the state water management agency. Whether the trained engineer or other specialist in the employ of the state is better able than the illiterate peasant to judge how much water to apply and when is arguable. But, for our present purpose, it should be emphasized that in no Middle Eastern country here considered is there anything remotely approaching irrigation management by users, such as, for example the irrigation district of the United States. In the latter, landowners and farmers form associations, under special legal powers granted by the states, to assume substantial responsibilities in providing irrigation water and drainage. In some cases, such districts have constructed reservoirs and built dams; in others, they have negotiated with public agencies that have built the works on such matters as repayment contracts; and typically the irrigation districts have operated and maintained the irrigation system, even when it was built by a federal agency. In the Middle East, in contrast, all these functions are performed by a government agency; the farmer has water available at places, times, and in amounts determined by the agency.

What goes for irrigation, goes equally for drainage. Since the drains must be constructed by the state, and maintained by it in Egypt as well as in Iraq or Syria, again the farmer's dependence on his government's activity becomes crucial. Infrastructure becomes very vital. All too often in recent years, drains have neither been constructed nor, if built, maintained, and the farmer has been worse for it.

What matters greatly are not only the societal arrangements under which irrigation and drainage works are constructed and maintained (though, as several Asian countries have demonstrated, private initiative can be a powerful originator of economically performing irrigation facilities), but also the conditions under which its use takes place. Thus we must look briefly at the economics of water use.

In the Middle East, by and large, farmers getting water from state irrigation enterprises do not pay for the water *as such*. Those who operate their own pumps obviously pay for their water, and pay in proportion to the amount of water they use. The farmer in the Nile Delta and Valley does not pay for his water directly; he does pay taxes in one form or another, and these, with other sources of revenue, provide the state with funds to construct and maintain the irrigation and drainage works. Israel is something of an exception to these statements, since farmers there do pay a water charge (although water costs have been subsidized, especially in areas of very high water costs, such as the Jerusalem corridor). It may be argued that, looking at the economy as a whole, it does not matter how the revenues are raised with which the state constructs and maintains irrigation and drainage works, as long as the income or benefits generated by the works exceed their cost. But the absence of direct water charges means that the farmer has no *financial* incentive to economize on water use, and this does affect even the cost-benefit calculations of the economy as a whole.

Various administrative controls on water use have therefore been instituted, in a trade-off between economic and administrative controls. For example, in considerable parts of the Delta and Nile Valley, the canals have been constructed below the level of the ground surface; a common system of canal management there is to provide a period of days (often five) when they are kept so full of water that gravity diversions to the farmer's land are possible, followed by an equal period when water level in the canal is so low that he must pump the water to his land, followed by a third period of equal length when there is no water in the canal. Though water pumps are coming increasingly into use, much of the pumping is done, as discussed in Chapter 4, through simple devices such as the water screw or the shadouf, powered in most cases by human muscle. Designed as a means of rationing water, the system has the further effect of keeping the peasant almost constantly in the water, thus countering strenuous efforts to control bilharzia or other water-borne diseases. Whatever its merits or demerits on other grounds, the system appears to have been ineffective in achieving its objective of limiting water use to an appropriate amount and, as studies elsewhere have shown, very costly.[9]

---

[9] Recent studies in India, for example, reveal truly large cost savings in the transition to mechanically powered water pumps. It is only where the farmer's alternative employment opportunities are estimated to be at or near zero and added leisure is equally judged without value that such labor can be judged to be economically effective. Under this aspect, the controversy of the degree of underemployment in Egypt and other Middle Eastern countries becomes anything but an academic pastime, and the authors are inclined from what evidence they have seen to judge that farmers' opportunity costs are substantially above zero.

Throughout the world, irrigation farmers, if given the chance to do so, use too much water. In almost every location where they have been made in countries where irrigation is practiced, careful experiments have established that the economically optimum water application is less than the application in fact made by the farmer. Exceptions exist in some parts of California and Arizona, where very high water costs make the economic rewards and penalties of careful water application patently obvious.

The most direct effect of too much water will be a reduction in the crop yields of the offending farmer; but its real effect is often upon others, or is delayed, or both. Prolonged excessive water application results in water-logging and consequently greatly reduced crop yields, unless, of course, adequate drainage is provided. Competent irrigation water-management programs will include occasional flushing applications of water, where a considerable part of the water applied quickly runs off or drains away; but continuous application of too much water either raises the water table ruinously or places an excessive burden on the drainage system, or both. Moreover, water savings of small proportions but over a large area would postpone the time when recourse has to be had to one or the other costly scheme of water augmentation from saline water, cloud-seeding, etc.

There is probably no other place in Middle Eastern agriculture where an institutional innovation, at a low cash outlay, would pay off better in terms of increased agricultural production, than in some system of irrigation water charges which would be technically and administratively practical and would provide a tangible and significant incentive to the farmer to apply water in reasonable amounts. The difficulties in the way of such a radical departure are not to be underestimated. The lack of water charges is not due to an oversight or caprice, but is deeply imbedded in the religious and cultural heritage of the countries. But perhaps an increasing realization that the measures that have been used to substitute for it are basically inefficient evasions of the issue can open the way towards a changing view of the matter and a different ranking of priorities in which the preservation of the land's integrity would rank high.

Cadastral Surveys and Land Records

In contrast to the developed countries, and in common with other less developed countries, land in the Middle East is not only valuable property, but the basis of economic life and personal security for most of the population. Through the centuries, the possession and use of land in the Middle East has changed and shifted, not only after wars but also on personal and political grounds. In the face of these changes and pressures, titles to land were not secure; land records were largely absent or were meaningless where they existed, or both. Possession was indeed nine points of the law, but was readily lost in the presence of greater force.

Wickwar traces the development of land registration during the later days of the Ottoman Empire in the Middle East; although no single date can mark a beginning, some steps were taken in the early 1800's and more, by the middle of that century.[10] Land registration necessarily involved some form of cadastral land survey, to delimit land ownership tracts on the ground, and involved as well as some system of land records, so that title to land could pass from owner to heir or could be sold or foreclosed upon. The land-registration system established often operated to the advantage of the large landowners—indeed, it may well have been devised to do so. They were in a far better position to press their claims before the government than were the peasants: cash costs of registration often discouraged the latter; sometimes large landowners were asked by peasants to register lands in their name, with the expectation that the land title would later be turned over to the peasants, only to have the large landowner retain it; and in other ways the larger landowners were able to turn the land registration to their advantage. The improvement in land titles also operated to the advantage of creditors, often foreigners, especially in Egypt; land could then be taken in satisfaction for failure to repay debts.

As Wickwar says: "These first efforts at the secular recording of deeds had also a great administrative defect; they were not accompanied by a general survey and demarcation of land rights . . . . In Egypt, a test in two villages about 1917 indicated that, of the property covered by registered deeds, about 4% could be exactly located, and about 37% could be located approximately by reference to land-tax registers and by supplementing these with other local evicence, while the remaining 59% could not be located" (p. 64). After World War I, under the guidance of French and British colonial administration, cadastral surveys were made throughout the Middle East, and land-record systems established. In general, the Torrens system (originally established in South Australia), under which the state guarantees and defines the land title, was adopted. Various means were established to resolve conflicting claims to land.

While all of this was not completed—and where completed was perhaps not always above criticism—yet by the end of World War II the countries of this region had land-title systems which included a major part if not all of the cultivated land. When the postwar agrarian land reforms began, these countries could therefore build upon an existing land-record system. It was incomplete in some areas, and still is—in Iraq and perhaps elsewhere—as to some lands; and the agrarian reforms have required additional surveys under such situations. But, on the whole, the Middle East has had far better land records than many countries which have sought to initiate land reform.

Agricultural Cooperatives

Great stress has been placed on the advantages of cooperatives in agriculture, in every country of the Middle

---

[10] W. Hardy Wickwar, *Modernization of Administration in the Near East* (Beirut: Khayats, 1962).

East. There were some in Egypt, for instance, before World War I.[11] By 1938, there were 730 societies in Egypt with nearly 70,000 members—but this was probably not over 2 percent of the number of all farms at that time; by 1952, before the Revolution, there were over 1700 societies with nearly half a million members—impressive growth, but still a rather small fraction of all farmers. The early cooperatives were largely concerned with credit and with cotton marketing, and operated more to the benefit of the larger than of the smaller farms.

The agrarian reform of 1952 placed great reliance upon agricultural cooperatives.[12] All new landowners were required to join a local cooperative; these cooperatives were the source of agricultural credit, and of other production inputs; they helped to organize agricultural production, especially the operation of the land in three large blocks, within each of which the farmer had a plot, so that each block could be devoted to a single crop, such as cotton; crops were marketed through the cooperative; and it assumed responsibility for many social services to its members. Since 1962, strong efforts have been made to extend the same cooperative system to all farms. Moreover, making the cooperatives the exclusive source of credits, supplies, and markets has forced most farmers to join; more than 75 percent of the total cultivated area in Egypt is now estimated to be included in agricultural cooperatives and land ownership carries with it a declining share of rights.[13]

The other Arab countries have tended to imitate Egypt in the promotion of agricultural cooperatives, as they have in their agrarian reforms generally. However, cooperatives have included far fewer farmers in other countries, at least so far, and are a less powerful force in the whole agricultural scene.

To those acquainted with agricultural cooperatives in economically advanced Western countries, the agricultural cooperatives of the Middle East raise a number of disturbing questions. Agricultural cooperatives in the Middle East have developed under strong government promotion, not to say pressure. A strong exponent of this view puts it as follows: " The main objectives of the Directorate General of Agricultural Cooperations is to organize, guide, supervise, and advise the agricultural cooperatives and audit their accounts at least once yearly . . . . the Directorate has organized the following

agricultural cooperatives and efforts are being made to form new ones as soon as the necessary preliminary studies have been carried out."[14] In 1962, Issawi, speaking of the agricultural cooperatives in Egypt that were developed for the land reform villages, wrote: "There seems little doubt that the new cooperatives have performed valuable services to their members and have reduced some of the disruptions which the removal of landlord control might have created. There is equally little doubt that these societies are cooperative in little more than name; in fact they are strictly controlled by government officials. Moreover, they could quite easily and imperceptibly be transformed from cooperatives to collectives, and mark the beginning of the end of private ownership in the one important sector where it is still important. But, in view of the crushing burden which the government assumed in the other sectors of the economy, and considering the manifest failure of collectivized agriculture from East Germany to China, it is perhaps not unreasonable to hope that such a step will not be taken in the immediate future."[15]

It is true that agricultural cooperatives have been stimulated by government action in many economically developed countries, including Denmark, Ireland, and the United States as prime examples. But the essence of the true cooperative is that its *members* manage it; they manage it jointly, for the good of each. Agricultural cooperatives may operate under special laws, as in the United States, or have special privileges of various kinds, or may be subject to varying degrees of governmental supervision, all without compromising significantly the right and the power of the members of the cooperative to decide for themselves. It seems fairly clear that the agricultural cooperatives of the Middle East are not yet, by and large, at the stage where their members in fact do run them. Even in Israel, where cooperatives have considerable independent economic and political strength, they are still heavily dependent upon government credit. Some progress may have been made in Egypt, in spite of Issawi's gloomy commentary. It may be that some agricultural cooperatives in other countries also have an important degree of self-government; but, by and large, one cannot class these as genuine cooperatives in the traditional sense of the term. Whether they represent a necessary step in the evolution from traditional to modern agriculture is a question to which the answer will become available only in time.

---

[11] National Bank of Egypt, "Agricultural Cooperation in the UAR," *Economic Bulletin,* No. 4 (1966).

[12] Gabriel S. Saab, *The Egyptian Agrarian Reform, 1952-1962* (London: Oxford University Press, 1967). See especially Chap. IV.

[13] National Bank of Egypt., *op. cit.*

[14] *The Iraqi Revolution in Its Fourth Year,* Issued by the High Committee for the Celebrations of the 14th July, Baghdad, 1962 (Baghdad: Times Press, 1962).

[15] Issawi, *op. cit.* (above, n. 6).

PART III

Development Potential of Middle Eastern Agriculture

THE purpose of Part III is to outline the agricultural development possibilities of the Middle East, and to consider how they might be realized. This is the part for which the earlier parts were an unavoidably lengthy prelude, and we should like to think will be the payoff of the entire study. The projections and speculations about agricultural development possibilities of the region are based upon the consideration of the natural resources of the region, (Part I) and upon the present agricultural development of the region, (Part II). The quality of a projection of the future is in large measure conditioned by the quality of the factual and conceptual foundation upon which it is based; this, and only this, is the justification for having dwelt at great length on the information presented up to this point.

The essence of agricultural development in the Middle East is the application of a package of *separate but closely interrelated* programs, technologies, and processes; it is their interrelationship which is truly significant. While the exposition that follows necessarily considers a topic at a time, we shall repeatedly emphasize the package nature of these programs. Any single program may have limited and sometimes even negative effect, if taken by itself; but may be highly productive if combined with other programs in proper proportions and proper timing.

Since the word "potential" is so freely used in this study, we must make at least an attempt to define it, for it is subject to a wide variety of interpretations. As used here, it denotes a level of output and income that could be reached under competent soil and water management and other modern farm practices; access at reasonable cost by all producers to services supplied to agriculture from other parts of the economy, including prominently the short- and medium-term capital market; availability to the country as a whole of foreign capital, by borrowing or otherwise; a level of health and education that permits effective and near full-time utilization of manpower; and stable and efficient government and administrative facilities that can launch and support long-term undertakings in the national interest.

Our idea of the potential of the Middle East does not reach to the far-off or futuristic, or depend upon radical departures from practices or technologies as they now can be found in many national agricultures. Transfer and adaptation to local conditions are sufficiently radical in themselves to keep one from going beyond.

For the major crops, these "potentials" approach physical potential, though again far from laboratory conditions. For the minor crops, our output projection cannot help but reflect a measure of economic constraints—i.e., traditional cropping patterns, rotation requirement, and size of markets have influenced our judgment.

While it might be physically possible to plant a large portion of the land to one or the other minor crop, a projection along such lines would be quite meaningless. In this sense then potential does not translate into "free of constraints," but only free of constraints preventing the emergence of an agriculture that is more productive of output *and* income.

There is finally, a question of time dimensions. Is this a potential that can be achieved in five years, in twenty-five, in fifty? There is no good answer to this legitimate query. Some of the investments that are discussed below will have to be spread over considerable time. Conventional wisdom tells us that in a year one can reasonably schedule and accomplish drainage installation in 200–250,000 acres. Thus new drainage in 15 million acres of Iraq would take considerably more than the rest of the century.

Other physical measures would take much less time. But then, what about social and institutional reengineering? And what about the changes in attitudes, values, etc., that are needed? Until recently, there has been an inclination to put very long time tags on these. But recent events in India and Pakistan, where rising numbers of farmers are rapidly embracing new techniques and habits lead one to question whether perhaps this inclination is wrong. More about this is offered in the concluding chapter. Suffice it to say here that the implied time horizon for reaching the potential is closer to thirty than to ten years, and is conceived more in terms of approaching than achieving it. This is not the stuff of which computer programs are fashioned. Assumptions and definitions are too open-ended to support a neat end product. But they convey sufficiently, we believe, what we mean and what we do not mean when we speak of the potential.

So defined, and anticipating the later conclusions, the Middle East has a great agricultural development potential. No single dramatic change emerges as a panacea, but the cumulative effect of many smaller but significant programs or inputs can be very great. At the same time, the difficulties to be overcome are very great, both because they are deeply rooted and many, and because they have persisted for long periods in the past. They will yield only to well-conceived programs carried out continuously for considerable periods of time. Progress must be sought along basic, and for the most part, unglamorous lines; there is no magic by which, in a single great dramatic step, the potential goals might be achieved.

Part III has been prepared in the conviction that a long-range projection of the agricultural development potentials of the region—imaginative without becoming fanciful—and a serious consideration of the obstacles to their achievement will prove valuable to the countries of the region, to other nations interested in this region, and to the various international organizations engaged in developmental efforts. If our analysis is correct, it reveals both the promise and the problems of agriculture in the region—problems which any country or group must face, promise which any country or group may, in some measure, achieve.

The greatest problems of the region are not primarily matters of resources, technology, nor even of economics in the narrower sense; they are institutional, social, and human, subject to solution in more than one way, and with help in their solution coming from more than one source.

CHAPTER 13

Land and Water Development Potential

W E begin consideration of the future of agriculture in the Middle East with a discussion of land and water development, especially of land and water not now used or not now used as intensively as they might be. In a region where these natural resources have been used—much of the time intensively—for thousands of years, one cannot expect to find unused land and water resources except in situations where new technology and/or additional amounts of capital will permit the development and use of some resource otherwise unavailable. Nor would one expect to find easily exploitable unused resources. In fact, as this and later chapters will show, there are only limited unused land and water resources in the Middle East; moreover, their development will be costly, nor can it alone provide the basis for much expansion of agriculture. Above all, the mere technical feasibility of developing certain hitherto unused or underused resources does not prove that it is economically advantageous to do so.

Development and Use of Natural Water [1]

A consideration of the possibilities of developing additional natural sources of water necessarily has to proceed by river systems or groundwater basins; we start with the best prospect, and gradually move to other water sources where little if any further water development is possible.

As was noted in Chapter 3, the Tigris River is international in the sense that it heads in Turkey, from whence comes about 40 percent of its total flow; it touches on Syria, and reaches the sea in Iraq, where lies by far the greater portion of the usable land that might be irrigated. Its flow has a marked seasonal pattern, as well as considerable variation in flow from year to year, so that storage reservoirs on the river or its tributaries could provide additional water for irrigation (capturing much of the peak flow that is now wasted), could provide a substantial degree of flood protection (especially to Baghdad), and could be the basis for generating hydroelectric power. The best combination of reservoirs, and the best method of operating such reservoirs as may be built, requires careful and detailed engineering study; and a large number of such studies have been made.[2] We do not attempt even to summarize all of them, much less to add their detailed analysis.

Reservoirs have been constructed at Dokan and Derbendi-i-Khan in Iraq and a major barrage or diversion dam (Samarra) built in Iraq to divert flood flows into the Wadi Tharthar depression and providing flood relief amounting to 7.1 million acre-feet (8,800 million m^3).[3] The best reservoir sites for the future are located at Eski Mosul, Bekhme, and Fatha in Iraq; these might be constructed to different heights with costs varying from perhaps 150 million dinar ($420 million) upward, and operated to produce varying proportions of irrigation water, flood protection, and power. Without storage, the Tigris may produce about 14 million acre-feet (17,000 million m^3) of usable irrigation water; with storage, the maximum economic supply for irrigation would be about 31 million acre-feet (allowing for evaporation losses, flood flows that could not be fully used, and the like) and assuming that irrigation usage predominated over other uses. Taking this storage step would utilize most of the total flow of the river and its tributaries, which (as noted in Chapter 3) is about 37 million acre-feet (45,000 million m^3). The additional supply of irrigation water thus obtained would permit substantial increases in irrigation water application, provided drainage is developed and modifications made in the diversion and canal systems—but would *not*, however, result in extensions of total irrigated area, as will become apparent later in this chapter.

At about its point of largest flow, the average annual flow of the Euphrates, allowing for upstream diversions, is about 28 million acre-feet (34,000 million m^3). It is more truly an international river than the Tigris. Nearly all of its water originates in Turkey, with the Keban Dam already under construction in that country, and considerable use of the water is planned (though originally for power production only, since a recent soil survey revealed the existence in Southeastern Turkey of over two million acres of land irrigable with Euphrates water, though at considerable distances, now for irrigation as well). Syria uses some of the water for irrigation, has begun construction of the Tabqa Dam, and plans further water use; and much water is used for irrigation in Iraq, thus far without any storage of Euphrates water in that country. The natural flow of the Euphrates varies seasonally, similarly to that of the Tigris, so that important additional amounts of water could be made available for use by storage.

Information about present water use of the Euphrates is at best approximate. Annual diversion is said to be 14.5 million acre-feet (18,000 million m^3) in Iraq; not quite 3 million acre-feet (3,500 million m^3) in Syria; and little, so far, in Turkey—for a total use of 17–18 million acre-feet (21,500 million m^3). One should add at this point that both Turkey and Syria reject the estimate of Iraq's current use as up to twice what it actually is, and regard it simply as a ploy to establish a position as a major claimant.

---

[1] Additional detail may be found in Appendix B.
[2] There have been 287 separate reports on one aspect or another of irrigation and drainage development in Iraq, up to 1964.

[3] Since so much of the water literature in this region is cast in the metric system, we give, in this chapter, both measures to permit easier identification by those familiar with the subject.

Be that as it may, current development plans amount to an additional 5 million acre-feet (6,000 million m^3) in Iraq, 9 million acre-feet (11,000 million m^3) in Syria, and 8 million acre-feet (10,000 million m^3) in Turkey, or a total of 22 million acre-feet (27,000 million m^3). Total future use would thus add to nearly 40 million acre-feet (48,000 million m^3), compared with an average flow per year (including present diversions) of about 28 million acre-feet (34,000 million m^3), which would be reduced by perhaps 10 percent through evaporation from new reservoirs.

Iraqi plans, some made as a result of studies by consulting firms from different countries, provide for a relatively large dam, the Haditha, on the Euphrates, and for a considerable number of smaller irrigation works and canal systems to use water from either the Tigris or the Euphrates River or both. Their combined costs would be in the general order of 250 million dinar ($700 million), but perhaps not all of these would be economical, especially if Turkey or Syria continue development of the river upstream.

However, when Turkey's Keban Dam has been built, the danger of floods to Baghdad and other locations along the Euphrates will be greatly reduced and a major motivation for additional waterworks in Iraq eliminated. At the same time, improvement in current water-use efficiency may become the more attractive alternative, given the large outlays that a new dam would call for.

Complex problems of management for irrigation, power, and flood protection exist on the Euphrates; complex problems of upstream versus downstream benefits and programs also exist, compounded by the international character of the river. Fair, equitable, and efficient development of an international river, even with the friendliest relations between the countries concerned is never easy, as the United States and Canada have learned in connection with utilizing the Columbia River.[4] Despite the obvious need for agreement, negotiations, begun back in 1946, that have involved only two of the three riparian powers at any one time, have been fruitless so far. As indicated, both current as well as future usage is in contention, thus adding to the normal difficulties in situations of this kind.

A recent engineering study of the Euphrates, concerning the current water use and planned additions, comes to somewhat different conclusions than this study has; but finds equally a great excess of eventually planned use over available flow. The relevant figures are in millions of cubic meters:

	Turkey	Syria	Iraq	Total
Current Use	1,500	2,980	12,860	17,340
Planned Additions	12,140	4,610	6,040	22,790
Total Future Use	13,640	7,590	18,900	40,130

As will be seen, the above estimates report a lower current use in Iraq and a lower future use in Syria, but a higher present and future use in Turkey than have been reported earlier in this chapter. But either set of estimates suggests that a radical cutting back of proposed projects would be in order if all three countries were to share in the river's expanded use. Furthermore, except in Turkey, the development of additionally usable irrigation water would at best involve only moderate, if any, extension of the irrigated area, but would result in better irrigation of the better soils now under irrigation—though project plans in each of the countries do provide for newly irrigated land in addition to more water for land currently in irrigated areas. In Syria, water from the Taqba project reportedly would eventually supply over 750,000 acres not now irrigated. In the worst of all contingencies—the development of all planned schemes by all three powers—there can be no doubt that both Syria and Iraq would find themselves with irrigation installations far in excess of what available quantities of water would justify.

With the completion of the High Aswan Dam, the Nile River in Egypt is now fully under control; no additional water supply can be developed,[5] but the present supply is ample for the Valley lands, and additional areas of reasonably good soils to which such additional water could be applied are not easy to identify. However, the management and use of the available water is highly important (see Chapter 15). The possibility of scouring action of the Nile River, which will in the future drop its silt load in Lake Nasser and will therefore tend to pick up silt from its channel below the dam, may create new serious problems for the downstream dams or barrages. Riverbed modification will, however, be slow to emerge, and some studies of particles picked up by the nearly siltless river as it leaves Lake Nasser suggest that the bed may stabilize itself. Only time and careful study will reveal changes in the regimen. Increased summer use of irrigation water will almost surely intensify the already serious drainage problem, to which we return below. Moreover, now that storage behind the dam will no longer permit the flooding of the Nile during its period of high flow, the natural flushing action of the river floods will also cease. Unless adequate drainage and proper water management are provided, the Nile Valley would accumulate salts from the evaporation of the irrigation waters, in the same way that the Mesopotamian Plain has been accumulating salts for several thousand years. This is the more serious as evaporation during the water's transit in Lake Nasser will raise its salt content. At the same time, more clay will form and as this water covers the land, the soil will become not only saltier but less permeable and infiltrable. While this situation can be remedied, it warrants watching and management.

Elsewhere in the region, only extremely limited opportunities exist for additional water development. First of all, the rest of the region simply does not have much water—the Jordan, Litani, and Orontes are very small rivers, and other springs and streams are still smaller. Groundwater supplies are limited. Some locally important additions to usable supplies may be possible but they will be small in regional terms.

---

[4] John V. Krutilla, *The Columbia River Treaty—The Economics of an International River Basin Development,* (Baltimore: Johns Hopkins Press, 1967).

[5] At least not without some wholly new project outside of Egypt, such as drainage of the Sudd swamplands.

As for groundwater, one can write off Israel, where it is now fully used. Judgment is less easy elsewhere, above all because interest in groundwater, especially in deeper layers, has recently been on the increase and has had some encouraging results in parts of Asia. Saudi Arabia, for example, has funded a very ambitious search for groundwater, in which several large international engineering firms are participating.

There may be more groundwater in Syria and even in Jordan than had hitherto been suspected. Egypt has groundwater in the Western Desert which it is trying to exploit. There is also groundwater in the coastal strip that extends west from Alexandria to the Libyan border and has since 1965 been the subject of a U. N. Development Program (UNDP) preinvestment survey; results so far suggest that the amounts are moderate, even if they could be economically extracted, which remains to be seen. The area of soils suitable for cultivation along this strip is estimated at 250,000 feddans, of which in good rainfall years about half is now grown to barley.

On the other hand, the amounts in the sources located so far, as in the Western Desert, are problematical on a sustained yield basis. Drying up of wells has occurred and counsels cautious appraisal. Origin of the water is largely undetermined and until it is known., evaluation of the potential remains informed guesswork. In many instances too, these smaller water sources have been more readily exploitable, with the capital and technology at hand, hence less is left unused than on the large rivers. The deep water in desert or near-desert areas certainly will require appropriate technology and considerable capital to reach. There may be surprises here, but at this point in time there is little hard evidence to go on.

One of the most knowledgeable students of water, and especially groundwater, in the Middle East recently put forward some ranges of cost in Saudi Arabia which illustrate at least the cost one may have to face in producing groundwater at the present time.[6] These costs, at the wellhead, are:

from flowing wells	$125 per acre-foot
from pumped shallow wells	$125 to $250 per acre-foot
from pumped deep wells	$250 to $370 per acre-foot

It is interesting to note that the top of the range in the third category is about the same figure as the current cost of desalted water in small installations of the kind used in many locations in the Carribbean—i.e., $1.00 per 1,000 gallons.

Desalting Seawater[7]

Of the means available to augment natural sources of sweet water, desalting has aroused perhaps the greatest interest in recent years. The possibility of truly large-scale programs for desalting sea water to produce fresh water for large-scale commercial agriculture is thus worth examining in some detail, especially as it has been most frequently associated with use in the Middle East. General Eisenhower did much to publicize this idea,[8] there have been two engineering studies made at the request of or by the U. S. government,[9] and there has been a host of popular and semipopular writing on the subject. There is unquestionably an unlimited supply of seawater, and its desalting is technically feasible; the catch is that for the present and the foreseeable future the cost of the water delivered to the land where it can be used, on a time schedule reasonable for agricultural use, is a large multiple of—perhaps ten times or more—the value of the water for such use.

Desalting seawater for large scale commercial agriculture involves three major closely interrelated components: (1) a source of energy; (2) a process for producing sweet water out of seawater; and (3) means for transporting water, at the right time, to the place of its application for the growth of crops.

Any source of energy will do for this process; most attention has been focused on nuclear power, due to a large degree to the great upsurge of nuclear power plants ordered by U. S. electric utilities in 1966, 1967, and 1968. But whatever one's view of nuclear competitiveness in the United States, there are some special drawbacks to its use in the Middle East. It will cost more to develop nuclear power in technically less advanced countries than it does in the United States; oil and gas are far cheaper in the Middle East than in the United States, and are probably a cheaper source of energy than nuclear power; and the scale on which a plant would have to be built to produce at lowest cost is such that outlets for the electric power generated would be hard to find for a long time to come. Egypt probably has from the Aswan Dam all the power it can use efficiently in the near future; the other Middle Eastern countries lack the capacity to absorb the excess power from a large nuclear power-desalting complex. Even if nuclear power should some day be much cheaper than is now in prospect, this would help the economics of desalting only if such cheap power were not also generated elsewhere, in which event it would simply operate to reduce the subsidy which desalting would receive from power sales. The most ambitious current proposals include the building of large agro-industrial complexes that would themselves utilize a large part of the power generated in the nuclear reactors—in associated chemical and metal industries—and thus avoid or lessen the problem of having to dispose of a large power surplus, at the cost, however, of losing the implicit subsidy.[10]

---

[6] *Hydrological Control of Development in Saudi Arabia*, David J. Burdon and Galik Otkun (XXIII International Geological Congress, Vol. XII [*Prague, 1968*]), pp. 145–153.

[7] This section is based upon the authors' article on the subject in *Science*, June 6, 1969. Since then a new study, under the direction of Professor Paul MacAvoy, has been published that reviews and analyzes all cost estimates: *Review and Analysis of the Costs of Desalted Sea Water 1969* (Washington, D. C.: Office of Water for Peace, U. S. Department of State).

[8] Dwight D. Eisenhower, "A Proposal for Our Time," *Reader's Digest*, June 1968.

[9] Kaiser Engineers and Catalytic Construction Co., *Engineering Feasibility and Economic Study for Dual-Purpose Electric Power-Water Desalting Plant for Israel*, (January 1966 and later revisions); and Oak Ridge National Laboratory, *Nuclear Energy Centers, Industrial and Agro-Industrial Complexes* (ORNL-4290, UC-80, Reactor Technology [Oak Ridge, Tenn.: Oak Ridge National Laboratory, Nov. 1968]).

[10] See *Nuclear Energy Centers, Industrial and Agro-Industrial Complexes, ibid.*

Although desalting of seawater by flash evaporation processes is now well established, the plants would encounter some obstacles because of the large size proposed (several hundred million gallons per day), obstacles which have plagued even the smaller plants now in operation. Large-scale desalting plants must be located at seashore, but land to be irrigated inevitably lies inland and often at considerably higher elevation; the proposed desalting plants would operate continuously, except for planned maintenance or breakdown shutdowns, but farming requires water in a considerably different time sequence. Transferring the water from the seashore to the place of use, and in some way storing water from periods of surplus to seasons of heavier demand, would be expensive; a major item in the cost would be a considerable loss in amount of water. The Middle Eastern candidates currently most often mentioned for desalted water—Israel and Egypt, primarily—lack good surface reservoir sites, and storage in underground acquifers presents many difficult technical problems; in either case, less water is recovered than is stored, and quality declines as well.

Accepting the engineering studies at face value, but including costs of water transportation from desalting place and time, to farmland place and time, and using more realistic interest rates (which for Israel and Egypt should not be less than 12 percent), we conclude that the annual cost of irrigation water at the farm will for the foreseeable future be in the general magnitude of $130 per acre-foot.

The desalting plants proposed would produce nearly 100,000 acre-feet (125 million m^3) annually in the case of the Kaiser study, and ten times this volume in the case of the Oak Ridge study. Such large volumes of water would necessarily have to be used primarily to produce staple crops such as grains and cotton; while some specialty crops such as flowers and winter fruit and vegetables could be grown, they could not begin to use all this water. For instance, although Egypt grows winter tomatoes for export, in its best year such export production has required only 700 acres; while the water required for even a greatly enlarged area would be only a small fraction of that available. Expansion of fresh specialty crops production on the required scale presently involve excursions into markets of wholly unknown demand-and-supply characteristics, and require a marketing competence that is acquired only over many years.[11] As for staple crops, it seems probable, on the basis of American experience, that they cannot pay more than $30 per acre-foot annually for water, and that $10 per acre-foot would be a more realistic figure.

Apart from our judgment that desalting of seawater for commercial agriculture is costlier for the foreseeable future than even its proponents judge it to be, additional considerations may be mentioned for the Middle East. Israel has indeed fully developed its available water supply; in its coastal plain and in the northwestern Negev there is a considerable acreage of good soil which could be irrigated if water were available at reasonable cost. This land lies some miles inland and at elevations mostly of 500 feet or more.

Adjacent to it, in the northeastern corner of the Sinai peninsula, is some acreage of similar land. At the right price, moderate amounts of water could be absorbed here. Although proposals have been made for irrigating additional areas in Egypt, west of Alexandria and in the depressions of the western desert, these soils have severe limitations. West of Alexandria the soils are mostly sandy and would have high water and nutrient supply requirements. The soils in the depressions that might be irrigated are mostly of very fine texture and are usually rather high in soluble salts. Moreover, as we have noted, Egypt does not lack water; the Nile, now regulated by the High Aswan Dam, has as much as can be used efficiently in the Nile Valley—indeed, taking some of this water out of the valley would most likely improve agricultural production in the valley over the long run.[12] While it would probably be costly to take Nile River water to the western areas near the coast, it is hard to conceive that it would be as costly as desalting sea water and moving it inland—the distances and the elevations between source-and-use points are not so greatly different as to offset the high costs of the desalting process.

In summary, we must conclude that the possibility of desalting seawater on a large scale, as a basis for additional commercial-scale agriculture, will not provide a practical solution to the economic problems of the middle East for the foreseeable future. Such proposals may even be looked upon as a distinct disservice to the people of the region, for they arouse hopes which cannot be fulfilled and distract attention from quite practical and urgently needed measures in the management of soil and water, and of farm practices and policies generally—measures which are the precondition of improved agriculture under any conditions, including use of desalted seawater.

Drainage

In contrast with the somewhat limited possibilities for development of more natural water and the economics of desalting seawater that for a long time to come appear highly unfavorable, the drainage of irrigated and irrigable lands is practical and economical on a large scale in the Middle East. Moreover, it is indispensable for growth.

The irrigation of desert soils almost inevitably creates drainage problems. Such drainage problems are often discussed as if the need for drainage arose from *over*irrigation; application of too much water does indeed lead to a saturated soil too close to the surface, and this in turn greatly impairs plant growth. But the need for drainage would arise, sooner or later, even if water application were at the minimum neccessary level. All irrigation water, even of the best quality, contains some dissolved salts; when this water evaporates from the soil surface or transpires from

---

[11] See Chap. 11, pp. 95-98.

[12] Just for perspective, it is interesting to note that a one-billion-gallons-per-day desalting plant—an undertaking several orders of magnitude above any now in operation—would at 80% availability yield annually the equivalent of 1.3% of the average annual flow of the Nile River. The conclusions forced upon one by such a comparison regarding the benefits to be reaped by a less wasteful utilization of river water need no elaboration.

the plants, these salts are left behind, and in time will accumulate to harmful and finally disastrous levels unless flushed out by periodic or occasional appplications of water removed through appropriate drains. Irrigation almost inevitably requires associated drainage; the world, including the United States, is full of unhappy examples of neglected drainage in irrigated areas.

To recapitulate briefly what has been said on the subject in Chapters 2, 3 and 4, the Mesopotamian Plain and the Nile Valley present each in its own way almost a classic example of quite different drainage conditions. The Mesopotamian Plain has been irrigated for 7,000 years at least, perhaps longer. In 2,400 B.C., wheat yields were as high as 30 bushels per acre (2000 kilograms per hectare) and it was at about this time that the first record of excessive salinity was noted.[13] Although the water quality in the Euphrates, the Tigris, and the tributaries of the latter is generally good (mostly under 400 ppm.), yet the plain lacks good surface drainage, so that much water applied in irrigation or flowing out from the streams in floods has evaporated there over the centuries. The salt accumulation has been large—averaging more than 600 metric tons per hectare (or 268 short tons per acre), and the surface of the land is white with salts in many areas. Extraordinary cropping practices are required for obtaining even a minimum crop on such lands. (See Chapter 4.)

In the Nile Valley, the natural flushing action of the floods and their drainage back to the river once was perfect for a primitive agriculture. Modern agriculture, of the past 100 years or more, has included summer irrigation, especially for cotton, and this has led to a serious rise in the water table. In extensive parts of the valley, the water table is within one meter (40″) of the surface, and in other areas is deeper but within two meters (79″) of the surface. Such a high water table causes drastic yield reductions for many crops, especially for naturally deep-rooted ones. In contrast with the Mesopotamian Plain, the Nile Valley has not yet generally accumulated salts in the soil at or near the surface in harmful concentrations—the period since the floods were restrained is too short for that.

The Mesopotamian Plain thus represents salt accumulation at about the extreme; the Nile Valley represents water saturation of the soil, bad if not at the extreme, but without as yet serious salt accumulation.

Serious as the situation is in each area today, time will make it worse, unless remedial action stops and reverses the trend. Salt accumulation will continue in the Mesopotamian Plain unless drainage is undertaken. In Egypt, the High Aswan Dam, by enlarging the supply of water in the summer, will worsen the waterlogging problem—increased cropping notwithstanding—and the elimination of floods, unless offset by careful soil flushing programs, will over a long time create serious salt accumulation problems.

The lack of adequate drainage for most of the irrigated lands of Egypt is a severe handicap for agricultural progress; for Iraq, it is a national catastrophe. In Egypt, as noted in Chapter 8, crop yields cannot be called high if one considers the favorable climate, soil, and water conditions; the present relatively low level is in considerable part a result of poor soil-and-water conditions and management; future progress in increasing crop yields will be disappointingly small unless or until the drainage is provided.

In Iraq, as noted in Chapter 4, the salt accumulation on and near the soil surface in many areas is so extreme that irrigation is practiced to flush the salts down into the soil a modest distance; a crop of barley (which is relatively salt-tolerant) is grown, with no more water than necessary, and yields of the order of ten bushels per acre obtained; thereafter land lies fallow for a year, while native weeds grow and exhaust the soil moisture to some depth; then the cycle is repeated. Such low crop yields, in alternate years only, from irrigated land, are about as marginal as crop production can get and still continue. Improvement is nearly impossible until the drainage problem has been corrected. While not all irrigated land in Iraq is that bad, virtually all of the land located in the flood plain of the Tigris and Euphrates Rivers would benefit greatly from provision of drainage.

These areas differ greatly from some old lake bottoms and swamps for which drainage has been attempted in the United States, with very poor results. Soils in both the Mesopotamian Plain and the Nile Valley have moderate to good internal drainage; it is wholly practical to drain each and to obtain good results from such drainage. The accumulated salts are readily soluble. Moreover, the costs of drainage are quite reasonable.[14] The capital cost of underground tile or plastic drains would be on the general order of $85 per acre; even for 7 million acres in the Nile Valley (on the estimate that all of it will ultimately require drainage even if some does not at present), this would mean a capital investment of about $600 million. In Iraq possibly 8 million acres will need drainage, if the additional water supplies described above are developed; at the same cost per acre, this would call for an investment of not quite $700 million. While these sums may seem large, the Iraqi budget has in recent years allocated $100 million annually to irrigation and agricultural investment, and capital sums of this magnitude, spread over many years, are certainly not impossible to obtain internally or externally. If the investment per acre is $85, a maintenance charge, amortization allowance, and interest on investment of, say, 15 percent annually would result in an annual cost per acre of perhaps $13. For a barley yield of 10 bushels per acre, this is utterly prohibitive; for a wheat yield of 100 bushels per acre or a cotton yield of 2 bales per acre—possible if the land were well drained *and* if the package of other inputs and practices were adopted—such a cost is tolerable.

It may be practical in the Nile Valley to undertake drainage by pumping from tubewells, as has been done successfully in West Pakistan. However, to date there has been no

---

[13] J. H. Boumans *et. al., Reclamation of Salt Affected Soils in Iraq* (Soil Hydrological and Agricultural Studies, Publication 11, International Institute for Land Reclamation and Improvement [Wageningen, The Netherlands, 1963]). This in turn quotes many sources for these and other statements.

[14] Boumans, *ibid.; Pilot Project for Drainage of Irrigated Land, United Arab Republic—Final Report* (FAO/SF:30/UAR [Rome: FAO, 1967]).

experience with this method. In the Mesopotamian Plain this method of drainage would be impossible until the high salt content of the water has been lowered. In saying that drainage is practical, we do not mean to minimize the seriousness of some of the problems that must be met. In Iraq, the water which will drain through the soil and into the underground drains will be highly saline in content, with in some cases several times as much salt as seawater. Such water cannot be put back into the Tigris and Euphrates Rivers without serious damage to present farm areas downstream, including the highly productive date groves and other irrigated areas along the Shatt al Arab. Some of this salty water might be put, for some years, into nearby desert depressions outside of the plain, and allowed to evaporate there, but there are limits to how much water can be so diverted. The ultimate solution must be discharge of these drainage waters into the sea, by special canals or conduits. These are costs additional to the local drainage costs cited above. Their magnitude is not at this time on the public record but might be high.

In each country, a successful drainage program must be undertaken on an adequate scale, which though ultimately needed for the entire irrigated area, might proceed by steps, with continuity in maintenance and operation of the drains to keep the investment from being wasted. Drains from field to sea must be viewed as a *system*, with close interrelationship of all the parts, and must be carefully coordinated with the irrigation program.

In Egypt, where good irrigable land is scarce, underground tile or plastic drains could replace the now prevalent open drains. The latter make unuseable for crops a substantial area of land, generally estimated at about 13 percent. The ability to use 13 percent more land would offset the cost of installing underground drains.[15]

In both Egypt and Iraq, while future agricultural progress is absolutely dependent upon adequate drainage of all irrigated land, such drainage should be given top priority in any agricultural program for the region; its value, however, if not accompanied by associated improvements, would be limited.

Other irrigated areas—in Syria, Jordan, Lebanon, and Israel—May also require some drainage. Syria has already undertaken drainage of the Ghab area, and some drainage has been installed elsewhere. A specific drainage problem may well develop in Syria on the cotton producing areas close to the Euphrates River when higher lying lands are irrigated. But the much smaller scale of the various areas that may need drainage now or in the future makes their problem greatly more simple than those of the Nile Valley and the Mesopotamian Plain.

Area of Irrigated Land

In considering the possibilities for irrigated area in the future it is necessary to distinguish clearly among three concepts or measures: (1) total area over which the irrigation network is extended, which might be irrigated at some time (often called "command area"); (2) acreage actually irrigated each year; and (3) length of the irrigation season, adequacy of water supply, etc.—the factors measuring the *intensity* of irrigation on the lands actually irrigated. Different situations exist among the countries of the Middle East for each of these three measures.

In Egypt, the irrigation network is spread over nearly all the valley and the delta, a total of 6.6—6.7 million acres. In the future, this might increase, some think by 300,000 acres, some by perhaps 1 million acres. Poorly drained areas in the delta can be diked and drained, and small additions made along the valley sides. The larger of the two estimates of added land could be reached only through expanding irrigation onto the desert. For reasons given in Chapter 2 such a development is not considered feasible in this study.

For the 1960—64 period the area of crops harvested annually (all of which are irrigated) was about 10.6 million acres (Chapter 8); the cropping ratio was thus 1.6. In Chapter 14, we shall point out ways in which this ratio in Egypt can be raised at least to 2.0, in which event the crop area actually irrigated each year would reach about to 14 million acres (illustrating, incidentally, how much more significant will be the results of more intensive use of land now in cultivation than extensions to new lands). At present, there is some irrigation all year round in Egypt, except for about six weeks from late December to early February, when the canals are closed for their annual cleaning; but often the application of water has been limited in the summer, especially to corn, for fear there would not be sufficient water for the cotton and rice crops. In the future, there should be plenty of water for optimum applications in all months, and there should be some increase in amounts applied and hence in yields of some crops— assuming that the necessary drainage is constructed and properly operated.

There is a serious lack of accurate information about irrigation water supply, irrigation water use, irrigated land-use, and related matters in Iraq; various figures are quoted in various reports, but all are nonauthoritative estimates. The figures used below are more or less a consensus gleaned from reading rather than from independent estimates. The total area of land to which irrigation has been extended, including pumped water supply, is in the general order of 15 million acres; the area irrigated each year, in recent years, is roughly half as great—7½ million acres. The very large gap between the area that could be and the area that is irrigated is due to the fact that the very bad drainage situation necessitates a crop-fallow rotation even on the irrigated land. The amount of irrigation water applied—probably in the neighborhood of 20 million acre-feet—is relatively low, if only because primarily winter crops are grown; for these the water requirements are much lower than they would be for summer crops. However, the efficiency with which the

---

[15] The beneficial results of underground drainage upon crop yields are difficult to quantify. In Egypt, they were recently put as high as 50% for rice, 45% for wheat, 47% for cotton, and 67% for sorghum (*Report of Regional Water Resources Development Officer*, cited in Item 7e of *Provisional Agenda*, Ninth FAO Regional Conference for the Near East, Baghdad, Iraq, September 21—October 1, 1968).

water is applied is also very low, so that more water is used than would be necessary under conditions of improved efficiency.

If all the additional water supplies discussed above were actually developed in Iraq, this would provide a maximum of something like another 20 million acre-feet or would roughly double the present supply. The amount of water that should be applied for most efficient crop production depends in large part upon the cropping pattern—more particularly, upon the relative area of winter crops which require comparatively little water and of summer crops which require a great deal more. Under any circumstances, some water will have to be used to flush out the salts—a great deal in the early years because of the magnitude of present salt accumulations, and some indefinitely into the future.

In any event, there will be little room and no need for a significant net increase in the area of land to which the irrigation network extends; some small additional areas of good soils might be included that are not now served and some may be eliminated; but there is no large area of land awaiting irrigation. Irrigation every year of all suitable land served by the irrigation network would be possible, or nearly so, if water development were pushed to its maximum. Overall, 40 million acre-feet for 15 million acres leaves little if any room for expansion. Irrigating what is now in the command area would also be desirable as a cost-saving factor; for irrigation investment, and to a considerable extent operation costs of the irrigation system, are a function of the gross area served rather than of the area actually irrigated. Increasing the latter, without modifying the former, therefore, leads to much lower irrigation costs per unit of water or of land. Moreover, as we shall point out in the next chapter, in the future a substantial part of the irrigated area in Iraq could and should grow two crops a year, a condition that would further concentrate the available water supply on a smaller area of land.

Irrigated agriculture in Iraq can indeed be greatly expanded, but the means should be more intensive use of good soils, not any significant expansion of the irrigation network to new land; and such intensification if it is to contribute to a much larger output of agricultural commodities must be part of a package for agricultural advancement. Moreover, while some storage would be helpful (and some may be provided as an incident to the completion of the Keban Dam in Turkey), we would not give investment in irrigation a high priority. That should go to additional drainage.

Elsewhere in the region, only modest increases in area irrigated are possible. Syria has plans for additional use of Euphrates River water and the Tapqa Dam is now under construction, designed to extend the irrigated area by close to one million acres, as well as to supplement existing irrigation. However, the basis for such additional irrigation development is not at all obvious, quite apart from the fact that the project is being constructed by Soviet planners and little is known about it. Even if Syria has additional Euphrates River water, or gains it through a treaty among the nations using water from this river, the area of good

soils reasonably accessible to the river for irrigation, is rather limited. Some past proposals have envisaged irrigation of soils which because of their poor characteristics, such as shallowness, would almost certainly be abandoned after a few years as unirrigable, for example. Altogether, it is wise to suspend judgment on both the availability of water, given the rival claims, as well as on the feasibility of substantial additions to irrigated acreage.

In the remaining countries of the region there are only very small amounts of unused water. Whatever development may occur will be important locally but of little moment in regional terms. Some development might derive from the fact that the efficiency of water application now is relatively low in many areas. If raised, it would permit either irrigation of additional areas or increased output from present areas.[16]

Improvements on Irrigated Land

Considerable opportunities exist for improvements on irrigated land in the Middle East; generally speaking, their costs will be moderate, they will consist of a large number of small activities on a local scale, but their long-run consequences will be considerable.

Foremost among such improvement to irrigated land is the leveling of the land wherever water is applied by surface irrigation methods—which is nearly everywhere outside of Israel. In the term "leveling" we include finishing to an exact level plane and also the more usual and generally better practice of shaping to a smooth plane with a predetermined slight slope. The latter facilitates furrow irrigation, makes mechanization easier, and results in 10 to 15 percent more complete utilization of the land area in comparison with bund irrigation.

Land leveling is basic to efficient application of water at the micro or field scale. Perhaps the crucial advantage of leveling is the avoidance of high and low spots, however small; the result is that less water is required, much better crop yields are secured from the irrigation, and usually less labor is required. To put it simply, low spots in a field will be covered by too much water and high spots by too little water. As a result, yields of most crops will be reduced. This reduction can be considerable from a depression or a high spot no more than three inches out of the plane of

---

[16] It is encouraging to witness the emerging emphasis on concentration of available water on promising areas, as, for example in the following excerpt from a recent FAO document:

> Given the limitations in water, the tendency in the past has often been to maximize returns to water, thus attempting to spread water thinly over large areas. With the break-throughs in High Yielding Varieties accompanied by high levels of fertilizer use and other inputs, the emphasis should be placed on optimal levels of water use, considering the need to create a moisture environment in which the productive capability of all productive inputs can be utilized.

(*Ad hoc* Consultation to Review UNDP/FAO Land and Water Use Projects in the Near East, Amman, Jordan, 12–18 May, 1969, *Brief Inventory of the Main Land and Water Use Problems in the Near East.*)

most of the field. A really good job of land leveling is basically a one-time capital investment;[17] once made, the soil can be restored to the same smooth plane with a minimum of cost for correction of unequalities brought about in crop production. Leveling of irrigated land is especially important in Egypt and Iraq, but is of significance also in other countries. The small farms of the Middle East—and, even more, the small fields into which the small farms are so often divided—constitute at best a handicap to good irrigation and especially to good land leveling, including the restoration of the surface to a smooth plane after cropping; but land leveling is important on even the smallest fields.

The canals in irrigated areas could also be improved. In some cases, especially in some sandy land areas in Iraq, they should be concrete-lined to reduce water losses through leakage and to simplify drainage. This requires a considerable initial investment, but if well done, maintenance is relatively inexpensive. In areas where irrigation water is applied by surface methods, better headgates or devices for diverting water from large to small canals and from canals to fields would be helpful. Also, more and better water-measuring devices, and better records on water use, are essential for efficient use of water. In Egypt, over the years canals should be raised so that water can be diverted to the fields by gravity or motor-driven pumps made available at a cost the farmer can pay. Replacement of human labor would be justified in terms of cost-saving alone. A recent study on India reports a drop from 495 rupees to 60 rupees per 10 acre-inches at a 40-foot lift, when hand-pumping is replaced by a diesel engine.[18] Replacement of human labor might also be a factor in the timeliness of water supplies, which will become a more critical element as other parts of the improvement package—such as increased fertilization, higher-yielding, but less drought-resistant varieties, and multiple cropping—spread to larger acreages.

On the other hand, the problem must be faced that availability by gravity would most likely lead to even greater excess in use, unless accompanied by some kind of incentive to use water sparingly.

A related matter, therefore—not strictly an improvement to the land but perhaps as basic to efficient land and water use—is the introduction of some institutional arrangement, or a variety of arrangements, which will provide irrigation farmers with a real incentive to reduce or eliminate excess water use. There are great possibilities in modified pricing systems; a basic amount of water may be available free, for example, with high marginal costs of additional water; or some other arrangement might be fashioned that is compatible with prevailing ethics and at the same time conducive to economic behavior. As it is now, the administrative power of the government official is often pitted against the desire of the user to obtain more water; and the latter all too often wins, frequently to his own detriment and

nearly always to the detriment of the larger area in which his excessive use of water creates drainage problems. Farmers must be educated and convinced about proper water use, but they also need clear immediate incentives in a form they can understand and respond to. It is possible that such social innovations will not fall on a receptive ground until modernization of farming practices has brought the Middle Eastern farmer much more deeply into accepting the basic equity as well as the national necessity for a water-rationing scheme through some kind of a charge. But its significance in the framework of resource management is such that it should receive very serious attention from both scholars and administrators, as well as from the most progressive farm operators themselves.

Land Use in Rainfed Areas

There is no possibility of a significant increase in the area of rainfed crop land in the Middle East; on the contrary, the use of rainfed land for crops is now overextended and some of the land now used for annual grain crops would be much more productive if converted to permanent grasses for livestock production.[19] The technology and the capital available to Middle Eastern countries over the past few decades have readily permitted the cropping of all rainfed land that could produce crops economically—and of some that cannot. As in all other parts of the world where rainfall is low and highly variable, a succession of relatively wet

[19] The case is quite forcefully put in the FAO document (*Report of Regional Water Resources Development Officer*) cited in footnote 16 above:

> The main land use in low rainfall areas is grazing, and much of the improvement of the whole agriculture of these regions needs to start with improving grazing use of land. Improved watering points and other facilities have often led to further pressures on grazing lands when not accompanied by some form of regulation of grazing methods and number of livestock. The problems of nomadic grazing are well known and the solutions to them have to be technically and economically attractive for the pasturalists before social and political solutions can be found for their settlement.

On the same occasion, there was a review of a UNDP pilot project in the Greater Mussayib Area, and the following comments deserve quotation:

> This basic problem . . . rises from the fact that the growing demand for food is generally met by the expansion of highly extensive methods of farming instead of greater intensification where the factors of production are favourable. Usually, the expansion of cultivation onto land that was formerly used for natural grazing is not followed by a more intensive livestock production pattern. Livestock raising is simply pushed to poorer and more distant areas with a consequent relative drop in productivity of both the cereal crops and livestock production. In Iraq, a good deal of the land that can be seen idle near the settled areas and that might be considered arable, is land that has been cropped at one time and then left unused due to the ravages of an inefficient fallow system. On the Mussayib, production from irrigated land is much less than should be obtained from the resources being used and fallow is much too common a practice and summer irrigation too limited in its use.

---

[17] Costs will vary so greatly with local conditions that it is not feasible to cite dollar figures. One estimate we have come across in the case of Jordan is about $10 per acre.

[18] John S. Balis, *An Analysis of Performance and Costs of Irrigation Pumps Utilizing Manual, Animal, and Engine Power*, (Delhi: USAID, June 1968), p. 4.

years has led to exaggerated notions about the crop potential of certain areas; in the inevitable dry years, crop failures have resulted.

In Chapter 14, we present some details about land-use practices on the rainfed lands, including estimates of the area that would best be converted from annual grain crops to permanent grasses or other forage crops. We will show that use of only the better (higher rainfall) lands for crops can, if accompanied by a package of improved practices, produce a great deal more grain than is now produced from the entire area. The poorer lands, from an annual cropping viewpoint, can nevertheless be fairly good grazing lands. While annual forage crops might be grown on them, the same factors that make annual grain crops uneconomical are likely to make the cultural and other necessary operations of annual forage crops too costly in relation to the results. Perennial grasses, on the other hand, might be seeded; if properly managed, stands of such grasses continue more or less indefinitely; in dry years, the grasses live over, if they have not been grazed too severely, and in wet years they respond to the available moisture by greater forage growth. While these lands might over a long time return to natural grasses, the process is very slow; and hence

planned reseeding might be economical. In the United States in the past twenty years, several million acres of generally similar lands have been sucessfully reseeded; and interestingly, the chief species seeded is crested wheatgrass, which was imported to the United States from the same general part of the world as the Middle East—more specifically, from the Caucasus. The costs of range reseeding are rather high in relation to the amounts of forage produced—costs of $10 per acre are rather common in the United States, and annual forage yields are often low. This means that regrassing of such land is economical, if at all, only when two conditions are *both* met: (1) the lands are thereafter managed to preserve the grass stand obtained, so that production can continue indefinitely without the necessity of frequent reseedings; and (2) reseeding is part of a total package of improved livestock management. The grazing animals of the Middle East are generally not only grossly underfed but often diseased, frequently of poor breeding that cannot utilize forage most effectively, and in other ways poor producers. Provided the other deficiencies are cured, then more and better livestock feed could be highly productive, especially if it supplied feed at seasons when the forage from grazing land is inadequate.

CHAPTER 14

The Long-Run Productive Force of Modern Farm Practices

In Chapter 13 we explored the possibility of land and water resource development in the Middle East, particularly the possibility of using land and water resources at present unused. Implicit in that discussion was the assumption that it would, in fact, be possible to develop any resources which it was economical to do—that the necessary capital would be forthcoming, that the public and private entrepreneurial capacity would be adequate, that the necessary supporting services would be forthcoming, etc.

In turning to the future possibilities of using modern farm practices and techniques in the Middle East, we explicitly assume that no factors, be these inputs or infrastructures, are limiting. That is, we assume that the necessary capital, professional expertise, governmental structure, institutional framework, and all the rest of it, are fully adequate to carry out any farm practices which promise to be economical and efficient in the Middle East for the next few decades. These obviously, but intentionally, unreal assumptions are further examined in Chapters 18 and 19, after we have here considered the full measure of growth that modern agriculture might offer to the Middle East.

While, as an expository matter, we must necessarily consider specific situations one by one, we have, throughout this chapter, very much in mind the package approach. We do not envisage the use of improved crop varieties without the use of better soil moisture conservation measures, without the use of more fertilizer, without better weed control, etc.; nor do we envisage the use of more fertilizer without improved crop varieties, better weed control, better moisture conservation, and the like.

With this as a preface, we turn now to a number of specific crop and resource situations in the Middle East, and suggest what modern farm practices might accomplish in an area in which they have as yet made little inroads.[1] The crop yields envisioned are deliberately set high; but in no case have we knowingly postulated yields that are not now attainable with the best known technology and in the absence of institutional constraints. These are, in other words, targets to shoot at—beyond reach now but not in a time span of the balance of the century.

Egypt

Wheat Production

The maximum yield of wheat that we know about—a yield of just over 200 bushels per acre—was produced on an ex-perimental area of six irrigated acres in the state of Washington. Some farm yields under irrigation in that state reach about 150 bushels per acre each year; and harvests in excess of 100 bushels per acre are general in the irrigated area. The all-farm average for the Columbia Basin Irrigation Project is nearly 100 bushels per acre. While the very high-yielding Gaines wheat that we are referring to is rather specifically adapted to the Washington, Oregon, and Idaho area where it was developed, it does illustrate yields that are attainable by adaptive research that develops varieties for specific areas. The Gaines wheat has short stiff straw and will tolerate high concentrations of plant nutrient without lodging.

At the International Maize and Wheat Improvement Center in Mexico (CYMMIT from its Spanish initials) other short-straw wheats have been developed that are more suited to the temperatures and latitudes of the Middle East. Some of these wheats have produced as much as 120 bushels per acre in Mexico and a bit above 100 bushels per acre on experimental areas in Egypt. There is little doubt that general yields of 100 bushels per acre are attainable with adequate inputs on well-drained Egyptian Nile Valley and Delta soils, where present average yields are about 40 bushels per acre. The increase in yield would come partly, perhaps 20 to 30 percent, by improved drainage; partly from improved irrigation techniques; and mainly from the higher yield potential of the newer varieties at much higher levels of fertilizer use (particularly nitrogen) than now used. A wheat crop requires about 2.7 pounds of nitrogen for each bushel of wheat grown. Thus a 100-bushel wheat crop requires about 270 pounds of nitrogen. While the soil will supply some of this, most of it must be added as fertilizer. Although Egypt is a heavy user of manufactured fertilizers, particularly nitrogen, application rates customary at present are not high enough to produce 100-bushel wheat. Moreover, we should point out that the use of such large amounts of nitrogen on present varieties would drastically reduce yields: they would soon lodge under their weight. In order to obtain sustained high yields, new high-yielding varieties must be developed, which to be raised successfully will call for a whole new package of practices. While the increase represents a rise of 150 percent, we do not believe that it is too unrealistic to set it up as a potential. It is less of a climb than the one that is contemplated, for example, for West Pakistan: from average yields of wheat of 15 bushels to 50 bushels by the end of the century. And

---

[1] "The growing of most field crops in the Middle East has not changed to any degree for many years" is the opening sentence in W. W. Worzella's recent report entitled *Cultural Studies on Yield and Quality of Field Crops in Lebanon* (Faculty of Agricultural Sciences, American University of Beirut, Publ. No. 30 [Beirut, February 1968]).

the genetic material for 100-bushel wheat yield—and better—is now in hand.

Corn (Maize) Production

Corn is the second most important grain crop in Egypt and is the first in acreage. It is the basic food of a large part of the rural population. In the past, because of limitations on water supplies in the early summer, planting of corn has sometimes been restricted until enough water was assured for other summer crops, primarily cotton. Hence this crop has usually been planted somewhat late and may have been limited in the amount of water applied to the land for its growth. Now that water is available on a year-round basis, except for the canal closure period in mid-winter, corn production can be greatly improved. Improved drainage alone has been found to raise yields from 40 to 75 percent. With adequate fertility levels, earlier planting times, and with adaptive research on varieties specifically suited to the Nile Valley and Delta, the country average could be raised to 100 bushels per acre or more. This is only one-half the yields obtained on several varieties under experimental conditions in Lebanon. It is true that to date no country has reached this national average, although some states in the United States have. But neither does any major corn producing country have all of its corn acreage on good quality irrigated land, where climatic factors are as favorable as in Egypt. If our figure of a potential of 100 bushels per acre for Egypt is subject to criticism, the criticism should be that it is an under- rather than an overestimate.

Some may say that corn is too low in economic value to make it profitable for use as a summer crop on expensive irrigated land; and in terms of world prices for corn as feed for animals this view, at most times, is perhaps true.[2] But this is not the burden of the issue here. Rather, the question is how much can be produced with the resources and know-how now available. And it should be remembered that corn is now a principal food crop for humans in Egypt.

Millet and Grain Sorghum Production

The food grains known as millet and grain sorghum together occupy about one-third of the acreage of corn in Egypt. They are summer crops competing with cotton, corn, rice, and vegetables for summer acreage. These two crops have experienced rather important improvements recently and give evidence of being equal or superior to corn for production on irrigated land in Egypt. They compete especially well under conditions of moisture stress. Thus they are important in India for using unused monsoon water in the soil following an early-harvested crop. But Egypt has adequate water supplies for her land, if properly used; thus sorghum and millet will compete with corn on the basis of equal potential for yield but involving shorter

growing seasons and less damage from hot winds at the critical plant-growth period.

Rice Production

The rapid advances in rice breeding and variety selection in the last few years make estimating a yield potential twenty or thirty years hence a pleasantly difficult job, owing in large part to the accomplishments of the International Rice Research Institute. Maturity time has been shortened so that three crops of rice have been grown on a single plot of land in one year.[3] The summer yield of these three crops from the same land was 20 metric tons of paddy rice per hectare—or over 18,000 pounds per acre. Equally amazing is the production of more than 20 metric tons of grain from one crop of rice and two crops of grain sorghum on the same land area in one year. These yields were produced at Los Banos, the Philippines, where rainfall is plentiful and and temperatures are high the year round. The new shortstraw rices can stand high rates of application of nitrogen fertilizers without lodging.

Yields in Egypt are about 4,500 pounds of rough rice per acre for the first eight years of the sixties. In the Mediterranean, Spain currently ranks highest among rice growing countries with 5,600 pounds per acre; and Australia heads the recorded world list with nearly over 5,700 pounds and most recently, 6,400. The new variety IR 8 has the potential of producing more than 8,000 pounds of paddy rice per acre in a single crop. In view of this and the present yields in Australia, we feel that rice yields in Egypt can over the years be raised to 7,100 pounds per acre. This would be just short of double the average yield of the last three years of about 3,900 pounds. The anticipated increase in yield would be due to a combination of new varieties, higher rates of fertilizer use, and to better management practices. Rice is a good crop to grow where some removal of salt from the soil profile is desirable. It does, however, require much water during the summer season.

Cotton Production

In terms of value cotton is Egypt's most important crop and is second only to berseem clover in acreage occupied. While at first glance the present yields of about one bale (500 pounds of lint) per acre compares favorably with the production of most countries, including the United States as a whole, the yield is not good for the excellent quality irrigated land found in Egypt. The current production in Israel is more than twice this yield and so is the yield on the irrigated areas of California and Arizona, where many of the large irrigated farm areas now produce about four bales per acre. Where drainage has been improved in Egypt, yields have increased by almost 50 percent.

These comparisons must be qualified by the fact that the

---

[2] It should also be noted, however, that in several states of the United States, corn was selling at higher prices than wheat for part of the 1968/69 crop year and that 200 million bushels of wheat were estimated to have gone into livestock feed.

[3] Robert F. Chandler, Jr., "The Case for Research," in *Strategy for the Conquest of Hunger,* Proceedings of a Symposium convened by the Rockefeller Foundation, April 1 and 2, 1968, at the Rockefeller University, New York, N. Y. p. 97.

extra long fiber cotton grown in Egypt is thought to have only about three-fourths the yield potential of the upland shorter staple cotton in most countries. However, the Egyptian cotton yield could certainly be double its present one-bale level. Nearly one-half of this improved yield could, for the country as a whole, come from improved drainage and the balance, from improved management practices—including timeliness of land preparation and planting, and better weed and insect control.

Berseem Clover Production

Berseem clover occupies the largest acreage of any single crop in Egypt. In the past it has had two main functions: one to furnish nitrogen to succeeding crops and the other to furnish forage for the work and milk animals. It is primarily a winter crop that competes with wheat. Sometimes it is grown as catch crop between other crops for a short time only. It matures rapidly and produces rather large amounts of nutritious feed.

With the increased use of chemical fertilizer, the need for berseem clover for fixing nitrogen from the atmosphere for succeeding crops has disappeared, though as a by-product of producing feed for the animals it has value. Thus the future acreage of berseem clover will depend to a considerable extent on the degree of mechanization of farm operations. One consideration here is that although the ubiquitous buffalo gives milk in addition to pulling the farm implements and providing manure for cooking the family meals, she is not a very efficient producer of milk for the feed eaten, once the work in the fields is no longer required.

A reduction in the need for draft animal feed could free acreage for the production of more fruits and vegetables, or more wheat, barley, beans, lentils; or it could be used to raise more animals for meat. If the latter were done, specific provision would have to be made for production of animal feed for use during the summer when the berseem clover acreage is mostly in other crops. Both corn and grain sorghum can produce very large tonnages of meat animal feed when well watered and fertilized. On a per acre basis, corn silage furnishes highest production of total digestible nutrients.

While berseem clover will continue to be an important crop in Egypt for a long time to come it is unlikely that it will occupy a larger percentage of the acreage than it does now.

Israel

Wheat and Barley Production

Wheat and barley together occupied 74 percent of the field-cropped area in 1964/65. Wheat yielded an average 36.5 bushels per acre, while barley with somewhat less poundage per bushel gave 40 bushels per acre. None of this acreage was irrigated in full though it is possible that some had supplemental irrigation. It is rather difficult to give a

good estimate of the yield potential of these grains. However, it is certain that large increases can be obtained by the use of the new short-straw wheats combined with higher rates of fertilization, particularly nitrogen. We believe that these steps together with excellent management could double the yields by raising them to 75 bushels to the acre. The acreages cannot be materially increased for lack of land having reliable rainfall.

Other Grain Production

Next to barley and wheat, sorghum is the most extensive grain crop in Israel, grown on about one-fourth the acreage devoted to wheat and barley. It is a summer crop of which about 10 percent is irrigated. The current yields are about 80 bushels per acre on irrigated and 27 bushels on dry land. There is a small acreage of irrigated and dry farm corn for grain, with yields comparable to those of grain sorghum. Excellent management could perhaps double these yields. However, these are summer crops that compete for scarce irrigation water. So long as the water can be used on crops that produce more income than these feed grains, the acreage is not likely to increase materially and feed imports, which form the greater portion of Israel's feed grains—mostly as maize— are likely to decline correspondingly.

Hay, Pasture, and Silage

Hay is produced on about 70,000 acres of land: of this acreage only 2,500 acres are irrigated. Another 60,000 acres are used for fodder, pasture, and silage. These materials are almost all destined for dairy cows. It is difficult to say what yields could be obtained under our assumption of "no constraints." Perhaps 50 to 75 percent above present levels would not be unreasonable.

Cotton Production

Cotton was produced on 22,000 acres in 1966. The present yield of 2.1 bales of lint cotton per acre is similar to yields obtained in California and Nevada under irrigation. That yields could be further increased is shown by relatively large acreages in the United States that do produce as much as 4 bales per acre. Perhaps it is reasonable to assume that over the next few decades the yield in Israel could be raised by about 50 percent to 3.2 bales per acre. The cotton that Israel produces is of medium fiber length, as is most of that grown in California and Arizona; it is not comparable with Egyptian long staple cotton.

Vegetables and Commercial Crops

Potato yields in Israel are quite good, with an average yield in 1966 of 360 bushels per acre. However, it is unlikely that this production can be raised more than 50 percent even by using best adapted varieties and excellent management practices. Tomato yields likewise are quite good at a 1966 average yield of over 18 short tons per acre. This production, too, might eventually be increased by 50 percent. The total acreage of vegetable and commercial crops in Israel is between 60,000 and 70,000 acres. Thus, while there is nothing on the horizon right now to cause one to assume that all of the crops produced on this acreage could have their yields increased in the long run by more than 50 percent, even using the best possible management practices, 50 percent does appear to be a reasonable "no constraints" level.

Fruit Production

In 1966, Israel had 220,000 acres in fruit plantations—a little more than one-half of it in citrus, and one-fourth in olives, pome, and stone fruits. Grapes occupied 14 percent of the fruit land and the balance was in bananas and miscellaneous fruits. Again it is difficult to say how much increase in yields could be obtained. We assume that over the long run, yields could be increased by 50 percent on the average.

In summary, Israel's agricultural output could be almost doubled in grain production and raised by perhaps 50 percent in vegetables, hay, fodder, silage, and fruits as a whole. Vegetable and fruit acreage could be increased at the expense of other crop acreage, and output increased further, if demand warranted. Some of the gain would come from better practices and varieties on all of the farms. But much would come from great improvement on those farms not being managed as well as the best farms are now managed. Needless to say, raising the level of production would require additional inputs such as fertilizer, insecticides, weedicides, and better seeds, as detailed in Chapter 17.

Lebanon, Jordan, Syria, and Iraq

Fallow: Its Uses and Abuses

To the people of Lebanon, Jordan, Syria, and Iraq, bread is truly the staff of life. Thus it is not surprising that wheat and barley occupy more area than any other crop. These grains, grown as winter- or fall-planted crops, fit well with prevailing climatic conditions. The rains come during the growing season and do not occur very often during harvest. Too, the principal growing season comes in the spring before very hot weather brings high water requirements for plant growth.

For a number of reasons the area, however, is less than ideal for winter grain production. Most of Jordan, Syria, and Iraq is desert and too dry for any appreciable agricultural production. In the areas that do get enough precipitation, on the average, to grow grain crops, large year-to-year variations cause a two or threefold range in the per acre yields between dry and wet years. Moreover, there are at times series of consecutive good or bad production years. Nor is the distribution of precipitation always ideal. In some years the rains are late in coming, and hence land preparation and seeding are late. Rains may be so intense in the midwinter months that the surface soil is puddled and sealed, so that water runs off rather than soaking into the ground where it is needed.

Another limitation on management of the area's soils for rainfed agriculture is that few are deep enough to permit the storing of rainfall from one season to the next. Experience in the United States has indicated that there should be a rooting depth in the soil of about six feet in order to make this practice feasible. Yet fallowing one year between grain crops is the general practice in our study area. In the fallow year, however, a crop of weeds is grown, a practice which would scarcely conserve moisture even if the soil depth were adequate, though the breaking of grain cropping with a year of weeds has undoubtedly had some benefits in reducing disease and pest problems.

Much of the weed growth during the fallow year is grazed by sheep or other animals. But the same soil, in the same time, with the same moisture, could produce a much better forage crop. We are suggesting that in all this area the wild plant or weed fallow be replaced by a purposeful crop. In the better land, wheat could be grown successfully for two or more years in succession, broken at intervals by another crop, even a grain, such as oats, which would make excellent forage and feed grain. Winter legume crops either for hay or for the seed as human food are also good fallow replacements. So is grain sorghum, which can be planted as early in the spring as possible and can make a good crop of both grain and fodder.

For much of the poorest yielding grain land, the best use in the long run may be as permanent grasses, even though they often cannot yield as much per unit of area as can sown forage crops. However, perennial grasses have two main advantages: (1) once seeded, if properly managed, they require little maintenance or reestablishment, hence are cheaper per unit of area than are planted forage crops; and (2) they are able to make effective use of moisture in wetter years and because of their deep-rooting habits, can survive a dry year. Harvesting of grasses is normally by livestock grazing, hence harvest costs, too, are low. The kinds of grasses to be raised would, of course, have to be determined locally, but should be given careful thought, given their permanency.

Planning the production of animal feed instead of relying only on grazing and aftermath would be new to these countries. But at present the animals are grossly underfed; much too high a proportion of the consumed feed goes into sheer maintenance of the life of the animals; and as a result meat prices are high. This problem is more fully discussed below, but the purposeful raising of feed certainly merits recognition at this point.

Some Special Situations

Lebanon has enough water in the Bekaa Valley to provide some irrigation for wheat and to produce a considerable acreage of irrigated corn and grain sorghum. Experimental areas have produced 200 bushels of corn and grain sorghum per acre. Reasonably managed farms now produce about one-half of this yield. Also, when the whole stalk and ears of corn are cut for silage, this perhaps furnishes the most total digestible nutrient per acre of any of the crops and represents especially good feed for milk cows.

The amount of water from the Euphrates River that will be available to Iraq in the future is uncertain, as we noted in Chapter 13. Upstream claims by both Turkey and Syria could limit the available water to that presently diverted. For this reason we do not estimate a large increase in summer crops. Cotton, corn, and especially rice have very high water requirements because of the very hot summer months. In a four-month summer period, it is said, the crop will require more than four times as much water diversion as is needed for a four-month period on a winter crop.

Because of the more efficient utilization of water by a wheat crop, there would be considerable advantage to growing one crop of wheat, followed by a moisture cleanup crop of sorghum each year on most of the acreage. Millet too, a good food-and-feed grain, is frequently so used in India and could also be used in Iraq.

There seems to be less room for a significant acreage of rice in either Syria or in Iraq. True, rice is doing well in Iraq relative to other crops, especially since the Government launched a special new varieties and fertilizer campaign in 1967,[4] with very encouraging effects on yields; but the water use is several times that required for a wheat and sorghum crop while the yield potential is only about 50 percent more than that of wheat.

Cotton production in Syria has been developing very well during recent years. A modest increase in acreage and a doubling of yield per acre is entirely feasible. In Iraq we suggest a moderate potential of 250,000 acres of cotton and a yield of 900 pounds of lint per acre, equal to that for Syria. That this is not an unreasonable goal in that part of the world is indicated by the fact that Israel yields now are above this level.

It is difficult to put figures on the potential for producing fruit and vegetables in Lebanon, Jordan, Syria, and Iraq. Perhaps it is sufficient to say that, in our judgment, present yields can be greatly increased—perhaps by 50 percent or more—and that acreage can shift between these crops and the grain crops as the demands develop. Actually, within a given diet pattern—which may already comprise a good deal of easily grown horticultural products, people do not substantially increase their consumption of these products until they have enough bread to eat. Increases in grain yields by even a considerable portion of the potential we have estimated could open up opportunities for a large increase in demand for a variety of fruits and vegetables, though growth pattern in the developed countries may be misleading guides in this respect.

A few areas, such as the warm, below sea-level Jordan Valley, are particularly favored for the production of early fruits and vegetables for the winter market both in Western Europe and in other parts of the Middle East, and perhaps at some future date in Eastern Europe as well. While this type of production could be quite important locally, especially in terms of furnishing employment and in earning of

foreign exhange, it cannot, for reasons outlined in Chapters 11 and 18, be large enough to be a major use of land and water resources for the region.

Livestock Production

Successful livestock production, no less than successful crop production, requires a package approach. If animals are to grow at reasonable rates, attain reasonable weights when full grown, and acquire a reasonable degree of fat prior to marketing, then they must have enough to eat—and good nutrition for animals means a well-balanced diet as well as enough bulk. But there is little to be gained in feeding an animal better if it is disease-ridden; and, even if the diseases are controlled and the diet is fully adequate, the type of animal which has been most successful under the relatively harsh conditions of the past may not make the best possible gains on the better diet—any more than the present crop varieties may be able to take advantage of more fertilizer.

Moreover, there is little point in having good types of livestock, reasonably free of disease and well fed, unless the meat and other products can be marketed effectively, reach the consumer in good shape, and produce a remunerative income for the producer. The parallels between the livestock package and the crop package are many and close; in each case, measures that would be highly productive as part of a package may have very limited, or even negative, value if taken singly. A program of importing improved livestock breeds and propagating them, for example, is of little value if disease cannot be controlled and better diets provided. The proposed Near East Animal Health Institute, a UN sponsored project, could be a long step in this direction.

Above all, however, to produce a significant output of meat or milk on a sustained basis, adequate amounts of animal feed must be produced. Yet, as pointed out above, outside Egypt and Israel there is little purposeful production of livestock feed. No significant amount of hay is produced and stored for future use in Lebanon, Jordan, Syria, or Iraq. The animals, to a large extent, find their feed in the deserts, mountains, field borders, weed fallows, and in the grain stubble. As a consequence of being poorly and erratically fed, a large portion of the animals are only maintained rather than being productive, and the number of animals slaughtered is very low for the total animal population—all of which makes meat very costly in terms of inputs per unit of output, certainly for cattle.

Lack of feed for livestock in the area need not be a continuing situation. There are good resources and opportunities for planned production of feed and forage, as discussed above. Part of the failure to develop these has been due to historic and ethnic reasons, above all the fact that, by and large, the animals are not owned by the people who grow crops. One prerequisite of successful livestock grain farming is that the farmer have an interest in the welfare and maintenance of the livestock. This interest is, for the most part, absent in the area, outside Egypt and Israel; nor is there a shortcut to changed conditions.

---

[4] Isam Al-Najar, *Introduction and Breeding on New Varieties with High Yield Potential Improvement of Rice Production in Iraq* (FAO, International Rice Commission, Working Party, Twelfth Seminar, September 9–14, 1968.

Pasture Mismanagement in the Middle East

Equally deep-seated as a problem is the matter of pasturage. Perhaps no sector of agriculture in the Middle East is so badly managed as are the natural grazing lands. For example, Van der Veen[5] has estimated that the number of sheep now on the range land in Syria is three times its carrying capacity. Probably the same holds for Iraq and to a lesser extent for Jordan and Lebanon; some areas in Israel, on the other hand, are not grazed as much as they could and should be.

The malady can be more readily diagnosed and remedies more easily stated in general terms, than practical and locally feasible steps can be suggested. For example, in order to reestablish good native ranges to their productive capacity to use the indigenous rainfall, one would ideally want periods of from one to three years of total exclusion of the animals. This could perhaps be accomplished, slowly to be sure, on one small area after another, provided proper policing could be furnished; or perhaps exclusion in the early spring months would be sufficient and easier to accomplish. Above all, marginal lands which have been plowed and then abandoned, or are still cropped but should be withdrawn—and these run into the hundreds of thousands of acres—might be the first to be reseeded and made available for grazing, to free currently grazed areas for the same treatment. And there may be other approaches. But two things are clear. First, no recipe for change can be provided in a study of this kind; only indication of direction, and these indications will not be new to nor satisfy the expert. And secondly, none of this will bring improvement without measures of controlled access. Again, however, it would not be enough to establish or reestablish a good vegetative cover of grasses and other forage plants; even if successfully established to the limit of the capacity of the soil and moisture of the area to produce such plants, such lands would have to be managed purposefully and skillfully in the future to prevent the present depleted condition from recurring. One basic element—but not the only one—of good range or grazing management is to leave ungrazed a substantial proportion of the plant growth each year. The precise proportion depends somewhat upon the plant species and also upon the necessity for building plant vigor so that the plants can survive one or a series of relatively dry years. In many cases, as much as 50 percent of the average annual plant growth must be left ungrazed if maximum annual production over the long run is to be achieved.

If livestock operators are to practice this type of careful range management, at least two factors are necessary, neither of which exists today in the Middle East. First of all, they must be convinced that this type of range management, including leaving substantial forage ungrazed each year, in the long run will produce more forage than will grazing the plants into the ground each year. It has taken

American ranchers a long time to learn this lesson, and some may not yet realize it fully. Today in the Middle East, the standard grazing management is to take every inch of plant growth that the animals can bite off; therefore, the nomads and other graziers must simply acquire entirely different concepts of grazing and use—and this will require changed habits akin to a cultural revolution.

But, even if such livestock operators realized better how to manage the land for maximum long-run output, the second essential factor is missing: the ability of a livestock operator to control use on the land where his animals graze. Unless an operator can control use—which means keeping other people off the land he grazes—then his forbearance in grazing will mean only that someone else will get the feed he leaves. As long as livestock can wander or be driven over any and all land where any livestock feed exists, and can graze as much of that feed as they are physically able to do, maximum long-run output from such land is impossible. This, too, is a lesson which took a long time to be learned in the United States. Unless some form of grazing land tenure or land management can be installed and made fully effective, increased forage production from the native ranges of the Middle East—and often, from the rainfed crop lands as well—will be extremely difficult if not impossible to attain. Perhaps in no other aspect of Middle Eastern agriculture is the clash between conditions as they exist today, and the social, cultural, and institutional prerequisites for increased output, sharper than for grazing. The physical potential exists for greatly increased output, and the technical requirements for achieving it are well known and not especially difficult; it is the human problems which are so intractable.

Potential and Problems of Rainfed Agriculture

Reference has been made on more than one occasion to the management issues in rainfed agriculture. Because of its importance, the following paragraphs, at the cost of some duplication, summarize the principal points that have been and need to be made.[6]

Somewhat more than one-half of the cultivated land of the study area is in rainfed agriculture. In addition, there are large areas of land used only to pasture grazing animals. Though this half of the agricultural area contributes less than half of total output, it is an important segment of agriculture. In Jordan and in Syria, and to a somewhat lesser extent in Lebanon, rainfed agriculture is dominant. The prevailing time-honored cropping pattern is that of one year of grain—mostly wheat—followed by a year of weed fallow, followed by another year of grain. Two principal reasons have been assigned for the fallow year. All have said that the land must be "rested," and some have said that

---

[5] J. P. H. Van der Veen, *Range Management and Fodder Development*, Report to the Government of Syria (Rome: FAO, 1967).

[6] Worzella, *op. cit.* (above, n. 1), puts the matter in proper perspective: "Cereal and forage crops produced under dryland conditions are grown in the absence of summer rains and are dependent in large part on the accumulation of winter precipitation in the soil. Under these conditions cultural practices are of major importance to economic yield since they, to be successful, must balance all factors of production with available water."

fallowing conserved a year's moisture for the succeeding grain crop.

There is a modicum of truth in the first, but practically none in the second proposition. In the absence of fertilizer use, it may well be that the soil could not produce a crop of grain every year. And the change in crop from grain to weeds undoubtedly helps to hold down the incidence of diseases and insect pests on the grain. However, little if any moisture is carried over from a weed fallow to the subsequent year's grain crop.

Recently there have been some great improvements in some parts of the world in the use of rainfed land that is marginal with respect to adequacy of moisture for production of grains, primarily wheat. The most important advances are (1) in the area of improved varieties and (2) in the conservation of the moisture that comes primarily as winter rain or snow. In the United States the development is most spectacular in eastern Washington and Oregon and in northwestern Idaho, areas in which average annual precipitation range and rainfall pattern are quite similar to that of the Middle East, except that the season of precipitation here is somewhat longer than in the Middle East and year-to-year total rainfall is somewhat less erratic.

The improved wheat varieties that are tailored to specific soil and climatic conditions are characterized by a short, stiff straw that does not weaken even with very large amounts of nitrogen fertilizer. They have the capacity to tiller profusely (produce a number of stalks from a single seed) in accord with amount of fertility and moisture available. Thus the planting rate can be much less than one bushel per acre. Varieties exist that will produce 150 bushels per acre under irrigation where water is not limiting and 100 bushels per acre or even a bit more occasionally with 300 mm. (12 in.) of winter rainfall with no irrigation. Though only the best farmers practicing a rigidly disciplined farm operation are getting these high yields, they can be and are being attained on many thousands of acres.

Some varieties recover well after a late spring freeze kills back the developing plants. Others recover very poorly. If there is no hazard of srping or late winter freezes, one does not need to consider this ability to recover. But if the hazard is there, the ability to recover may mean the difference between a good crop and a very poor one, though perhaps in no case is there a complete recovery and yields equal to what would otherwise have been attained. Varieties of barley are being developed that can be sown in the spring and that use the stored winter rainfall and give good yields in a very short maturity time. Some varieties are developed for late planting after the winter rains come. These do not generally have the yield potential of the varieties that are planted earlier in the year and get well established before winter.

Our purpose here is not to catalogue the various kinds of grain varieties that have been and are being developed, but to call attention to the necessity of knowing the soil and the climatic conditions before selection of a variety of wheat to plant. The new Mexipak wheat, for example, runs into difficulties when taken too far north in the northern

hemisphere, but this should not be a problem in the Middle East as we define it. The problem is that this variety of wheat does not respond enough to lengthening days and takes too long to mature under northern dry land conditions. The water required for maturing the crop earlier in the season is less than required later on, when transpiration per unit weight of plant growth is much greater. Thus early maturity is desirable, even though the plant with the later maturity date may have the potential to produce greater yields if water was adequate.

The second advance in the use of rainfed land that is marginal with respect to adequacy of moisture has to do with conservation of the moisture that comes as rain and snow. If soils are properly tilled, mulched, and contoured, as much as 80 percent of the rainfall may be carried over from one season to the next crop year and thus provide nearly two years' rainfall for producing one crop. This permits good yields of wheat in alternate years on land having an annual precipitation of no more than 200 mm. (8 in.). In central Washington, where climatic conditions are similar to those of much of the Middle East, competent farmers average 35 bushels of wheat per acre on land with no more than this amount of precipitation. However, for this moisture carryover to be effective, the soil rooting depth must be enough (usually six feet) to permit the storage of the rainfall from both years' precipitation. Unfortunately, most Middle Eastern soils are shallow to rocks, gravel, or to an impenetrable horizon or layer of calcium carbonate or gypsum, and, therefore, do not have this depth of root zone available for moisture storage. Thus, when the rooting depth of the soil is so shallow that an average year's precipitaion wets it down to the obstruction, there can be no effective carryover of moisture from one year to the next, with one exception that, under some circumstances, could be useful: moisture could be carried over from a fallow year (without weeds) so that the grain crop could be planted and start growth prior to the initial fall of early winter rains.

The loss of grain production in the Middle East due to competition from *weeds* is pathetic. In many of the fields—particularly the lower yielding ones—weeds use half or more of the limiting available moisture. The grain yield is cut in half again by the fact that there is usually, though not always, a year of weed fallow between crops. The grain therefore gets little more than one-fourth of the moisture available for plant growth.

Weeds can be controlled, both mechanically and by use of chemical weed killers. Control is a matter of organization and costs. On the latter we have only episodic data, suggesting orders of magnitude; in the Bekaa Valley of Lebanon, for example, a spray program to kill wild mustard could be applied for the cost of about two bushels of grain per acre—a cost much less than the increased yield that could confidently be expected to result. Moreover, on experimental areas where such a program has been followed for several years, the weeds are eliminated and spraying is no longer required each year.

There is much inequality between the countries of the study area and even within these countries with regard to *tilling and planting equipment and methods.* The most advanced and most highly mechanized areas are perhaps the marginal wheat and barley areas of Jordan, Syria, and Iraq. There, farming is the extensive kind and mechanization units are large. In other parts of the same countries much wheat and barley is still produced by primitive methods. In order to make full use of the inputs required to produce high yields, a degree of mechanization is necessary. For example, some of the new wheats will not give good stands of plants under the traditional practice of throwing seeds on the ground and then plowing them under. Instead, a grain drill must be used. Similarly, the application of the large amounts of fertilizer required calls for mechanized application for best return for the money invested. So does weed control: weed killers must be sprayed on the foliage or on the soil, and weeds and grass must not be allowed to grow on fallow land in fall and winter when the moisture from winter rains is to be used to grow a spring crop of forage or legume.

Timeliness of land preparation, planting, and harvest is important to high productivity, as is pointed out below. Thus mechanical equipment, which in developed countries is often thought of exclusively in terms of labor-saving, actually is a necessary part of the "high potential package" to which the farmer must have access for timely and satisfactory land preparation, fertilizing, and planting.

The necessary tillage operations require not only proper implements, but also adequate power to pull them, understanding of moisture retention and movement in the soil, and rigid discipline in the timing and precision of the tillage operations. Moisture can be stored and kept near enough to the surface so that wheat can be seeded in the moist layer before winter rains begin. In this way good growth can be obtained before cold weather sets in (where this is a factor). Drills are now available that will with precision seed in 16-inch rows at a depth of about 5 inches (about 11 centimeters). In the Columbia basin of Washington, 40- to 50-bushel wheat yields are common in alternate years in the area of 9-inch annual precipitation, with timely tillage operations and complete weed control.

The basic equipment needed includes chisels, used in the fall to loosen the ground, improve moisture penetration, and to break up any plow pans; sweeps, to be used in the spring near the end of the rainy season ahead of weeders; and rod weeders, to fill air pockets and to make a mulch to prevent moisture evaporation except at surface mulch. Rod weeding is continued through the summer season to maintain the mulch and to prevent weed growth. Sweeps are pulled ahead of the last rod weeding operation before fall planting. If these operations have been properly carried out, a moist layer close to the surface is available for planting by the deep drill aready mentioned so that a good stand can be obtained before the fall and winter rains come.

Closely linked to methods of tillage is the timeliness of its application. For each combination of soil, climate, and variety there is a rather narrow range of planting time for grains that will give the best possible yield. Usually for each week or even each day the planting is delayed beyond this

time, there is a reduction in yield. In some good wheat farming areas in the United States, when planting starts it is continued on a 24-hour a day basis until finished. Thus, maximum use is made of the expensive items of equipment needed to get the job done on time. The knowledge of best planting date is gained from a combination of research finding and actual on-farm experience with the particular variety. Although effort and a certain amount of self-discipline are required to get planting and other time-important operations done at the right time, the rewards in higher yields are very great indeed.

Mention must also be made of the losses due to blowing when the soil is dry, due to runoff when there is an excessive rate of rainfall, and of ways of minimizing them. In the first place, soils that are so steep that they cannot be farmed to winter grain crops without severe erosion should be left in close-growing vegetation for grazing purposes. One of the best control measures on the cropped land is a good cover of growing grain plants during the winter months when the rains come. Too, the condition of the soil, such as size of clods left in seedbed preparation, materially influences the rate of intake of rainfall and retention of snow. Some vegetation such as straw stubble from the previous crop may be of great help here.

It is very important in fallowing operations to keep some of the stubble showing above the soil surface. This helps to cut surface wind velocity and thus reduce soil blowing. The chisels and sweeps mentioned earlier are also effective in this regard, as well as the rod weeders which tend to return to the surface crop residues that have been partially buried near the surface. At times windbreaks of trees help, but of course a row of trees requires water for their growth.

Fertilizer Needs and Practices

Nowhere in the Middle East can continued high yields of wheat and other grain be obtained without adequate application of plant nutrients. This is particularly true because soil organic-matter levels are very low and hence only small amounts of nitrogen and phosphorus are made available each year for plant growth by microbial decomposition. The nitrogen required for each bushel of wheat or barley produced if not furnished by the soil must be supplied as animal manure or added as purchased mineral fertilizer. A portion of the total nitrogen metabolized during the growth of the crop will be returned to the soil by the straw and roots of the plants if nothing but the grain is removed in harvesting. But in converting a low-producing wheat and barley area to one of high production, large amounts of fertilizer nitrogen will be necessary—although, because of the microbial fixation of nitrogen, less will be needed if a legume crop is grown between grain crops.

Nitrogen is the first limiting nutrient. The shortage of phosphorus is usually not quite so limiting, but needs to be overcome for sustained high yields. Potassium is usually available in reasonable amounts, but it is likely that some application may become necessary if high yields of grain are taken off the land for long periods of time.

Coordination of water with fertilizer use is an important aspect of the practices package. According to data obtained at the Agricultural Experiment Station at Pullman, Washington, each bushel of wheat requires 2.7 pounds of nitrogen, and each inch of soil moisture transpired through the wheat plants can produce seven bushels of wheat per acre. If one does not match the available water and the nutrients applied, a loss of one or the other may result. Shallow soils cannot store enough moisture during the rainy season to produce a large crop, where rainfall stops well before maturity. On such soils it is obviously wasteful to fertilize for obtaining a crop yield that the soil moisture is incapable of producing—hence, the necessity of knowing the important soil characteristics. In general, the soil depth must be of the order of 60 inches to 80 inches for useful amounts of moisture to be carried over from one rainfall year to the next. For soils that have less than 40 inches depth, annual cropping with the use of appropriate amounts of fertilizers is a far better usage. The hazards of plant diseases and pathogens, produced by growing the same crop on the same soil every year, can be controlled or suppressed by breaking the cycle with a different crop or by the development of varieties that are resistant.

In areas of high winter rainfall or coarse-textured soils it is usually desirable to add only a portion of the needed nitrogen fertilizer at planting time, with the balance applied in either one or two applications in the spring. With some of the fine-textured soils all of the needed fertilizer may be applied at planting time without much loss in efficiency and with consequently lower cost of application.

Fertilizer alone, even in relatively large quantities, not only would in many cases not materially increase production but might reduce it in some. Present native varieties are very well suited to growing under the conditions of low-fertility level that have prevailed for many hundreds or even thousands of years. To produce the high yields envisioned in our estimates of production potential, then, increased fertilizer use must be accompanied by such other inputs as new varieties, better soil preparation, timeliness of planting, and control of weeds. The absence of any of these necessary inputs is likely to reduce yields and may even cause the loss of the entire investments involved in the others.

What the preceding sections have spelled out in detail is that it is of critical importance that the major farm operations be done in the best way at the right time. Failure to do a good and timely job on any one factor may result in near-failure altogether. Particularly in dryland areas, high-quality farming requires a high degree of discipline on the part of the farm operator. Within this context, the extreme importance of interaction among the various factors can hardly be overemphasized.

At the same time, there must be a relationship between costs and prices that stimulates the adoption of improved technology. Fertilizers are a case in point. The smaller the quantity of farm products needed to purchase a given tonnage of fertilizer, the more likely that the purchase will be made, provided, of course, that the varieties planted will allow the desired production response. In terms of such cost-price relations, conditions vary widely. Estimates made on the occasion of a recent conference[7] show that in Egypt (in 1964/65), it took 5 kg. of wheat and 6 kg. of rice to purchase 1 kg. of (unspecified) fertilizer. In Lebanon (1965/66), it took less than 2 kg. of wheat, and Syria (1965/66) only 4.8. (It took only 1 kg. of rice in Pakistan, in 1965/66—an interesting statistic in view of that country's recent advances in agricultural production). In this connection, one must note that the Egyptian government's support price for wheat (in 1966/67) was nearly 20 percent *below* the average import price. In Lebanon, it was nearly 40 percent *above* the import price. Thus, price policy is a crucial ingredient in the evolution of farm practices.

---

[7] "Economic Incentives and Marketing for Expanded Agricultural Production," Item 7b of *Provisional Agenda*, Ninth FAO Regional Conference for the Near East, Baghdad, Iraq, September 21-October 1, 1968.

CHAPTER 15

The Output Potential

IN Chapter 13, we considered the resource development potential of the Middle East, and in Chapter 14 we considered the productive potential of the adoption of modern farm practices; in each, either explicitly or implicitly, as explained at the beginning of Part III, we assumed that there were no limiting factors other than resources and technology. To the extent that these productive potentials were realized in practice, agricultural output in the Middle East would rise: implicit in our discussion throughout the preceding two chapters was increased output. In the present chapter, we quantify those increases without abandoning the simplifying assumption of "no constraints." Indeed, here we must stress that one factor under this assumption—no concern over market outlets for the commodities that might be produced—has not hitherto been especially relevant. In Chapter 18, we explicitly consider the matter of market outlets, as in Chapter 19 we consider more fully the nature of the social and institutional limiting factors. Moreover, in the present chapter we do not consider, explicitly, the economics of producing what the resources and the technology are capable of producing: i.e., we do not here assert that the production increases that we posit as achievable will maximize agricultural income in the Middle East. We do judge that these increases in output, or part of them, are likely to prove profitable in the sense that output will increase far more than the costs of inputs, and that the increases we suggest as possible are, by and large, "practical" in the broad sense of the term. But—and this point is crucial to a proper understanding of the chapter—what is shown is neither a plan nor a program, but only an outline of potentials, in line with what has been said earlier in the introduction to Part III. Each of the crops dealt with could increase more or less than shown, absolutely or relatively. What is of primary interest is the direction and scale of change.

We have not, thus far, explicitly considered the matter of the time required to achieve the potentials we have outlined. Under the very best of circumstances, one can hardly envisage all the changes in agricultural production outlined in Chapters 13 and 14 being achieved in less than two or three decades. It takes time to build irrigation works, to construct drainage canals and facilities, to develop tailor-made crop varieties, to reestablish grass on denuded grazing lands—even when there is a great will to do so, ample professional manpower to do the job, and adequate investment capital. How much time it takes to change institutional and social arrangements is perhaps less easily answered today than in times past. The rapid course in parts of Asia of agriculture's breaking out of its traditional patterns gives hope that perhaps these transitions are less difficult than one had long assumed

them to be. Whether this is true also of the countries of the Middle East here considered remains to be seen; and until the area's capacity to change on a sufficiently large scale is more amply demonstrated, one must remain cautious. Our time schedule in the two foregoing chapters and in the present chapter is thus unavoidably vague; we can simply say that it would be surprising to see the increased output being achieved in full before 1990 or 2000, although this would not exclude large increase much sooner through rapid advances of certain segments of the agricultural economy.

Crop Potentials of the Middle East

In this section an attempt is made to estimate, country by country, crop production of the Middle East as it might follow from full utilization of the resource development potentials of Chapter 14. For easier comparability with local statistics, as well as with various plans, metric units are shown throughout the chapter. "Present" conditions refer roughly to the mid-1960's.

Egypt

In estimating the output potential of Egypt (Table 15-1), the following assumptions were made:

(1) As a result both of the additional supplies of water from the High Aswan Dam and of other modern farm practices, two crops per year are possible on all cultivated land; as an average, that is, the ratio of crops harvested to crop land will rise to 2.0.

(2) By use of the best farm-management practices, with all needed purchased inputs and with the new crop varieties now available, the yields of crops will rise to or near levels now deemed achievable with best and closely controlled practices.

(3) Acreages of wheat, corn, and rise will increase somewhat, but no increase will occur in the acreage of sorghum and millet.[1]

(4) A small increase will occur in the acreage of cotton and a small decline in that of berseem clover.

(5) An 80 percent increase will occur in the acreage of fruits and vegetables.[2]

---

[1] This assumption is debatable. It is based on the reasoning that while the potential yields of corn, sorghum, and millet are similar, corn was the more desirable food grain in Egypt. A factor in favor of sorghum and millet is their somewhat shorter growing periods and less damage from hot winds at critical growth periods. However, it is assumed that an increase in sorghum and millet acreage would be at the expense of corn; no overall difference would be made in the total grain production.

[2] A larger increase is technically quite possible, but as explained in the introductory section to Part III, it makes no sense to estimate

Table 15-1

Egypt: Present and Potential Production of Main Crops

(All Crops Irrigated)[a]

Crop	Present Yield (mt./ha.)	Potential Yield (mt./ha.)	Present Crop Area (1,000 ha.)	Potential Crop Area[b] (1,000 ha.)	Present Production (1,000 mt.)	Potential Production (1,000 mt.)	
Wheat	2.7	6.7	605	800 (900)	1,620	5,400	(6,750)
Corn	3.6	6.3	661	700 (750)	2,360	4,400	(4,700)
Sorghum and millet	3.9	6.3	220	220 (250)	860	1,400	(1,600)
Rice	4.1	8.0	486	700 (750)	2,000	5,600	(6,000)
Total food grains	N.A.	N.A.	1,972	2,420 (2,650)	6,840	16,800 (19,050)	250% of present production (280%)
Cotton	0.59	1.2	781	1,000 (1,100)	462	1,200 (1,320)	260% of present production (285%)
Fruit and vegetables			330	600			180% of present production
Sugar cane	84	100	50	50	4,200	5,000	120% of present production
Berseem	N.A.	N.A.	1,063	1,000	N.A.	N.A.	95% of present production
Fodder and all other, incl. beans, lentils	N.A.	N.A.	124	490 (510)	N.A.	N.A.	400% of present production
Total			4,320	5,560			

[a] For the potential acreage there are about 57.5 milliards m³ of water. This allows 10,340 cubic meters per hectare per crop or 3.4 acre-feet per acre per crop. This should be adequate even allowing for inevitable losses.

[b] The area underlying this column is 2.9 million ha., or 7 million acres, possibly a conservative estimate (see Chap. 4, above). Another 350,000 ha.–or 875,000 acres–is the target set by the Government. It may be noted that a similar result could be achieved by raising the intensity, not to 2.0 as assumed here, but to 2.1. The figures in parentheses suggest the potential based on 7,875 million acres.

(6) Any remaining acreage of irrigated land will go into fodder crops for animals or into the "all other" category, which includes beans and lentils.

On the basis of these specific assumptions, substantial increases in crop output per acre are estimated (Table 15-1). By and large, with some exceptions, crop yields are estimated at double the present or recent past; in addition, some increases in acreage of crops will be possible, as the cropping ratio rises, in spite of a relatively fixed area under irrigation.

Israel

A different set of assumptions is made for Israel. Specifically, it is assumed that there will be no appreciable change in the area of cultivated land or in the supply of water for irrigation. Israel has already developed virtually its entire usable supply of irrigation water, and has extended crop production to all lands capable of continued crop production at a profitable level with present water supplies. However, many of the present farmers, especially those settled since 1948, have not yet attained the level of production which characterizes the older farmers; and over a period of years it may reasonably be assumed that these newer farmers will catch up with the older ones.

potential of *minor* crops without some reference to current practices and production mixes. Since one of the vehicles of income growth in agriculture is the transition to higher-valued crops, the subsequent valuation of these potential patterns may be on the conservative side.

We do assume that new, short-straw wheats, capable of much higher yields, will be introduced and widely used and be supplied the needed fertilizer. Still further, a general improvement in farm management practices and a more careful use of land, including borders and field edges, will take place.

The potential production increases differ considerably according to the crop. Except for wheat and barley, they tend not to exceed 50 percent above present levels (Table 15-2). Although we have not estimated an increase in the acreage of fruits and vegetables, such an increase, as in Egypt, is technically quite possible if the demand warrants.

Lebanon

The plowable land in Lebanon and that under permanent crops, other than meadows, is about 225,000 ha. (630,000 ac.). Of this area, 204,000 ha. (502,000 ac.) is plowable land not in permanent crops. Of the plowable land used for annual crops, about 56,000 ha. (138,000 ac.) are fallow each year. In our calculations we have projected the elimination of the fallow. Also we have assumed that crops such as irrigated wheat can be followed by another crop, such as a vegetable for forage crop, and that there can be more double cropping in the irrigated orchard areas. Thus the cropped area could be raised from the present 216,000 hectares (534,000 ac.) to about 300,000 ha. (742,000 ac.). All this would permit increasing the acreages of wheat, corn, sorghum, legumes for grain, and vegetables; but we assume that the industrial crops, sugar beets, tobacco, and peanuts would retain their present acreage but with higher yields. For example, sugar beets presently yield about 6.5 tons of sugar per hectare. This yield can perhaps be raised to 8 or 10 tons. Peanut yields, now about 1600 kg./ha. can certainly be doubled. By the general use of adequate amounts of fertilizer; by the adoption of new crop varieties, particularly the short-straw wheats; and by weed control and other necessary associated practices described above, crop yields can be greatly increased.

As a result, grain production would rise sharply; vegetable and fruit production would rise considerably, but less dramatically (although still further increases in acreage are technically possible); and increases in industrial crops, not shown in Table 15-3, could also be significant.

Jordan

In making our projections for Jordan, it is assumed that 90,000 ha. (225,000 ac.) of the driest land previously used for crops would be reseeded or allowed to revert to grass. Because of the problems of moisture and weed control on fallow land, as outlined in the preceding chapter, the alternate crop-fallow system on the remaining unirrigated crop land would be abandoned in favor of annual cropping. A food crop such as beans, peas, or lentils—or a forage crop such as oats, sorghum, or vetch—would be substituted for fallow. This will necessitate the use of fertilizers, especially nitrogen. It is further assumed that the short-straw wheats,

Table 15-2

Israel: Present and Potential Production of Main Crops

Crop	Present yield (mt./ha.)	Potential yield (mt./ha.)	Present crop area (1,000 ha.)	Potential crop area (1,000 ha.)	Present production (1,000 mt.)	Potential production (1,000 mt.)
Wheat	2.0	4.0	80	100	160	400
Barley	1.0	2.6	50	30	50	78
Sorghum						
rainfed	1.7	2.5	30	30	50	75
irrigated	5.1	7.5	3	3	15	22
Hay	3.6	4.5	30	30	110	135
Green fodder						
rainfed	17	20	5	5	85	100
irrigated	72	72	16	16	1150	1150
Sugar beets	43	43	6	6	258	258
Cotton	1.2	1.5	22	22	26	33
Vegetables[a]			29	29		
Fruit[a]			88	88		
Maize crops[a]			41	41		
Total			400	400		

[a] Data could be calculated from total tonnage and area, but such aggregates considered not useful.

Table 15-3

Lebanon: Present and Potential Production of Main Crops

Crops	Present yield (mt./ha.)	Potential yield (mt./ha.)	Present crop area (1,000 ha.)	Potential crop area (1,000 ha.)	Present production (1,000 mt.)	Potential production (1,000 mt.)
Wheat	1.0	3.4	66.5	84	67.8	285
Barley	1.2	3.2	13.4	16	15.8	51
Corn	1.8	6.3	3.0	26	5.2	164
Sorghum	.9	6.3	1.3	24	1.2	151
Legumes for grain	2.1	3.0	13.8	22	29.2	66
Vegetables	11.4	15	30.5	40	348	600
Fruits	8.7	10	74.6	75	650	746
Industrial			13.2	13		
Total crops			216	300[a]		
Fallow			56	none		
Total			272	300		

[a] Includes 28 from double cropping legumes or corn after wheat.

which can stand high fertility levels even in dry years and which make efficient use of available nitrogen, will come into general use.

These changes will not eliminate the variations in crop yields arising out of dry and wet years; but yields of all crops can be at a much higher level, with variation around that higher level. Such additional irrigation as permitted by availability of both water and good soils is assumed to be carried out. The area will not be large and is assumed to be used for relatively high value crops, rather than grains. Controlled grazing and other good management practices are assumed on the reseeded grasslands.

With these changes, wheat production would increase more than threefold; barley, somewhat less than double; lentils and other legumes, more than fivefold; and forage crops, substantially (Table 15-4). Only modest increases in acreage of fruits and vegetables are assumed, but larger increases are technically possible if demand warrants.

Syria

In Syria, it would be reasonable to reduce the cultivated area from 6 to 5 million ha. (15 million to 12½ million ac.) by converting 1 million ha. (2½ million ac.) now in marginal wheat and barley to grass, either by natural reversion or by reseeding. The remaining rainfed cropland—consisting of the wetter areas—could then be cropped continuously, rather than on a crop-fallow basis as at present, for reasons discussed earlier. Some of this increased crop area would go into wheat, but much—perhaps 800,000 ha. (2 million ac.)—would also go into legume and forage crops. Since grazing of grasses would generally be available in the winter and spring, these feeds could be stored as hay or in pit silos and used during the summer and fall dry season when natural grazing is exhausted. We assume that only little additional irrigation (100,000 ha. or 250,000 ac.) will be possible, based on using water flow from the Euphrates; in part, this conclusion rests less on the probably limited availability of water than on the certainly limited availability of good irrigable land.

At the present time corn is not a major crop in Syria. However, this crop could be grown on the irrigated land of the Ghab and in the Euphrates Valley at high production levels. We have not considered this possibility in our table for Syria. Another crop that could be grown on these irrigated acres is soybeans. This crop may deserve serious consideration because it is a good source of protein for human consumption. Again, as in the case of corn, we are not attempting to predict its acceptance as a major crop.

The substantial increases in forage from the reseeded grazing lands plus the large increases in harvested forage from croplands will make possible a large increase in livestock production, which in large part should come from increased output per animal, based on better feeding. Specific estimates of livestock production possibilities are presented later in this chapter.

The consequences of these assumptions, combined with achievement of good farming practices practices as in the preceding instances, are shown in Table 15-5. Wheat production in Syria would increase to more than fivefold, barley production would more than double, cotton production (from only a modestly larger acreage than at present) would increase two-and-one-half fold and a great deal more feed for livestock would be produced. Only modest increases in fruit and vegetable acreage are estimated, but, as for the other countries, further increases are technically possible.

Iraq

For the rainfed land in Iraq, we assume the same kind of adjustments that have been discussed for Jordan and Syria, and for the same reasons. That is, acreage of rainfed cropland would be reduced somewhat, to about 2 million ha. (5 million ac.), with some of the drier lands diverted to grass, and with annual crops on the wetter rainfed lands continued in production—wheat on the wetter and barley on the drier lands.

The greatest output potential lies in the irrigated lands, partly because production per acre is extremely low. While these offer some possibilities for increased irrigated areas

Table 15-4

Jordan: Present and Potential Production of Main Crops

Crops	Present yields (mt./ha.)	Potential yields (mt./ha.)	Present crop area (1,000 ha.)	Potential crop area (1,000 ha.)	Present production (1,000 mt.)	Potential production (1,000 mt.)
Wheat	0.8	3.0	235	200	188	600
Barley	0.8	2.5	89	50	71	125
Lentils & other legumes for grain	0.7	1.5	20	50	14	75
Vegetables			38	40		
Fruits			70	75	assume 50% increase in yields	
Forage & other feed crops		3.0	25	135		400
Fallow			167	none		
Total			640	550		

Table 15-5
Syria: Present and Potential Production of Main Crops

Crop	Present yield (mt./ha.)	Potential yield (mt./ha.)	Present crop area (1,000 ha.)	Potential crop area (1,000 ha.)	Present production (1,000 mt.)	Potential production (1,000 mt.)	Remarks
Wheat	0.9	2.8	1,200	2,000	1,049	5,600	530% of present prod.
Barley	0.9	2.6	645	500	589	1,300	220% " " "
Cotton	0.5	1.0	239	300	123	300	240% " " "
Dry legumes	1.1	1.6	193	600	218	960	440% " " "
Vegetables			122	150			125% of present hectarage
Fruits			277	280			no change
Winter legume & fodder crops for animals		3.3		800		2,640	new crops
Other crops			81	370			
Total			2,757	5,000			

Table 15-6
Iraq: Present and Potential Production of Main Crops

Crop	Present yield (mt./ha.)	Potential yield (mt./ha.)	Present crop area (1,000 ha.)	Potential crop area (1,000 ha.)	Present production (1,000 mt.)	Potential production (1,000 mt.)
Barley	0.71		1,169		832	
rainfed		2.7	580	1,000		2,700
irrigated			590	none		none
Wheat	0.48		1,737		826	
rainfed		2.7	1,300	1,000		2,700
irrigated		6.7	400	2,000		13,400
Sorghum after irrigated wheat		3.0		2,000		6,000
Rice	2.2	8.0	141	100	309	800
Cotton	0.25	1.0	39	100	10	100
Sorghum and millet	1.1		15		16	
Summer vegetables			130	200		
Other crops			75	100		
Total			6,176	6,500	2,000	25,600
Total grain					2,000	25,000

as discussed in Chapter 13, the future production possibilities in Iraq presented in Table 15-6 rest on the assumption of only the present estimated water diversions. Among the reasons are: (1) unless and until a treaty is worked out between Iraq, Syria, and Turkey, the availability to each—and most certainly to Iraq—of additional water from the Euphrates River is uncertain; and (2) a low priority on additional irrigated areas in relation to other measures to increase agricultural production. In fact, the assumption underlying this table is that the area to which irrigation is extended will decline from 3.3 million ha. (8 million ac.) to 2.5 million ha. (6 million ac.), but that the area of irrigated land cropped each year would rise substantially due to a minimum of *one* crop each year on *all* irrigated land, and two crops each year—wheat and sorghum—on the irrigated wheat land. Since rice uses a great deal more water than wheat, and uses it in summer when water supply is often limited, we assume the area of rice grown in Iraq would be reduced somewhat below the present level. Since recent successes in increasing rice yields

make it possible to get the rice needed on a smaller acreage, the area devoted to the growing of cotton and vegetables in summer can be increased, to utilize the available summer water supply. It has been assumed that about 7 acre-feet of water per acre (21,000 m^3 per hectare) would be required for these summer crops. For several years, extensive flushing operations would be required to reduce the salt content of the irrigated lands; during those interim years, barley may have to be continued as a major crop on the irrigated lands. After the salt content has been lowered, however, the higher-yielding wheat would replace barley on the irrigated lands. We estimate that 4 acre-feet of water per acre (12,000 m^3 per hectare) would be applied to winter wheat and the following crop of sorghum. This is a heavier rate of application than at present, but it would permit the growing of a crop of sorghum or millet on the same land in the same year—and these crops could provide excellent forage and grain for livestock. On the basis of present irrigation water use, grain output could be increased many times, as Table 15-6 shows; so could the output of cotton and vegetables readily be increased as far as the markets would provide outlets.

In the assumptions that underlie Table 15-6 are included the best possible land-, water-, and crop-management practices, including weed and pest control; fully adequate drainage; lined irrigation canals where necessary and other changes in the engineering features of the irrigation system from its present condition that would allow conveyance for a continuous cropping regimen; good seedbed preparation and good harvesting practices; and all the other assumptions involved in the "no limitations" general assumption underlying this chapter. This, we realize, is assuming a great deal.

On the other hand, if one wishes to assume that additional irrigation water will be developed on the Tigris and its tributaries and on the Euphrates, then the estimated production of Table 15-6 could be increased considerably—possibly even doubled, were all irrigation water development possibilities realized. Where such production would go is open to question. This additional grain would depend on Iraq's ability to find foreign markets, or to find ways and means of greatly raising the domestic consumption of livestock products. In any event, since the stipulated yields could not be achieved without large purchased inputs, the development of markets of one kind or another would be crucial, though its oil income would make it less urgent for Iraq than for some other countries to find foreign customers as sources of foreign exchange. (See also Chapter 18.) Additional water might well go more toward summer crops, including more rice. Corn, cotton, sugar cane, and other summer crops also have real possibilities if summer irrigation water is available in larger quantities.

The options opened up in the preceding paragraph obviously reveal the limits of the "potential" approach. Many crops can "potentially" be raised in terms of output, but which of them will in fact be grown, will depend on the constellation of domestic and foreign markets at some future time. For that reason, we have generally abstained from postulating too radical departures in local output patterns. At the same time, we readily admit that the feasibility of projecting more nearly present output patterns into a distant future also proves little as to advisability. But then it must be recalled that the potential we are discussing and seeking to quantify is, up to this point in any event, more a way of showing capability rather than feasibility other than technical.

The Region's Grain Supplies

In many respects, it is useful to speak of "grain" rather than to consider wheat, barley, rice, maize, and other grains separately. For the most part, all grains now grown in the Middle East are for human rather than for livestock consumption, hence one kind of grain is largely substitutable for another. Even if more grain is produced as a livestock or poultry feed in the future, considerable substitution possibilities will still exist.

In the Middle East, as here defined, at present about 11 million (metric) tons of all kinds of grain are produced; the region is also a net importer of grains—wheat and flour for human use, maize and other grains for poultry, etc.—hence its present consumption of all grains is about 14.5 million (metric) tons, or about a third above regional production (Table 15-7). Were the production potentials we have described in the foregoing pages all to be realized, regional production would rise to about 50 million (metric) tons—or 4½ times current production and 3½ times current consumption. Since twenty to thirty years would be required to reach the potential even under the general assumption of no limiting factors,[3] regional consumption would also greatly rise by that time. It is noteworthy that every country in the region, except Israel, has a potential grain output well in excess of present domestic grain requirements, though not above end-of-the-century requirements. On the latter score, population is likely to double before the end of the century, since it takes an annual growth rate of only 2.4 percent for a doubling in thirty years, a rate now equaled or exceeded in most of the area. In addition, incomes are likely to rise. The combined effect would be more than a doubling of consumption, but probably a great deal less than needed to absorb the production of 50 million tons—which is 3½ times present consumption, but on a regional basis.

Table 15-7 takes a very rough stab at the quantities involved. Consumption by the end of the century is assumed to rise by 250 percent, except in Israel and Lebanon, where rising incomes will lead to little if any increase in per capita consumption. The figures are grossly rounded to avoid any impression of precision. Moreover, since the base consumption figures do include grains used for feeding, it is implicitly assumed that grain consumption for livestock feeding will rise *pari passu* with human consumption. Nonetheless, the results are revealing in that, on these assumptions with regard to both production and consumption, Iraq emerges as the overwhelming surplus grain producer of the region, followed at a great distance by Syria. Jordan would be in equilibrium, and all the other countries in a deficit position. More than proportionate increases in grain fed to livestock would reduce the surplus of Iraq and Syria and increase the other countries' deficits.

Livestock Production Potential of the Middle East

At the beginning of Chapter 9, we pointed out that a great deal less is known about livestock production than about crop production in the Middle East. We also pointed out how crop production is much more important, in terms of income, than are livestock; and how livestock operations are generally separated from crop production. If our knowledge of present livestock production is severely limited—and it is—then our knowledge of livestock production potentials is even more severely limited. There is every reason to believe that livestock output could be increased—

---

[3] Assuming that time remains a constraint that can be relieved by capital infusions, but not beyond a certain point.

Table 15-7
Grain Production Potential, Regional Summary[a]
(1,000 mt.)

	Present		Potential		Potential production compared with:	
	Production	Consumption	Production	Consumption	Present consumption	Potential consumption
Egypt	6,840	8,600	16,800	21,500	+8,200	-4,700
Israel	275	940	575	1,900	-365	-1,325
Jordan	128	267	725	700	+458	+25
Lebanon	95	487	592	1,100	+105	-508
Syria	1,640	2,350	6,900	5,800	+4,500	+1,100
Iraq	1,970	2,000	24,000	5,000	+22,000	+19,000
Region (rounded)	10,950	14,650	49,600	36,000	+34,900	+13,600

[a] Should Egyptian acreage reach Government goals (see Table 15-1 and associated text discussion), then grain potential would rise from 16,800 to 19,050 for Egypt and from 49,600 to 51,850 for the region. The deficit for Egypt would be reduced from 4,700 to 2,450, and the regional excess rise from 13,600 to 15,850.

Table 15-8
Livestock Production Potential

	Present situation			Potential for future		
	Numbers (million)	Rate per head (kg.)	Total production (1,000 mt.)	Numbers (million)	Rate per head (kg.)	Total production (1,000 mt.)
Meat						
Beef, veal & buffalo	6	43	254	6	100	600
Mutton, lamb & goat	24	6	147	30	15	450
Pork			4			5
Subtotal, animals			405			1,055
Poultry			181			500
Other, including edible offal			26			50
Total			612			1,605
Milk, all kinds			2,397			(see text)

Source: Based on data in Chapter 9.

and intuitively one feels that very great increases are possible—but solid evidence is hard to come by.

Only one thing is certain. If livestock output in the Middle East is to rise, a number of things must occur: animals must be better fed, in quantity and balance of rations; they must be reasonably disease-free; present strains of livestock, while perhaps well suited to survive and to reproduce under the difficult conditions under which they must live, generally lack the capacity to make efficient use of better feed supply and must be upgraded; livestock products must be marketable through efficient channels, and at prices attractive to producers; and so on. Above all, present attitudes and institutional arrangements, which in most cases are deeply imbedded because of many generations of experience, must be modified greatly, as has been pointed out in the discussion of grazing land management in Chapter 14.

But the productive potential of livestock in the Middle East depends not only on livestock management practices, but also upon land management and associated cropping practices. In this section, we assume not only that a con-

siderable part of the productive potential of the land is realized, but more particularly, that the additional forage and grains produced on farms, and available for livestock feed, will somehow actually be provided to the livestock— that the present almost complete divorcement between livestock and crop production will somehow be ended.

As more grass and forage for livestock become available, a natural tendency will be to increase livestock numbers. In general, we feel that this would be a mistake, at least until the present livestock numbers had been brought up to a high level of productivity. A large proportion of all feed consumed by animals now merely maintains them, leaving little for their growth. A much more efficient system of livestock production would be a much higher level of nutrition, and a much more intensive operation generally, with present numbers of livestock—but with animals of much better breeding.

On the basic assumption then that livestock numbers would remain relatively constant, but that output per head would be increased greatly, total meat production in the Middle East would rise greatly (Table 15-8). We assume that meat production per head (of all animals in the herds and flocks) would rise from 42 kg. for cattle and buffalo at present to 100 kg. (in the United States today it is over 140 kg.); for sheep and goats, we assume it would rise from 5 kg. today to 15 kg. (in the United States today it is 22 kg.). No change in pork production, insignificant in that part of the world in any event, is suggested. Although we have estimated in Table 15-8 that poultry meat production would rise nearly three times, there is no physical limitation that restricts it to this output level. If the region's grain production estimated as possible earlier in this chapter were actually realized, poultry output could be several times higher than here suggested.

The estimate that total meat production would rise by about 250 percent is based upon holding livestock numbers constant and greatly increasing output per head; more probably, numbers would increase somewhat and output per head would increase less than we have estimated, and this might lead to a somewhat smaller total output.

Table 15-8 shows present milk production from cows, buffalo, sheep, and goats as one total. While there is no entry in the "potential" column, it is clear that from the viewpoint of physical limitations on output, this production could increase very greatly—severalfold, if it were desired to do so. But a much higher consumption of milk would raise a number of problems we are not prepared to trace through in the context of this study. Many groups in the Middle East today do not have a strong tradition of milk consumption, especially by adults; and changing their food habits might be difficult, and perhaps not even advisable—for questions may well be raised about increased milk consumption from a dietary viewpoint as well as from one of sound production economics. Apart from the likelihood that unless production, marketing, and consumption situations change greatly, milk in the Middle East is as likely to be a source of food infection as of improved nutrition, it is a fact that there are other ways in which the nutritional needs of the people might better be met.[4]

In this section, as in the preceding one on crops, we are postponing the question of market outlets for the potential production until Chapter 18. But we might point out here that in most of the world—and we believe the Middle East is no exception—meat and milk consumption per head is quite responsive to average real income per capita. If the countries of the Middle East succeed in developing their economies and restraining their population growth, so that per capita incomes double, treble, or more over the next generation, then certainly one way in which some of this increased income will be used is in the purchase of more, and better quality, meat.

---

[4]It might well be argued that we are here abandoning strict observation of the "no restraints" approach, for it is no more daring to assume the existence of appropriate milk-handling facilities than that of many other requisites of growth. However, the climatic conditions combined with the absence of a milk-consumption habit have made it seem preferable to "go easy" on the milk potential. In all honesty we must, nonetheless, admit that the matter should not end there.

Gross Agricultural Output, If Production Potentials Are Realized

By way of price assumptions it is possible to convert all the foregoing estimates of agricultural production potentials to value terms and arrive at an estimate of gross agricultural output. Since the various commodities have not been estimated to increase at the same rate, this conversion combines their various rates into a single overall measure or index.

On this basis then, the potential output of Middle Eastern agriculture is 2.63 times the present output. No great significance can be attached to this precise figure, except that we can say, with some confidence, that total agricultural output can be more than doubled in a generation, if all of the necessary assumptions are met—which is a tall order, indeed. The relative increase differs somewhat by commodities, but a general increase for all major commodities is evident. It is doubtful, however, if much reliance can be put on the differences in rates of increase among commodity groups; for instance, although the increase estimated is less than a doubling for fruits and vegetables, in considerable degree the size of this increase measures possible market opportunity only—for surely the increase in vegetables could be far greater than this.

It may sound quite encouraging to say that agricultural output in the Middle East may be more than doubled over the next generation, even under the heroic assumptions made. But we pointed out in Table 10-5 that the value added per worker in agriculture in the regions as a whole is now $358; doubling this would be a great achievement, and

Table 15-9

Gross Income Potential of Middle Eastern Agriculture* (million $)

	Present[a]	Potential[b]
Cereals, cotton, other field crops	1,642	4,500
Vegetables	385	900
Fruits	352	900
Other crops	15	40
Milk	328	900
Eggs	91	250
Meat of all kinds, including poultry	568	1,500
Other livestock products and miscellaneous	84	150
Total	3,465	9,140

[a] Taken from Table 10-1.

[b] For the grains and livestock relying on grazing of forage, these represent physical potentials at present prices; for vegetables, fruits, milk, and eggs, where output could be increased greatly by acreage diversions or by use of imported foods, respectively, these represent physical potential increases per acre or per head, plus modest increases in acreage or numbers of livestock which are in turn approximately related to market outlets, in each case at present prices.

*See text for numerous underlying assumptions.

would surely improve the welfare of rural people as well as contribute greatly to national income. But it would still leave value added per worker at about $750—hardly an amount which anyone can consider satisfactory for the long run. Moreover, this assumes that the number of agricultural workers will remain constant, even though agricultural output doubles. If physical output per worker in Middle Eastern agriculture should remain as low as it now is, or should be no more than doubled by reason of increased production, rural incomes will still be low. If the increased output could be produced with less labor input, then value added per worker would rise, but so might unemployment in the country concerned. And, as we shall see in Chapter 16, the chances of a stable rural labor force riding the crest of a rapidly rising nonfarm demand for labor are exceedingly slim.[5]

____

[5] In this perspective, the choice of prices for computing the value of gross output becomes less important. One can argue at length over the appropriateness of such a broad and, on its face, simplistic assumption as constant prices, the selection made here. But the alternatives vastly exceeded the authors' research framework. The search for "reasonable" price levels three decades from now would simply involve a new study that extends to the future of world markets for major farm products. Table 15-9 thus merely sets a base. It shows, for example, that if prices across the board were to decline by, say, 20%, aggregate value would only double, and per capita value remain constant for a doubled farm population. Even such speculations, it has seemed to us, are instructive enough to excuse the luxury of simple assumptions.

CHAPTER 16

Employment Implications of Reaching the Agricultural Potential

FOR several reasons, particular importance attaches to the employment possibilities in agriculture in any lower income country, including countries of the Middle East.

(1) A great many persons in such countries depend upon agriculture now as their source of livelihood. The Middle East (except Israel) is about half agricultural, as far as employment is concerned (Chapter 6).

(2) Agriculture in such countries is a way of life as well as a means of earning a livelihood. That is, the people engaged in it, and to a lesser degree the entire populace, regard agriculture in a very special light, with special virtues in addition to its capacity to produce a living for its workers. Many statements, some strongly emotional in tone, have been made in support of this special role for agriculture.

(3) Farming as an occupation and land as a form of capital are often the best possible economic security for the mass of people in such countries. Poor people in such countries seek to obtain farms and farm employment as the most practical way, or at least the way they know best, to improve their lot in life.

(4) The rate of natural increase in population among the present farming populations of the countries of this region is high. Health programs since World War II, among other less easily identifiable factors, have greatly lowered death rates, especially in children, while birth rates have stayed relatively unchanged. Rural areas now, and for the foreseeable future, will produce a substantial annual excess of births over deaths; as these children reach maturity, they will seek employment.

(5) It is most improbable that urban employment can expand fast enough to provide jobs for the growing numbers of workers from both urban and rural areas. Even if the rate of urban employment grows at a relatively rapid rate, urban employment is too small to expand fast enough to absorb all the increase in the labor force—and still more, too small if employment in agriculture should fail to expand or even decline.

(6) As pointed out in the previous three chapters, much of the progress in agricultural production involves a substantial degree of mechanization. While such mechanization is production-improvement and not labor-saving oriented, the latter is likely to be a by-product, whether wanted or not, and if not right away, then in the long run.

Furthermore, in considering the employment outlook for agriculture in the Middle East, it is helpful to distinguish two situations: (1) Farming—or the operation of a farm as owner or tenant, or working for someone else who is a farm operator—in order to grow crops and produce livestock products; and (2) Agribusiness employment—or employment in the production and distribution of farm inputs such as fertilizer; in the provision of specialized services related to agriculture, such as tractor repair and maintenance; and in marketing and processing the agricultural output.

In countries with a technologically advanced agriculture, the trend has been for farmers to become increasingly specialized, in the sense that they buy more and more goods and services as inputs into the production process on the farm. Conversely, the range of activities performed by the farmer narrows. One measure of this is the relation of the cost of goods and services, as inputs into agriculture, to the value of the output. The great contrast between the United States and the Arab countries of the Middle East was noted in Chapter 10: whereas the cost of inputs to the farmer in the United States is close to 70 percent of the value of the gross output, in the Arab countries of the Middle East it is mostly less than 25 percent. An implication not then pointed out is that while the inputs cost the farmer a substantial share of his gross income, they do provide significant employment elsewhere.

In this chapter then, we are asking: if the productive potential—defined as before—of Middle Eastern agriculture is realized, what is likely to happen to employment in agriculture, in both farming and in agribusiness? However, before seeking to answer this question, it is important to look quickly at population trends as a whole, for they, after all, will provide the setting, in which the agricultural labor force is only one of several aspects.

With due caution as to their accuracy in reflecting reality, annual growth rates are currently believed to be as follows:

Egypt	2.8%
Israel	3.4%
Lebanon	2.5% dropping to 2.3% after 1980
Jordan	3.0%
Syria	2.9–3.0%
Iraq	2.9%

Source: FAO, Near East Commission on Agricultural Statistics, Fourth Session, 1968. For Israel and Iraq, *United Nations Statistical Yearbook,* 1967.

Thus, considering the relative size of the various countries, the joint growth rate per year is somewhere around 2.9 percent. At this rate, the population would double every twenty-four years. We have neither the information nor the capacity to guess at the direction in which the rate will move in the next decade or two. But even if it should begin to decline—and this is not only dependent on future birth rates, but also on mortality rates that still have a long way to go down in that part of the world—the decline will hardly be rapid. Thus, population pressure on employment opportunities will remain a potent force.

Agricultural Technology and Farm Employment

Available data suggest, with few exceptions, that every country which has had a rapid and sustained increase in agricultural output in the past generation or two has decreased the farm labor input per unit of agricultural output. It may indeed be argued that this is one way in which one can define the term "agricultural revolution," or that it is at least a necessary ingredient that labor input per unit of output decline significantly and continually. Increased crop yield per acre or increased output per head of livestock, reduces labor input per unit of product; for much of the labor input is a function of crop area or of livestock numbers, irrespective of their yield—a kind of overhead labor. For instance, the time required to plow a field is not influenced significantly by the size of the later crop harvest; even harvest labor probably increases less than proportionately to the increase in volume harvested. This is true, even when the methods of crop and livestock production remain essentially unchanged except that yield is increased.

In fact, however, methods of crop and livestock production have not remained unchanged. The methods commonly employed to produce the sustained and rapid increase in agricultural output which one could call an agricultural revolution have at the same time greatly reduced farm labor input *in total*. In the United States, for instance, while gross output has nearly doubled since 1940, total labor input into farming declined by nearly two-thirds. These two changes are deeply related. The same measures which were employed to increase output—and which would probably have been necessary, even if a deliberate effort had been made to avoid them—also reduced total labor input continuously and sharply. We see no reason to think the experience will be different in the Middle East.[1]

Farm mechanization, especially the substitution of tractor for animal power in field operations, is a good example. This process is often discussed—especially in literature dealing with agriculture in technologically not advanced agriculture countries—as if its chief purpose was laborsaving. Tractors will require a lot less human labor to plow a field than would use of bullocks, and in this sense are indeed laborsaving. More importantly, if properly used

and if supplemented by all the other parts of the package needed for high crop yields, the tractors will save a great deal more labor by increasing output per acre. By increasing output per acre, they are also land- and water-saving. The kind of moisture conservation that is possible by tractors and proper machines on rainfed lands can be the basis for a large increase in wheat yields, if supplemented by other parts of the package, such as better varieties and more fertilizer. And, also importantly, the land devoted to raising feed for draft animals becomes available for higher-value uses. In the absence of the tractor and its associated machines, however, the rest of the package may produce little more than do present methods of operation.

This sort of example could be multiplied greatly. Mechanization, and other improved practices, might be judged primarily in terms of their effect upon agricultural output, and secondarily in terms of their effect upon labor input—but the result would often be closely similar to that which would result if they were looked upon primarily as a means of reducing labor input.

Since labor productivity is simply a measure of the relationship between product output and manpower input, it follows that it will rise steeply under the conditions depicted, and, as a corollary, that to produce a *given volume* of output will require a smaller amount of labor.

To the extent that the productive potentials of Middle Eastern agriculture (outlined in Chapters 13, 14, and 15) are realized in practice, actual labor input on farms will almost surely decline. The very forces that are involved in the process of achieving the potentials will result in less labor input. If there were large additional areas of land suitable for irrigation in the Middle East, by traditional or present methods, or if there were large additional areas of rainfed land that could be brought into cultivation, the crop areas could be increased and in this manner total output increased, with labor input increased more or less proportionately.[2] But as we have tried to demonstrate, this is not the case. On the contrary, under proper management, land used for crops (both rainfed and irrigated) is likely to shrink in total extent. Crop acreage harvested will rise, it is true, as fallowing is reduced or eliminated and as more than one crop is grown in the same year on irrigated land. But this latter can generally be accomplished by use of present workers during a longer working season; i.e., there will be little demand for additional workers, though each worker is likely to be more fully employed.

In Chapter 15 we noted that the value of gross agricultural output could increase about two-and-one-half fold over the next generation or so; but even this increase would not raise farm family incomes to a very acceptable long-term level, and would raise them very little, if farm population doubled in the next twenty-five years. If, on the other hand, as part of the modernization process, labor input were halved, then returns per worker could be twice the two-and-one-half fold increase. This amount of increase would begin to approach a more acceptable level.

---

[1] There is some evidence that in the early stages of the agricultural revolution in Asia and other developing countries—referred to often as the "green revolution"—this tendency is hidden by the need for intensification of all stages of production required in order to exploit the genetic potential of the new varieties as well as by the greatly enlarged opportunities for multiple cropping. But it is not likely that, even if this is true initially, the obverse tendency toward a decline in labor per unit of output will not reassert itself in the long run.

[2] Though it is likely that the opening up of new land involves especially new methods from the outset and thus is less labor-intensive than the cultivation of land already in use.

In view of the uncertainty in the rate at which agricultural output will in fact rise, little purpose is served by making specific projections as to labor input at selected future dates. Also, various public programs might seek to retard the displacement of farm labor without inhibiting the increase in agricultural output. Though the tendency toward reduced labor input is strong, as output increases under that combination of measures necessary to increase output, yet there might be some ways in which labor displacement could be minimized. One of these is a deliberate effort to favor labor-intensive crops that yield the highest value with the highest possible labor input. Fresh produce of high enough quality to be exported is generally in this category. But the question of market outlets and of marketing arrangements, discussed in Chapters 11 and 18, would appear to put rather narrow limits on such opportunities, in terms of land and labor so employable. In any event, our concern is more with the force and the trend than with specific numbers.

Because Iraq has substantial volumes of unused water and has large areas of land irrigated only in alternate years or even more randomly, some people have felt that Iraq could provide rural settlement opportunities for large additional numbers of people. This might be possible, but it would be difficult. The utilization of such land and water for greatly increased output would largely mean more intensive use of present—or indeed, of reduced—land areas; there would be little opportunity to settle large additional numbers of people and at the same time realize the agricultural progress we postulate as possible. Iraq's land reform program has experienced severe difficulties, in part because tracts are now smaller while the old methods of crop production persist. The kind of technologically more advanced agriculture we describe for Iraq has little if any place for more farmers, though, as cropping patterns have changed, present farmers might be productively employed more days during the year. Effecting any significant increase in rural population density would represent a deliberate trade-off between increasing agricultural labor productivity—and probably productivity as related to *all* inputs—and the cost of maintaining the potential settler in another segment of the economy. While this might recommend itself under some conditions, it does not form part of the package for achieving the country's agricultural potential, but would rather detract from it.

Employment in Agribusiness

Although farm employment is likely to decrease as more advanced technology is applied to agriculture, employment in agribusiness, which in national income accounting and other statistics will mostly show up in categories other than agriculture, will increase, but probably not enough to offset the decrease in farm employment.

In Chapter 17 we shall present some estimates of farm machinery, fertilizer, and other chemical inputs that will be required if the production potential is to be realized. Substantial increases in these inputs will be needed—larger,

proportionately, than the increases in output because so little input is, as a rule, used now. That is, at present a considerable part of the crop output depends in large part upon fertility in the soil and receives no fertilizer; the increased outputs, however, will require fertilizer in sufficient quantity to provide the essential nutrients needed for the increased output, and so with other inputs. All of these increased inputs will require some labor, somewhere, for their production. However, the employment effect in the Middle Eastern countries will almost certainly be less than the loss of employment in farming, mainly because these added inputs are efficient, in the sense that less labor is required to produce them than is saved by their application in farming. Generally speaking, specialization increases efficiency and less labor will be employed under new than under old methods of agriculture, adding the direct farm labor to that expended for the production of inputs. In this connection, our earlier suggestions about the possibilities for developing agricultural supply industries on a regional scale should be recalled. The Middle East, on a regional basis, might well develop factories to produce the tractors—or some parts of the tractors, the other agricultural machines, the fertilizers, the other chemicals, and all the other inputs needed, for the farmers of the Middle East. If done on a regional scale, advantage could be taken of any expertise in any country, and the volume would probably be large enough to be economical. But from an employment point of view, the effect would be even more adverse to greater employment opportunities.[3]

Agribusiness also includes handling the farm output of agricultural commodities. The volume of output depicted as the potential for the region will require greatly increased handling. This increase will be steep, since a much larger share of the additional production must be marketed than of the current output. Thus, a farmer who has marketed 10 percent of his production might come close to doubling his marketing, if he increases his output by 10 percent, provided he retains but little of the new output for himself. And even if he retains a significant share, the relative rise in the amount marketed will still be much greater than in the amount produced. At the same time, the amount of processing per unit of product is also likely to rise (though this depends to a considerable extent on a concomitant rise in per capita income). Certainly, that has been the experience in the United States and in other high-income countries. For a long time to come, one would not expect the farm commodities of, say, Egypt to be processed to the same degree as they are in the United States, nor perhaps for so many services to be provided along with the food commodities, to save the time and work of the housewife; but some changes in this direction might well occur. In many technologically less advanced countries of the world, a substantial quantity of food is wasted annually through losses in storage and in marketing. To the extent that such

---

[3] If tractors were largely imported, the offsetting nonfarm employment would be in those parts of the economy that produced the foreign exchange to pay for the tractors. But to trace this flow leads beyond the scope of this study.

losses can be reduced, more labor (and capital and management) used in the marketing process can be quite productive from a national standpoint.

It would be instructive to quantify the magnitude of employment in agribusiness and farming, now, on a country-by-country basis, and then speculate on how that relationship might change. But data do not even permit the first step, let alone the second. One is thus limited to hypothesizing that although the gains in employment in agribusiness activities would be unlikely to offset the associated losses in farm equipment, they would go some way toward mitigating them, especially as the agribusiness employment would probably be better paid than farming. On the whole, more training and more skills are required for agribusiness jobs, and wages are generally above those in farming. This would not change for quite some time, if ever.

National Policy Issues

The employment-decreasing effects of adoption of an advanced agricultural technology pose several serious national policy issues:

(1) What is the optimum trade-off for a country to choose, between advancement of agricultural production and diminution of agricultural employment? Or, to put the same issue differently, what should be the trade-off between efficiency and welfare? There exist many situations in the Middle East in which it would be possible to reorganize the land into many fewer farm units, develop much larger farms, mechanize them, and operate much more intensively, with the package of technologically advanced practices we have described repeatedly, and as a result materially increase total agricultural output in a relatively few years. But this process, in addition to requiring a managerial competence that is often scarce and not quickly producible, would displace many people from their present farms; and it is difficult to see where else in the economy these people could be absorbed, even if they were willing to be relocated. There is every reason to believe that the countries of the Middle East will increasingly be confronted with this situation.

Land reform poses this problem sharply. Most large farms and estates in many countries, prior to land reform, were both technically backward and resulted in a high disparity in income between landowner and agricultural worker. But it may be questioned if, in technical terms, new small farms are a satisfactory answer; certainly they pose serious problems of efficient farming even today and more seriously in the future. This is not to quarrel with the necessity of breaking up the big estates of Egypt, for instance, and of destroying the political stranglehold such estate owners had on the country; but one must raise the issue of new one-hectare farms as a serious obstacle to increased agricultural output. The extensive use of cooperatives in Egypt and other countries, especially for land reform areas, has been one measure to overcome this

problem. The countries may have done as much as anyone could do, yet the problem still remains.

(2) How far should a country seek to promote agricultural development, and how far should it seek to promote industrial and other nonfarm employment and output? An easy, but misleading, answer might be, to push both. But capital for investment, and even more, for managerial talent—whether public or private—is limited in every country. How much of it should be channeled toward agriculture and how much elsewhere? With given amounts of investment capital and given amounts of managerial competence, where would national income be stimulated most and rise fastest? Who would bear the cost of accelerated development in his economic sector, and how would the gains from greater output and efficiency be distributed? Are the long-run or secondary costs and benefits the same as, or different from, the immediate ones? These are some of the difficult issues of national policy, which the potential of advanced agricultural technology tends to raise or at least to sharpen.

(3) What can a country do with the surplus of new young workers arising in the farming families, with their annual surplus of births over deaths? As was stressed early in this chapter, the population trends in every Middle Eastern country produce such an annual surplus of workers over available jobs in the country as a whole and, undoubtedly, in the rural sector, and will continue to do so until population-control measures halt the trend. Should a country try to absorb as many of these new workers on its farms as possible? If so, how? As earlier parts of this chapter have stressed, such an adjustment is contrary to the direction that agricultural development is taking and will take. Should a country try to employ this rural surplus labor force in its cities? If so, at what? Although industrial and other urban employment has been rising rapidly in some countries, it may be extremely difficult to absorb the rural surplus labor force. In meaningful terms, some of the countries may have little farm surplus labor today in the sense that much could be removed without affecting the operation of agriculture as presently organized (Chapter 6). But whatever the magnitude of any current surplus, a substantial one is almost bound to develop in the next several under the twin impact of a continued increase in population and a declining need for labor as agricultural technology advances.

The countries of the Middle East are now in considerably differing situations, as far as nonfarm vs. farm employment of labor is concerned. As we saw in Chapter 6, agricultural employment in Israel is already down to about 12 percent; young men and women, growing up on farms in Israel, have been flocking to the cities for jobs, much as they have in the United States. With the present larger size of the labor force in nonagricultural employment in Israel, further reductions in the agricultural labor force that advancing agricultural technology may create can be absorbed therefore without major difficulty. The situation in Lebanon is not the same as in Israel, but is nearer that of Israel than is that of any other Arab country. Beirut has grown enormously since World War II and has provided a large number of jobs; while there are many problems in its further growth and continued prosperity, yet the nonfarm employment prospects here—especially in real rather than artificially inflated service industries—are relatively better than in most countries. Egypt, on the other hand, with its very large rural population, its very small farms, and its high rate of population increase, faces an extremely difficult job of absorbing its growing labor force in urban employment. Processing industries, perhaps located in rural areas, would act as a buffer, but the product mix would have to change substantially toward products that lend themselves to processing (other than cotton) in order to make this a quantitatively significant way out. Servicing of new inputs (tractor, pump, and engine repair and maintenance stations, fertilizer distribution points, etc.) is another potential employer; but one would have to know much more of the relative magnitudes to appraise the net balance. Some thorough research in that direction would be a valuable contribution.

We conclude this chapter by pointing out that resources and technology provide the basis for a great potential in Middle Eastern agriculture. But the achievement of this potential poses a very sobering problem of truly excess manpower—the more so in that the requirements for achieving greater production themselves will depress the size of the needed labor force.

CHAPTER 17

Inputs to Reach the Output Potential

IN this chapter we have estimated the inputs that must be used by Middle Eastern farmers to achieve the production potential postulated in Chapter 15. These estimates we believe are reasonable but they are not, of course, precise. The uncertainties include, for example, our lack of knowledge of the potassium requirements of soils for sustained high levels of production of various crops in Middle Eastern areas where the soil potassium supply is adequate for present production levels and where no appreciable potassium fertilizers have ever been used. Yet we know from the experience gained on similar soils in other parts of the world that potassium fertilizers will be required if the high levels of production postulated are to be reached and maintained over long periods of time. All of the nutritious food-and-feed crops of the Middle East remove a great deal of mineral elements from the soil, therefore a high level of mineral nutriments in the soil must be maintained if sustained high production is to be achieved. Another uncertainty is the costs that we have assumed for inputs. It is true that demand of the size we have postulated for the Middle East would far exceed the input supplies now available for sale there. But as and if the need develops, these purchase inputs are most likely to become available either by development of sources within or without the region. While large, they would still be only a small fraction of swiftly expanding world supplies. Whether the prices assumed are reasonable depends so much on matters extraneous to the Middle East that we must plead extenuating circumstances and justify them as leading at least to rough orders of magnitude.

Fertilizer Needs

As pointed out earlier, for each bushel of wheat produced (60 lb. or 27 kg.), approximately 2.7 lb. or 1.22 kg. of nitrogen must be available to the plant root system. All of this becomes a part of the wheat plant, but all is not removed in the grain. (Approximately 1.5 lb. of nitrogen are found in the protein of a bushel of wheat, and approximately 1.2 lb. of nitrogen in the plant roots and in the wheat straw.) In estimating the nitrogen needed to reach the production potential of the Middle East, we have provided about two lb. of nitrogen for each bushel of wheat produced. This allows for removal of the wheat and perhaps most or all the straw. In addition, the roots of the wheat plants will decompose by microbial action and make a portion of the contained nitrogen available for producing subsequent crops. In appraising the total amounts assumed we should recall that most of the cultivated soils in the Middle Eastern area have been in cultivation for a long time and the hot summer temperatures accelerate decomposition of soil organic matter. Regardless of the cause, we know

that the native nitrogen in the soils of the area is very low indeed. Hence the nitrogen needed for producing large sustained yields must be added to the soils as fertilizers. The exception to this is the small amount, perhaps a few pounds only, of fixed nitrogen that is added to the soil each year in the precipitation and that amount fixed by legumes. Perhaps, therefore, the needed amounts stipulated are a little on the conservative side.

In estimating the needs for phosphorus and potassium we may also have been conservative. Most if not all of the soils producing grain have high p^H values (low acidity), and this reduces phosphorus availability. Thus it is necessary to add more phosphorus than the plant will use. Over long periods of time the phosphorus content of the soil will build up so that eventually the need for this element may decrease somewhat. However, this would be a long time ahead. For the potassium needs, therefore, we have estimated that the K_2O would equal in weight the P_2O_5 added per unit of area and that both of these would be required in one-half the amount of the nitrogen.

We have previously remarked on the good supply of available potassium in the soils of the Middle Eastern area. However, it has been found that where crops are grown, the very high levels of nitrogen required to produce a high yield of grain also require high levels of available potassium in the soil. It may be that we have overestimated the potassium needs particularly for the near term. But we doubt this. Though admittedly we have little evidence to back our projected potassium needs, particularly for grains, we believe that the amounts postulated are not far from what would be needed in the long run.

So far we have talked about only one grain—wheat. Most of the other grains will have similar requirements. We have allowed somewhat higher levels of phosphorus and potassium for rice. The allowances for cotton have been based on U. S. experience in California and Arizona areas where soil conditions are not too different from those found in the Middle Eastern area. Allowances for vegetables and for fodder crops have been in accord with our best judgment from experience elsewhere.

Perhaps the recommendation based on least evidence is that for fruits. But here again we have the experience of growers in the Southwestern United States. Our recommendations are similar to best practices there. For olives, acreage or which is included in fruit acreage, recent experiments in the Middle East indicate responses to complete fertilizers. Since improvement in the olive production in the Middle Eastern area is badly needed, we believe that fertilizer usage should be one item in a package of practices necessary for accomplishing the increase in production.

The fertilizer prices that we have shown in each table

Table 17-1

Egypt: Fertilizer Requirements and Costs per Year of Development Potential, by Crops and Nutrients

I. Requirements

Crop	Metric Tons		
	N	P_2O_5	K_2O
Cotton	120,000	60,000	60,000
Wheat	180,000	90,000	90,000
Corn	146,000	73,000	73,000
Rice	220,000	150,000	150,000
Millet & Sorghum	46,700	25,000	25,000
Sugar Cane	50,000	35,000	35,000
Berseem Clover	None	100,000	100,000
Fodder and other crops	25,000	50,000	50,000
Total	787,700	583,000	583,000

II. Costs

	Price (U.S. Cents per kg.)		Present Production		Potential Production	
	Present	Potential	Requirements (thousand metric tons)	Cost (million U.S. $)	Requirements (thousand metric tons)	Cost (million U.S. $)
N	31	25	250	78	788	197
P_2O_5	18		55	10	583	105
K_2O	11		1	negligible	583	65
Total			306	88	1,954	367

are approximately those shown in Table 5-7. Mostly these are the prices at the farm. Hence the cost to the country—in foreign exchange—is lessened by the cost of hauling and profits along the chain of distribution. Costs could of course be lower when the fertilizers are bought in much larger quantities and perhaps profit margins could be cut by buying through farmers' cooperatives. It should be noted that the highest prices were in Jordan and Syria. To some extent these may reflect transport costs inland from Beirut across two mountain ranges over inadequate roads. Needless to say production of the amounts of grain we have postulated would require better transport and storage facilities for fertilizer than are now available.

How to utilize present fertilizer prices—disregarding whatever inaccuracies may affect these statistics—to estimate outlays far in the future raises serious problems. There is, to begin with, the likelihood of lower fertilizer production cost around the world, based on improved technology. Advances in ammonia production and attendant price declines are a recent example. There is the effect of scale of production and marketing generally as a price-reducing factor. There is the likely reduction in losses and waste when marketing attains a greater importance. But there are also substantial investments to be made, as near-costless, primitive practices are modernized. And present depressed price levels in the industry are not likely to persist for very long, thus presaging an upswing in prices.

If it is difficult to evaluate how all this affects present fertilizer prices in Middle Eastern countries, it is many times as difficult to make predictions about the future. On balance, it is more likely, however, prices will decline rather than rise or remain constant. This applies above all to nitrogen. As a nod in that direction we have therefore (for each country), chosen a somewhat lower price of nitrogen, to the farmer, than appears currently to prevail. If this is still too high, it will help balance any conservatism in our estimates of requirements.

Farm Machinery Requirements

Of the input needed to sustain high levels of agricultural production, adequate farm power and implements are an important part. There are many reasons for this. One of the important prerequisites for high crop yields is a good seedbed; but this cannot be prepared, especially in fine-textured soils, without adequate power for plowing, harrowing, smoothing, and pulverizing the soil. These operations cannot all be adequately done by hand labor or by animal power in a short period of time. Yet if the land is to grow two or three crops a year, a possibility opened up especially by the new high-yielding grain varieties with substantially shorter growing seasons and less photosensitivity,[1] any delay in soil preparation is lost time for growing crops and lost income. Timeliness of preparation may be equally important, even when only one crop a year is grown. This is particularly true in the rainfed wheat areas, where land preparation by man or animal

____

[1] The new IR-8 rice, for example, matures in 120–125 days, as against 180 days in the case of traditional, local varieties.

Table 17-2

Israel: Fertilizer Requirements and Costs per Year of Development Potential, by Crops and Nutrients

I. Requirements

| | Metric Tons | | |
Crop	N	P_2O_5	K_2O
Cotton	3,310	1,660	1,660
Wheat	12,000	6,000	4,000
Barley	2,340	1,200	600
Sorghum	2,910	1,400	700
Hay	1,400	1,000	500
Green fodder	3,200	3,000	2,000
Beets	1,500	1,000	1,500
Vegetables	5,800	5,800	5,800
Maize	5,000	4,000	4,000
Fruit	12,000	8,800	8,800
Totals	49,460	33,860	29,560

II. Costs

| | Price (U.S. Cents per kg.) | | Present Production | | Potential Production | |
	Present	Potential	Requirements (thousand metric tons)	Cost (million U.S. $)	Requirements (thousand metric tons)	Cost (million U.S. $)
N	26	22	25	6.4	49	11
P_2O_5		18	10	1.7	34	6
K_2O		11	5	.5	30	3
Total			40	8.6	113	20

Table 17-3

Lebanon: Fertilizer Requirements and Costs per Year of Development Potential, by Crops and Nutrients

I. Requirements

| | Metric Tons | | |
Crop	N	P_2O_5	K_2O
Wheat	7,000	3,500	3,500
Barley	1,250	625	625
Maize	5,000	2,500	2,500
Sorghum	4,500	2,750	2,750
Dry Legumes	1,100	2,200	2,200
Vegetables	8,000	8,000	8,000
Fruits	10,000	7,500	7,500
Industrial Crops	2,600	2,600	2,600
Totals	39,450	29,675	29,675

II. Costs

| | Price (U.S. Cents per kg.) | | Present Production | | Potential Production | |
	Present	Potential	Requirements (thousand metric tons)	Cost (million U.S. $)	Requirements (thousand metric tons)	Cost (million U.S. $)
N	27	22	12	3.0	39	9
P_2O_5		18	6	1.2	30	5
K_2O		12	2	.3	30	4
Total			20	4.5	99	18

Table 17-4

Jordan: Fertilizer Requirements and Costs per Year of Development Potential, by Crops and Nutrients

I. Requirements

	Metric Tons		
Crop	N	P_2O_5	K_2O
Wheat	18,000	9,000	9,000
Barley	3,750	1,800	1,800
Dry Legumes	2,500	5,000	5,000
Vegetables	4,000	4,000	4,000
Fruits	10,000	7,500	7,500
Forage and Feed Crops	13,500	13,500	13,500
Totals	51,750	40,800	40,800

II. Costs

	Price (U.S. Cents per kg.)		Present Production		Potential Production	
	Present	Potential	Requirements (thousand metric tons)	Cost (million U.S. $)	Requirements (thousand metric tons)	Cost (million U.S. $)
N	35	27	2.0	.7	52	14
P_2O_5	24		.8	.2	41	10
K_2O	13		3.0	.4	41	5
Total			5.8	1.3	134	29

Table 17-5

Syria: Fertilizer Requirements and Costs per Year of Development Potential, by Crops and Nutrients

I. Requirements

	Metric Tons		
Crop	N	P_2O_5	K_2O
Cotton	30,000	15,000	15,000
Wheat	168,000	84,000	84,000
Barley	39,000	20,000	20,000
Dry Legumes	30,000	60,000	60,000
Vegetables	15,000	15,000	15,000
Fruits	35,000	28,000	28,000
Winter Legumes and Fodder Crops	50,000	50,000	50,000
Other Crops	15,000	15,000	15,000
Total	382,000	287,000	287,000

II. Costs

	Price (U.S. Cents per kg.)		Present Production		Potential Production	
	Present	Potential	Requirements (thousand metric tons)	Cost (million U.S. $)	Requirements (thousand metric tons)	Cost (million U.S. $)
N	31	25	10.0	2.9	382	96
P_2O_5	25		2.0	.6	287	72
K_2O	21		.4	.1	287	60
Totals			12.4	3.6	956	228

power is nearly impossible prior to the softening action of the first rains. This delay in planting time means lost time for growth of the wheat early in the winter. The result: lower yields.

The payoffs from machinery have been well categorized in a recent article: "Mechanization can ... permit the completion of tasks with more precision, accomplish more work quickly, develop resources not presently utilized, and accomplish tasks not possible with traditional techniques."[2]

---

[2] Lyle P. Schertz, "The Role of Farm Mechanization in Developing Countries," *Foreign Agriculture* (U. S. Dept. of Agriculture), VI, No. 48 (November 25, 1968).

Table 17-6

Iraq: Fertilizer Requirements and Costs per Year of Development Potential, by Crops and Nutrients

I. Requirements

Crop	Metric Tons		
	N	P_2O_5	K_2O
Cotton	20,000	10,000	10,000
Wheat	483,000	240,000	240,000
Barley	80,000	40,000	40,000
Sorghum	180,000	90,000	90,000
Summer Vegetables	40,000	40,000	40,000
Other Crops	15,000	10,000	10,000
Total	818,000	430,000	430,000

II. Costs

	Price (U.S. Cents per kg.)		Present Production		Potential Production	
	Present	Potential	Requirements (thousand metric tons)	Costs (million U.S. $)	Requirements (thousand metric tons)	Cost (million U.S. $)
N	31	25	4.8	1.5	818	205
P_2O_5	18		1.8	.3	430	77
K_2O	11		.2	negligible	430	47
Totals			6.8	1.8	1,678	329

Table 17-7

Summary of Fertilizer Requirements and Costs per Year of Development Potential, by Countries and Nutrients

I. Requirements

Country	Thousands of Metric Tons		
	N	P_2O_5	K_2O
Egypt	788	583	583
Israel	49	34	30
Lebanon	39	30	30
Jordan	52	41	41
Syria	382	287	287
Iraq	818	430	430
Totals	2,128	1,405	1,401

II. Costs

	(million U.S. $)	
	Present	Potential
Nitrogen	92	532
Phosphorus	15	275
Potassium	1.3	184
Total	108	991

Tractors

The most important single item of machinery for the small and medium-sized farm in the Middle East is the tractor, specifically one that has sufficient power to pull the implement of heaviest draft. This implement is usually the plow or chisels used as the initial step in seedbed preparation. Small, two-wheeled tractors are usually not adequate in soils of heavy draft.

The question of what constitutes adequate minimum tractive power for general farming has been discussed recently by G. W. Giles.[3] He concluded that the minimum power requirement for high-level crop production was in the range of from 0.5 to 0.8 horsepower per cultivated hectare (including animal and manpower). The only country in the Middle East that meets this criterion is Israel, with about 0.85 horsepower per hectare. For the subsequent calculations, we are assuming that an adequate level would be 0.75. We also believe that an area with predominantly fine-textured soils such as the Middle East would be underpowered at 0.5 horsepower per hectare, and that 1.0 horsepower per hectare would represent overpowering, although understandable perhaps in terms of convenience. According to this assumption, all countries other than Israel must be considered underpowered.

In calculating the available power for cultivation of the soil, human and animal power as well as the tractors operating on fossil fuel should be counted. The power from a man or from an animal depends, of course, on many factors, including health and size. Table 17-8 attempts to show the available horsepower for crop production in a number of areas and countries. Of the countries of the Middle East, only Israel and Egypt are included.

---

[3] *The World Food Problem*, A Report of the President's Science Advisory Committee, The White House (Washington, D. C., September 1967), III.

Table 17-8

Available Horsepower per Hectare of Arable Land and Land Under Permanent Crops[1]

	Africa	Asia	India	Latin America	Taiwan	Oceania	U.A.R.	Israel	Europe	U.S.A.	U.K.	Japan	World
Tractor[2]	.03	.02	.008	.18	.080	.33	.181	.815	.81	1.0	1.57	.004	.268
Garden Tractor[3]		.03			.063	.006		.007	.02	.014	.03	2.06	.024
Animal	.01	.10	.145	.07	.063	.011	.065	.010	.08		.02	.088	.044
Human	.01	.05	.056	.02	.113	.001	.12	.016	.02	.002	.09	.148	.024
Total	.05	.20	.21	.27	.27	.35	.37	.85	.93	1.02	1.71	2.30	.36

[1] Calculations based largely on values taken from FAO's 1965 Production Yearbook.

[2] Defined by FAO as wheel and crawler tractors developing over 8 H.P. and used in agriculture.

[3] Defined by FAO as tractors developing under 8 H.P. and/or weighing under 850 kgs and used in agriculture.

From G. W. Giles, "Agricultural Power and Equipment," in *The World Food Problem,* a Report by the President's Science Advisory Committee, The White House, Sept. 1967 (Washington, D.C., September 1967), III, 178.

Table 17-9

Tractor Requirements for Middle East Agriculture at Potential Output Level

					Additions Needed	
Country	Projected cultivated area[a] (millions of ha.)	Horsepower required (millions)	30 H.P. Tractors required	Present Tractors	Number of tractors[b]	Cost[c] (million U.S. $)
Egypt	2.78	2.09	70,000	11,000	59,000	212
Israel	.40	.30	10,000	11,000	none	none
Lebanon	.30	.23	7,700	3,500	4,200	15
Jordan	.55	.41	13,700	2,100	11,600	42
Syria	5.0	3.75	125,000	7,400	117,600	420
Iraq	3.3	2.48	83,000	2,400	80,600	290
Totals	12.3	9.26	309,400	37,400	273,000	979

[a] Note: this is physical area, not sum of cropped areas.

[b] This assumes that present tractors average 30 hp.

[c] At $120 per horsepower.

The most economic tractor size for the Middle East differs according to different soil and terrain situations. Most of the soils are fine textured, with consequent heavy draft. Some have pans that must be loosened for best production. In steep terrain situations a large tractor cannot be used, particularly with vines and tree crops. For the fine-textured soils where cultivated areas are large enough, a 40–45 horsepower tractor will do a much better job. We have used the 25–30 horsepower-range tractor size as a basis of calculations. The fact that this size may be either too large or too small, depending on specific situations, does not appreciably affect our overall calculations of costs per hectare of land calculated. At the present time manufacturing costs would permit sale to the farmer at about $120 per horsepower. On this basis, Table 17-9 shows the need of the Middle Eastern area for additional power as well as the cost that will meet the needs for attaining the production potentials we have deemed possible. The additional power needs are projected on the basis of our latest knowledge of the number of tractors being used. Since the number in use is probably increasing, the additional requirements are likely to be somewhat lower than here calculated.

Machinery Other Than Tractors

While the tractor is usually the largest single item of farm expense, the grain combine, serving more area than the tractor, is a more expensive piece of machinery. In our consideration of the cost of plows, disc harrows, drills, planters, fertilizer distributors, and other equipment, we have assumed that equipment can be used to its full potential—i.e., irrespective of the size and distribution of farms. In the example just cited, one grain combine may harvest the area served by five 30-horsepower tractors. Thus the farm units must be rather large or else large machines must be shared by a number of farms, as is beginning to be done in some countries like Thailand under custom-hiring arrangements. We assume further, in our context of maximum potential production, that some form of sharing will prevail. It should be noted that some items, such as turning plows, match the tractor in numbers. But many do not.

Giles[4] has given study to the needs in Asia, as well as some other areas, for the implements and accessories

[4] G. W. Giles, "Agricultural Power and Equipment," in *The World Food Problem, ibid.*, p. 178.

needed for high-level production agriculture. In spite of the objections to such intercountry analogies, we do not believe that we can do better than follow his calculations. We therefore use his data but eliminate items that seem inappropriate to the area. Little is to be gained from a detailed calculation of the numbers of plows, planters, and drills that might thus be needed. Rather, we consider the total costs of all these items and relate them to the cost of the required tractors. It then turns out that the cost of equipment other than tractors approximates the cost of tractors. With elimination of a few items, the cost of tractors in Giles' table represents 56 percent of total equipment cost. The major item missing in these calculations is transport equipment such as trucks for hauling grain and other produce. The net outcome of the calculations—based on 0.75 horsepower per hectare, $120 per tractor horsepower, and tractor costs representing, say, 60 percent of total equipment cost—is total equipment investment of $150 per hectare, or $61 per acre.

In the past, the cost of maintenance and repairs of tractors and other major farm machines in the Middle East as in developing countries generally has been higher than in the countries where such machines were manufactured, in part because of the cost necessarily associated with import of spare parts, in part because of lack of trained and experienced repairmen, and in part because maintenance and repair have often been neglected until more costly repairs were necessary. This situation, it is hoped, will improve in the future—especially as such machines are used in larger numbers—providing both economies of scale in servicing and putting more machines of the same make in each area.

The foregoing analysis suggests the need for a new investment of about $1,000 million in tractors (Table 17-9, last column) and a similar amount in all other farm machines. In U. S. experience, the annual replacement, repair, and operating costs of this lot of machinery should be somewhat under 20 percent of the cost of such machines new, or something less than $400 million annually. This is, at best, only a rough estimate, since the amount of use the machines will get, the skill with which they are used and maintained, the cost of spare parts and of fuel and other operating costs, and other factors may well vary greatly in the future. A rough benchmark figure of $500 million will serve.

Pesticide Use for Production Potential

In Chapter 5 attention was called to the necessity for using chemicals to control insect and fungus pests and to suppress or eliminate weeds. The necessity of the use of these chemicals increases greatly with the very high levels of production that we have postulated, and so does the payoff.

The kinds of chemicals used are constantly changing as newer and more effective compounds are developed; as newly developed strains of pests become immune to a chemical that has been used in the past; and as awareness rises of the possible harmful effects of persistent residues

Table 17-10

Projected Pesticide Needs and Costs per Year

	Total nutrients required (1,000 mt.)	Pesticide required (1,000 mt.)	Costs at $1.25/lb. (million U.S. $)[a]
Egypt	1,955	57.8	159.5
Israel	113	3.4	9.4
Lebanon	99	3.0	8.3
Jordan	134	4.0	11.0
Syria	956	28.7	79.2
Iraq	1,678	50.4	139.2
Totals	4,935	147.3	406.6

[a]The cost in foreign exchange would, of course, be much smaller.

left in the soil from previous applications. Thus one cannot project with certainty how much and what kind of pesticides will be required for high-level production in the Middle East at some time in the future. However, it is possible to give a fairly good idea of how much it will cost. W. B. Ennis et al.[5] have prepared a detailed study of the need for additional pesticide inputs for increased agricultural production in a number of parts of the world, particularly in the less developed areas. These authors noted that in the countries of Israel, Japan, and the United States over a period of years there was a relationship within and between countries showing pesticide use at about 3 percent of the amounts of fertilizer nutrient applied. Of course, some crops, such as cotton, entail much larger outlays for pest control than do other crops, such as wheat.

Currently available statistics are too weak a reed to offer support for calculating average costs per pound of pesticide. Definitions and sources differ, and so does the product mix. For example, one might deduce from pesticide statistics regularly published in much product detail by FAO and from summary-value statistics published until 1965 in Egypt's official statistics, that the country was buying pesticides at an average cost of a little less or a little more than $1 per pound. But this average is heavily affected by large volumes of low-cost sulfur and mineral oils. It hardly could be representative for the future. Analogous or greater difficulties affect statistics for the other countries.

A better way for calculating average costs for pesticide is perhaps to take recourse to calculations recently made in the United States that suggest the cost of a pesticide at the formulating plant, ready for use, at about 70 cents, based upon an imported technical pesticide price of 50 cents.[6] Adjusting the cost of the latter to 70 cents, to be representative of a wider range, and allowing for a markup after the product leaves the plant, might result in a cost per pound of perhaps $1.25.

This assumption would result in the following figures for the six countries (Table 17-10), if we assume the above-noted linear relationship between fertilizer use and pesticide use given by Ennis et al. (e.g., 30 kilograms of pesticide per metric ton of fertilizer).

[5] *Ibid.*, p. 130–175.
[6] *Ibid.*, p. 159.

Miscellaneous Inputs

The volume of agricultural production that a realization of the productive potential would entail, would require a considerable amount of miscellaneous inputs—for such things as packaging and marketing materials, seeds and planting stock, and others. Transportation within the farm or rural area, as well as from it to the larger markets, is another requirement to consider. Some items rather small in cost can be essential to production also. We do not attempt to estimate the costs of these items, or even to enumerate all the kinds of inputs that would be needed; their costs would not, however, be large in comparison with the items which we have listed. Any organization of agriculture which could realize the productive potential could take care of such miscellaneous items.

Capital Requirements for Resource Development

The drainage and irrigation possibilities outlined in Chapter 13 would require a considerable investment. There would be some annual maintenance or upkeep and some annual operating costs for such facilities, but this would not be particularly large; a far more serious matter would be the annual fixed charges that should be levied against these investments. The total investment for all the drainage discussed in Chapter 13 is estimated to be of the general order of $1.5 billion; this would no doubt be spread over a considerable number of years, and must be accorded priority over further investment in irrigation facilities, except for those that are necessitated by new drainage installations. If the full irrigation possibilities of the Tigris and Euphrates, within the region of study, were developed, this would call for an investment in the general magnitude of $1 billion; although this too would be spread over some years. Not only would such installations have associated facilities for considerable electrical power (which the countries concerned could absorb only gradually), but there may be some doubt as to the wisdom of complete irrigation development of all available water supplies. Gradual development would allow periodic reviews.

Considering the general lack of capital in these countries, a full annual charge (for maintenance, operation, depreciation, and capital use) should probably be not less (though possibly more) than 12 percent of the original investment cost of these developments. At 12 percent, an annual charge of $180 million would arise from drainage, and one of $120 million from irrigation; but this level of costs would be reached only gradually. Part of these costs would not be cash payments but accounting costs.

Summary of Inputs and Outputs

The annual cost needed above present cost to support potential output, as estimated earlier in this chapter, would be about $1,000 million for fertilizer, and about $500 million for machinery; about $400 million for pesticides[7]

---

[7] Obtained as difference between potential requirements of $400 million and an estimated $50 million now being spent in the region.

and miscellaneous items might account for another 100 million annually. Total added cash outlay annually then would be $2,000 million. As compared with the better than $5.5 billion estimated annual increased gross output (Table 15-9), this is a modest input or conversely, a relatively high output from the input. While these are cash costs, the increase in output would also be a cash income. The drainage cost would be $180 million annually, much of which would not be a cash cost each year. The irrigation investment would amount to an annual charge of $120 million, much of which would also not be in cash. Adding both charges at full value still would leave the ratio of added inputs to added outputs relatively favorable—$2,300 million to $5,500 million—without considering the ways in which full irrigation would lead to a potential output greater in value than was estimated in Table 15-9.

These extremely rough calculations do, perhaps, serve to emphasize two facts: (1) as agricultural technology advances, the cost of inputs in relation to the value of outputs will necessarily increase—it will simply be impossible for Middle Eastern agriculture to increase its total production on the present ratio of identified inputs to outputs of about 25 percent (Chapter 10); and (2) very careful consideration should be given to costly investments in drainage, and more particularly in irrigation, to assure (a) that the needed complementary farm programs are really carried out, (b) that the investment is not premature, and (c) that the long-run economics are favorable.

Other Inputs

There are other inputs, previously discussed, that do not lend themselves to even the rough approximation of costs attempted above. Credit facilities, research and extension, a well-functioning input supply mechanism, and others. What is to be said about their role in advancing the potential has been said elsewhere, specially in Chapter 11, and nothing is added here except that most of these costs will have to be borne by the economy as a whole rather than by the producer alone.

Economic Potentials in Supplying Agricultural Inputs

We may close this chapter by pointing out again that the supply of needed agricultural inputs provides opportunity for expansion elsewhere in the economy of the region. There will be an employment effect, in manufacturing and in distributing the various inputs, as well as in marketing, processing, storing, etc., the outputs. While this effect may be less than the probable shrinkage in farm employment, it is nevertheless highly important and deserves attention. There are considerable possibilities in regional cooperation in the manufacture of fertilizers, machines, pesticides, and other inputs. If such production could be planned to operate on a regional rather than a national basis, its economic efficiency would surely benefit.

CHAPTER 18

The Market Possibilities for Increased Agricultural Output

In the previous chapters of Part III, we have assumed that there were no limiting financial, social, or institutional factors to achievement of the physical production possibilities that the natural resources of the Middle East and modern agricultural technology could permit.

In this chapter, we will examine what limits if any might in fact be set by market outlets, one of the factors hitherto assumed as nonlimiting. How much, in other words, of the important agricultural commodities can be sold, where, and at what prices, and how do the amounts that can be marketed to advantage compare with the amounts that might be produced?

In making this brief analysis, a few major facts or relationships about the Middle East should be recalled.

(1) The Middle East as a whole is a region of very rapid population growth. At a minimum, food needs will rise in proportion.

(2) The Middle East as a whole is a region of low per capita incomes with a relatively low present consumption of food, especially of the more expensive kind. As real incomes per capita rise in the future, one would expect per capita expenditures on food to rise. In part this would spell a greater quantity but also increasingly a shift to higher-priced protective foods.

(3) As described in Chapter 11, the Middle East as a region imports a substantial proportion of its foods, especially cereals, and exports a few commodities, of which cotton and citrus are two of the major ones (Table 18-1). About half of all exports, by value, are cotton, and nearly all of this is Egyptian cotton. Another major product is citrus, especially from Israel and Lebanon. With the exception of exports from Jordan and Lebanon, most of the trade is with countries outside the region; intraregional trade is by and large a small proportion of the total foreign trade of the various countries (see Tables 11-4 and 11-5) Cereals and oilseeds are the two large classes of agricultural imports, and they nearly all come from outside the region; while cotton and citrus, the major exports, do not stay in the region. Of items not shown in Table 18-1, the region is also a considerable net importer of meat.

The Region as a Market

While for the most part we use our own projections of supply possibilities, the projections of demand, made as part of the FAO Indicative World Plan,[1] offer an excellent

starting point for consideration of future markets for agricultural commodities. In contrast to Chapter 11, when we were concerned with describing the current situation, we now focus our attention on the comparison between 1962, the base year for FAO data in this study, and 1985, the final target year. This period for demand analysis is substantially shorter than the time interval we judge will be needed for the production potential to be realized, but it may be that the same general order of increase will apply to a longer period. In any event, the amounts projected should be roughly indicative of those that will be demanded.

The data in the FAO study indicate roughly a doubling of population between 1962 and 1985 (Table 18-2). A doubling of population in twenty-three years is a very high rate of natural increase, by world standards, but we have no information that would lead us to quarrel with the FAO projections.

The FAO study also projects a rise in per capita incomes (measured in constant prices) which ranges from an increase of about 50 percent to a near doubling among the different countries. If population doubles and per capita income rises 50 percent, then total income for the region will be three times higher in 1985 than in 1962, with larger increase in per capita incomes associated with greater total income in 1985. Because a smaller proportion of the added income is likely to be spent on food, expenditures on food will presumably rise somewhat less than increases in total income, but a substantial rise in total outlay for food can be taken for granted.

The population effect is to double the consumption of grains; but because the income elasticity of grains in the Middle East is very low, the income effect on grains is negligible. For the starchy foods—sugar, and pulses and nuts—the income effect is somewhat greater, leading to as much as 20 percent more consumption than the population effect alone would require. The income effect is somewhat more for vegetables, and much more for fruit, meats, eggs, and milk. While these projections are based upon observed or assumed relationships between income and per capita consumption of the present or recent past, the function used to project them into the future is not always the same, but provides for varying degrees of continuity or change.[2] And while the projected increases differ between countries as well as between the major food products, a general rela-

---

[1] The study for the Near East excludes Israel; we have attempted to supply more or less comparable figures for Israel, but the comparison may not be precisely correct.

[2] The methods used are briefly but well outlined in *Agricultural Commodities—Projections for 1975 and 1985* (Rome: FAO, 1967), II, xxiv, xxv. This volume also contains the income elasticity assumptions made for the projections, and while these may differ somewhat from those used in the *Indicative World Plan* they are

Table 18-1

Intraregional Trade in Agricultural Commodities, in Relation to Total Exports and Imports, 1961-63 Average

(million U.S. $)

	Exports				
Product	6 countries excluding Israel	Israel	Total 7 countries	Intraregional	Imports
Cereals	60	[a]	60	12	264
Pulses				4	
Vegetables	} 75	[b] 2	} 155	7	} 50
Fruits		[b] 78		11	
Oilseeds	11	[b] 2	13	4	61
Cotton	[b] 341	[b] 2	343	[c]	
Total enumerated items	487	84	571	38	

[a] Less than 0.5.

[b] Average 1962-63.

[c] No information; assumed negligible for raw cotton.

Sources: (1) *Indicative World Plan, Agricultural Development 1965-85—Near East,* (Rome: FAO, 1966), *II: Explanatory Notes and Statistical Tables.* (2) *UN Yearbook of International Trade Statistics 1965.* (3) U. S. Department of Agriculture.

tionship between projected consumption and projected population, modified by some differences in per capita consumption, is evident in Table 18-3.

The great importance of cereals—as a user of crop land, as a basic part of the diet of lower income people, and as an expenditure for import—requires a closer look at the cereal balance in the region (Table 18-4). In the 1960—62 period, total cereal production within the region averaged 10 million tons; in favorable years it will be higher, in dry years, lower. For the same period, imports of all cereals averaged nearly 4 million tons annually. Of the total supply

probably very close. They are of sufficient interest to show for a selected list of items:

Demand Elasticities for Selected Foods

	Egypt	Israel	Lebanon	Jordan	Syria	Iraq
Wheat	0.3	−0.4	0.1	0.2	0.2	0.4
Rice	0.3	0.3	0.3	0.6	0.6	0.6
Sugar	0.9	0.3	0.6	0.6	0.7	0.5
Pulses and nuts	0.5	0.0	0.3	0.3	0.4	0.6
Vegetables	0.3	0.2	0.3	0.2	0.6	0.4
Citrus	1.0	0.1	0.4	0.8	1.0	1.2
Beef and veal	1.0	0.6	0.7	1.0	1.2	0.9
Mutton and lamb	1.0	0.5	0.8	1.2	1.2	1.0
Poultry	1.6	0.2	1.0	1.5	1.6	1.6
Eggs	1.2	0.2	1.0	1.0	1.2	1.2
Fish	0.7	0.5	0.5	0.8	1.0	0.8
Milk, excl. butter	0.8	0.1	0.9	1.0	1.0	0.8
Vegetable oils	1.0	0.2	0.3	0.4	0.4	0.7

Source: FAO Agricultural Commodities—Projections for 1975 and 1985 (Rome: FAO, 1967), II: *Methodological Notes, Statistical Appendix.*

of nearly 14 million tons, nearly 10 million tons were used for human food. Although the region is a heavy net importer of cereals, it does export some—especially rice from Egypt, and, in good years, barley from Syria.

The nonfood uses of cereals amount to about 3.25 million tons annually; a little of this is for seed, some is for dairy cows and chickens, especially in Israel, but it is not clear what all of it is used for in Iraq and Syria.

The outlook in Table 18-3 was for a need by 1985 of 17 million tons or more of cereals for direct use as human food. Since the demand for meat and livestock products will rise relative to the increase in population, some additional grain will be needed here. As a preliminary estimate, we suggest that 8 million tons of cereals will be used domestically for nonfood purposes. The 25 million tons of cereals required (disregarding exports) are half of the potential grain production of nearly 50 million tons, as shown in Chapter 15. If we exclude rice from the production potential, on the grounds that much of it will be exported, the remaining 44 million tons is still significantly above the projected demand for cereals. Finally, if we project consumption to the end of the century—as we have done very roughly in Table 15-7, and on ground rules somewhat different from those used here—we have a requirement for human and feed consumption of 36 million tons, or ten million more than in 1985. This projection also would leave a large balance beyond needs so calculated, whether or not one deducts rice from both production and consumption.

What is equally impressive in Table 15-7, however, is the fact that Egypt, Israel, and Lebanon would still show a grains deficit—in Egypt of nearly 5 million tons; that Jordan would only just break even; and that substantial surpluses beyond domestic consumption would emerge only in Syria and Iraq, with the latter in a sense the only one of the six countries that would show surpluses of such magnitude as to raise serious questions of disposition. Indeed, the region without Iraq would remain a cereals-deficit one, despite the sharp increases in production here projected.

This situation highlights, for one thing, the role of regional cooperation; for Iraq would find in the neighboring countries customers for about one-third of its cereal surpluses, or somewhat less when Israel is excluded. Secondly, the countries could find full satisfaction of their cereal needs long before they reach the level of production termed "potential," but again, only on a regional basis. Thirdly, if it proved difficult to compete in other markets in such volume as here indicated (after all, 14 million tons are nearly equal to the total wheat exports of the United States in the 1968/69 marketing year), alternative production patterns would be called for, either in the direction of more grains for livestock feeding for domestic or export markets, or diversion of crop land to pasture, or other variations. In any event, the notion of the "potential" here provides a useful tool for previewing policy alternatives.

Because it is income-elastic—some of it highly so—the domestic demand for fruits and vegetables is projected to increase by much more than the projected doubling of population. Indeed, the projected increase in demand is significantly greater than the increase in yield per acre which was projected in Chapter 15; but as has been indicated there, the necessary increase in output can be achieved by using more land for these crops. Since the land potentially suitable for vegetables is vastly more extensive than the need, the demand can readily be satisfied by expansion of acreage. For fruits, the climatic requirements are somewhat more exacting, but the increased demand projected can be met by a combination of expanded area and increased yield per acre. Neither would notably cut into the acreage of the major crops.

The projected demand for meat and other livestock products (Table 18-3) is roughly of the same magnitude as the projected production potential of Table 15-8. This projection assumes that no difficulties would arise in domestic production replacing imports of the commodities involved.

Export Possibilities

An attempt to estimate probable or possible agricultural exports from the Middle East must rest as much on political as on economic assumptions. Nearly half of present exports from the region consists of Egyptian cotton, now largely committed to the U.S.S.R. in exchange for other commodities or as service on loans. Presumably some of this cotton is consumed in the U.S.S.R. and some is sold to other countries. How much more Egyptian cotton might be disposed of to the U.S.S.R. in the future? Within some limits, the answer will depend upon political relationships between the two countries: a larger volume might be taken, or the present volume might be reduced; and some or much might be sold in third markets, eventually in competition with cotton from Egypt itself. Can Egyptian cotton regain, or obtain, direct export outlets in other countries? As we have noted in earlier chapters, Egyptian cotton has special qualities, especially in fiber length; while these make it a somewhat different commodity than shorter staple cotton, Egypt does not have a monopoly on this type of cotton nor is any type of cotton wholly free of competition from other cottons and from synthetic fibers, at least for some uses. Within some limits, relative prices influence the relative uses. There is a real question whether Egypt will be able to place her cotton abroad without substantially lowering its price.

Briefly, at least three factors incline one to view future cotton outlets and prices cautiously. These are (1) the large additions of cotton acreage in several countries and

Table 18-2

Population, per Capita Income, and Food Consumption, 1962, and Projections for 1985

	1962				1985			
Country	Population (million)	Income per cap., $[a]	Total calories per day	Total proteins per day, grams	Population (million)	Income per cap., $[a]	Total calories per day	Total proteins per day, grams
Egypt	27.2	148	2390	69.7	53.1	288	2560	75.3
Israel	2.3	700[b]	2791	84.5	4.3	1580	3000	90.0
Lebanon	1.9	385	2420	68.5	3.5	671	2630	76.9
Jordan	1.7	236	2230	62.2	3.9	360	2480	69.3
Syria	4.9	210	2350	68.4	10.7	307	2510	74.0
Iraq	6.6	298	2140	62.1	14.4	451	2390	70.9

[a] Gross domestic product at factor cost.
[b] Net domestic product at factor cost.

Source: (1) Except for Israel, *Indicative World Plan for Agricultural Development, 1965-85—Near East* (Rome: FAO, 1966), II: *Explanatory Notes and Statistical Tables.* (2) Israel: 1962 figures—*The Statistical Abstract of Israel, 1968.* 1985 figures prepared by the Agricultural and Settlement Planning and Development Centre of the Ministry of Agriculture and the Settlement Department.

Table 18-3
Consumption and Trade in Food Products, 1962, and Estimated Consumption, 1985

| | 1962 | | | | | 1985 Consumption | | |
| | Consumption | | | | | | Total | |
Food group	Per cap. (kg./yr.)	Total (1,000 t.)	Imports (mil. U.S. $)	Exports (mil. U.S. $)	Net trade (mil. U.S. $)	Per cap. (kg./yr.)	1962 rate (1,000 t.)	1985 rate (1,000 t.)
EGYPT								
Cereals	168.2	6,419	116	28	−89	162.0	10,911	10,512
Starchy products	11.2	292				12.4	591	655
Sugar	15.8	384	9	4	−5	20.5	758	983
Pulses and nuts	10.8	289				14.3	573	759
Vegetables	89.4	3,121	} 7	} 27	} 20	113.2	6,418	8,124
Fruit	66.5	1,035				97.2	2,155	3,150
Meat, inc. offal	9.6	256	6	a	−6	15.5	409	660
Eggs	1.1	30				2.0	58	106
Milk (liquid eq.)	42.9	1,167	4	a	−4	62.7	2,277	3,329
Fats and oils	5.8	158	20	3	−17	6.7	308	356
ISRAEL								
Cereals	117.4	266	45	1	−44	90.1	502	385
Starchy products	33.6	76				36.2	143	154
Sugar	32.1	73	4	a	−4	46.9	137	200
Pulses and nuts	9.1	20				10.5	39	45
Vegetables	109.9	249	2	3	1.0	130.8	469	558
Fruit	144.8	328	1	61	60	163.7	619	700
Meat, inc. offal	40.8	92	6	a	−6	53.9	174	230
Eggs	19.9	45		9	9	23.6	84	100
Milk (liquid eq.)	134.0	326	1	a	−1	145.7	572	622
Fats and oils	19.4	44	8	3	−5	23.4	83	100
LEBANON								
Cereals	121.6	300	25	1	−24	119.7	568	559
Starchy products	15.6	29				16.9	55	60
Sugar	23.3	42	4	0	−4	29.7	81	103
Pulses and nuts	13.4	25				16.3	48	58
Vegetables	103.2	246	} 13	} 17	} 4	125.6	462	562
Fruit	158.5	247				193.0	461	561
Meat, inc. offal	31.2	59	20	0	−20	43.2	100	139
Eggs	2.7	5				3.8	9	13
Milk (liquid eq.)	49.8	95	8	a	−8	77.5	168	261
Fats and oils	12.5	24	6	a	−6	13.6	44	48
JORDAN								
Cereals	135.9	286	15	a	−14	148.6	668	730
Starchy products	10.4	18				12.0	41	47
Sugar	23.2	42	5	0	−5	26.4	93	106
Pulses and nuts	10.2	1.7				11.7	40	46
Vegetables	117.8	267	} 7	} 6	} −1	13.3	636	730
Fruit	112.3	130				137.7	302	370
Meat, inc. offal	11.9	22	5	0	−5	15.0	40	51
Eggs	1.8	3				2.3	7	9
Milk (liquid eq.)	37.1	63	1	a	−1	47.6	145	186
Fats and oils	10.0	17	2	1	−1	11.2	39	44

a Less than 0.5.

Sources: Except for Israel, and except as noted below, *Indicative World Plan for Agricultural Development, 1965-85–Near East,* (Rome: FAO, 1966). These data are confusing because: (1) for cereals, total consumption includes the weight of the whole grains, whereas the per capita figures for cereals deduct milling losses of bran, etc.; and (2) total consumption of vegetables *includes* some products (possibly melons) which per capita data *exclude,* whereas for fruits the same products are *excluded* from the total but *included* in the per capita figures; these

Consumption and Trade in Food Products, 1962, and Estimated Consumption, 1985

| | 1962 | | | | | 1985 Consumption | | |
| | Consumption | | | | | | Total | |
Food group	Per cap. (kg./yr.)	Total (1,000 t.)	Imports (mil. U.S. $)	Exports (mil. U.S. $)	Net trade (mil. U.S. $)	Per cap. (kg./yr.)	1962 rate (1,000 t.)	1985 rate (1,000 t.)
SYRIA								
Cereals	157.6	879	17	24	7	159.6	2,020	2,047
Starchy products	8.9	42				9.8	95	105
Sugar	16.7	79	6	0	−6	21.3	162	207
Pulses and nuts	12.8	61				15.0	136	160
Vegetables	33.4	527	} 10	} 6	} −4	43.3	1,045	1,355
Fruit	142.0	345				163.2	800	919
Meat, inc. offal	13.7	65	3	4	1	19.6	131	188
Eggs	1.5	7				2.3	16	25
Milk (liquid eq.)	65.7	322	2	1	−1	94.1	703	1,007
Fats and oils	9.5	47	2	7	5	11.0	102	118
IRAQ								
Cereals	129.6	1,020	20	8	−11	141.4	2,349	2,562
Starchy products	3.8	22				4.3	57	64
Sugar	29.5	207	16	0	−16	33.5	440	499
Pulses and nuts	5.7	36				7.2	79	100
Vegetables	56.8	516				75.5	1,151	1,530
Fruit	70.9	313	5	19	14	93.8	712	942
Meat, inc. offal	17.5	129	a	a	a	26.0	245	364
Eggs	1.1	7				1.8	16	26
Milk (liquid eq.)	75.5	498	2	0	−2	107.4	1,088	1,547
Fats and oils	3.9	26	9	1	−8	5.1	56	73
TOTAL								
Cereals		9,170	238	62	−175		17,018	16,795
Starchy products		479					982	1,085
Sugar		827	44	4	−40		1,671	2,098
Pulses and nuts		448					915	1,168
Vegetables		4,926	} 45	} 139	} 94		10,181	12,859
Fruit		2,398					5,049	6,642
Meat, inc. offal		623	40	4	−36		1,099	1,632
Eggs		97		9	9		190	279
Milk (liquid eq.)		2,471	18	1	−17		4,953	6,952
Fats and oils		316	47	15	−32		632	739
TOTAL, EXCLUDING ISRAEL								
Cereals		8,904	193	61	−131		16,516	16,410
Starchy products		403					839	931
Sugar		754	40	4	−36		1,534	1,898
Pulses and nuts		428					876	1,123
Vegetables		4,677	} 42	} 75	} 33		9,712	12,301
Fruit		2,070					4,430	5,942
Meat, inc. offal		531	34	4	−30		925	1,402
Eggs		52					106	179
Milk (liquid eq.)		2,145	17	1	−16		4,381	6,330
Fats and oils		272	39	12	−27		549	639

respective differences about offset each other for each country. For the 1985 total consumption at the 1962 rate of consumption, we modified the reported 1985 total consumption at the 1985 per capita rate by the same percentage that the reported 1962 per capita consumption deviated from the reported 1985 per capita consumption, so that the data in the last two columns are fully comparable. For Israel: Statistics prepared by the Agriculture and Settlement Planning Development Centre of the Ministry of Agriculture and the Settlement Department.

continents; (2) the continuing heavy inroads of synthetics; and (3), most recently, a U. S. policy of favoring longer staple cotton growing at home rather then importing it. Whether these factors will be offset by the demand pressure from sheer population growth, or by the slackening of the pace at which new producers have entered the field is most uncertain. The answer would likely be "no" for the next decade or so, but might be "yes" for the longer run. Perhaps most serious is the competition from man-made fibers, not only in the developed countries, but more recently also from elsewhere. In South Korea, for example, cotton's share of the fiber market between 1964 and 1967 dropped from 80 to 65 percent. In Taiwan between 1963 to 1968, it dropped from 83 to 75 percent. In Thailand, a polyester plant with an annual capacity of 20 percent of present consumption is planned to operate in 1972.[3] If developments in the United States and Europe are a precedent—and they probably are—then these are the beginnings of a rearrangement in the fiber-consumption mix.

We have suggested in earlier chapters that a modest amount of cotton (something in the general order of 400,000 bales annually) could be grown in Syria and Iraq, beyond the amounts now grown. This would all be medium and short staple cotton, not the long staple Egyptian kind; as such it would be competitive with cotton grown in many countries around the world. By the same token, the world price would not be very sensitive to modest increases in output from these countries. While there are many crosscurrents in the world cotton market, the production of

Table 18-4

Production, Trade, and Consumption Balance for Cereals, 1960–62

(1,000 Metric Tons)

Country	Production	Imports	Total Supply	Local Utilization		Exports
				Food	Nonfood	
Egypt	6,151	1,629	7,780	6,419	954	407
Israel	99	[a]660	759	[b]361	[c]398	0
Lebanon	88	310	398	300	86	12
Jordan	141	214	355	286	63	6
Syria	1,377	367	1,744	879	647	218
Iraq	1,902	306	2,208	1,020	1,045	143
Saudi Arabia	236	320	556	518	38	0
Total	9,994	3,806	13,800	9,783	3,231	786

[a] 1965 only; from *UN Yearbook of International Trade Statistics, 1965.*

[b] Wheat, rye, and rice; data from *UN Statistical Yearbook, 1967.*

[c] Difference between total supply and estimated food consumption.

Source: Except as noted above, *Indicative World Plan, Agricultural Development, 1965-85–Near East* (Rome: FAO, 1966), *II: Explanatory Notes and Statistical Tables.*

---

[3] These examples from *Foreign Agriculture* (U. S. Department of Agriculture), September 8, 1969.

Syria and Iraq might well be marketable without undue difficulty provided—and this is a more pressing problem—they can grow cotton at costs which produce a fully competitive product.

The competitive situation would perhaps be more serious in grains, of which, as has been shown above, the region, and specifically Iraq, would have a sizable surplus, if the potential were achieved. Grains, like the common types of cotton, are sold in a world market. But in contrast to cotton, the volume that could be produced in the Middle East is large enough to affect world prices significantly, depending upon how quickly the potential production came onto the market; though some grain, at least, could probably be sold without too serious effect upon prices. Among the more serious problems are (1) that Middle Eastern grain growers must bring their costs down to become fully competitive with the grain exporting countries, which, it must be assumed, will not stand still in their efforts to become more competitive, so that each will be shooting at a moving target; and (2) that most likely such exports would have to penetrate markets now served by other countries; this calls for development of a market strategy and market promotion.

While one must guard against being overly impressed by contemporary changes in outlook, nonetheless the notable advances in yield and production in the grain deficit areas of the world and the pressure of continually large supplies of grains from traditional exporters do not have the aspect of a brief interlude. Speaking about rice, for example, a recent survey judged that ". . . by 1980 it is expected that the world rice situation will have adjusted from the recent pattern of relatively high prices and scarce supplies to one of generally adequate exportable supplies and significantly lower prices..." (*Foreign Agriculture* [U. S. Dept. of Agriculture], August 11, 1969). When Japan is seriously attempting to reduce its rice acreage and even considers diversion to nonfood uses, it will not be easy to enter the cereals export market, or even to maintain one's position, unless one is a low-cost producer.

Citrus is another important agricultural export from the Middle East, especially from Israel and to a much smaller but still important extent from Lebanon. Since 1948, Israel's principal market has been Western Europe, especially the United Kingdom and more recently West Germany. Between them, the Common Market and the European Free Trade Association absorb about 90 percent of Israel's citrus exports. While it is true that Jaffa oranges can at times be seen in U. S. grocery stores, the volume amounts to less than one percent of the export volume. Lebanon's strength, on the other hand, lies more in lemons than oranges or grapefruit: one-third and sometimes more in volume of its exports—and a somewhat higher proportion in value—consists of lemons. Of these, a sizeable percentage goes to Eastern Europe (about 50 percent in 1965), and most of the remainder to neighboring Syria and Jordan. Of the oranges, the great bulk goes to these two countries and Saudi Arabia.

The future competitive situation is difficult to evaluate. Israel is even now in competition with other Mediterranean countries, such as Spain and Italy,[4] as well as with North Africa and at times South Africa. Periods of oversupply and low prices have been common in the past and suggest a market that is not very price-elastic. Population growth is low in Israel's citrus markets and cannot be counted on as much of a demand booster, but rising per capita income should help. If the Eastern European countries would ever open up to citrus from Israel and elsewhere, large volumes might be marketed there in time; but this is a very speculative possibility, indeed. Altogether, the outlook must be considered as not dynamic—in the sense that, barring a breakthrough in Eastern Europe, the odds on a greatly enlarged market are rather long.

While Lebanon has a historical foothold in the Soviet bloc, speculation about the degree to which consumption there might expand is beyond our competence. On the other hand, the Arab countries of the region could absorb substantially increased volumes at rising per capita incomes (see p. 154 for demand elasticity estimates). The same goes for Lebanese apples, of which countries of the region at present take the great bulk, with the U.S.S.R. a modest "in-and-out" customer.

It is perhaps not too sweeping a statement to say that the hesitation one feels in predicting a very rapid growth in export sales for citrus holds for fresh fruits and vegetables in general. A somewhat closer look at the whole subject is therefore worthwhile, particularly as the possibility of such exports in the winter season to both nearby Arab countries and to Europe has recently gained much prominence, not least because exports of such high-priced items would justify the use of new sources of high-cost water. The prerequisites for such exports have been described in Chapter 11. But in addition to developing the managerial capacity and the physical plant for becoming exporters in a field that makes exceptionally high demands on managerial skills, there are unanswered questions as to the competition and the size of the export markets at prices that will allow such sales. Some studies have recently been undertaken, especially in the FAO framework,[5] which tend to be relatively short-run projections that are not necessarily indicative of conditions twenty-five or thirty years hence. Nonetheless, their conclusion that penetration of Western European markets on any large scale, is beset by very substantial obstacles, deserves attention. Among these obstacles are the preferences accorded to members and associates of the Common Market and, in specific instances, new types of competition. An interesting illustration of

the latter is the spread of production under cover, mentioned in connection with a recent FAO recommendation to establish a UN-assisted Advisory Service in this field for the Near East:[6]

> As regards the production of out-of-season vegetables for export, it should be stressed that the advantages of many Mediterranean and Near Eastern countries—mild winter climate, cheap manpower—are not sufficient by themselves to ensure the success of the endeavor. The glasshouse vegetable industry is developing in Europe. In the "classical" area of glasshouse production, i.e. North-Western Europe, areas under glass increased by 45 percent between 1956 and 1964, the most marked increase—75 percent—being in The Netherlands. But the new and striking feature of the last ten years is the tremendous increase in areas under protected cultivation in Southern Europe (especially in the south of France and Italy) due mainly, but not exclusively, to the use of plastic sheets.

On the same subject, according to a recent Organization for Economic Cooperation and Development (OECD) report that made short-term projections,[7] the European members of OECD were estimated to have an increase in fresh tomato export availabilities from 471,000 tons in 1961-64 to 680,000 tons in 1970, largely through increased "undercover" production in Spain, Belgium, and the Netherlands. A much smaller growth in import needs, from 632,000 to 700,000 tons, would leave an import need of only 20,000 tons in 1970. At the same time, there would be a steep increase in the excess of export availabilities over import needs of processed tomatoes, of the order of about 800,000 tons, in the same countries. It is factors such as these that need the most careful analysis, if both capital and human investments are not to be wasted.

As for Eastern Europe, which, with such exceptions as Lebanese lemons, now is a negligible customer, the future is impossible to foresee, since decisions will be dictated by considerations of a political rather than an economic character. And as for intraregional sales, one of the striking discoveries one makes as one inquires into this aspect is that more than one of the countries involved figures both as a would-be exporter in its own development plans and as a projected importer in those of another. Again, regional-scale planning would help in correcting potential collision courses of this kind.

In summary, the wide range between lowest and highest per capita consumption seems to indicate ample room for substantial increases in exports. But the realities of the technical and managerial capacity to export perishable, high-quality commodities; the lack of knowledge of the nature of these markets, especially

---

[4] Italy's competitive position was in serious jeopardy in 1968/69, when the Government intervened on a large scale and acquired stocks of oranges to upport a sagging market. Export subsidies were introduced and sizeable tonnages reported to have been destroyed in Sicily.

[5] See, for example, *Outlook for Production and Trade of Selected Horticultural Products in Mediterranean Countries,* a project sponsored by the Commodities Division of FAO that culminated in a conference held at Bari, Italy, in September 1967, and published as *Conferenza Nazionale per l'Ortoflorofrutticultura, Prima Commissione di Studio sulla Analisi e Proiezione al 1970 e 1975 della Domanda e della Offerta dei Principali Prodotti Ortofrutticoli,* Bari, 29-30 Sett. 1967.

[6] G. Piquer, "Advisory Service on Production, Marketing and Export Trade of Fruits and Vegetables in the Near East and North Africa," Item 9 of *Agenda* of the Regional Commission on Horticultural Production in the Near East and North Africa, First Session, Beirut, Lebanon, October 7–12, 1968.

[7] *Tomatoes: Production, Consumption, and Foreign Trade of Fruits and Vegetables in OECD Member Countries; Present Situation and 1970 Projects.* (Paris: OECD, 1968).

with regard to price elasticity; the consequences of uncoordinated planning for expansion by many countries enamored of the same idea; and, finally, the effect of this and other national policies—including exclusionary regional blocks—on any one of them call for careful exploration and testing before plans ripen into commitments.

To illustrate these comments it is worth quoting a few findings from relatively recent FAO reports. One of them originated in an FAO conference in Beirut[8] and makes these points:

(1) Regarding exports to Europe, the obstacles cited were (a) the need for selection of more suitable varieties for longer distance transport; (b) the high transport cost in comparison with other supplies to European markets; (c) the various tariff and nontariff measures which affect the trade in fruit and vegetables in most European importing countries; (d) the need to reduce nonlabor costs, as rising incomes will raise wages and thus reduce the competitive edge, if any, of the conditions in the Near East, where low wages currently can offset the higher efficiency prevailing in countries like the United States or Israel; (e) the lack of processing facilities and of low-cost varieties that are suited for processing, in comparison with competing countries that are aiming specifically at the market for processed products.

(2) As for specific products, the FAO report states, for example, that a demand-supply equilibrium in peaches might in the future be reached only "at reduced prices"; that the outlook for table grapes "...was not very promising, especially for prospective newcomers..."—presumably because export availabilities showed a very large excess over import requirements, at least for 1970, which was the target year in this instance; that for apples, Europe "...showed a high level of self-sufficiency... which was not expected to show any decrease in the medium term future," and as a consequence, shipments from the Near East would exist only on an "...irregular basis"; that for orange and tangerine exports to countries outside the Near East "...rising competition could be expected from strong production increases in other areas..." and that "...strong short-term market disturbances might be more frequent in the future than hitherto"; that for tomatoes, the varieties presently cultivated in the Near East "...did not meet customer tastes of the European market" but that off-season shipments had possibilities and should be further investigated.

The FAO Outlook study mentioned earlier (n. 5) which extends the projections to 1975 suggests the likelihood of exportable citrus surpluses by 1975 which could be absorbed only "...if there should be further

decline in relative price".[9] It foresees that "competition on international citrus markets is likely to become very stiff, resulting in growing pressure on prices...." It takes a more optimistic attitude toward increased exports of early tomatoes and potatoes and of continuing exports of onions from Egypt.

It is interesting too how the export market is viewed by a strong competitor like Italy. In the same study as the one from which the preceding observations are cited, the view is expressed that the absorptive capacity of Italy's principal tomato markets would even by 1970 be inadequate and might adversely effect returns to growers. With regard to fruits, the report points out that the output projected for 1975 is already based on the "...recently encountered difficulties in disposing of fruit that is significantly exposed to the sharpening of foreign competition, which is related to the expansion of output in other Mediterranean countries in apples, pears, and oranges. Thus the export availabilities are very conservative and must be fought for." Commenting on oranges specifically, the author of this particular segment of the report states as follows: "It is logical to foresee that unless production techniques are improved, quality and product differentiation pursued so that the product becomes more acceptable in export markets, there may occur a profound crisis, graver than any in the past, in this sector of agriculture."

Price-support schemes and recourse to the above-mentioned destruction of modest tonnage of citrus fruit in 1968 testify to the accuracy of the above observation. Nor is the problem confined to Italy. France too engaged in destruction of surplus fruit in 1968[10]. Large government subsidies are required, for example, to permit exports of processed tomato products from Israel, against sharp competition;[11] and other examples could readily be cited.

Conclusion

We conclude this brief and little more than indicative review of market potentials by saying that there is room for markets of Middle Eastern agricultural commodities to expand but not without great effort and within limits only. Certainly, in the next twenty-five years or so, the available markets could not accommodate all the productive potential, as we estimated it. They could, however, absorb a good deal at prices at least moderately satisfactory. The greatest difference between potential output and prospective markets is for the cereals; and the exporter's means of narrowing that gap are perhaps smallest here—for there is little if anything

---

[8] *Marketing and Refrigeration of Perishable Produce in the Near East,* Report on the FAO Near East Regional Conference on the Marketing and Refrigeration of Perishable Produce held at Beirut, Lebanon, September 20–28, 1965 (Rome: FAO).

[9] FAO's Study Group on Citrus Fruit in its Fourth Session (April 1969) estimated that export availabilities of oranges and tangerines in 1975 would exceed import demand by 4.3 million tons—or about 13% of world supplies, an outlook that gives substance to the above statement.

[10] *Foreign Agriculture,* June 30, 1969.

[11] *Foreign Agriculture,* April 28, 1969.

that gives the Middle East a relative advantage over its competitors, most of which have traditional trading channels and markets. Thus the degree to which grain markets can satisfactorily absorb increased output depends basically upon the relative rates of increase in world output and demand. On the other hand, if population and per capita incomes increase as projected, and if the farmers of the Middle East can produce grain cheaply enough to replace imports, then a very substantial increase in grain production over a period of years at more or less constant prices (in relation to world grain prices) can be supported, irrespective of exports.

Success in increasing the level of fruit and vegetable—or other specialty—exports depends finally on the nature of the market, and here research is urgently needed; but secondly, increasing the level requires the development of facilities, organization, and discipline—from raising the right variety at the right time, to having it arrive in the specified condition at the right price. Recognition of marketing skills, in the widest sense, then is a prime prerequisite, and one, incidentally, that is more readily accessible to private entrepreneurs than to government bureaucracies. To take these matters for granted or to assign them a subordinate role can only result in disappointment. That is why the casual reliance on these crops to make a crucial contribution to agricultural growth, which one meets not only in the Middle East, is a dangerous trend.

CHAPTER 19

Priorities and Strategies

Even though we have in the immediately preceding chapters cast our ideas of the area's potential into specific numbers, we are in no way wedded to those ideas as "targets." Nor do they constitute anything that might be called a plan or a program. Their sole function is to convey in a manner that is at the same time pedestrian and dramatic the order of magnitude of possible change.

To do this, it seemed useful to be specific and to deal in acres, tonnages, gallons, and dollars. If these were taken, however, as anything but illustrative, we would be the first to regret having engaged in this exercise altogether. For to carry through any calculations of future output in defensibly realistic detail, whether by use of a sophisticated econometric model or by some simpler device, would require a large number of specific assumptions about acreages, yields, numbers and management of livestock, consumption patterns, foreign markets, capital availability, wage rates, prices, etc. Such a well-articulated and comprehensive projection, consistent internally between its components and externally with the expected growth of the remainder of the economy, has not been the aim of the study. The Indicative World Plan of the FAO represents a courageous step in that direction, and we have greatly benefited from its existence as well as from conversations with several of its authors. But our aim has been less ambitious. It has been to evoke a sense of what is possible, what stands in the way, and what it might take to scale the obstacles that block its achievement.

We have also kept in mind that the audience we most wish to reach is not so much the expert or technician as the policy maker on one hand and the "informed layman" on the other. Thus, our emphasis has been more on defining issues and prospects broadly, though wherever possible sharply, than on achieving a consistent model of the future agriculture establishment at a given date. The study, then, is more a guide through the problems of Middle Eastern agriculture than to its solutions, more a map for orientation than an itinerary to a specified goal.

We would, however, be subject to legitimate criticism, were we to close without at least sketching our thoughts on the strategies available to achieve the goals we have discussed, and on the major step in devising strategy, which is a sorting of priorities.

The Package Approach Revisited

It is all very well to stress the "package approach," as we have done repeatedly; but it is also possible to become paralyzed by the complexity of the components of the package. He who attempts to move evenly on all fronts with limited means may end up by moving on none. This is a problem the developed countries never had to face. Their

economies just grew—not out of a central authority's thought about how they might or should grow—but out of the combined action of vast numbers of individuals. They grew unevenly—with lags, lurches, and bounces, and with only moderate concern for the effects of development on individual fortunes—and at times with no concern at all. But grow they did—unplanned and uncoordinated, with private initiative bringing up the rear in supplying the missing links as their absence or inadequacy became evident, and with their growth there arose the opportunity for profitable investment.

It is precisely the total or partial absence of this spontaneous process that has led development theorists to speculate on the bundle of factors needed to accomplish growth. Once identified—and the lists coming off the typewriters of different researchers differ some in emphasis but little in composition—the resulting package in each case poses a task for action. And this raises the clear question of priorities and strategies.

If what we say in the subsequent paragraphs seems to be applicable not just to the Middle East, but to developing countries generally, this is hardly surprising. In its general dimensions the idea of the package approach has been so well described recently by W. David Hopper that one prefers quotation to the use of one's own words. There is room here only for citing a few incisive passages from the work in question. "It is now an accepted cliché that in approaching the farmer the stress must be on a package of practices. ...It is clear...that the package of farm practices is only one package of many that are and will be required for underpinning the movement toward a scientific agriculture. The contents of some of these packages must interact concurrently to initiate development; others must be added in sequence as the process of growth gathers strength and momentum. The priorities of action are dependent upon the stages of growth, and the phasing of investments must be sensitive to long- and short-term needs, to existing and anticipated requirements."[1]

In the subsequent paragraphs Hopper then describes briefly the various types of packages, such as production services, improved farm practices, and others. He goes on to show how each of these packages in effect is a container for several others "nested in it." Thus, the package of basic production services contains a

> ...pattern of organized transport, a system for receiving marketed farm produce and supplying produc-

---

[1] W. David Hopper, "Investment in Agriculture: The Essentials for Payoff," in *Strategy for the Conquest of Hunger*, Proceedings of the Symposium convened by the Rockefeller Foundation, April 1 and 2, 1968 (New York: The Rockefeller Foundation, 1968), pp. 106–8.

tion requisites, an administration for controlling water distribution and maintenance of irrigation works in areas where irrigation is involved, and so on. Indeed, if each of these nested packages were dissected, a basic package would eventually involve a descriptive analysis of the national economy, its government, and its social and economic institutions. The package of practices—responsive seed and fertilizer—is likewise composed of nested packages. Responsive seed is a product of research; fertilizer, of chemical engineering. Each, in turn, rests on a foundation of human enterprise and skill and on an allocation of investment resources. Again the lines can be traced to the structure and dynamic of national culture.

Because the phenomenon of change in agriculture, even in its earliest stages, ultimately involves the complex of national life, I have an inherent dislike of the package analogy. It often obscures important components that are nested deep inside; and there is a tendency that is not easily curbed to engage in a search for *the* package, the touchstone, the magic box, which, when opened, will transform poverty into abundance. Scientific agriculture will not come packaged. Its components are part of the variegated lattice of national life; to weave them together into a functioning network capable of sustaining a swelling volume of agricultural produce will require a sensitivity to the waves of unfolding events that is seldom found among package handlers.[2]

Hooper concludes his observations with the remark that instead of a "catalogue of musts," there are in fact a few "salient points that seem to set the environment for payoff."

Before coming to these salient points, with particular application to the Middle East, a double disclaimer is in order. For one, we make no pretense that the ideas we present are new, either among the governments and leaders of the Middle Eastern countries involved or among those on the outside who are engaged in assisting their development. But we believe that our comments will serve to inject into the preceding speculations a modicum of realism that would otherwise be absent. Thus, to many readers we merely recall rather than reveal. The second disclaimer is that by pointing to steps to be taken we do not imply that no such steps have been taken in any of the countries at any time to any extent. Quite the contrary. Elements of all of them are found in all of the Middle Eastern countries; and patient research could no doubt pinpoint or at least approximately suggest the degree of their achievement, country by country. But again this would be a book we did not set out to write.

From the beginning we have laid emphasis on the many factors that in combination contribute to the emergence of a productive national agriculture. Some of these factors have been discussed, others have not. Among those which we have treated in some detail are the importance of markets and marketing, agricultural credit, agricultural research and extension, education generally, availability of inputs, levels of management of soil and water, farming practices,

---

[2] *Ibid.*

etc. We shall not here repeat or summarize what we have said previously about these factors, except to say that as their presence favors and accelerates growth, so their absence inhibits or retards it.

The Role of Government

Some other factors have received only the barest mention. Among them is one that in a way sets the stage: government—its objectives, its determination, its view of agricultural progress, its own stability, and its efficiency in action. In this perspective, government has several features: continuity, so that something done today is not abandoned tomorrow; provisons for an orderly transfer of power from one set of national leaders to another, preferably through an elective process but by whatever means the country chooses; sufficient political strength in the ruling administration or group so that it can afford to undertake and continue programs whose payoff extends over a long period and which may lack in visibility; sufficient competence to carry out programs which are decided upon; and, crucial to the subject at hand, the will to assign a high priority to the development of an efficiently producing agriculture.

But the relation between government and agricultural development is not all one way. There is a certain reciprocity between agricultural progress and the nature of the country's government. A taste of substantially higher output, profits, better living, etc. is likely to affect the political process and eventually the kind of government that will be readily sustained. Responsiveness to the farmer, recognition of the role of agriculture in economic development, and a willingness on the part of the farmer to respond in turn to the government go hand in hand. In their absence, agricultural development programs are likely to go astray: either a multiplicity of objectives will dilute the forward thrust; or there will be concentration of effort on program aspects and on engineering works that lend themselves most easily to central control, that are in one way or another glamorous or spectacular (one may put a name on a dam but hardly on a drainage ditch or stretch of wellleveled land), and that involve the least amount of reengineering of attitudes and institutions. This emphasis is not only damaging but generally fatal to such development programs. Outside interests that wish to sell their services to the country have often played, unwittingly or otherwise, a significant role in these miscarriages. But if governments are secure enough of their tenure to profit from this more sophisticated insight, a major obstacle to successful agricultural development will be on its way to elimination.

Agriculture's Status

Another aspect of a positive government view toward agriculture is the recognition that a changed role for the agricultural specialist is necessary. It may well be, for example, that for the immediate future the pace of successful drainage in Iraq will be determined more by the availability of trained men than by limitations of capital. But the market for competent trained specialists in nearly every field today is an international one; such men can and do move to other countries, when employment opportunities are better than in their homeland. Men from Middle Eastern countries are no different in this respect from others. They may have some degree of emotional or sentimental or patriotic attachment to their native country, but this will hold them only for a limited time or to a limited degree when faced with much better job opportunities abroad. This is not to suggest that salary scale is the only determinant of a man's decision where to conduct his work. But unless the agricultural expert ceases to be low man on the totem pole and agriculture ceases to be considered as something to get away from rather than to build up, there will be no skilled manpower to turn plans into reality. This goes for standing among the professions, for recognition in government service, for participation in the international intellectual community, etc.

Production Versus Welfare Objectives

When we come more directly to agricultural strategies, two of the basic choices—it becomes clearer all the time—are (1) whether to give at least a moderate amount of priority to one or the other of the package components or to insist that all must be advanced at an even pace; and (2) whether to improve agriculture gradually across the board or to concentrate the effort in a specific area—be the criterion of selection geographic, product, type of farm, or any other unit.

On the first point one is probably safe in saying that even if it were humanly possible to quantify the degrees of advance in such diverse and hard-to-measure items as credit availability, transportation, or the technology of irrigation, it would still not be possible to accomplish coordinated growth. The catalogue, or whatever else one wishes to call it, is useful above all as a reminder that none of these items must be totally neglected, or that some may need additional attention if overall progress appears to be lagging.

In a way, the package approach must not be taken too inflexibly. Agricultural progress can be made in the absence or with inadequate provision of one or the other ingredient. There are countries with poor road systems that nonetheless show remarkable agricultural growth. Others have poor credit systems, but the country's social structure may allow the farmer to draw on sources larger than those of his own family, until he reaches a point where he becomes an acceptable credit risk. In still others, the absence of a well-developed extension system may be compensated for by proximity of successful farmers, or by a form of farm organization favorable to the transfer of technology.

But even in these cases it is normally understood that somewhere in the country rapid progress in production is being made in a form that leads others toward imitation. This calls for an affirmative answer to the question whether effort should be concentrated rather than dispersed. With limited capital resources and professional talent a certainty,

the Middle Eastern countries cannot expect to undertake every new program in every part of the country at the same moment.

Concentration of effort will be enormously more productive of results, even though it will unavoidably spell neglect for other areas. To the extent that the concentrated programs favor one geographic area over another, many people will be dissatisfied; and charges of favoritism, political discontent, and other adverse consequences may well flow. Undeniably, a clear choice between a production objective and a welfare objective is involved. If one favors the first, it is because truly spectacular advances can now be achieved by such a concentration of effort: when responsive seeds, appropriately formulated and distributed fertilizer, and other inputs can be made available where needed, the effect of dramatic improvements upon the rest of the farming community can be truly revolutionary. A doubling in crop yields is not just ten times a 10 percent rise; it is a quantum leap that breaks through the barrier of tradition. Recent events in India and other parts of Asia surely point in that direction. Concentration will be especially productive if it is not just geographic but also oriented toward the most progressive farmers and a limited and most promising number of crops. In the Middle East these will undoubtedly be the leading cereal crops: wheat, rice, and perhaps corn and barley. A production-oriented program in cereal crops, concentrated in the most promising zones—i.e., those having the most favorable conditions of soil, water, and other natural resources—is likely to yield the quickest and largest payoff and become a force in itself that will affect others.

Even when such a basic decision is reached, there are further ones along the road. How is one to proceed in this concentrated effort to exploit a technologically feasible production potential? Here the package may consist of such steps as determining

* the most economic level of fertilizer application
* the most efficient method and kind of fertilizer application
* optimum water requirements
* suitable soil preparation techniques
* proper dates, depth, and spacing of planting
* effective method of weed and pest control
* most profitable crop rotations
* other practices—e.g., most suitable farm implements and machines

Trial and error on demonstrations farms and in entire zones will gradually produce a body of knowledge that can be carried to the rest of the country, preferably with the help of those that have been directly involved in these initial phases of the process.

Concentration, however, must not be wholly at the expense of the more gradual aspect of agricultural programming in the Middle East. This is exemplified in matters such as drainage programs, which can perhaps be initiated in areas where the concentration takes place but which will necessitate continuing investments elsewhere in order to result in a balanced program. More often than not, local drainage works will have to be accompanied by canals or conduits that carry the drainage water to larger permanent outlets. Likewise, irrigation works often are part of a larger plan, taking many years to accomplish, even though the first and immediate impact might be in the areas that have been selected for rapid growth. Gradualism is also evident in other aspects of development in the sense that there will be a cumulative effect of works undertaken at any time. As the soil gets better drained, weed control becomes more important; and as weeds are brought under control, fertilizer becomes more important. As these practices get adopted, the stage is set for the introduction of an improved crop variety and this in turn leads to the need for more fertilizer again, even better weed control, and so on.

Incentives

If a third component of the package merits to be singled out, it is probably the relationship of input and output prices. This is usually dealt with under the heading "incentives," and it is one of the more perplexing pieces of policy-making for governments, even in developed countries. The urge to be internationally competitive, the desire to keep urban populations from rebelling against the high cost of food, the general lack of political influence of the bulk of farmers in developing countries, the small scale on which inputs are produced or marketed and the inefficiencies in that system—all these and others tend to depress the prices of farm products and raise those of inputs. The squeeze contributes heavily to making farming one of the less attractive occupations.

It has been contended—and recent experience seems to have borne this out—that whatever farmers are not capable of doing, they are surely capable of recognizing the profits that can be obtained as the difference between the cost of inputs and outputs. W. David Hopper has put the matter eloquently:

> Of all the instances of neglect seen in past development programs, probably the greatest has been the neglect of incentives . . . . For some countries the ultimate cost of the absence of incentives for even traditional investment has been fearful; in most cases, and in spite of improvements, it is still being paid. One can hope that a like cost will not have to be paid in other countries. But unless agriculture is made profitable on a continuing basis, society will pay a debt for this neglect.[3]

Because outlets—at home or abroad—for greatly increased production cannot be viewed as a certainty for the Middle Eastern countries we have discussed, and because inputs—often from abroad—are generally procured on a small scale and distribution channels are inefficient and expensive, the matter of output and input prices deserves special attention by the respective governments and a high priority on the list we are here suggesting.

A final word about the role of institutions, both with a

---

[3] *Ibid.,* p. 110

small and a capital "I," is in order. Societies generally shrink from measures that would strongly alter them and prefer recourse to new technology that helps to evade and avoid the issue. At the extreme, a nuclear desalting plant is more eagerly sought than the introduction of measures that would bring stream water savings totaling perhaps many times such a plant's output. We have pointed out the hopelessness of raising the countries' livestock enterprise without measures to control grazing, and the hopelessness of restoring much overgrazed marginal land to productivity without the removal of livestock from its surface. Fragmentation of much of the land calls for organization that permits efficient cultivation, increasingly by machine, and this means institution-building. So does satisfactory handling of irrigation and drainage works.

In all this, the role of government has been stressed, largely because in this day and age the center stage is taken by government plans and programs, frequently prepared as a basis for obtaining foreign loans or grants from private or, more often, public sources abroad; but also because the large cultivators have as a rule shown little interest in national objectives, and the small ones are too deeply entangled in the struggle to stay alive to have either means or vision for larger goals.

But in many parts of the agricultural economy, private enterprise can play an important role. This is eminently true of the marketing of both inputs and outputs, in which the heavy hand of bureaucracy can—and often does—fatally damage proper timing and in general lacks the flexibility of private entrepreneurs. Exhortation by government frequently achieves less than the patient prodding of a salesman intent on making a profit which can become permanent if the results he has pictured as possible do in fact materialize.

As long as government is in position to keep in check the drive for the one-time, quick profit and creates the environment for more continuous endeavors, there is much to be gained from keeping government out of other aspects of marketing. Here again, the concentration of effort, the effect of demonstrating dramatic change, can provide the focus for the injection of private marketing initiative.

Beyond these remarks we can hardly hope to venture. The specific circumstances of each country will dictate different paths of policy and program. In each country a different region, a different product, or a different set of producers will be selected to receive immediate attention designed to provoke a radical spurt in output. In each country different fundamental and long-term programs will recommend themselves to the government. One would hope that there would be drainage in Iraq and Egypt; introduction of greatly improved cropping practices on rainfed lands in Syria, Iraq, and the drier parts of the other countries; a greatly raised stature of agriculture as a branch of the economy, of farmers as members of society, and of agricultural experts as high-ranking professionals; and a general determination to give unglamorous, continuing tasks at least equal standing with, if not preference over, the flashy and spectacular. There is now a vast stock of available technology and practices at hand which, combined with the great natural resources of the Middle East, are adequate for doubling or tripling agricultural output in the next two or three decades, without recourse to miracles wrought by untried, revolutionary technologies and practices. The real miracle will come in the combination of these elements in a favorable political and institutional environment, not in the discovery of one magic approach that will quickly bring major change to the region.

Soils of the Middle East

THIS appendix and the 1:2,000,000 scale maps in the pocket of this book are included for those readers who have need for or interest in more information on the soils of the Middle East than is given in Chapters 2 and 4. The presentation, however, is not meant for the use of soil scientists. Rather, it is aimed at more general readers who would like some information about the soils of the region.

The map delineations were taken from 1:1,000,000 maps prepared by the World Soil Geography Unit, Soil Conservation Service, United States Department of Agriculture. It should be pointed out that most of the map preparation was done a number of years ago, with map legends and mapping-unit designations abstracted by us from some more complete and detailed ones prepared for a worldwide map that covers areas other than the one with which we are concerned. In shortening these legends and designations and in taking out some more technical language, we hope that we have not done too much violence to the purpose and intent of the originals. We are very grateful to the Soil Conservation Service for letting us use their data and maps, which we feel add greatly to the value of the book. The maps and soils information have not been previously published.

Some people have questioned our use of the older soil terminology at a time when new systems of classification and designation are being developed and accepted both in the United States and in other countries, as well as by a recent soil map project of FAO and UNESCO. There are several reasons why we have stayed with the older terminology. In the first place, the map and the descriptions of the map-delineation units were only available in terms of the older classification system. It was not possible to convert to the newer system without redoing the maps. This could not be done in the time available to us. In the second place, the new classification systems were not published in final form at the time this manuscript was prepared. Thus, though available to most soil scientists now and for some time past, they have not been available to most readers of this book. Moreover, perhaps, the older system, though not as elegant or informative as the new, would, we believed, give more useful information to our readers with less expenditure of effort—this apart from the fact that the authors, as with an old hat, felt more comfortable with the older terms.

We must emphasize that these maps, though useful, are schematic only. They were made from the best information that was available at time of completion some years ago, with many sources of information on geology, terrain, soils, and vegetation available and used. Although we do not show the degree of reliability of the map delineations and designations on our maps, it should be obvious that they will be more reliable in well-known areas such as the Nile Valley and Delta of Egypt, the coastal plain of Israel, and the Mesopotamian Plain in Iraq than, for example, in a remote portion of the deserts of southwest Jordan.

What the maps do show, at the Great Soil Group level—Alluvial, Solonchak, Terra Rossa, for example—is where the various soils occur, as well as their relationships to each other and to the climatic zones where they are found. Our shortened mapping unit descriptions tell something about the soils within the delineations and give some information on their suitability for producing agricultural products.

In discussing and describing the soils of the study region we have tried to minimize the use of technical terms that are not familiar to the general reader. But since there are limits to this effort, with the reader faced directly and inescapably with the Great Soil Group names used both in the text that follows and in the mapping-unit descriptions, there follows a very short and hence inadequate statement about each of the Great Group designations indicated in the text and soil mapping-unit descriptions. The aim is to convey to the reader some bit or bits of useful information about the soils included within each of the designated Great Soil Groups.

Great Soil Group Names, With Brief Descriptions

Map Symbol	*Name*
A	*Alluvial Soils* are formed from geologically young materials deposited by running water in stream valleys, deltas, or at the foot of slopes. Soils of the Nile Valley and the Nile Delta are examples.
BD	*Barren Desert Soils*, as the name implies, have little or no vegetative cover. They occur in our study area in the almost rainless areas of the Egyptian deserts.
BF	*Brown Forest Soils* were developed under a deciduous forest vegetation on calcareous or neutral colluvial materials under moderately high rainfall. They occur in mountain basins of Iraq.
CT	*Chestnut Soils* were developed under grass vegetation on calcareous materials under moderate rainfall. They are found in an area north of Damascus, Syria.

D	*Desert Soils* were developed in areas of low rainfall but have some desert-type vegetation. They occur in extensive areas in Syria and Iraq.
L	*Lithosol Soils* are thin with a high proportion of bare rock or very stony surface. They are soils of general occurrence in our study area.
NB	*Noncalcic Brown Soils* are nonacid soils that do not contain free calcium carbonate. The sandy coastal plain areas of Israel and Lebanon that have been leached free of calcium carbonate are the main occurrence in our study area.
RB	*Reddish Brown Soils* were developed under grass vegetation with moderate rainfall. They are found, for example, in northern Syria along the Turkish border.
RD	*Red Desert Soils* are soils in the desert areas with reddish color. They occur in the desert areas of Egypt, Jordan, Syria, and Iraq.
R	*Rendzina Soils* are highly calcareous dark or grayish-brown soils developed under grass or broad-leaf forest vegetation. Examples are the dark soils associated with the Terra Rossa in the hilly and mountainous areas of Israel and the West bank of Jordan.
RM	*Rough Mountainous Land* Soils are just that.
S–M	*Beach Sand–Dune–Marsh complex.*
S	*Sand Dunes.*
SR	*Sierozem Soils* are developed under desert-type vegetation with some short grass, under somewhat more rainfall than prevails in the deserts. Large areas of these soils are found in northern Syria between the true deserts and the areas of better rainfall towards the mountains to the north.
SK	*Solonchak Soils* are salty soils that occur within desert depressional areas where evaporation of runoff waters has left salt accumulations. Similarly, they occur in higher rainfall zones where salt accumulation is high. Numerous areas are found throughout the drier portions of our study area.
TR	*Terra Rossa Soils* are, as their name implies, red soils. They were formed primarily under broad-leaf forest from hard limestones and under moderate to high winter-type rainfall.
TRR	*Terra Rossa–Rendzina* complex is an intimate association of the Terra Rossa and Rendzina soils. (By "complex" we mean that the two soils occur within the map delineations in undetermined proportions and in undesignated locations.)

Soils of the United Arab Republic

The agriculture of the United Arab Republic is confined to the Nile Valley and the Nile Delta. The exception is a small amount of irrigated land in the several large depressions in the Western Desert, where land is watered by fossil water from the Nubian sandstone.

The Nile Valley and Delta comprise about 3.1 percent of Egypt's 247 million acres of total area. These 7.7 million acres of arable alluvial soil consist of an overlay of Nile "silts" several meters thick over layers of sand and gravel deposited in an earlier age. These Nile silts originate for the most part in the volcanic surfaces of Ethiopia.

The mineralogical composition of the heavy mineral fraction of the silt is in keeping with this origin. The principal part of the heavy mineral fraction—after the silts are cleaned by acid treatment—consists of augite, amphibole, and epidote. Sediments in the river at floodtime may also contain more than 50 percent clay-size particles. The clay may contain some kaolin, but is dominantly illite with a considerable component of montmorillonite. The proportions of clay and coarser materials vary according to the velocity of the water depositing the sediments. The highest percentages of clay are found in the sediments from still backwaters and in impoundments.

The textures of the soils of the Nile Valley and Delta are fairly fine. Perhaps the most usual are silty clay loams and silty clays. With the relatively high percentages of swelling montmorillonite clays, the soils are not easy to cultivate.

Not only do they require much power for tillage, but also there is a strong tendency to form clods, particularly if the moisture is not just right at the time of cultivation.

The Nile silts—or mud, as it is sometimes called—have been famous for their fertility throughout recorded history. Part of this fertility is undoubtedly due to the fresh, weatherable minerals that can serve as a source of inorganic nutrients required by the plants. Another contributing factor is the high content of organic matter of the silts when deposited.

According to Ball,[1] the average organic matter content of the suspended matter of the Nile River at Cairo in 1924–27 was 2.5 percent. This organic fraction had a carbon-to-nitrogen ratio of 11. Thus, good supplies of nitrogen needed for plant growth are deposited along with the mineral nutrients. One must conclude that the fertility of the Alluvial soils of Egypt is due to the loss of fertile soil mainly in Ethiopia. In the main the fertility was not developed subsequent to deposition. Here it should be pointed out that though the natural fertility of the Nile Alluvial soils is justly world famous, it is not adequate for the sustained high production of modern agriculture. Egypt now uses relatively large amounts of manufactured fertilizers and must use even more if the higher levels of production that are possible are to be attained.

The lack of adequate drains and the overuse of irrigation

---

[1] John Ball, *Contributions to the Geography of Egypt* (Cairo: Government Press, 1939).

water has led to high water tables in many of the Nile Valley and Delta soils. This problem has become more acute since it has become common to take more than one crop per year from the land. Balls[2] has given us a fine discussion of seasonal fluctuations and the consequences of these high water tables. Without question, a large portion of the Delta now has limited crop yields because of this situation. A massive effort at lowering the water table, particularly in the Delta, is needed. This should involve not only an adequate system of drains, but also restraint on overuse of irrigation water.

Soils of the Desert Areas

With the exception of a small barley growing area southwest from Alexandria with 200 mm. (8 in.) precipitation, there is no dryland cropping in Egypt. West from Alexandria there is some rough grazing along the Mediterranean Coast. Also, there is a small amount of irrigated agriculture in the oasis depressions of the Western Desert. Thus, outside the Nile Valley and the Delta and the named small-area exceptions, the whole of Egypt is an inhospitable desert. In the southern portions the rainfall averages less than 25 mm. (1 in.). In some years there is no rain—although occasionally there is a deluge.

Because these desert areas abut rather immediately on the Nile Valley and the Delta, Egypt has very little land for extending the area of cultivation. There are some cultivable areas in the desert depressions. But the water, both artesian and pumped, is fossil and is subject to depletion. Too, these soils are mostly salty and lie below the level of surrounding areas with no outlets. On a longtime basis, drainage will be a problem. The question of how long the water will last is, of course, a crucial one. Some wells from Roman times are still flowing, while some recent ones are now dry. Another serious hazard to agriculture in these depressional oases is the surrounding drifting sand dunes.

Soils of Jordan and Israel

The most universal characteristic of the soils of Jordan and Israel is their usual high content of calcium carbonate,[3]

---

[2] W. Lawrence Balls, *The Yields of a Crop* (London: E. & F. Spon Limited, 1953).

[3] Calcareous soils, as we use the term, are those that contain enough calcium carbonate or limestone to materially influence their properties. Usually this will be several percent or more. The ways in which calcium carbonate affect soils are many.

Because calcium carbonate or limestone react with soil materials and with rainwater to release calcium ions, these soils are always well supplied with calcium ions. Also, because most all limestones and other calcareous deposits contain a component of magnesium, the soils are adequately supplied with magnesium ions as well. Reasonable amounts of both calcium and magnesium ions are of course necessary for plant growth. The presence of free calcium carbonate has a beneficial effect on the physical condition of a majority of soils. It tends to make them more granular and loamy.

On the other side of the coin excessive amounts of calcium carbonate maintain a high pH—alkaline condition—in the soils. This suppresses the availability of some of the minor nutrient elements, causing a chlorosis or yellowing of the leaves. Citrus are particularly susceptible, as are many other crops. Too, if the percentage of calcium carbonate is very high, it acts as a diluent to the active soil components—clay fraction and organic matter, for examples.

which usually ranges from 25 percent upward. Except for the coastal area, only in a few soils of the highest rainfall areas—above 600 mm. (24 in.)—are even the surface horizons free of calcium carbonate. And even in these higher rainfall areas most of the soils are calcareous to the surface.

Another general characteristic of the soils of East Jordan is the very low content of soil organic matter. Organic matter does not persist in these areas of very high summer temperatures. As a consequence the soils are quite poor in native nitrogen. As a general rule nitrogen is the first limiting plant nutrient for crop production.

Soils of the Jordanian Deserts

Most of Jordan is desertic with an average annual rainfall of less than 100 mm. (4 in.). With the very hot summers that prevail and this low precipitation, very little vegetation is produced. Useful vegetation for grazing animals is produced only in the wadi (gully) bottoms and other drainageways where there is accumulation of some rainfall runoff from higher ground. Even here the grazing potential is very limited indeed. The soils are mostly of the Gray-Desert kind. They are of medium texture, calcareous throughout, and may contain some soluble salts. They occupy more than 50 percent of the total area of the country. Little or nothing can be done to improve the plant production in these desert areas.

In the south part of Jordan on the dissected plain and in the Araba Valley are sands and thin soils from sandstones that do not show soil profile development. These areas, too, have only a very limited use for grazing of the natural vegetation since there is very little to graze. Within these desert areas are two oases that have in the past furnished a limited amount of stable crop production from groundwater. The total production potential is quite limited, however.

In the northeast part of Jordan are large areas of desertic soils from basalt. Some of the soil material on these basalt fields has undoubtedly been brought in by the wind. As with areas covered by Gray Desert soils, the production of vegetation that can be utilized by grazing animals is very low. Some of the desertic areas in the northeast part of the country are highly gypseous (containing much calcium sulfate). In fact, in the Azraq area large expanses of almost pure gypsum.[4] Production of vegetation in these areas is very scanty.

Soils of the Steppes

Between the desert to the east and south, and the areas of higher rainfall to the west lies an area of what Moorman has called Yellow soils.[5] On our map we have designated them Sierozems. In this area the average rainfall is from about 100 mm. (4 in.) to 250 mm. (10 in.). These soils also are found in a similar rainfall belt between the higher rainfall areas of the mountains, and the drier Jordan Valley and Wadi Araba to the west.

---

[4] F. Moorman, *The Soils of East Jordan* (FAO Report No. 1132, [Rome, 1959]).

[5] *Ibid.*

The grazing potential of these soils is much greater than that of the deserts. Indeed, in some of the broad plateau areas and in some of the depressions having deep soils some extensive cultivation of barley is carried out. However, the best use of the major part of this area is for grazing, and at present the land is badly overgrazed.

Some water-spreading structures of about one-half meter height have been built in the region of Qasr el Hallabat. Where this has been done some fair wheat crops have been grown. However, it would seem that here is a real opportunity to use these water-spreading structures to greatly increase and to stabilize the forage supply for the animals—primarily sheep—of the area. Where the soils are of sufficient depth, soaking the ground with extra water from flooding during the winter rainy season would make possible the growth of large amounts of fodder for the summer feeding of the sheep. Of course, this would reduce to some extent the runoff of water in the wadis into the Jordan River Valley. But this local use could be a much more efficient utilization of the water than for irrigation further downstream.

This steppe area has been used much more intensively in the past than it is being used today. As evidence of this, in addition to the many one-stone high water-spreading structures, there are remains of semi-irrigation works of past ages. As pointed out by Moorman, it may be that there was more rainfall a few thousand years ago than now. Also, it is quite probable that a family then could be pleased with a lower amount of produce for its labor than today. In either case there is an opportunity to produce much more from the land and water than is being done today.

Soils of the More Humid Zone

The Red and Yellow Mediterranean soils of these areas have average annual rainfall of over 250 mm. (10 in.). These soils, like the others we have discussed, are highly calcareous. In this feature they may differ somewhat from many of the soils of the western Mediterranean area derived from limestones, such as the well-known Terra Rossas, which are not generally so calcareous in their upper horizons.

The bulk of the rainfed agricultural production in East Jordan is on these Red and Yellow Mediterranean soils. Generally, the soils of these areas of higher rainfall will produce crops each year. They have quite good water retention characteristics and plant roots penetrate well into the deeper horizons.

Perhaps the most common misuse of the soils of the higher rainfall area is in the usual year of weedy fallow that falls between crops of grain in alternative years. These weedy fallows are used for feeding of sheep. But this is a very inefficient use both of the land and the natural rainfall. True, the fallow may be serving some purpose in the control of plant disease. Wheat planted every year on the same land does run into problems. But even if continuous wheat production is not desirable, the alternate year could be used to produce a crop of forage or another grain crop such as barley or grain sorghum rather than the weed fallow. A weedy fallow that has enough vegetation to be grazed profitably is not a moisture-conserving practice.

Interestingly, it has been suggested also that under similar rainfall conditions of the Northern Jesireh in Syria, a mixed-farming system that made use of the land to grow a crop each year would be a desirable change from the almost universally practiced wheat-fallow-wheat farming system now in use.

In these areas of higher rainfall, winter grain is a natural crop to use because the growing season coincides with the rainy season (when moisture is in the soil and evaporation rates are low). It is true that the annual amount of precipitation is somewhat more erratic than in some other parts of the world that have a winter-type rainfall pattern. It will perhaps always be true too that there will be an undesirable annual fluctuation in the grain crops produced from this area. But much better use than presently can be made of the available moisture by control of weeds, timeliness of cultural and planting operations, and use of better adapted varieties. Both the soil and the water are capable of producing several times as much grain as now obtained.

While the more level areas are suited to grain farming on an extensive scale, the steeper areas are suited to the production of more intensive crops. Usually rock terraces are required. While these do require considerable labor to prepare, they can be relatively permanent. The tree and vine crops grown include olives, apples, and grapes. Where the soil is too shallow for these crops and where the areas between the limestone rocks are too small, good use can be made as grazing land.

To the west of the area of higher rainfall is the transition zone between this and the below-sea-level Jordan Valley. This zone is intermediate in elevation and in amount of precipitation between the higher rainfall area to the east and the very dry valley. The soils found here are on relatively steep topography; and usually are rather shallow, very calcareous, and not very productive. Except for some wadi bottoms that can be irrigated, the best use is for grazing land. Though this is the present use, this area—as is the case for all of Jordan—is badly overgrazed, with consequent increased soil erosion and less overall production than would be possible under better pasture management.

Soils of the Jordan Valley

The principal area of Jordan in which an irrigated agriculture can be developed is in the Jordan Valley. Recent estimates indicate net suitable irrigable land between 44,000 and 45,000 hectares. This amounts to only 4 percent of the arable land of Jordan and to less than 0.5 percent of the total area. But the full use of this area under intensive irrigated agriculture could make a very large contribution to the productive capacity and the financial stability of the country. Of this irrigable land, about 70 percent is on the east side of the Jordan River.

The soils of the Ghor terrace of the Jordan Valley are mostly underlain by saline marl (soft limestone). Over the marl are fluvial and colluvial materials moved down-slope by gravity and water, but not suspended in a stream—deposits of moderately fine textures—clay loams, silty clay loams, and loams. Where the overlays are thin, the salty marl comes to or near to the surface. The soils shallow to marl have less value because of the shallower root zone and the presence of salt. There are some areas of clays, but these only amount to about 3 percent of the total.

About one-third of all of the suitable soils of the Ghor terrace are salty enough to require drainage and appreciable preleaching before irrigated agriculture can be established. Although these soils have the handicap of the salt, its removal leaves a soil thought to be equal in potential productivity to the comparable nonsaline soils. Thus, removal of the salt by leaching removes the handicap permanently, provided good irrigation practices are subsequently followed.

The soils of the Zor—the lower terrace of the Jordan River—are formed from rather recently deposited materials that have somewhat coarser textures than the soils of the Ghor. These soils yield somewhat better than the soils of the Ghor. Perhaps this is due both to their more sheltered situation at lower elevation and to their somewhat coarser textures, making for better root penetration and water movement. At the present time, more than one-half of the area of the lower terrace is subject to annual occasional flooding from November to March. The proposed dams on the Yarmouk River should reduce the flood hazard, but there will still be a number of wadis that run directly into the Jordan from both the east and the west.

Soils of the West Bank of Jordan and Israel

On the west side of the Jordan Valley and Wadi Araba lies a belt of steep rocky Lithosols (shallow thin soils). In fact, most of the area is bare rock. For most of the area the exposed rock is limestone, but a small area adjacent to Lake Tiberias is mostly basalt.

In the south along the west side of the Dead Sea, rainfall is less than 100 mm. (4 in). Here there are only occasional patches of desert vegetation. There is very little grazing for animals. In the north near Lake Tiberias, the precipitation is about 400 mm. (16 in.). Here the grazing is, of course, much better.

As one goes up the escarpment to the west the rainfall increases with elevation, though still in the rain shadow. On the west side, this area of steep Lithosols is bordered by the Terra Rossa (red soils residual from limestones) and Rendzina (black calcareous swelling soils) complex area. This is the highest rainfall area of Israel and the West Bank of Jordan. The precipitation which comes November through April ranges from 500 to 700 mm. (20 to 28 in.). In spite of the relatively wet winter season, nearly all of the soils are calcareous in all horizons. The red soils have developed on the residuum for weathering of hard limestones and the black ones from marls (soft limestones).

These upland soils are mostly on hilly terrain with a large proportion of rock outcrop. Agriculture is carried out mostly on the areas that can be terraced, though there are some less steep rolling areas, particularly in Samaria. These are mostly on the marls. There has been a great deal of erosion here in the past. Many remnants of old broken stone terraces bear mute testimony of the labor that went into gaining a small plot for cultivation. The principal crops have been olives, vines, and grain. Now some tomatoes, tobacco, and vegetables such as squash and cucumber are grown.

That these soils have good water-retention characteristics and permit root penetration to good depths is shown by the fact that tomatoes and tobacco are planted in May after the last rain and still, with no rain and no irrigation, make a reasonable crop. Where some groundwater is within reach of pumps, supplemental irrigation, of course, gives greatly increased yields. But where these soils are of sufficient depth and where they are smooth enough, or can be held by terraces, they are good soils for the production of crops, particularly those that can grow early in the summer. The long, dry summer is, however, a handicap since, during it, no supplemental irrigation water can be had.

Between the hill country and the Mediterranean Sea in Israel is the coastal plain, covered by soils derived from sandy materials. In these soils the calcium carbonate has been removed by leaching to a considerable depth. The soils, in contrast to the soils mentioned so far, do not generally contain free calcium carbonate. They are about neutral in reaction. These noncalcic Brown soils—the name connotes lack of free calcium carbonate—are used intensively for the production of oranges, grapefruit, and many other kinds of produce.

The average annual precipitation of this coastal plain area ranges from 500 mm. (20 in.) in the south to 700 mm. (28 in.) in the north. This is adequate moisture for winter grains, but summer crops including citrus must be irrigated to produce well and to survive. The topography is that of a series of hills and ridges. This unevenness reflects the probable origin of the sands as windblown ridges and dunes—before the time of stabilization by vegetation. The soil textures are moderately coarse, consisting of sands and loamy sands.

To the southwest of Lake Tiberias is an area of Reddish Brown soils forming the Plain of Esdraelon and the Vale of Jezereel. These soils, derived from basaltic materials, are among the most intensively cultivated in all of Israel. They are calcareous throughout and are, with minor exceptions, well drained.

To the south of the area of noncalcic Brown soils in Israel lies an area of deep Loessial—fine-textured calcareous windblown material—soils. This area lies around but primarily to the west of Beersheba. The rainfall here ranges from 150 to 300 mm. (4 to 12 in.). The vegetation is that of a grassy steppe. The area grades on the south to somewhat coarser windblown materials that get thin toward the hills to the south and east. Also, the rainfall lessens in the same directions. The general topography of this area is smooth, with some stream channels cutting through. These could be good soils for farming if water were more plentiful. Some scanty crops of barley are occasionally grown, but without irrigation water the production potential is very poor.

Along the coast for the whole length of Israel are narrow belts of sand dunes, though they are not continuous in all places. Some of these are still in motion and may be a hazard to city areas.

The deserts of the southern part of Israel receive less than 100 mm. (4 in.) of rainfall a year. Some years there is no rainfall. Some of the area is sandy but most of it is rocky, with very little vegetation.

Soils of Syria and Lebanon

In rough proportions, almost one-half of Syria's 45.8 million acres is desert or rockland that has little or no agricultural potential. Over one-fourth is grazing land. The remainder is suitable to varying degrees for rainfed or irrigated agriculture. More than one-half of the country gets less than 250 mm. (10 in.) of annual precipitation on the average, and perhaps one-third of the total area gets 150 mm. (6 in.) or less. With the prevailing hot summers, the resulting climate is truly quite desertic in much of the country.

Soils of the Desert Areas

Muir[6] and Reifenberg[7] have described several kinds of soils in the Syrian Desert. Muir recognized Brown desert soils, solonchakous (salty) Brown desert soils, and undeveloped soils on alluvium. As he has described them, all of these soils have depths of the order of 75 cm. (30 in.). Actually, over most of the Syrian desert area the soils are Lithosols (thin soils), with a large proportion of bare rock surface. All of these soils have moderate amounts of clay—usually of the order of 20 to 30 percent of the fine-earth fraction (less than 2 mm. diameter). As is generally true in this part of the world, all of the soil-forming materials are highly calcareous, with a percentage usually between 25 and 50 percent. The calcium carbonate is disseminated throughout the whole soil mass in all of the size fractions. Gypsum (calcium sulfate) is generally present, especially in the lower soil horizons. In some of the areas this material is the dominant component of the geologic material underlying the soil. Some of the gypseous formations in the soil indicate that this mineral has dissolved and reprecipitated as crusts. These gypseous materials were originally deposited in inland seas during the Miocene geologic period—about 20 or 30 million years ago.

Where the Desert soils were formed on materials residual from the decomposition of hard limestones, there are many flint (a dense, fine-grained form of silica) rocks. These rocks, when exposed on the surface, get a dark-colored surface coating called desert varnish (this is usually a very thin coating of iron oxide with perhaps some manganese oxide). When these flint rocks predominate, the area is called a "flinty" desert. Characteristically, these desert areas are windswept and inhospitable at times. In many places a few inches of sand accumulate around the few sparse shrubs that manage to survive the hot, dry summers. In some places there are sand dunes.

The desert areas may furnish a very limited amount of grazing for camels and sheep. But the total rainfall is low and very erratic. Some years there may be no rainfall in a given location. In others there are occasional deluges. Thus,

pasturage for nomadic flocks is where the rains happen to fall. The nomads may hunt for spots where it has rained as a fisherman may have to hunt for schools of fish. At best, survival of man and beast is not easy. At worst, not too many survive.

As already stated, the soils of the desert areas contain much calcium carbonate and usually gypsum in varying amounts. Since rainfall is so low, chemical weathering is very slow. Thus, minerals are generally little weathered. The clay minerals are usually montmorillonitic (swelling clays that erode deeply when dried from a wet condition); but Muir has also shown that a fairly unusual soil clay mineral, polygorskite, is also an appreciable component of these Desert soils.

Soils of the Steppes

The steppes—from the Russian designation—are extensive treeless plains that lie between the desert and the areas having enough rainfall for a sustained agricultural production. We call the soils of the steppes Sierozems, also a Russian term. In Syria the Sierozem soils are light in color, low in content of organic matter.

The Sierozem soil area of Syria is to the north and west of the deserts, covering about 20 percent of the area of the country. The rainfall in the Sierozem area is generally between 200 mm. (8 in.) and 350 mm. (14 in.). Short-duration periods of frost in the winter are common here. The native vegetation probably consisted originally of grasses and shrubs. But now there is little grass.

Until relatively recent times this area was used almost exclusively for grazing. In recent years, however, some extensive, large-scale grain farming has been undertaken. This, together with the farming in areas to the north with somewhat more rainfall, constitutes the most highly mechanized farming in Syria. When rainfall is good, reasonably good crops—particularly of barley—are obtained. Certainly, though, under systems of management now generally practiced in Syria, the portion of this soil belt next to the desert is suited only for grazing. Moreover, although this Sierozem soil area comprises about one-fifth of the total area of the country, it can only produce under native vegetation about 0.3 million sheep-years of grazing. (By "sheep-year," we mean "furnish feed for one sheep for one year.") In other words, while the area comprises 80 percent of the total grazing area of Syria—exclusive of the deserts—it can furnish only about one-fifth of the total sheep-years of grazing, exclusive of that on cultivated land.

The Sierozem soils of Syria are generally shallow, high in calcium carbonate, and frequently high in gypsum. They contain little organic matter and have weak structure. They are of medium texture—generally 20 to 30 percent clay. Although the clay mineral composition is variable, montmorillonite and polygorskite together—with some kaolinitic components—are most common.

Potential of the Sierozem Belt in Syria

The production of grain in areas that are marginal with respect to rainfall is always hazardous, due to uncertainty of the amount of rainfall for any particular crop year. It

---

[6] Alex Muir, "1951 Notes on the Soils of Syria," *Journal of Soil Science*, II, No. 2 (1951).

[7] A. Reifenberg, "Soils of Syria and Lebanon," *Journal of Soil Science*, III, No. 1 (1952).

seems, to the farmer at least, that average years come infrequently—and in these areas when there is seldom enough rainfall for maximum yields—that one lives with a perpetual shortage of moisture for high crop production. But frequently low yields are due more to poor management practices than to a shortage of moisture. In the dry-farm—rainfed—areas of many if not all Middle Eastern countries, weeds transpire more moisture than the grain.

Van Liere[8] has said that lack of higher grain yields in years of average rainfall is not due to low soil fertility, soil depletion, nor insufficient moisture. Rather, it is due to irregularities (deficiencies) in soil management. It is the custom, for instance, in the rainfed areas of grain production in Syria to have a fallow year between the alternate years of grain—primarily wheat. It has been said that the fallow year is to conserve moisture and to "rest" the soil, as well as to provide pasturage for the sheep. Actually on these soils some nitrogen was perhaps built up for the next crop. But the weed fallow does not save any moisture for the next wheat crop and the weeds as pasturage are a very poor use of the water that came during the previous winter. A benefit that did accrue to the farmer because of the fallow, however, was some control of pests and pathogens that tend to become problems in a continuous monoculture.

A serious question can now be raised regarding any necessity for continuing the fallow year between grain crops. The nitrogen needed can be had from fertilizers and the continuous monoculture can be broken up by introducing a crop as a replacement for the fallow. Vetch, lentils, sorghum, and soybeans have been suggested as possibilities. The production of fodder crops to replace the fallow year would necessitate many adjustments. But, undeniably, the land and the water now available can produce far more food than now.

Recent developments in soil management and in water conservation, together with new wheat varieties, have made it possible and profitable in the United States to produce good crops of wheat in alternate years in a rainfall belt of the same magnitude as the marginal grain zone designated by Van Liere (200—300 mm.—8—12 in.). Usual yields under good management in this area of the United States are now of the order of 40 to 60 bushels per acre (2700 to 4000 kilograms/hectare). A wheat producing area in the State of Washington, furthermore, lying between the drier portion of the Columbia Basin and the Palouse country to the east, has average precipitation values about the same as the area in Syria under discussion (200—350 mm.—8—14 in.) Recent improvements in varieties of wheat and new knowledge of how to conserve and use moisture have made it possible to increase yield very greatly in the Washington State area, with yields of 40 to 50 bushels per acre (2,700 to 3,400 kilograms/hectare) commonly obtained in alternate years.

Two circumstances work against a comparable potential for the Syrian soils. One of these is the fact that the winter rainfall period is shorter for the total amount of precipitation in Syria. Another factor—general shallowness of the

soils—makes it impossible to carry appreciable moisture from one rainfall period until the next year. In order for moisture to be carried over, the Washington farmers need a depth of about 2 meters (6 feet). Few if any of the rainfed Syrian soils have such depths.

However, there is enough moisture each year to wet the soil to a reasonable depth, making it is possible to grow annual crops. Perhaps one of the great needs here is specific adaptive research aimed at the selection and development of varieties of grains and fodder crops that can get started early in the spring and mature early before the dry, hot summer season. Barley may be the best grain bet for early maturity and drought resistance. Fodder crops such as grain sorghum (also a widely used food grain) and vetch also deserve serious developmental research.

Reddish Brown Soils in Syria and in Lebanon

The Reddish Brown and associated soils of Syria and Lebanon are in a higher rainfall belt than the Sierozems and in a lower one than the Terra Rossa and associated soils. The average annual precipitation of this Reddish Brown soils area is generally in the range from 300 to 500 mm. (12 to 20 in.). All of this comes as winter rain or snow. Frost occurs in the winter but the duration of freezes is short. Summers are hot and dry.

Most of the soils are derived from calcareous materials—limestones or chalks. However, about one-fourth of the area covered is derived from basic igneous rocks—lava, ash, pumice. These, too, contain liberal amounts of calcium carbonate, except for the surface horizons of a few soils under the highest rainfall. The occurrence of calcium carbonate in soils and soil-forming materials residual from basalt and related rocks is authigenic—generated on the spot—from calcium, not leached from the weathered rock and the carbon dioxide in the air and rainwater.

The Reddish Brown and associated soils are used mainly for grain—mostly wheat—under natural rainfall. Where available, irrigation water is used on specialty crops such as fruits and vegetables. Some cotton also is grown with supplemental water. Soil textures are in the range from medium to heavy.

The soils derived from basaltic materials have a high clay content, with montmorillonite as the principal constituent. As a consequence, some of the soils crack widely and deeply when dried. The filling in of these cracks in the dry soil by sloughing from the crack sides, causes them to churn during long periods of time. These soils are called Grumsols or Vertisols. The rainfall on these soils is adequate to wet them down to their full depth each year. Thus, they are suited for production of grain that makes its growth in late winter and spring. Without irrigation they are not suited for production of summer vegetables and longtime maturing summer crops, such as cotton or maize. However, fodder crops such as sorghum could be grown in alternate years with wheat.

Terra Rossa-Rendzina Complexes and Associated Soils of Syria and Lebanon

Terra Rossa (red) soils are derived from hard limestones. The earthy soil-forming materials are the residium left when

---

[8] W. J. van Liere, *Report to the Government of Syria on the Classification and Rational Utilization of Soils* (Report No. 2075 [Rome: FAO, 1965]).

the calcium carbonate of the limestone has been dissolved and taken away in the leach water. The Rendzina soils (usually dark in color) are formed by the weathering of soft limestones. Clay contents are higher in the Rendzina soils and the proportion of montmorillonite in the clay fraction is higher than in the Terra Rossa. (By "complex" we mean that the two soils occur within the map delineations in undetermined proportions and in undesignated locations. In sufficiently detailed mapping the two soils would be separately designated.)

Moorman[9] has avoided the use of the name "Terra Rossa" for the red soils of East Jordan; he calls them "Red Mediterranean" instead. Primarily he does this because the term "Terra Rossa" is usually limited to the red soils of the Mediterranean area derived from hard limestones. Under the term "Red Mediterranean," then, he has included both the soils from limestones and those of similar color from basalt. On the other hand, Muir[10] emphasizes the differences in composition between the red soils from limestone and those from basalt. Perhaps the most important difference between them is the higher content of the swelling clay montmorillonite in the ones from basalt. Both varieties make excellent agricultural land when of sufficient depth and on suitable terrain. A third kind of associated Red Mediterranean soils is that derived from conglomerates, as for example the basin at Homs.

Actually the soils from basic igneous rocks and conglomerates tend to have a darker color than the Terra Rossa from hard limestones. Sometimes they are almost black. All of them are highly productive under good managment.

Soils of Iraq

Iraq is an irregularly shaped area lying between 29° and 38° north latitude, and 38° and 49° east longitude. The total area is about 172,000 square miles, or 110 million acres, and thus is a little larger than California. Only the northeastern one-third of the country gets more than 20 centimeters average precipitation. Thus, most of the country has a subtropical desertic climate with very hot, rainless summers.

The dominant physical feature of Iraq is the Tigris—Euphrates River system and the flood plain that is a common feature between them. Both rivers arise in and have their main water catchment areas in Turkey. Because the rainfall for these rivers comes mainly in the winter season, much of it as snow, both have a flood period in late spring and early summer. The Tigris River has an average flood peak discharge at the edge of the flood plain at Samarra, Iraq, of about 3,200 cubic meters per second or 113,000 cubic feet/second. The corresponding figure for the Euphrates is 2,100 cubic meters/second. The maximum flood peaks for both rivers yield much greater flows than these. For the Tigris, the peak is about fourfold greater.

It is these tremendous peak floods that have made the flood plains of the two rivers coalesce into a single geomorphic unit. The main or lower flood plain begins near

Samarra, Iraq, on the Tigris, and near Ramadi, Iraq, on the Euphrates. This plain—375 miles long and 125 miles wide—has been a center for irrigated agriculture for several thousand years. While it has been one of the world's most productive areas during ancient times, a great deal of the land is now salted up beyond possible use without reclamation measures. The salt has come from the small amount in the river water by evaporation from plant and soil surfaces during the several thousand years of use for growing crops. Most of the area, however, can be reclaimed and used for intensive agriculture.

Except for the relatively abundant water of the two rivers, this flood plain would be a part of the deserts to the north, west, and south that cover the greatest part of Iraq. The whole area of the deserts and flood plain has an annual precipitation rate of about 150 mm. or less (5—6 in.). Vegetation in the absence of irrigation from river or ground water is generally very sparse. A few shrubs and some grass grow in the bottoms of wadis and furnish scant grazing for nomadic heards following occasional heavy rains. There is underground water in some places, and a number of wells have been put down to get water for livestock and for the nomadic people who own them. It is thought that the groundwater originates far distant and that to some extent at least it may be fossil.

The northeastern 5 percent of Iraq is rugged mountain land with maximum elevations of about 12,000 feet. In this area of parallel mountain ranges, the precipitation is as high as 1,300 mm. (50 in.). Elevations and precipitation fall off to the southwest. Kirkuk, about 110 kilometers southwest of the border with Iran, has about 375 mm. (15 in.). Even in the mountains all of the precipitation comes in the winter and spring. The summer and early fall are rainless.

According to E. Guest (1933),[11] an ecological subdivision may be made of the mountain country with respect to elevation: lower mountain slopes (500—2,000 m.), mainly oak forest; higher mountain slopes (2,000—3,000 m.), mainly with grass vegetation; over 4000 m., an Alpine region. Buringh[12] indicates that the areas above 9,000 feet (2,500 m.) belong in the Alpine zone. The oak forest and scrub oak with grass is the vegetation on 80 percent of the total mountain area. The other 20 percent of the mountain areas are about equally divided between hilly and steep grassland with oak trees or brush, and the good valley land. Less than one-half of 1 percent of the mountain area is Alpine.

Between the mountains and the dry desertic and river flood plain areas lie the foothills and plains that grade between the relatively high rainfall of the mountain area and the low rainfall of the deserts. The foothills are fairly well-watered grassland, where not too steep. The plains near the mountains and foothills are a good winter grain area, that ranges from 14 to about 28 inches of precipitation. Southwestward this plains area grades into the steppes and then to the desert. This entire foothills and plains area is grazing land that ranges from good to poor depending on the rain-

[9] See above (n. 4), p. 169.
[10] See above (n. 6), p. 172.

[11] Quoted in P. Buringh, *Soil and Soil Conditions in Iraq* (Baghdad: Ministry of Agriculture, 1960).
[12] *Ibid.*

fall and the nature of the soils and surface configuration. The steep slopes effectively, of course, have less precipitation than the level land, primarily because of excessive runoff.

All of the Iraqi grazing lands, mountains, foothills, plains, and steppes long have been badly overgrazed. There are simply too many animals for the forage produced.

Soils

Most of the irrigable soils of Iraq are found in the flood plain of the Euphrates and Tigris Rivers. Parts of this flood plain have been cultivated for a long period of time. Since the rainfall is now and has always been too low for the production of crops from this source alone, irrigation with water from the two rivers has been carried on for as long as 6,000 years. The water from the rivers carried then, and still does, much silt and a modest amount of soluble salt. There is no reason to believe that the rainfall or temperature has materially changed in the last 6,000 years.

Most of the flood plain of the Euphrates-Tigris Rivers has been cultivated and abandoned many times. A legacy left by the irrigators is layers of river silt and mud to a depth of one to several meters and a content of salt that is the major problem in the continued use of the area for agriculture. The salt comes from the irrigation water rather than from underground sources. Thus, though the soils are very young in the sense that man has made them during historical times, they are old in terms of man's use and the salt he has left in them. The layering seen in these soils is depositional rather than genetic, as in soils that have been in place for a long period of time.

The river sediments are not materially different, for the most part, from the deep alluvial deposits that formed the flood plain prior to the activities of the early agriculturalists. In fact, the two river beds moved about the valley many times in the past. Some of the valley alluvial sediments are said to be as much as three kilometers thick. One of the most characteristic of all the properties of these alluvial deposits is their high content (usually 20–30 percent) of calcium carbonate. This carbonate occurs in all size fractions. It is an important factor in the reclamation of the soils that are too salty for good agricultural production, furnishing calcium ions that help to maintain reasonable permeability to leaching water.

Mention has been made of the high salt content of the soils that have been irrigated with the river water for long periods of time. From this, one might be led to suppose that the water is relatively salty. But that is not the case. Both the Euphrates and the Tigris have rather good water (200 to 400 parts per million). The cause of the salt accumulation is a complete lack of planned drainage of the irrigated land and the very high rate of evaporation of water during the hot summers, both from crops growing on the land and from the bare soil surfaces. Without adequate drainage for removal of groundwater, the water table rises high enough for capillary action to carry the water and dissolved salt to the surface, where the water is lost by evaporation to the air. The salt then accumulates, frequently as a white crust or as a surface coating.

Another important component of the flood-plain soils is the ubiquitous small percentage of gypsum or calcium sulfate. This slightly soluble salt, not harmful to plant growth in small amounts, appears to be formed by precipitation from the river water that soaks the soils during irrigation or during natural flood periods. The importance of this salt lies in its property of maintaining a high concentration of calcium ions in the soil solution, thus ensuring that the sodium ion does not become dominant in the soil during leaching for reclamation purposes. Soils dominated by sodium ions have low water permeabilities and may become highly alkaline.

Fortunately, the soils of the Mesopotamian Plain have a fairly good lateral permeability, partly due to the layering of the geologically recent deposited river silts. These also have a fairly good vertical permeability that is due to the high content of calcium carbonate and the modest content of gypsum. Too, the dominant clay mineral, polygorskite, does not swell and shrink nearly so much as soils with a comparable content of montmorillonite, which is a more common clay mineral. According to Buringh[13], almost all of the soils of Iraq are affected to some extent by salt. The exceptions are those found in the 10 percent of mountainous, hill, and upland plain country of the northeast that has precipitation of about 15 inches or more per year. All of the Tigris-Euphrates flood plain is affected by salt to some extent, with the area from somewhat north of Baghdad south seriously handicapped by their accumulations. These soils can, however, be desalted if properly designed irrigation and drainage systems are installed. There are a few areas of soil that would be difficult to reclaim, but in the main the salt-affected soils can be reclaimed at costs that are economical as a part of a complete system of irrigation and drainage. Enough extra water in addition to that needed by the plan must be made available for the leaching process. The troublesome soluble salts that must be washed out are primarily sodium chloride (common table salt), as well as lesser amounts of calcium and magnesium sulfates. When the excess of these unwanted salts has been removed, the soil has lost some of its needed nutrients, including nitrogen, potassium, and phosphorous. Thus, a soil that has been heavily leached to remove salts may need immediately more fertilizer than comparable soils not so leached.

Most of the soils in the flood plain have fine silty, fine loamy, or fine clayey textures. By this we mean that they are neither extremely sandy or heavy clays. There are some sand and pseudo-sand (aggregates behaving as sand-size particles) dune areas within the flood plain, which amount to about 3 percent of the area. These sand dune areas have little or no potential for agriculture. Near to the river channels are areas of well-drained, somewhat sandier soils that are now the most valuable soils in production. These, too, have rather fine textures of fine sand to silty clay loams.

Structure in a soil is due to a number of finer particles acting together as a unit. The structural units so formed increase aeration and permeability to water. The soils of the river flood plain do not have a well-developed, strong struc-

---

[13]*Ibid.* (above, n. 11).

ture. Hence, they are prone to slake, causing loss of permeability, when wetted by rain or by irrigation water. Organic matter usually contributes to the stability of structural aggregates. But the soils of this area have very little organic matter. This lack is primarily due to the very high temperatures of the soil in summer. At these high temperatures, microorganisms quickly decompose plant and microbial residues. Another contributing factor to the low level of organic carbon is the type of clay mineral dominating the fine fraction of the soil. The rod-shaped polygorskite does not so readily fix organic matter in a form unavailable to microorganisms as do the platy clay minerals, montmorillonite and illite, that are more common in occurrence.

Desert Soils

According to Buringh, 75 percent of the land area of Iraq is covered by soils that are submarginal for agricultural use. By this, he means that they are not suitable for growing crops, forestry, or grazing, either under irrigation or natural rainfall. Of this area, the far greater portion has rainfall of less than 200 mm. (8 in.). In this greater portion, with its hot, rainless summers, truly desert conditions prevail. There is essentially no vegetation. Bushes occur only where their roots can reach groundwater. To the north, the desert area grades to the steppes. On the fringe between these two areas, with rainfall at about 200 mm. (8 in.), there is some winter grass for sheep and camels. In a few wadi bottoms bits of dry-farming are found. But the total area is very small.

Actually the soils of the desert areas of Iraq have not been studied in much detail. However, enough work has been done to determine that a wide range of soils exist there. In some of the large areas the soils are underlain by rock that is primarily gypsum. In other areas the soils are from limestone. To the south there are desert areas that consist of sands and gravel of continental origin.

It is possible that some desert areas in Iraq have some soils that could support irrigated agriculture. However, since Iraq has far more potentially irrigable land in the plain than water presently available for irrigation, there is little urgency for evaluating these desert areas that have no probability of being irrigated in the near future.

Soils of the Uplands and Mountains
of Northeast Iraq

The mountainous and foothill areas in the northeast of Iraq get more precipitation than the desert areas to the west and south. However, the pattern of the precipitation is the same. All of the rain comes during the fall-winter-spring period, with a rainless summer that is a bit shorter than in the desert areas. The greater amounts of rain and snow allow for more percolation through the soil and for more chemical weathering of the minerals in the rocks and soil-forming materials. Thus, the soils are generally nonsaline; and they have, where geologic erosion is not extreme, developed genetic horizons over a long period of time. This condition contrasts both with the flood plain, where man-induced layers prevail, and with the deserts, where there has

been relatively little chemical weathering and biological activity. Along the streams there are, of course, some recently deposited alluvial sediments. Also, there is more soil organic matter, and as a consequence more moisture, more plant growth, and lower soil temperatures.

Rainfed Areas with Adequate Moisture for
Winter Crops—Most Years

The most important dry-farm or rainfed farming in Iraq is the Kirkuk-Erbil-Musel Plain, on the east side of the Tigris River. It has a rainfall from 400 mm. (16 in.) in the southwest to 700 mm. (28 in.) in the northeast part of the plain. This rainfall range corresponds to the precipitation in the Palouse area of Washington and Idaho. According to Buringh, the plain has a usable area of about 2.7 million acres, which is about the same potential crop area as the Palouse and Nez Perce Plains in Idaho, Oregon, and Washington. The rainfall is about the same too and the distribution pattern is similar. In Iraq, however, the dry summer period is somewhat longer and more rainless. Also, it is likely that the uncertainty of rainfall from year to year is greater in Iraq. Nevertheless, this is a good production wheat area.

The soils are calcareous with a zone of carbonate accumulation at a depth of from 12 to 20 inches. Salts, including gypsum, are not present except in a few small depressional spots. In all probability the potassium supply of these soils is reasonably good. Crop-yield responses to nitrogen and phosphorous under good management are certain.

Although these soils are quite good for winter crops such as wheat and barley, summer crops cannot be produced without irrigation. Wheat is sown in November or December and harvested in May and June. A good crop of wheat leaves the whole of the soil completely dry. Thus, after harvest and without irrigation, there can be no plant growth until the rains come in November and December.

Soils—Marginal for Dry-Farms

Between the good rainfed area and the deserts are areas that have marginal precipitation for dryland agriculture. According to Buringh, this area is somewhat smaller than the better rainfed area that has been discussed—about 1.6 million acres. The precipitation ranges from 150 mm. (6 in.) at the edge of the desert to 400 mm. (16 in.), or a bit more in some places. Not only is the precipitation extremely variable from season to season, but also the soils are not very good in much of the area. Highly gypseous beds underlie much of the area. Some of the soils contain as much as 60 percent of calcium carbonate. Sand-size particles of these soils are usually gypsum crystals. Such soils cannot be good either for dry-farming or for irrigated agriculture. Soils derived from the limestone materials rather than from the gypsum beds are better. Where the soils have sufficient depth—more than one meter—and are in the higher rainfall portion of the area, good wheat can be grown in the better rainfall years.

These soil areas have more relief than the previously described group. Consequently erosion is more of a problem.

Also, runoff from the sloping land reduces the moisture available to wet the soil. Hence the steeper soils are hardly suitable for anything other than pasture.

In recent years considerable areas of these marginal soils have been put to farming with use of large-scale mechanization. In the years of higher rainfall they may produce good crops. But in years of low rainfall crop failure is expected. These areas were in natural-steppe grassland vegetation prior to the introduction of large-scale mechanized farming, and were used for grazing by the animals of the nomadic people. Thus, the area of grazing for these herds has been greatly reduced. To get enough forage for the animals of the nomads is a real problem at the present time.

According to various studies that have been made, there is considerable land in the area that has a good potential for irrigated agriculture. Some areas can be irrigated with water from the Greater Zab and Lesser Zab rivers. However, this takes water from the Mesopotamian Plain. Therefore, this irrigation has to be considered as an alternative use rather than as additional irrigated agriculture for the country.

There are a number of reasons to suggest that the drier portion of this area should be left in grass. The higher rainfall deep-soil areas, on the other hand, may profitably be used for dryland agriculture under improved methods of water conservation. These soils are of medium texture.

Soils of the Mountains

The rocks of the mountains of Iraq are nearly all limestone. Only on the very highest elevations are there some areas of igneous rocks. All of the soils of the mountains, their valleys, and the foothills are calcareous. Only the larger mountain valleys have appreciable areas of land that can be cultivated. Buringh gives the area of these soils at 107,000 hectares, of which 70 percent is estimated to be usable. Thus, there are about one-quarter million acres which are usable. The winter rainfall here is between 900 mm. and 1,000 mm. on the average (35—39 in.). However, in a very dry winter season precipitation may be only one-half the average. Even so there is usually enough moisture to saturate the soil during the winter rains. These soils produce relatively good wheat crops. Under a high level of management they can produce very high yields. Crop responses have been shown to both nitrogen and phosporus in applied fertilizers. The soils are of medium texture—consisting of clay loams and silty clay loams—and have a reasonable amount of organic matter in them. That is, they have a few percent, in contrast to soils of the Mesopotamian Plain that have less than one percent. While these valleys are good wheat lands and will produce other winter crops, summer crops require irrigation. There is a flood hazard on some of the land.

Of the mountain areas other than the valleys approximately one-half is suitable for grazing and forests, while the balance is too bare, too steep, or covered by ice or water. Knowledge of the mountain soils, other than in the valleys, is scant.

Between the mountains and the plains is an area of foothills. This area of strong relief is cut by deep valleys that issue from the high mountains. The rainfall is generally quite good, ranging up to 900 mm. (35 in.) near the mountain masses. While some small area of it are fair for farming winter wheat and small patches can grow rice, the area as a whole is best suited to grazing. The potential of the area is low for cultivated crops.

The soils of this area are mainly formed from colluvium from the mountains. On the steeper slopes, where runoff is high, the soils are drier than those at the foot of the slopes or those on level ground. Thus their characteristics and their production of forage are those of a climate with lesser rainfall. In spite of the good total precipitation of the zone, the natural vegetation is grassland. Efforts at establishing forest have not been very successful.

Map Delineations and Symbols

The delineations shown on our maps are soil associations at a high level of generalization. This is to say that more than one kind of soil occurs within the boundaries. These kinds of soil may be similar or very contrasting. But at the scale of our maps and with our knowledge of the areas they cannot be mapped separately. The mapped soil areas are defined in terms of a dominant Great Soil Group and its associated soils as they characteristically occur on a specified land form from a specified kind of parent material.

Great Soil Group Symbols

The symbol used to identify each delineated area on the maps designates the dominant great soil group, the land form, and the parent material of the soil. However, it is best to consider the symbol as a whole unit rather than in terms of its component parts. It represents all of the implications of character and distribution of soils that are inherent in the factors represented.

Most of the symbols consist of three parts: (1) the symbol of the great soil group, (2) the symbol of the land form on which the soil occurs, and (3) the symbol of the parent material from which the soil was derived. The parent-material symbol forms the denominator and the land-form symbol the numerator of a fraction. Thus, $SR \frac{H}{S}$ designates Sierozem soils from sedimentary rocks on hilly terrain. Absence of the numerator of the fraction indicates level-to-undulating terrain.

Commonly on our maps, two distinct soil associations are included in a single delineation and are identified as a single map unit. In this case two kinds of symbolization have been used. In one of these each soil association is indicated by its respective symbol and the two joined by a hyphen. The more extensive association is shown as the left portion. In our area, then, A_N-L^S designates calcareous alluvial soils on level plains with steep lithosols on adjacent escarpments. The Nile River Valley south of Aswan is designated by this symbol.

The second way in which two soil associations in the same map unit are designated is by use of a multiletter single symbol. An example used on one of our maps is $TRR \frac{R}{LS}$ which designates Terra Rossa—Rendzina complexes from limestone, on undulating rolling terrain.

Terrain Types and Land-Form Designations

Terrain symbols on our maps are shown in the following listing together with a brief statement of what they indicate.

Symbol	Meaning
none	Level, gently sloping or undulating with slopes generally less than 5 percent.
C	Strongly sloping land on volcanic cones.
DF	Dissected fans at the base of mountains.
F	Fans at the base of mountains.
H	Hilly land with slopes mainly 15 to 30 percent.
M	Mature, rounded mountains, a high proportion soil-covered.
R	Rolling or sloping with slopes mainly 5 to 15 percent.
RH	Rolling-hilly complexes.
S	Steeply sloping land, mostly over 25 percent.
SU	Steep-undulating complexes.
T	Terraces, weakly dissected.
U	Undulating (frequently getting no terrain symbol).

Parent Materials Symbols

Consolidated Rocks

Symbol	Meaning
B	Basic igneous
BV	Basic volcanic
C	Undifferentiated
Ch	Chalk
Ls(or LS)	Limestone (may be with shale)
S	Sedimentary, undifferentiated
Sd	Sandstone
TS	Sedimentary, Tertiary, weakly consolidated

Unconsolidated materials

A	Alluvium
Ae	Aeolian sand
D	Diluvium—more properly colluvium
L	Loess
U	Undifferentiated
V	Volcanic ash; lava beds

Modifying Symbols

In addition to the symbols that have been used for the Great Soil Group, land form or terrain, and parent material, it has been convenient to use a few letters as subscripts modifying symbols. For our maps the principal ones are these:

Modifying Symbol	Meaning
B	Mountain basins and footslopes
C	Coarse-textured—sands to sandy loams
D	Dunes
f	fine-textured—silt loam to clay
g	gravelly
H	Hydromorphic

N	Calcareous-surface horizons
R	Stony
S	Sandy (not so used with Alluvial soils)
S	Saline. Salty enough to interfere with plant growth (so used only with Alluvial soils)
W	Subirrigated—oases in arid regions.

Symbol	*Meaning*
A_{CN}	Coarse-textured Calcareous Alluvial soils and associates, on level to gently sloping first bottoms or delta plains. Mapped throughout Iraq. Mostly along narrow bottoms where sand bars, narrow terraces, and adjacent bluffs are sometimes included. In southeastern Iraq mapped only on alluvial flats in association with sandhills. A few areas of Solonchak are included. Rainfall, 4 inches in south to 25 inches northeast. Used for grain and vegetables with and without irrigation according to rainfall.
A_{fN}	Fine-textured Calcareous Alluvial soils and associates, on level alluvial plains. In Syria a few areas are mapped along the Euphrates River north of Aleppo. Solonchak and Reddish Brown soils of terraces are common inclusions. Rainfall is 15 to 20 inches. Used for rainfed wheat and barley and for summer crops under irrigation. In Egypt found in the whole of the Nile Valley north of Aswan and in the Nile Delta. Soils harden and crack deeply when dried. When wet they are highly plastic. Soils were somewhat salty before irrigation. Salinity now depends on soil management and degree of drainage. Precipitation is two inches or less. Used for intensive irrigated farming for all crops of the area.
A_H	Hydromorphic Alluvial soils and associates. Mostly these are poorly or very poorly drained Alluvial soils with small areas of well-drained associates. Organic soils are common inclusions. In Syria an area is mapped in the Ghab Depression, where the soil was originally flooded for long periods of time during the winter months. Rainfall is from 20 to 30 inches. The area has been largely reclaimed and is now under irrigation. Both winter and summer crops are produced.
A_N	Calcareous Alluvial soils and associates, on nearly level first bottoms or delta plains. Mostly these are well-drained Alluvial soils with some local sports of Solonchak or marsh. Mapped in northern Syria along the Euphrates River. Mostly medium and fine textures and include some low sand dunes. There are a few isolated hills, some of which are volcanic. Included are some terraces with Reddish Brown or Sierozem soils. Annual precipitation from 6 to 15 inches. Used mainly for winter grains with and without irrigation. Mapped in Israel and Jordan south of Lake Tiberias and in the Bekaa Valley in Lebanon. Predominantly clay textures, but with some clay loams. Used for rainfed winter crops, and for winter and summer crops under irrigation.
A_S	Saline Alluvial soils and associates, on nearly level first bottoms or delta plains. Mostly Alluvial soils with sufficient soluble salt to interfere with but not prohibit cropping. Some nonsaline Alluvial, Solonchak, or Marsh included. Mapped in Syria along Euphrates River and its tributaries in desert areas. Four to six inches annual precipitation. Textures are mostly silt loams, but range is from clays to sands. Used for irrigated agriculture where water is available.
$A_H{-}A_N$	Hydromorphic Alluvial soils and associates with well drained Calcareous Alluvial soils, on nearly level first bottoms. Poorly drained Alluvial soils, commonly saline, with small inclusions of Calcareous Alluvial soils. Locally there are large areas of Calcareous Alluvial soils with a small proportion of saline and poorly drained nonsaline associates. Mapped in Israel and Jordan in the vicinity of Lake Tiberias and Lake Huleh. The Lake Huleh areas now largely reclaimed. Precipitation is from 20 to 26 inches.
$A_N{-}A_S$	Calcareous and Saline Alluvial soils and associates, in varying proportions, on nearly level flood plains or delta plains. Mapped in the Tigris-Euphrates Delta. The areas are well drained and usually do not flood, although there is a potential for occasional flooding.

Textures are mostly silty clay loams, silt loams, or silt. Clays and sand are uncommon. Annual precipitation, 4 to 8 inches. All crops grown under irrigation. A large proportion of the agriculture of southeastern Iraq is on these soils. Many are badly salty.

A_S–A_{CN} Saline Alluvial soils with coarse-textured Calcareous Alluvial soils and associates, on level to nearly level flood plains and delta plains. Occurs as mixtures of these two soils in varying degree with smaller amounts of Solonchaks, Sand Dunes, or Marshes. Mapped in Jordan in the northern end of Wadi Sirhan. Infrequent flooding. Very low precipitation. Mapped in Iraq in numerous wadis in western and southern areas. Textures range from sands to clays. Floods are infrequent. In some places water table is near the surface most of the time. Four inches rainfall. Used for occasional grazing by Bedouin-owned flocks.

A_N–$L^{\underline{S}}$ Calcareous Alluvial and associates with steep Lithosols, on adjacent escarpments. Mostly Calcareous Alluvial with lesser amounts of Saline Alluvial soils. Lithosols of short, steep slopes to adjacent deserts included. Mapped in Egypt along the Nile River south from Aswan. Nile silts and clays intermixed with sand. Barren Desert soils on either side of river. Intensive irrigated agriculture on the Alluvial soils.

BD_g Gravelly Barren Desert, on level to undulating plains, with some Shifting Sand Dunes, steep plateau escarpments, and steep-sided wadis. Mapped in Egypt adjoining the Kharga oases and along the Libyan border. One inch or less rainfall, little or no vegetation.

$BD_g\frac{U}{\underline{S}}$ Gravelly Barren Desert over sedimentary rock, on level to undulating plains with scattered flat-topped, steep-sided hills and plateau escarpments. Mapped in Egypt over portions of the Libyan Plateau. One inch or less rainfall, little or no vegetation.

BD_{LS} Stony Barren Desert over limestone, on level to undulating plains broken in places by nearly vertical slopes of isolated flat-topped hills, plateau escarpments, and walls of wadis. Mapped in Egypt over portions of the Libyan plateau north of Siwa. One inch or less precipitation, little or no vegetation.

BD_{Sd} Stony Barren Desert over sandstone, on level to undulating plains broken in places by nearly vertical slopes of isolated flat-topped hills, plateau escarpments, and wadis. In Egypt mapped west of the Great Sand Sea. Less than one inch precipitation.

$BD\frac{SU}{LS}$ Stony Barren Desert over limestone, on steep-undulating terrain complexes broken into small and large areas by steep slopes of a complex system of wadis, numerous flat-topped hills, or by a series of plateau escarpments. Mapped in Egypt on an extensive area north of Dakhla. Precipitation less than one inch. Vegetation generally absent.

$BD\frac{SU}{Sd}$ Stony Barren Desert over sandstone, on steep-undulating terrain complexes of level to undulating plains broken by steep slopes of a complex system of wadis, by numerous isolated flat-topped hills, and by a series of plateau escarpments. In Egypt marked on the Gilf Kebir Plateau and some other places. Less than one inch precipitation, little or no vegetation.

BD_g–$BD\frac{H}{C}$ Gravelly Barren Desert soils, on level to undulating plains with stony Barren Desert hills in groups or isolated. Mapped in Egypt in inextensive areas along the outliers of oases bounded by steep escarpments. One inch or less rainfall, little or no vegetation.

$BD_g\frac{U}{\underline{S}}$–$BD\frac{H}{C}$ Gravelly Barren Desert soils over sedimentary rocks, on level to undulating plains with stony Barren Desert hills which are the remnants of an eroded peneplain. Mapped in Egypt on the limestone plains and hills of the Libyan Desert. Less than one inch precipitation, little or no vegetation.

$BD_g\frac{U}{\underline{S}}$–$BD_S$ Gravelly Barren Desert soils over sedimentary rocks and Barren Desert sands, on level to undulating plains. Mapped in Egypt on the limestone plains of the Libyan Desert. Less than one inch precipitation, little or no vegetation.

BDS—BD$\frac{H}{C}$ Sandy Barren Desert soils, on level to undulating plains with stony barren Desert hills in groups or isolated remnants of an eroded peneplain. Mapped in Egypt on the Libyan Desert. One inch or less rainfall, little or no vegetation.

BD$\frac{U}{gS}$—S$_{BD}$ Gravelly Barren Desert soils over sedimentary rocks, on level to undulating plains with numerous isolated or groups of Sand Dunes. Mapped in Egypt in the Libyan Desert. One inch or less rainfall, little or no vegetation.

BD$_{LS}$—S$_{BD}$ Stony Barren Desert over limestone, on level to undulating plains broken by scattered Sand Dunes and by occasional flat-topped hills and steep-sided wadis. Mapped in Egypt over the northern part of the Libyan Desert. One inch or less rainfall, little or no vegetation.

BD$_S$—S$_{BD}$ Sandy Barren Desert, on level to undulating plains with isolated Shifting Sand Dunes in hills or long ridges. Mapped in Egypt south of the Qattara Depression, in the south near Sudan and other smaller areas. Generally less than one inch annual precipitation, very little vegetation. Gravelly Barren Desert soils are commonly included where the bedrock outcrop is near the surface.

BD$_{Ls}$—SK   Stony Barren Desert over limestone, on level to undulating plains with Solonchak in depressions. Mapped in Egypt over portions of the Qattara Depression. Rainfall generally one inch or less. Vegetation generally absent but there are some salt-tolerant plants on the Solonchaks.

BF$_B$ Brown Forest and associated soils of basins and piedmont plains, on level to sloping terrain with moderate dissection in some places. Mapped in Iraq in a few basins in the Kurdish Mountains of the northeast. Mainly they occur in association with Lithosolic Brown Forest and Terra Rossa of the uplands. Small areas of poorly drained soils are included. Annual precipitation is from 25 to 40 inches in winter season. These soils are used for wheat, barley, fruits, cotton, and for grazing.

BF$\underline{M}$ Lithosolic Brown Forest and associated soil. The gradient of the slope ranges between 15 and 50%. Mapped in Lebanon and Syria in association with Brown Forest soils. Precipitation is from 35 to 60 inches. The natural cover is mostly trees. Mostly the areas are used for summer grazing, but some of the lower slopes have been terraced and are used for production of fruits, vegetables, and grains.

CT$\frac{H}{TS}$ Chestnut and associated soils from weakly consolidated materials, on hilly terrain. These gravelly soils occur between 5,000 and 6,000-feet elevation in Syria in area north of Damascus. There are many small inclusions of shallow soils. Precipitation is between 20 and 30 inches. Part of the area is used for rainfed crops of the area.

D$_{Ch}$ Desert and associated soils from chalk, on level to undulating plains. Clays and loams are the dominant textures. Extensive areas mapped in eastern and southern Syria. These soils are mostly 18 to 24 inches deep to rock, and in some places are as shallow as 12 inches. In a belt along the Euphrates River the underlying rock is mostly gypsum. Annual precipitation is mostly five inches or less. In Iraq mapped west of the Euphrates River. Soils similar to those in Syria. Suitable only for occasional light grazing. Not good soils for irrigated agriculture, though some may be so used.

D$_g$ Desert and associated soils, on gravel plains that are level to undulating. Several areas mapped in southern and western Iraq. Most of the gravel is at the surface of these soils, which are only two to three feet deep. Associates are Lithosols, Solonchaks, or Sandy Desert soils. Two to four inches rainfall, only occasional sparse vegetation. These plateau soils are not suitable for agriculture of any kind.

D^M Lithosols of desert moutains that are steep and rocky. Mostly they are devoid of vegetation. Mapped in the Sinai Peninsula and in the eastern desert of Egypt. Precipitation generally less than one inch. No agricultural value.

$D\dfrac{R}{Ch}$

Desert and associated soils from chalk, on rolling plains with usually less than 15 percent slopes. These soils are dominantly shallow, usually less than one foot. Clays and loams are the dominant textures. Lithosols and Solonchak are included. Mapped in southeast Syria and in western Iraq north of the Euphrates River. No agricultural use except light grazing. Four or five inches annual precipitation.

D_s

Sandy Desert and associated soils, on rolling plains with some local areas of Solonchak. Annual precipitation about seven inches. Only agricultural use is grazing by nomadic herds. Mapped in northern Iraq west of the Tigris River.

$D\dfrac{T}{A}$

Desert and associated soils, on level to undulating terraces along rivers. Solonchak, Solonetz, and saline or calcareous Alluvial soils are principal associates. Mapped in Syria and Iraq along the Euphrates River in the area of Desert soils. There is a sparse stand of desert shrubs. Silt loam textures are dominant, but the range also includes sandy and gravelly. A few small areas are irrigated by water lifted from the Euphrates River by pumps for the production of wheat and barley. Annual precipitation is four to five inches. There is a sparse stand of desert vegetation.

D_U

Desert and associated soils from unconsolidated materials, on undulating to rolling plains with local flats and small hills. Soils are mainly from alluvium, commonly reworked. Locally, some sandy materials. The texture range is from gravelly loam to clay. Extensive areas are mapped east of Deir eg Zor in Syria and these continue across into Iraq. The underlying materials are gravels, sands, and clays. Average annual precipitation is four or five inches. Vegetation is a sparse stand of desert shrubs. There is some light grazing by nomadic herds. Locally there are some spots and flats that are salty. No agricultural use other than very light grazing.

D_B–SK

Desert and Solonchak soils and associated soils of basins and piedmont plains, on level basin bottoms and adjacent sloping coalescing fans. Somewhat more than one-half of the area is Desert soils and most of the remainder is Solonchak. Associated soils of lesser extent are Solonetz and Lithosols. The main area is mapped in Middle Syria east of Palmyra at an elevation of about 1,500 feet. Textures are mainly loams and clays. Four to five inches rainfall, sparse vegetation. There is light grazing occasionally. No agricultural value beyond this.

D_{TS}–SK

Desert and associated soils from weakly consolidated sedimentary rocks with Solonchak, on level to rolling plains. Mapped in western Iraq. Estimated 60 to 80 percent Desert soils and 20 to 40 percent Solonchak. Underlying rocks are thought to be largely chalk and gypsum with admixtures of sandstone, shale, and unconsolidated material. The Solonchak occurs in numerous shallow depressions, many of which contain ephemeral lakes. These have a fine texture. The Desert soils are variable in texture and depth. Vegetation is sparse on the Desert soils, but the Solonchaks generally support some salt-tolerant grasses. Rainfall averages five to eight inches. There is some grazing by nomadic herds. No other agricultural potential.

$L\dfrac{H}{C}$

Lithosols and associated soils from undifferentiated consolidated rocks, on hilly terrain in arid and semiarid regions. Mapped east of the mountains in the west central part of Syria. Much of the area is bare rock, mostly basalt and limestone. Where there is some soil the vegetation is grazed. Average annual precipitation is from 12 to 16 inches. Agricultural value for grazing only.

   A number of areas are mapped in the eastern desert of Egypt. The rough terrain is essentially devoid of vegetation. Mean annual precipitation generally less than one inch. No agricultural value.

$L\dfrac{H}{Ls}$

Lithosols and associated soils from limestones, or hilly terrain in arid and semiarid regions with a high proportion of bare rock or very shallow and stony soil. Some deeper soils on gentle slops included. Solonchak and Solonetz occur in small depressions. These areas are almost barren of vegetation. In Israel and in Jordan an area is mapped on the West Bank between the Terra Rossa of the mountains to the west and the Ghor (upper terrace of the

Jordan River) on the west side of the Jordan and the Dead Sea. Rainfall is from 8 to 16 inches per year. The area is underlain by soft marls and harder limestone. Vegetation is rather sparse. Area is used for grazing. On the East Bank of Jordan there is a large continuous area from near Amman south to Maan. The underlying rocks are flinty limestone and marl. Rainfall range is from about 4 inches up to 16 inches.

In northeast Syria this unit is mapped on and around a hill that rises to 1200 feet above the level of the surrounding plain. Much of the underlying rock is chalk. Rainfall averages 16 to 20 inches. In spite of this good level of precipitation vegetation is not good. The area is used occasionally for grazing. There a few small pockets of irrigated fields. This area has very little agricultural potential.

In other and drier portions of Syria and in Iraq this unit is mapped where the surrounding areas are deserts. In this situation vegetation is exceedingly scant. Production of forage is very low indeed. These soils have no other agricultural potential.

$L\frac{H}{S}$

Lithosols and associated soils over sedimentary rocks, on hilly terrain with a high proportion of bare rock or very shallow and stony soil. Some areas of Sierozem and Alluvial soils are included in the areas mapped in Middle Syria. Underlying rocks are sandstones and clays with isolated beds of limestone. Average annual precipitation is about ten inches. The only agricultural use is occasional grazing.

In northern Iraq a few areas mapped in association with Desert, Sierozem, and Reddish Brown soils. Underlying rocks are limestones, sandstones, shales, and conglomerates, as well as some local unconsolidated material. Average annual precipitation here is from 8 to 18 inches. Most of the areas are barren except for local accumulations at foot slopes and in wadis where shrubs and grass grow. In these there is some cultivation and grazing.

$L\frac{H}{Sd}$

Lithosols and associated soils over sandstone, on hilly terrain with a high proportion of bare rock or very shallow and stony soils. Mapped in southeastern Iraq over red sandstone. Locally sandy Red Desert soils are associates. Precipitation is between eight and ten inches annually. The hills are largely bare, but desert shrubs persist in the local areas of Red Desert soils. These furnish a small amount of grazing. There is no other possible agricultural use of these soils.

In Egypt some areas are mapped in the eastern hills. The hills are dominantly bare of vegetation. Mean annual precipitation is generally less than one inch. These soils have no agricultural value.

L_{LS}

Lithosols and associated soils from limestone, on level to undulating plains in arid and semiarid regions with a high proportion of bare rock or very shallow gravelly and stony soils. Less shallow soils occur locally in protected sites. Solonchak and Solonetz occur in some shallow depressions.

In Jordan large areas of this unit are mapped in the west and southwest. Characteristic of these soils in Jordan is the large amount of pebbles and gravels covering the surface and comprising the principal portion of the upper horizon. These generally windswept deserts are locally knows as "Hammadas." Annual precipitation two to four inches. Except for a few favored sites these soils are bare of vegetation. They have little or no agricultural value.

In southern Syria several areas of these soils are mapped where there is a higher proportion of flinty limestone than there is in middle and eastern Syria. These areas are at the northern edge of the "Hamad," or stony desert. However, there are some areas of Desert soils from chalk or marl that have a greater depth of soil than these Lithosols and that are relatively free of stones. Also included are numerous small shallow depressions with Solonchak soils. Average annual precipitation is four inches or less. The soils are largely bare of vegetation except for a few favored sites such as the fringes of some of the depressions. These soils have no agricultural value.

This unit is the principal one mapped in southwestern Iraq. Level-bedded limestones are at or very near the surface over most of the area. Locally there are some standstones, shales, conglomerates, and unconsolidated materials. The soils are dominantly shallow, stony, and gravelly; but included are many shallow depressions with deeper soils—mainly Solonchaks. Also there are numerous wadis with narrow bands of gravelly and stony soils.

Locally there are some clayey Alluvial soils that are always high in lime and frequently high in soluble salts. Average annual precipitation is about four inches. Some local spots are occasionally grazed. These soils have no other agricultural value or potential.

L<u>M</u>

Lithosols and associated soils, on mountains of arid and semiarid regions. In Syria the principal areas mapped in this unit are two mountain ranges northeast of Damascus. In the main, these are bare limestone rock. However, many small fans of wash from these mountains are included. There is some cropping and grazing on these fans. Otherwise there is no agricultural value to the areas.

A few areas are mapped in mid- and west-northern Iraq. Most of these areas consist of rock exposure. There are some areas of soil at the base of slopes and in narrow wadi belts. On these there is some vegetation which is grazed. Except for these there is no agricultural value.

L<u>RH</u>
B

Lithosols from basic igneous rocks, on rolling to hilly terrain complexes in arid and semiarid regions. There are large rolling areas associated with hilly areas of steeper slopes. The rolling and hilly areas are not separated. The main areas mapped are in southern Syria with a continuation across the border south into Jordan. These areas are adjacent to the lava flow areas. The area is mainly boulders, stones, and cobbles with numerous exposures of bedrock. Small areas of Sand Red Desert and Solonchak soils are included. Average annual precipitation is under eight inches in most places. The area is almost bare of vegetation except in favored sites. The areas have no agricultural value of consequence.

L<u>S</u>

Steep Lithosols of arid regions from a variety of rocks in a range of climatic conditions. In Egypt this unit is mapped on numerous scattered escarpments and steep hills. The soils are stony and dominantly barren of vegetation. The underlying rocks are primarily limestones. The precipitation is two to three inches in the north part of Egypt and less than this in the interior deserts. No agricultural value.

In Israel and the West Bank of Jordan this unit is mapped on the steep irregular slopes in the east adjacent to the Jordan Valley, the Dead Sea, and the Wadi Araba. The area is nearly devoid of vegetation except for small pockets of smooth Sierozem soils on which there is some vegetation. These comprise no more than 2 or 3 percent of the area. Underlying rocks are basalt near Lake Tiberias and limestones and marls near the Dead Sea. Except for some grazing on the pockets of deeper soil, the area has no agricultural potential. Rainfall eight inches or less.

In East Jordan there is a long continuous area of these soils adjacent to the Rift Valley. Mostly the underlying rocks are limestones, but in the south there is some granite. The area is quite steep and irregular. Rainfall is mostly four inches or less. Toward the upper part of the slopes rainfall increases and there are some patches of Reddish Brown or Terra Rossa soils. These are used for crops where terraces have been built to hold the soil in place. Other than these there is little vegetation and little or no agricultural value.

In Syria this unit is mapped in numerous small areas in the arid and semiarid parts in association with Desert, Sierozem, and Reddish Brown soils. Underlying rocks are mostly limestones and marls, but in some areas basalts occur. In some places these are mixed. The areas are largely barren of vegetation, but where the rainfall is highest there is some vegetation. The precipitation ranges from about 4 to 18 inches. The vegetation is grazed occasionally, but otherwise there is little or no agricultural value. In Iraq a few areas are mapped on the border with Iran, in Middle Iraq, and in the north part of western Iraq. Average annual precipitation ranges from 8 to 20 inches. The areas are practically barren of vegetation except for some deeper soil areas at the foot of slopes and in narrow wadis. In these there is some vegetation that is used for grazing. As a whole there is little agricultural value.

L_S

Lithosols and associated soils from undifferentiated sedimentary rocks, on level to undulating desert plains in arid and semiarid regions. These shallow, stony, and gravelly soils are mostly from limestone, sandstone, and shale, but frequently from a mixture of these. Average annual precipitation is about four inches. The mapped areas are in southern Iraq. A sparse desert vegetation covers much of the area. Light grazing is the only agricultural use.

$L\frac{SU}{LS}$

Lithosols and associated soils over limestones, on dissected plateaus with undulating plateau remnants in arid and semiarid regions. These shallow soils dominate on the steep slopes of the V-shaped valleys and also on the level to undulating intervening areas. The proportion of bare rock is high. In Egypt mapped in an extensive area south of Cairo and another on the Sinai Peninsula. Dominantly barren of vegetation. One inch or less annual rainfall. No agricultural value.

In Iraq this unit is mapped in several areas in the southwest. Included are many wadis with narrow bands of Calcareous Alluvial and Saline Alluvial soils. Small playas are common in some areas. In some places there are some deeper Desert soils. On these and in the wadis there is some vegetation. These areas are used for occasional grazing. No other agricultural value.

$L\frac{SU}{S}$

Lithosols and associated soils over sedimentary rocks, on dissected plateaus with undulating plateau remnants in arid and semiarid regions. These are dominating shallow soils on steep V-shaped valley walls and on level to undulating intervening areas with a high proportion of bare rock. Mapped in the arid part of Syria, including the bluffs on either side of the Euphrates River. The underlying rocks are limestone, marl, or gypsum. Gypsum is the main rock along the Euphrates River. Average annual precipitation is from three to ten inches. The areas are largely barren of vegetation, but there is some desert vegetation in favored places. This is grazed occasionally. No other agricultural value.

In Iraq this unit is mapped on several areas in the west and south. Limestones are the most extensive underlying rocks but there are some sandstones and shales. There are many narrow wadi beds with Calcareous Alluvial soils. Average annual precipitation is about four inches. Vegetation is scanty but there are some desert shrubs. These are grazed occasionally. No other agricultural value.

L_V

Lithosols of lava beds in arid regions, on undulating to rolling plains with numerous steep slopes at irregular intervals. In Jordan this unit is mapped on a rather large area, especially in the north-central portion of the country. The surface cover ranges from a few inches of unconsolidated material to bare lava rock. The lava is rather fresh and unweathered. Average precipitation is less than four inches. The areas are not used to any appreciable extent for grazing. No agricultural value.

In Syria the main areas of this mapping unit are mapped in the south, but some areas are also found in the north. In some areas the lava is broken into large boulders with some soil in between. Some vegetation, including herbaceous plants and small trees, grow in these pockets. Average annual precipitation is from 8 to 20 inches. The areas are used for grazing but are very difficult for the animals to cross. No agricultural value other than the small amount of grazing.

In Iraq a few areas are mapped at the border with Saudi Arabia and Jordan. Annual precipitation is about four inches. Areas are almost bare of vegetation. No agricultural value.

$L\frac{S}{}-A_N$

Steep Lithosols of arid regions with Calcareous Alluvial soils, on strongly dissected uplands with moderately broad Alluvial flats. The units consist of Alluvial soils on first bottoms too narrow to show separately on the map and of Steep Lithosols on the escarpments of the adjacent uplands. In Syria most areas of this unit are found in the east on either side of the mountains. Rainfall varies greatly depending on the geographic location. The Calcareous Alluvial soils are used to a large extent for crop production. In many places they are irrigated. In some places the underlying rocks are limestone in some places and basalt in others. Although subject to some flooding, the Calcareous Alluvial soils are good. The Steep Lithosols have little agricultural value.

In western Iraq several areas have been mapped along wadis cut into the western part of the desert plateau of south-central Iraq. The soils are used for occasional grazing along with the associated Desert soils. No other agricultural value.

$L_{Ls}-RD_f$

Lithosols from limestone, on level to undulating plains with fine-textured Red Desert soils from unconsolidated materials on level to undulating plains. Mostly shallow stony soils with lesser areas of the Red Desert soils from alluvium in depressions. Minor areas of Solonchak and Solonetz. There are a few steep slopes of wadis and isolated flat-topped hills. In Egypt this unit mapped as an extension of the Libyan Plateau. Mean annual precipitation is two inches or less. Vegetation is sparse. Not used for grazing. No agricultural value.

$L\underline{S}$–RD$_S$

Lithosols on steep terrain with sandy Red Desert and associated soils, on level to undulating plains. The steep slopes are on hills or ridges with intervening level to undulating areas. Mapped in Egypt along the southern border of Israel as an extension of the main areas in Israel. The underlying rock is limestone. Sandy alluvium may constitute up to 20 percent of the areas. Most of the area is steep, with shallow soils with the Red Desert comprising a lesser percentage. Average annual rainfall is from three to six inches. There is essentially no vegetation. No agricultural use of value.

$L\dfrac{SU}{Ls}$–RD$_f$

Lithosols from limestone, on steep-undulating terrain complexes with fine-textured Red Desert soils from unconsolidated materials on level to undulating plains. The greater amount of the area has the shallow soils, and the mainly clay-textured Red Desert soils from alluvium makes up a lesser part. Minor areas of Solonchak and Solonetz occur. In Egypt this unit is mapped on the chalky limestone plateau south of Alexandria. This is an area of stony soils within which occur numerous clay flats and depressions. Annual rainfall is two or three inches. Vegetation is scarce and not used for grazing. No agricultural value.

$L\dfrac{H}{Ls}$–SR$_U$

Lithosols from limestone, on hilly terrain with Sierozem and associated soils from undifferentiated unconsolidated materials on level to undulating plains. Shallow soils on the hills occupy most of the area. The lesser components of Sierozem soils are on fans, in basins, and on low plains adjacent to the hills. Several areas are mapped east of the mountains in west-central Syria, where there are narrow ranges of hills with narrow valleys or basins. Average annual precipitation is about ten inches or less. Vegetation is sparse and furnishes some grazing. In a few places irrigated crops are grown. The agricultural value of the Lithosols is very little and that of the Sierozems hinges on availability of water for irrigation.

NB$\underline{M}$

Lithosolic Noncalcic Brown and associated soils, on mountains. The mountains are steep and rough. The Noncalcic Brown soils are on the less steep rolling portion of the areas. This unit was mapped on a few areas in northern Syria as extensions from larger areas in Turkey. Since the areas are underlain by basalts or limestones and the rainfall in Syria is only about twenty inches, there is some doubt that these soils are really noncalcareous. The area is used for grazing.

NB$\underline{S}$

Lithosolic noncalcic Brown and associated soils from consolidated rocks on steep hills with narrow valleys between. A few small areas are mapped in northwest Syria adjacent to Turkey. The underlying rocks are largely sandstone or igneous rocks that weather to sandy material. Average annual precipitation is about 30 inches. Pines grow well on this unit. Can be used for forestry or grazing.

NB$_S\dfrac{H}{Ae}$

Noncalcic Brown and associated soils from aeolian sands, on hilly terrain with dunelike or modified dunelike topography. A few areas with steep slopes are included, as are a few Reddish Brown soils on terraces. Textures are from sands to sandy loams. These soils occur on the Coastal Plain in western Israel and Syria. They are developed from sandy materials that have been stabilized with vegetation long enough to have lost their calcium carbonate content by leaching. They are now slightly acid to neutral throughout. In Israel these soils are used extensively for production of oranges and grapefruit. In Syria they are used for producing a variety of fruits and vegetables. Nearly all of the areas in both countries are suitable for cultivation. Annual precipitation of these soils in Israel is 20 to 24 inches. In Syria it is about 30 inches.

$R\dfrac{R}{Ch}$

Rendzina and associates from chalk, on undulating to rolling plains. Associates are Alluvial soils. Textures are silt loam to silty clay loam. An area is mapped over chalky marl along the coast in northwestern Syria. Terra Rossa soils are common inclusions. The Rendzina soils are shallow and in some places may have a thin overwash. The area is generally eroded. Average annual precipitation is from 20 to 30 inches. These soils are used for growing grain and for grazing.

RB$_B$ Reddish Brown and associated soils, in basins and on piedmont plains. Mapped on gently sloping fans at the base of mountainous slopes. Associates are Solonchak, Solonetz, and Alluvial soils constituting less than 20 per cent of the areas. A few areas are mapped in the western slopes of the Anti-Lebanon Mountains in Lebanon and in Syria. Textures are coarser and soils more stony near the mountains. Average annual precipitation, 12 to 15 inches.

RB$_{Ch}$ Reddish Brown and associated soils from chalk, on level to undulating plains with clayey or loamy textures. Alluvial soils and Solonetz are common associates. Several areas are mapped in northern Syria in the vicinity of Aleppo. These soils have a zone of precipitated calcium carbonate in the rooting zone of plants that is called "caliche." Farmers are able to break this layer up with special tillage implements. Precipitation is about 15 inches. These soils are used for the production of winter grains without irrigation. (There is some evidence that the parent material of these soils is unconsolidated calcareous materials rather than chalk).

RB$\frac{F}{D}$ Reddish Brown and associated soils from chalk, on level to undulating plains on coalescing fans at the foot of mountains or high hills. Alluvial soils, Solonetz, Solonchak, and Lithosols are the principal associates. Textures range from coarse to fine but are dominantly medium. A precipitated carbonate layer is probably present in places. Mapped in Iraq on either side of Jebel Sinjar in the northwest. These soils grade into Sierozems on the south. Annual precipitation is 12 to 15 inches. These soils are used mainly for grazing.

RB$_f\frac{R}{TS}$ Fine-textured Reddish Brown and associated soils from weakly consolidated sedimentary rocks, on undulating to rolling plains. Associates are Solonchak, Solonetz, and Alluvial soils. Textures are predominantly clay loams, silty clay loams, and clays. Underlying rocks are mainly soft limestones. Several areas are mapped in Iraq between the Tigris River and the Kurdish Mountain in northeastern Iraq. Rainfall is from 12 to 20 inches. Mainly the soils are used for grazing but there is a significant amount of dry-farming, with wheat and barley as the crops.

RB$_f\frac{R}{U}$ Fine-textured Reddish Brown and Associated soils from undifferentiated unconsolidated materials, on undulating to rolling plains. Solonchak and Solonetz are the main associates. Textures range from silt loams to clays. The parent materials are various kinds of sediments, mostly several meters thick over gravel. Near the hills the sediments may be quite thin. Several areas are mapped east of Mosul in northern Iraq. Average annual precipitation is between 13 and 20 inches. These soils are used for production of wheat and barley under natural rainfall. These are among the most important soils in Iraq.

RB$_g\frac{F}{D}$ Gravelly Reddish Brown and associated soils, on weakly dissected sloping colluvial fans at the base of mountains. Calcareous Alluvial soils and Solonchak are common associates. Mapped in the northern end of the Bekaa Valley in Lebanon. Underlying material is a weakly cemented conglomerate that is high in calcium carbonate. These soils are used mainly for grazing but with some local production of winter grains. The water-holding capacity of these soils is low, hence plants suffer from drought late in the growing season.

RB$\frac{H}{C}$ Reddish Brown and associated soils from consolidated rocks, on hilly terrain. Dominantly hilly sandy loam to clay loam Reddish Brown soils, mainly over limestones and marls. Rolling Reddish Brown soils comprise a part of the area. Precipitation is from 15 to 20 inches. Mapped in northern Syria and Iraq as extensions of the main areas occurring in Turkey. These soils are used mainly for grazing but some winter grain is produced.

RB$\frac{H}{Ls}$ Reddish Brown and associated soils from limestone, on hilly terrain from hard limestone, chalk, and limestone conglomerate. The soils are generally shallow and somewhat stony or gravelly. Barren Lithosols are common associates, as are also Solonchak and Calcareous

Alluvial soils. Textures are clayey or loamy. The precipitated limestone horizon, caliche, is of common occurrence. Several areas are mapped in northern Syria. Precipitation is between 15 and 20 inches on the average. These soils are used for grazing and for grain production. The small included areas of Alluvial soils are frequently irrigated.

$RB\frac{H}{TS}$ Reddish Brown and associated soils from weakly consolidated materials, on hilly terrain. Hilly sandy loam to clay loam Reddish Brown soils predominate, but locally there is as much as 30 percent rolling Reddish Brown soils. Solonetz and Solonchak are associates in small percentage in isolated areas. Several areas are mapped in Syria and northeastern Iraq west of the Kurdish Mountains. Underlying materials consist of conglomerates, shales, limestones, sandstones; and in some places the material is unconsolidated. The range in texture and depth is wide, but heavy-textured soils are quite extensive. Average precipitation is from 15 to 25 inches. The main use of these soils is for grazing.

$RB\frac{H}{U}$ Reddish Brown and associated soils from unconsolidated materials, on hilly upland terrain. Perhaps one-third of the area is rolling rather than hilly. Two small areas are mapped in southwestern Israel. They border on being Sierozems. The parent material of these soils is a clayey material washed from adjacent limestone uplands and deposited as fans or terraces. There has been severe erosional cutting of the areas. Rainfall is from 8 to 16 inches. The land is used for citrus and grapes under irrigation, and sorghum and grazing under natural rainfall.

$RB\frac{R}{B}$ Reddish Brown and associated soils from basic volcanic rocks, on rolling plains that are slightly to moderately dissected. Textures are mainly silt loams and silty clay loams. In Israel three areas are mapped near Lake Tiberias. The soils are calcareous throughout. Rainfall is 20 to 24 inches. These soils are nearly all used for general crops of the area. Part are irrigated.

In Jordan there are two areas mapped in the northwest near the Syrian border. These have numerous boulders over the surface but not enough to prevent cultivation. Precipitation ranges from 12 to 24 inches. There are some dry-farm and some irrigated crops. The soils have fine textures and are calcareous.

In Syria some areas are mapped east of the mountains in the northwest, but the main area is between Damascus and Lake Tiberias. The soils are fine-textured. There are some stones and boulders, however. These are among the most important soils in Syria and are used mainly for the production of winter grains. Rainfall is between 15 and 20 inches.

$RB\frac{R}{Ch}$ Reddish Brown and associated soils from chalk, on undulating to rolling plains. Caliche is common but is generally soft enough to be broken up with special implements. Associated soils are Lithosols and Calcareous Alluvial soils. Several areas are mapped in northern Syria. Rainfall is about 15 inches. Winter grains are produced under dry-farming, but many smaller areas are irrigated for vegetable production. The parent rock underlying these soils may be unconsolidated calcareous deposits rather than chalk.

$RB\frac{R}{Ls}$ Reddish Brown and associated soils from limestone, on undulating to rolling plains. Common associated soils are Solonchak, Solonetz, and Saline and Calcareous Alluvial. Textures are loams and silt loams. Several areas are mapped in Jordan on the plateau east of the Dead Sea. Some steeper areas are included, and here there are some Lithosols. Precipitation is from 12 to 20 inches; most of the area is dry-farmed to winter grain. There is a short continuation of this area into Syria, where conditions are comparable to those in Jordan.

$RB\frac{R}{TS}$ Reddish Brown and associated soils from sedimentary rocks, on undulating to rolling plains. Textures are loams and silt loams. Associated soils are Solonchak, Solonetz, and Saline and Calcareous Alluvials. A narrow area is mapped in Syria on the border with Turkey. The underlying rocks are limestone or chalk. Average annual precipitation is 16 to 24 inches. The soils are used for winter grain production.

$RBR\frac{C}{BV}$ Stony Reddish Brown and associated soils from basic volcanic rocks, on slopes of volcanic cones. The soils have numerous basaltic gravel, stones, and boulders on the surface and in the soil. There are some Lithosols consisting of piles of boulders on bare rock surface. There are also small areas of Calcareous Alluvial and Solonchak soils. These soils are too stony for ordinary cultivation, but a few places have been cleared of stones and are used for crops; most of the area is used for grazing. Precipitation is from 12 to 20 inches. Crops are wheat and barley. In Jordan is a short continuation of the main area, which is located in southern Syria.

RB_{RB}^{H} Stony Reddish Brown and associated soils from basic igneous rocks, on hilly terrain. Mapped in Syria north of Lake Tiberias, these soils have gravels, stones, and boulders on the surface and in the soil that prohibit ordinary cultivation. Basaltic outcrops are common and the soils are generally shallow. There are some Lithosols consisting of piles of boulders or bare rock. There are also some Solonchak, Solonetz, and Calcareous Alluvial associates. A few fields have been made by clearing the rocks, but most of the area is only used for grazing. Precipitation is from 10 to 20 inches annually. Some fruit is grown where irrigation is possible.

RB_{A}^{T} Reddish Brown and associated soils, on weakly to moderately dissected terraces. Most of the soils are on terrace remnants some thousands of yards wide. Lithosols are found on the steep sides of the dissecting gullies. There are small areas of coarse-textured alluvium, as well as a few areas of hydromorphic soils. These soils occur in the Plain of Esdraelon and the Vale of Jezereel in Israel. They are among the most intensively used soils in the country. Average annual precipitation is 20 to 24 inches. The underlying materials from which the soils are derived is mixed limestone and basaltic colluvium. The textures are fine. Calcium carbonate occurs throughout the soil profiles.

A small area is mapped in Syria along the Euphrates River near the Turkish border. Average annual precipitation is between 10 and 15 inches. This area is used for production of winter grains.

RB_U Reddish Brown and associated soils from unconsolidated materials, on level to rolling plains. Textures are mostly sandy loams and loams. Locally there are small areas of hilly Reddish Brown soils and some Solonchak and Solonetz in small depressions and flats. In Syria these soils are continuations of areas occurring in Turkey. Calcium carbonate horizons occur in places. Textures are mainly medium to fine. Rainfall is from 12 to 16 inches annually. The areas are mostly used for grazing and for growing winter grain. Many small areas are irrigated from wells.

In Iraq several small areas are mapped in the north. Parent materials are alluvium over conglomerates, sandstones, shales, marls, clays, and sands. The alluvium appears to be quite old. Medium and fine textures predominate. Rainfall averages from 10 to 18 inches. These are very important agricultural soils in Iraq. They are used for grazing and for the production of wheat and barley.

RB_{A}^{T}—A   Reddish Brown and associated soils, on weakly dissected terraces, with Calcareous Alluvial soils on moderately broad first bottoms. In Israel these soils occur near the coast between the sand dunes and the steep upland to the east. For the most part the soils are developed from material washed down from the limestone uplands. About half of the area is on recent alluvium that is still flooded occasionally. The remainder is on terraces that are now above overflow. Textures range from loamy sand to silt loam. Rainfall is 20 to 24 inches. These soils are mostly under irrigation but there is some rainfed grain production. The agricultural use is intensive.

In Lebanon these soils are mapped in the Bekaa Valley, where they are used both as dry-farm and as irrigated agricultural land. In Syria they are mapped adjacent to the Ghab. In these areas there is a higher proportion of Reddish Brown soils on the terraces than in Israel. Parent material of the alluvium in the Bekaa Valley is mostly from basalt uplands, while the material near the Ghab is from limestone uplands. Texture of the Reddish Brown soils is mostly loam, while the Calcareous Alluvial soil ranges from loam to clay. A higher proportion of these soils is used for grain in Lebanon and Syria than in Israel, though production of other crops, particularly fruits and vegetables, has been increasing. Rainfall averages from 20 to 30 inches on this mapping unit in these two countries.

RB$_{Ch}$–L$\frac{H}{Ls}$ Reddish Brown and associated soils from chalk, on level to undulating plains, with Lithosols from limestone on hilly terrain. Reddish Brown soils occupy half or more of the areas. They are of clayey or loamy texture. In Syria several areas are mapped in the north. Precipitated limestone horizons are prevalent in the Reddish Brown soils, but they can be broken up by the farmers. The Lithosols are largely barren. The Reddish Brown soils are used for the production of winter grains. Numerous inclusions of Calcareous Alluvial soils are irrigated. Annual precipitation is from 15 to 20 inches.

RD$_g$ Red Desert and associated soils, on gravel plains that are level to undulating. Textures are gravelly sand loam to gravelly loams. In Egypt numerous areas are mapped on the coastal and alluvial plains and along the wadi beds of Sinai and the eastern deserts. The wadi beds are commonly bouldery on and beneath the surface. Average precipitation is two to three inches. Vegetation is sparse. This unit has no agricultural potential.

In Jordan an area occurs south of the Jafr Depression, which is in the south-central part of the country. This is brown flint gravel-covered, windswept desert almost barren of vegetation. Annual precipitation is less than four inches.

In Iraq areas occur along the border with Saudi Arabia. These areas are windswept gravel pavements with unconsolidated materials beneath the gravel concentration at the surface. The very scant vegetation furnishes a very little amount of grazing for camels. Average annual precipitation is about four inches.

RD$\frac{H}{U}$ Red Desert and associated soils from unconsolidated materials, on hilly uplands with characteristic desert land forms. This unit mapped in southern Iraq as an extension of a larger area in western Kuwait. Some undulating and rolling areas included. Textures are dominantly gravelly and sandy, but locally can be clayey. There are some Solonchak soils in small depressions. Parent materials are mainly sands and gravels but there is some sandstone. Annual precipitation is about four inches. Vegetation consists of some low desert shrubs occasionally grazed by nomadic livestock. This unit has no other agricultural value.

RD$_S$ Sandy Red Desert and associated soils, on level to undulating plains. Textures are dominantly loamy sands and sandy loams. Dry sands, shifting and stabilized sand dunes are found as the main associates. Solonchak, Alluvial soils, Gravelly Red Desert soils, and Lithosols are also present. In Egypt discontinuous stretches of this unit are mapped on both sides of the Nile River. Solonchaks are found in depressions as are some gravelly flats. Mean annual precipitation is two to three inches in the north and less than one inch in the south. There is some scrubby vegetation. This unit has no agricultural value.

The largest area of this unit mapped in Jordan is found in the Jafr Depression in the south. Parent materials in this depression are from alluvial materials washed from flinty limestone and sandstone of the adjacent region. There are some sand dunes. Wadi beds are frequent. There is some sparse vegetation. Average rainfall is less than four inches. This area has no agricultural value other than occasional light grazing.

In Iraq this unit is mapped mainly west of the Tigris River in the southeast on flat plains consisting of broad coalescing fans and on some old alluvium deposited by the Tigris River. Annual precipitation is six to seven inches. These soils support some desert shrubs and grasses that are used for grazing. A few small areas are irrigated for crop of wheat and barley. There is little agricultural value.

RD$_U$ Red Desert and associated soils from unconsolidated materials, on nearly level to undulating plains. Textures are dominantly loamy sand to loam. Locally there are some clayey soils. Solonchak and Solonetz occur mainly as small areas but may occupy a considerable percentage of some areas. Although the average annual precipitation is only about four inches, there is some vegetation for grazing animals. This unit is mapped mainly in southern Iraq. There is no agricultural value other than the limited grazing.

RD$_W$ Red Desert and associated soils of oases, on level to undulating plains and basins. This unit is mapped in Egypt in the oases of the western deserts—Siwa, Bahariya, Kharga, Dakla, and Farafra. The surface areas may be gravel, windblown sand, or clay. These undifferentiated surficial materials generally overlie a thick stratum of red clay. Solonchaks are common inclusions. The cultivated lands in these oases are scattered and not continuous. They are located where successful wells have been put down. Mean annual rainfall is generally less than one inch. The value of these soils for agriculture depends on their texture, depth, and saltiness. The continuity of a supply of water is questionable if too much area is put into agriculture.

$RD_g–L\frac{H}{C}$
Gravelly Red Desert and associated soils, on level to undulating plains, with Lithosols from consolidated rocks on hilly terrain. The area of each of these two components of the unit may be about equal or they may vary in either direction from this proportion. Desert sands are commonly included. In Egypt this unit is mapped over portions of the eastern desert and the Sinai Peninsula. Boulders are found on and in the Red Desert soils, where there is some scattered desert shrubs. The Lithosols are dominantly bare with numerous areas of rock outcrop. Average annual precipitation is one inch or less. This unit has no agricultural value.

$RD_S–L\frac{H}{C}$
Sandy Red Desert and associated soils, on level to undulating plains, with Lithosols from consolidated rocks on hilly terrain. The proportions of the two components is quite variable but may vary in either direction from equality. In Jordan there is an area of this mapping unit northeast of Asraq in the central northern part. This area consists of an intricate pattern of lava beds, sandy deposits, and salt flats. Much rough hilly bedrock is exposed. Annual precipitation is less than four inches. The land produces little vegetation and is of no appreciable agricultural value.

In Syria, adjacent to Jordan, is a similar area except for a greater proportion of Clayey Red Desert soils. Solonchaks are extensive inclusions. Again, precipitation is less than four inches. The area has no appreciable agricultural value.

$RD_S–L\frac{H}{Ls}$
Sandy Red Desert and associated soils, on level to undulating plains, with Lithosols from limestones on hilly terrain with isolated hills, or with small groups or rows of hills with intervening level to undulating plains. Sandy Red Desert soils and associated Solonchak, Solonetz, and Alluvial soils may constitute one-half the areas—or more or less than this. The Lithosols with areas of bare limestone rock make up the balance. Annual precipitation is less than four inches. The unit has no agricultural value. The only area mapped is in the tip of southern Israel.

$RD_S–L\frac{H}{Sd}$
Sandy Red Desert and associated soils, on level to undulating plains, with Lithosols from sandstone on hilly terrain. Isolated hills with intervening level to undulating plains. In Jordan an area is mapped as a continuation of a larger area in Saudi Arabia. Rainfall is four to six inches. There is some grazing. Other than this there is no agricultural value.

$RD_S–L_{Ls}$
Sandy Red Desert and associated soils, on level to undulating plains, with Lithosols from limestones on level to undulating plains. Principal associates of the Red Desert soils are Solonchak and Solonetz. In Egypt one area mapped south of the coastal sand dunes and marshes adjoining Alexandria. Precipitation is two to three inches. Sparse vegetation. Not used for agriculture.

In Iraq several areas are mapped in the southwest near the Saudi Arabian border. Included are some gravel plains and narrow wadi beds with saline Alluvial soils. Annual precipitation is three to four inches. Areas support scant vegetation that is grazed occasionally.

$RD_g–L^{S}$
Gravelly Red Desert soils, on level to undulating plains, with Lithosols on steep terrain. There are isolated hills or groups of steep hills, with intervening level to undulating plains or wadi beds of gravel. In Egypt mapped over extensive areas of the eastern deserts. The Red Desert soils occur on the level to undulating areas between the hills. Annual precipitation is one inch or less. There is no agricultural value.

$RD_s–L_S$
Sandy Red Desert and associated soils, on level to undulating plains, with Lithosols from sedimentary rocks on level to undulating plains. It is thought that limestones predominate; but sandstones, shales, and conglomerates occur. Some gravelly areas are included. Solonchak and Solonetz are principal associates. In Iraq a few areas are mapped

in the south. Annual precipitation is three to four inches. Sparse vegetation is occasionally grazed. There is no other agricultural value.

RD_S-L_{sd}

Sandy Red Desert and associated soils, on level to undulating plains, with Lithosols from sandstone on level to undulating plains. Solonchak and Solonetz are the principal associates. In Jordan one area is mapped as a continuation of an area in northern Saudi Arabia. Annual precipitation is three to six inches. The Red Desert soils support a sparse desert vegetation, and some additional vegetation comes after the winter rains. Used for light grazing.

$RD_S-L\frac{SU}{S}$

Sandy Red Desert and associated soils, on level to undulating plains, with Lithosols from sedimentary rocks on steep-undulating terrain complexes. There are numerous hills and ridges with steep slopes and nearly level tops. One large area mapped in Central Jordan. The underlying rock is horizontally-bedded flinty limestone, with the exposed surfaces covered by flints. Annual rainfall is eight inches or less. Most of the sparse vegetation is found in the wadi beds. The grazing supports a scattered population.

RD_g-S_D

Gravelly Red Desert and associated soils, on level to undulating plains with shifting sand dunes. Solonchak is an associate. In Egypt an extensive area is mapped adjoining the coastal sand dunes of the Sinai Peninsula. This is a gravelly Alluvial plain with sand dunes. There are some open sand flats. Annual precipitation is two to three inches. Vegetation is sparse. There is no agricultural use.

In Israel and in Jordan these soils are mapped in the Wadi Araba south of the Dead Sea. There are alternate areas of sand dunes, salt flats, Gravelly Red Desert soils, and Sandy Red Desert soils. The parent soil materials have mostly been washed from adjacent steep slopes and deposited as fans in the Rift Valley. Annual precipitation is less than four inches. The vegetative cover is very thin. There is no agricultural value.

RD_S-S_D

Sandy Red Desert and associated soils with sand dunes, on level to rolling plains. Associates are soils in local depressions. In Egypt one area of minor extent is mapped along the Mediterranean coast between Solum and Sidi Barrani. Rainfall is three to six inches. The sandy soils are cultivated where wells furnish irrigation water.

Mapped in middle Iraq. Small areas of Solonchak occur in depressions. Average annual precipitation is five to six inches. There are some desert shrubs that furnish limited grazing.

RD_S-SK

Sandy Red Desert and associated soils with Solonchak, on level to undulating plains. Several areas are mapped in southern Iraq. The areas are low-lying flats with materials that are at least partly water-deposited. Some areas are subjected to flooding and have high water tables. Annual precipitation is four to five inches. Vegetation is relatively more abundant than on most desert areas. These areas furnish more grazing than most of the areas of comparable rainfall.

RM_{AM}

Rough mountainous land of Alpine Meadow soil regions in rugged mountains. Steep Lithosols and bare rock with Alpine Meadow soils, on concave slopes and on mountain summits. Some areas mapped in northeast Iraq between elevation of 6000 and 9000 feet. Precipitation over 24 inches. Much snow in winter but little precipitation in summer. The areas are used for summer grazing.

RM_D

Rough mountainous land of Sierozems and Desert soil regions. In Egypt one extensive area mapped in southern Sinai. The mountains are rough and stony. There is little vegetation. Rainfall is generally less than one inch. No agricultural value.

RM_{TRB}

Rough mountainous land of Terra Rossa—Brown Forest vertical soil zones. Mapped in the high mountains of Lebanon, Syria, and Iraq. Trees were the dominant vegetation at one time but much of the area now has no trees. Annual precipitation is twenty inches or more. These areas furnish some summer grazing.

S_{BD}

Shifting sand dunes of the Barren Desert region. Low to high dunes in systems of rounded hills or long, broad ridges with intervening troughs. In Egypt mapped in the area known as the vast Great Sand Sea, as well as in numerous smaller areas. Annual rainfall is less than one inch. No agricultural value.

S_D

Shifting Sand Dunes. In Egypt mapped west of Alexandria adjoining the Libyan Plateau and adjacent to the coast. Rainfall is about six inches or less. Also mapped along the Sinai coast, where rainfall is less than two inches. Solonchaks are in the small depression. No agricultural value.

In Israel mapped along the seacoast and a few miles inland in the south. In the south vegetation is sparse while further north where rainfall is higher there is more vegetation, including some trees and shrubs. Little or no agricultural use.

In Lebanon and Syria a few narrow areas are mapped along the coast. Rainfall more than twenty inches. No agricultural use.

In Iraq the main areas are mapped southwest of the Euphrates. Some Solonchaks are in depressions. Rainfall about four to six inches. No agricultural value.

S—M

Beach—Sand-Dune—Marsh Complex. Sand Dunes along the beach, with salt or fresh water marsh between the dunes. Mapped in Egypt in one small area along the Mediterranean coast at the mouth of the Nile River. Rainfall three to four inches. No agricultural value.

S_D—SK

Sand Dunes with Solonchak. A complex of low Sand Dunes with Solonchak in depressions or troughs between dunes. In Egypt an area is mapped west of Alexandria. The dune ridges are dominantly stabilized by vegetation. Narrow interdune flats and broad depressions from dried-up lake beds constitute the Solonchak. Rainfall three to four inches. Little or no agricultural value.

In south Iraq an area is mapped as a narrow belt southwest of the Euphrates. The area is low-lying and receives water from the wadis of the desert to the southwest. The dunes are largely barren, but there is some salt-tolerant vegetation on the flats between the dunes. Annual precipitation is four to five inches. The area is used for grazing. No other agricultural value.

SK

Solonchak and associated soils, on broad depressions or benches. Associates are Solonetz and other local inclusions. In Egypt an extensive area is mapped in the Qattara Depression, as well as in numerous smaller areas elsewhere. These areas may be continuously wet during the whole year or they may be wet in winter and dry in summer. Mean annual rainfall is less than one inch.

In Israel and in Jordan these soils occur at both ends of the Dead Sea. There are three main components: one is the actual flood plain; one is the old marine sediments that are very salty; and the third is fan materials from adjacent uplands. Where the latter are deep enough, salts can be leached and the soils farmed. Fans may comprise as much as 70 percent of the total area. Annual rainfall is less than four inches. In Jordan other areas are found throughout the country.

In Syria there are a number of areas in the north and east, as well as small areas in the south. Average annual precipitation is from 4 to 12 inches. There is some vegetation, but most of the areas are bare.

In Iraq several areas are mapped in inland depressions. Small areas of Sierozem, Deserts, or Red Desert are included. Annual precipitation is from four to seven inches. There is some salt-tolerant vegetation, but some areas are barren. There is some grazing.

SK_A

Solonchak and associated soils from recent alluvium, on level flood plains or delta plains. Associates are Saline Alluvial and Alluvial soils. In Israel and in Jordan these soils are mapped at both ends of the Dead Sea. They occur just above the level of the Dead Sea and are muddy the whole year as a result of a high water table and the flood waters of streams that flow into the Dead Sea. Annual precipitation is four inches or less.

In Syria an area is mapped in the northeast in a broad alluvial flat. Average annual precipitation is from 12 to 15 inches.

In Iraq several areas are mapped in the Tigris-Euphrates delta. The areas are marshy in winter but are mostly dry in the summer. Some areas flood regularly, others flood only infrequently. Average annual precipitation is from four to eight inches. Reeds are the main vegetation. The soils are not used for agriculture except for dates in a few places.

SK–A$_S$

Solonchak with Saline Alluvial soils and associates, on lowlands and basins with numerous ponds and lakes. Solonchak predominates, while Saline Alluvial soils occur in lesser amounts.

In eastern Jordan a number of areas are mapped on narrow to wide wadi bottoms. Salty lakes occur in some of the broader flats. Textures are probably silty or clayey. Average annual precipitation is less than five inches. There is some grazing.

In Middle Syria a few areas are mapped adjacent to streams. Average annual precipitation is from 8 to 13 inches. The areas are used mostly for grazing but there is some irrigated farmland.

In Iraq this is one of the most extensive units mapped in the Tigris-Euphrates Delta. Some areas are also mapped in middle Iraq. Textures are predominantly silt loams, silty clay loams, and silts. Large areas are irrigated and produce a crop every other year. The salt content is in large measure the result of long irrigation without provision for drainage to remove salts. If properly drained and managed, these can be first-class soils. Average annual precipitation is four to eight inches.

SM

Salt-Water Marsh, in low-lying areas along coasts. In Egypt numerous small areas of tidal marsh are mapped at the edge of the Nile Delta. Sand flats and dunes are commonly included. Annual rainfall is three to four inches. No agricultural value.

In southern Iraq several areas are mapped along the coast. The areas are largely mud flats and may be either barren or reed-covered. Some coral reefs and sandy areas are included. No agricultural value.

SR$_B$

Sierozem and associated soils, on basin and piedmont plains.

Mapped in east-central Syria, the largest area being the Oasis of Damascus. Textures are mainly loam and silt loam with some clay. Solonchak is a common inclusion. The Damascus basin is perhaps the most productive area in Syria. It has been irrigated for thousands of years. It is very productive of fruits and vegetables. Where not irrigated there is a sparse stand of shrubs and annual grasses. Average annual precipitation is from 9 to 13 inches. Nonirrigated areas are used for occasional grazing.

In southeastern Iraq areas are mapped on the Iran border on plains formed by coalescing fans. Textures are probably medium. Average annual precipitation is from 8 to 11 inches. Mostly these areas are used for grazing, but some small areas are irrigated for the production of wheat and barley.

SR$_{Ch}$

Sierozem and associated soils from chalk, on level to undulating plains. Mainly loamy and clayey Sierozems with associated Solonchak and Lithosols. Extensive areas are mapped in northern Syria over formations of chalk, marl, or gypsum. These soils are generally shallow to bedrock. Precipitation is from 8 to 14 inches, with vegetation not so sparse as on the deserts. The areas are mostly used for grazing, but some small areas are irrigated.

SR$\frac{DF}{D}$

Sierozem and associates, on dissected colluvial fans at the base of mountains. Sloping to moderately steep sandy loam to silty clay Sierozems are on the valley sides. Medium and fine-textured Sierozems with Solonetz occur at the base of slopes on colluvial materials. In southeastern Iraq some areas are mapped where precipitation is from 8 to 12 inches. Vegetation consists of shrubs and grasses. The soils are used mainly for grazing, but small areas are irrigated for production of barley and wheat.

SR$_g$,

Sierozem and associated soils, on gravel plains that are level to rolling with occasional steep escarpments. A few areas are mapped in west-central Syria, where the underlying material is weakly cemented conglomerates. These rocks are high in calcium carbonate. The soils are gravelly and low in productivity. Precipitation is between 9 and 13 inches. The soils are used for grazing.

SR$\frac{H}{S}$

Sierozem and associated soils from weakly consolidated sedimentary rocks, on hilly terrain. Mapped in eastern Iraq over red sandstone, silts, and marls. Some of the soils are reddish and hence not typical Sierozems. Textures range from sandy loam to clay. Some of the soils are shallow and stony. Precipitation is from 8 to 12 inches annually. Vegetation is denser then on the deserts. The soils are used mainly for grazing.

SR$\frac{R}{Ch}$

Sierozem and associated soils from chalk, on undulating to rolling plains. Textures are mainly clayey and loamy. Extensive areas are mapped in northern Syria over formations of chalk, marl, gypsum, or unconsolidated calcareous deposits. These soils are generally shallow to bedrock. Annual precipitation is between 8 and 14 inches. Some areas are irrigated, but most of the areas are used for grazing.

SR$\frac{R}{L}$

Sierozem and associated soils from loess (windblown silts), on rolling plains. In Israel this unit is mapped in the vicinity of Beersheba and consists of a gently rolling area covered with several feet of loess. The area is cut by stream channels to some extent but for the most part is relatively smooth. To the south the loess gets coarser and thins out at the junction with the steep uplands. Annual precipitation is from 4 to 12 inches. The vegetation is that of a steppe. These would be good productive soils if water were available.

Sr$\frac{R}{Ls}$

Sierozem and associated soils from limestone, on rolling plains. In western Jordan two areas are mapped. The bedrock is horizontally-bedded flinty limestone and marl. A considerable portion of the area is covered by shallow soils or Lithosols. Some soils can be cultivated by tractors, though production of a crop is very uncertain because of erratic rainfall. Average annual precipitation is four to eight inches. Altitude is from 3000 to 5000 feet. Mainly the area is grazed, though the vegetative cover is sparse.

In southern Syria a few areas are mapped where rainfall is a little higher and vegetation more abundant than on the deserts. Average annual precipitation is eight to ten inches. The soils are used for grazing.

SR$\frac{R}{TS}$

Sierozem and associated soils from weakly consolidated rocks, on rolling plains. Textures are predominantly sandy loam to clay loams. Some hilly Sierozems are included. Several areas are mapped in north-central Iraq. The underlying materials are chalk, limestone, gypsum, sandstone, shale, and some unconsolidated sands, gravels, and clays. Short, abrupt slopes leading to wadis are common. Average annual precipitation is from 7 to 12 inches. The soils are used mainly for grazing, but some areas are irrigated for production of wheat and barley.

SR$_R\frac{H}{B}$

Stony Sierozems and associated soils from basalt, on hilly terrain. In Jordan and Syria there is an area of this unit mapped. Most of the area is in Jordan, with a smaller continuation into Syria. The soil is mostly covered by basalt boulders. Local lava flows and cinder cones occur. Annual precipitation is less than eight inches. Vegetation consists of scattered shrubs and grasses. The area is used for grazing.

SR$_S$

Sandy Sierozem and associated soils, on level to rolling plains. Solonetz and Solonchak are associates. In northern Iraq an area is mapped where the evidence indicates that the soils are prevailingly sandy. Average annual precipitation is from 7 to 12 inches. The soils are used for grazing.

SR$_U$

Sierozem and associated soils from unconsolidated materials, on level to undulating and locally rolling plains. Associates are Solonchak, Solonetz, and some Sandy Sierozem. In northeastern Syria a few areas are mapped over a mixture of gravel, sand, and clay, with sandstone, conglomerate, and marl occurring locally. Lithosols are common on the consolidated rocks. Average annual precipitation ranges from 7 to 12 inches. The soils are mainly used for grazing.

In northwestern Iraq several areas are mapped. The underlying material is variable. Considerable Solonchak is found in the depressions. Annual precipitation is from 7 to 12 inches. These soils are used mostly for grazing, but a few areas are irrigated for production of wheat and barley.

$SR\frac{T}{A}-A_{CN}$

Sierozems and associated soils, on nearly level to gently sloping terraces with coarse-textured Calcareous Alluvial soils. Solonchaks are common associates. This unit is mapped in eastern Iraq. As delineated it includes some low mountain outliers and isolated hills. The surface soil is dominantly loamy, but the substratum is gravelly. The average annual precipitation is about eight inches. These soils are mostly irrigated and are used for the production of wheat, barley, tobacco, and fruit.

SR_S-S_D

Sierozems and associated soils, on rolling plains with sand dunes. A large area is mapped in southeastern Israel and extends into Egypt. These soils are developed on windblown materials that are coarser in texture than loess. Annual precipitation is from two to four inches. Vegetation is scant. Some of these soils could be used for agriculture if irrigation water were available in plentiful amounts.

$SR-SK$

Sierozem-Solonchak complexes and associates, on lowland plains. The topography is nearly level to gently undulating. This unit is mapped in basins and on low flats in northwestern Syria. The Sierozems constitute perhaps 70 percent of the areas. Solonetz is an associated soil. The parent material is alluvium and colluvium with some inclusions of marl. Average annual precipitation is about 13 inches. Much of the unit is irrigated and is used for the production of grain, as well as some fruit and vegetables.

TR_U

Terra Rossa and associated soils from undifferentiated unconsolidated materials, on level to rolling plains. Common associates are Rendzinas, Calcareous Alluvial, and Brown Forest soils. A few areas are mapped near the coast in Lebanon and in Syria. The parent materials are mostly alluvium, colluvium, and marine deposits. The alluvium and colluvium originated in the limestone mountains to the east. Some sandy coastal deposits are included. Average annual precipitation is 30 to 40 inches. These soils are intensively used for growing fruits and vegetables.

$TR\frac{R}{Ls}-RB_U$

Terra Rossa soils from limestone and Reddish-Brown soils from unconsolidated sediments, on level to rolling plains. The Terra Rossas are on the more rolling portion of the landscape. Calcareous Alluvial soils and Lithosols are common associates. This unit is mapped in northeast Syria around Homs. Average annual precipitation is from 20 to 30 inches. These soils are used for growing grains and fruits.

$TRB\underline{M}$

Lithosolic Terra Rossa—Brown Forest soils vertically zoned on mountains. This is the main unit mapped on the low mountains in northeastern Iraq. Elevations are mainly from 2,000 to 4,000 feet. Oak forests occupy or have occupied much of the area. The soils are highly calcareous. Average annual precipitation is from 30 to 45 inches. The soils are used for grazing, with some crop production in small valleys.

$TRR\frac{H}{Ls}$

Terra Rossa—Rendzina complexes and associated soils from limestone, on hilly terrain with karst-like (sinkhole) topography. In Israel and the East Bank of Jordan these soils occur on the uplands. The Terra Rossa soils have developed on the hard limestones while the Rendzinas have developed on marl. Annual precipitation is dominantly in the 20- to 28-inch range. The soils are used for olives, grapes, and for cereals where level enough. The area was once covered by an oak forest, and there has been much erosion over thousands of years. Where not cultivated the areas are covered by grass and shrubs. Cultivation is mostly on areas that can be terraced.

Several areas are also mapped along the coast of Lebanon and in northwestern Syria. The soils here are similar to those in Israel and Jordan except that precipitation is mainly between 30 and 40 inches. Some Brown Forest soils are probably included. The soils are generally shallow and stony with numerous outcrops of limestone. Here too many of the areas have been terraced and are used to produce fruits and vegetables.

In Iraq an area is mapped near the entrance of the Tigris River into Iraq. The area is used mainly for grazing, though there is some cultivation. Precipitation is from 25 to 40 inches a year.

$TRR\underline{M}$

Lithosolic Terra Rossa and Rendzina complexes and associated soils, on mountainous karst-like topography. A few areas are mapped in northwestern Syria, mainly just east of the Orontes River. Elevations are mainly between 700 and 2500 feet. There is much bare rock. Annual precipitation is between 20 and 30 inches. The areas are used mainly for winter grazing, though a few included areas of Alluvial soils are cultivated.

$TRR\frac{R}{Ls}$ Terra Rossa-Rendzina complexes and associated soils from limestones, on undulating and rolling terrain. An area mapped in western Syria. Lithosolic Terra Rossa and Rendzina soils are associates. There is much bare rock. Annual precipitation is between 20 and 30 inches. The soils are used for fruit and vegetable production.

$TRR\underline{S}$ Lithosolic Terra Rossa—Rendzina complexes and associated soils from limestone, on steep hills. Several areas of this unit are mapped in northwestern Syria. Underlying rocks are mostly hard limestones, but there are some marls and chalks. The terrain is rugged and there is much bare rock. Annual precipitation is mostly from 20 to 30 inches, though it is as much as 40 inches in some places. Scrubby trees, shrubs, and grass grow where there is soil material. The areas are used for grazing. Where alluvial soils are included they are usually cultivated. A few of the steep hillsides are terraced with stone and are used for fruit and vegetable production.

In northern Iraq an area is mapped near the Tigris River where it enters Iraq. The soils are dominantly covered with scattered trees—mostly oak. Average annual precipitation is from 25 to 40 inches. The area is used mainly for grazing.

Water Supply of the Middle East*

The Nile River

General Description

IN length of its channel and quantity of its flow, the Nile River ranks among the great rivers of the world. It is the dominating feature of northeastern Africa and affects all or parts of Egypt, Sudan, Uganda, Ethiopia, Kenya, Tanzania, Rwanda, Burundi, and the Congo (Figure B-1). The sources of the Nile, long unknown to many who made use of its waters, are found in the Atbara River in Ethiopia, the Blue Nile with its origin in Lake Tana in Ethiopia, and the White Nile with its many branches in the vast Sudd of Southern Sudan, as well as in the Lake Plateau region of Lake Victoria, Lake Albert, Lake Edward, and Lake Kyoga.

While no appreciable amount of its flow originates in Egypt, the greater part of development and use of Nile water has been within Egypt. Moving upstream from the Nile delta, the first tributary of any consequence is the Atbara in Sudan, which flows into the Nile at a point about 1,700 miles upstream from the Mediterranean Sea. In fact, from the Atbara River to the Mediterranean Sea, the Nile flow is steadily reduced due to evaporation and diversion for irrigation. Rainfall along the Nile Valley in Egypt is practically nonexistent, averaging about 32 mm. (1.27 in.) per year at Cairo and even less proceeding upstream toward Sudan. Consequently, the Nile serves primarily as a channel to conduct the runoff from the upper basin areas toward the sea. The river carries a heavy load of sediment, especially in times of flood discharge, which it is continually depositing in the river channel. This process has formed the huge delta at the mouth of the river, which accounts for most of the agricultural area of Egypt. The process has also been responsible for the raising of the river channel bed to the entent that the gradient in the flood plain is actually *away from* the channel rather than toward it as is normal in river beds receiving runoff from the surrounding terrain. This creates drainage problems. Hurst[1] reports estimates of sediment buildup in the river bed amounting to three to four feet in a thousand years.

Geology of the Nile Basin

The country surrounding the Nile River is markedly different on the eastern side from the western. The eastern desert area gives evidence of erosion by water in times past, probably in the Pleistocene period, when the climate was very different from the aridity exhibited today. The sharp relief of the eastern desert is not present on the west side of the river channel, where all signs of large drainage systems have been removed by wind action, which has shaped the western desert areas over the past half million years. From a report of the Food and Agriculture Organization of the United Nations[2] the following is quoted:

Sometime during the late Tertiary period the present Nile Valley was a sea gulf which was gradually filled with marine and terrestrial deposits during Pleistocene and recent times. The sediments were in part deposited by the Nile, but the greater part consists of fan and piedmont deposits brought down from the adjacent highlands in flash floods by steep torrential streams. The majority of these sediments consist of unconsolidated gravels, cobbles, and sands, except in the coastal zone west of Alexandria, where in texture they are mainly clay loams and clays. As many as five different terrace levels have been observed, both in the river deposits and in the rubble deposits of piedmont plains. In many places the terraces have been eroded or covered by fan deposits; large areas are covered by wind-blown sand.

The higher desert plateaus are geologically older, ranging in age from Cretaceous to Pliocene; these formations include sandstones, shales, and limestones, and also gravel, rubble, and sand deposits in various stages of consolidation. Most of the desert areas are denuded and rocky.

In Upper and Middle Egypt from Aswan to Cairo the Plateau usually rises in steep, rocky escarpments to a considerable elevation above the valley floor; as much as 100 meters for the Nubian Sandstones in the Aswan-Isna area. In Lower Egypt the transition between the fringe lands and the desert is not so clearly marked and the older stony desert lands grade into the Pleistocene landscapes of the alluvial plain.

The semi-detailed surveys. . . .are located within the recent Nile delta, which was built up during the Holocene and where the deposition of clay and silt in the lagoon lakes and adjacent swamps is still going on.

Most of the agricultural lands in Egypt exist on the river deposits described above.

The River Flow

The flow of the Nile River has been measured for many years at a number of locations. The measurements at Aswan provide an excellent basis for determining the amount of water flow into Egypt. The gauge is located at the Aswan Dam, which is just a few kilometers downstream from the

---

* Compiled by Dean K. Fuhriman, Brigham Young University, Utah.

[1] H. E. Hurst, *The Nile* (London: Constable Publishers, 1952).

[2] *Soil Survey Project, United Arab Republic, Vol. I, General Report* (FAO/SF: 16/U.A.R. [Rome: FAO, ca. 1966]).

MED. SEA
CAIRO
SINAI
JORDAN
SAUDI ARABIA
Asyut
EGYPT
Esna
Nile
Aswan
Lake Nasser
RED SEA
River
LIBYA
CHAD
Atbara R.
KHARTOUM
SUDAN
Blue Nile R.
White Nile R.
Lake Tana
ETHIOPIA
ADDIS ABABA
Bahr El Arab
Lol R.
Bahr El Ghazal
THE SUDD
Sobat R.
Bahr el Jebel
CENTRAL AFRICAN REP.
CONGO REPUBLIC
Lake Albert
Lake Kyoga
KENYA
UGANDA
Lake Edward
Lake Victoria
RWANDA
BURUNDI
TANZANIA
THE NILE RIVER
0 100 200 300 400
KILOMETERS

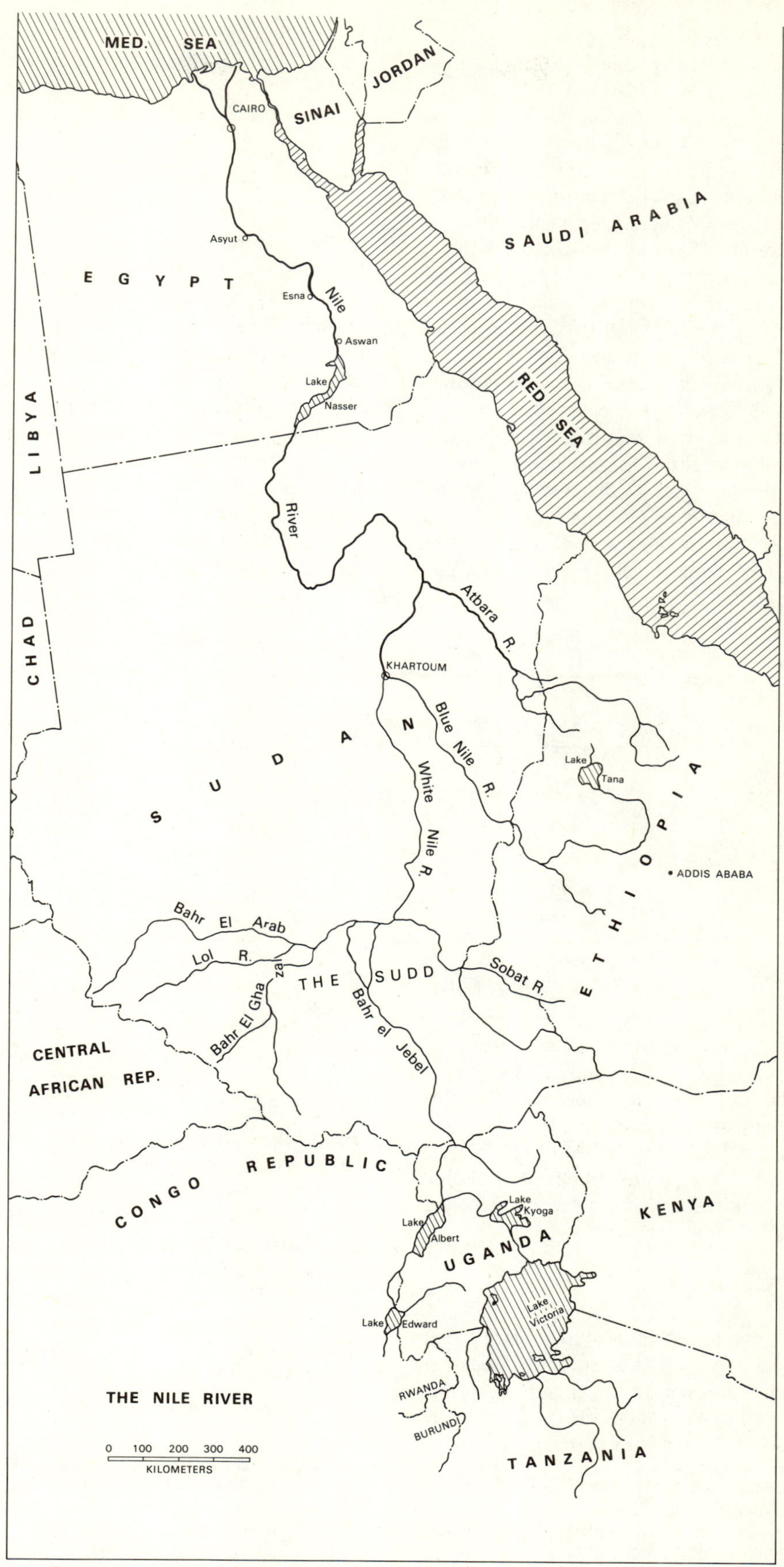

Fig. B-1

new High Aswan Dam (Sadd et Aaali). A record of these measurements is reported by Hurst, Black, and Simaika for the years 1869 to 1945,[3] 1946–58[4] and 1959–65.[5] A summary of these records showing monthly and annual runoff for the period 1871 to 1965 is shown in Table B-1 (end of Appendix B). It should be noted that there is some regulation upstream which affects this record, due to diversions in Sudan and storage in Sudan and some upstream reservoirs. However, the net effect in comparison with the total flow of the river is small and the difference is not significant to the purpose of this report. Since the allocation of waters between Egypt and the upstream countries is on the basis of the "natural" river, the figures given in Table B-1 could not be used for an exact determination of water available for development in Egypt. Hurst and his colleagues tabulate both computed "natural" river flows (at Aswan) and measured Aswan flows for the period 1946 to 1958. In order to show the general magnitude of difference between the two, the figures for the year 1951 (the year of lowest flow in the 13-year period) and the year 1954 (the year of highest flow in the 13-year period) are tabulated here:

natural flow, with only limited storage during the low-water period. This has been dramatically changed by completion of the High Aswan Dam, which provides a vast storage reservoir for the regulation of Nile streamflow in Egypt.

Periods of Flood

In the past there have been a number of years of serious flooding on the Nile, and therefore in the Nile delta a system of levees has been constructed to hold back the water in time of flood. The completion of the High Aswan Dam will change the picture here, too, and it is unlikely, assuming efficient flood control operation of the reservoir, that the delta will ever again be in danger of serious flooding. The waters are now being retained in Nasser Lake and the flood control features of the project are already a reality. It should be noted that the years 1874, 1878, 1879, 1894, and 1895 were years of excessive flooding. In 1878, the peak flood was about 13,400 cubic meters per second, probably the maximum on record. The year 1946 was a year of serious flooding in the Sudan; especially from the

Nile River Flow, Low (1951) and High (1954)[6]

Year	Jan.	Feb.	Mar.	Apr.	May	June	July	Aug.	Sept.	Oct.	Nov.	Dec.	Annual total
1951													
Measured	2.75	2.92	2.41	2.43	2.92	3.66	3.83	14.9	17.5	10.1	4.35	4.30	72.3
Natural river	3.39	2.24	2.27	2.34	2.15	1.18	2.85	15.4	17.8	12.0	7.79	4.93	74.4
1954													
Measured	2.55	2.40	2.24	2.30	2.76	4.04	6.79	23.2	27.0	19.7	6.85	3.82	104.0
Natural river	2.87	1.83	2.55	2.27	1.82	1.18	5.63	24.8	28.3	20.5	8.55	5.05	105.0

To provide a physical concept of the amount of water represented in the figures above, it may be noted that 1.0 in the tabulated material (representing one milliard cubic meters) is an amount of water which would cover 810,710 acres of land with a one-foot depth of water. Stated another way, the total water storage capacity in Lake Mead (the reservoir formed at Hoover Dam on the Colorado River) is equal to 31.1 million acre-feet (about 38 milliards of cubic meters) and the Nile flow in the below-average-runoff year of 1951 would fill Lake Mead twice.

Periods of Reduced Supply

Table B-1 shows the fluctuations in available water supply from the Nile River in Egypt. The months of December through May represent the period of minimum runoff. There has been relatively little storage on the river in the past, and the use of the Nile has been restricted to the

Atbara River. The effects of this flood downstream were minimized because of the limited but useful regulation which was possible on the Nile upstream from the Atbara and also by the unintentional flooding of large areas in the Sudan, the total volume of which is estimated by Hurst, Black, and Simaika[7] to have been of the order of 3.0 milliards of cubic meters. The storage capacity created in Lake Nasser by the construction of the High Aswan Dam is 157 milliards of cubic meters (125 million acre-feet),[8] or more than the maximum year runoff for the period shown in Table B-1. By reserving a predetermined volume of the capacity to receive floodwaters during the flood period (August, September, and October), the magnitude of flow downstream can be regulated within tolerable limits quite easily. Hurst[9] indicates the "time of travel" of flood from Roseires (confluence with the Blue Nile) to Aswan varies

---

[3] H. E. Hurst, R. P. Black, and Y. M. Simaika, *The Nile Basin* (Cairo: Ministry of Public Works, 1946), III.

[4] H. E. Hurst, R. P. Black, and Y. M. Simaika, *The Nile Basin* (Cairo: Ministry of Public Works, 1959), IX.

[5] H. E. Hurst, R. P. Black, and Y. M. Simaika, *The Nile Basin* (Cairo: Ministry of Public Works, 1966), X.

[6] Figures are in milliards (billions) of cubic meters. One milliard cubic meter = 810,710 acre-feet. From Table B-1.

[7] H. E. Hurst, R. P. Black, and Y. M. Simaika, *The Nile Basin* (Cairo: Ministry of Public Works, 1959), IX.

[8] Aswan High Dam Authority, *Aswan High Dam*, brochure published in May 1964.

[9] Hurst, *The Nile, op. cit.* (above, n. 1).

between ten and thirty-five days, depending on the volume of flow, with time decreasing as the flows increase. Most of the floodwaters of the Nile originate on the Blue Nile or the Atbara. For flood control at the High Aswan, this should then provide a warning time of at least ten days to prepare to receive the flood flow at the reservoir.

Silt in the Nile Water

All natural streams carry suspended and dissolved material—a maximum in time of flood. The Nile silt has been deposited on irrigated lands for many years and has no doubt contributed considerably to the soil-building process on these lands. According to Hurst,[10] the concentration of suspended matter generally ranges up to about 2500 parts per million during the flood season; and the annual total is about 100 million tons—about 30 million tons of fine sand, 40 million tons of silt, and 30 million tons of clay.

Groundwater in Egypt

Since the Nile River represents the only surface water supply available to Egypt, there have been a number of attempts to develop groundwater outside of the Nile Valley. In the vast sandy desert expanse of Egypt, there are only relatively few areas which show any promise for the development of either land or water beyond the valley of the Nile. From ancient times there have been a limited number of oases, where springs or wells have provided groundwater to the thirsty traveler and his animals. Some agricultural development and use of the waters at these oases have taken place, and some investigations have been undertaken to determine the water-development potential in these areas. The Egyptian General Desert Development Organization has been working toward such development under the New Valley Project. All of the investigation has been in the Western Desert (west of the Nile River). Some studies have also been conducted in the oases at Bahariya, Farafra, Dakhla, and El Kharga. These represent the larger and more important potential development areas of the general region known as New Valley. This region is generally underlain by Nubian sandstone, at varying depths, depending on topography, which is the source of most of the groundwater. According to Roberts,[11] the cropped area irrigated from groundwater (in 1962) in the Dakhla, Kharga, and Farafra areas of the Western Desert was distributed as follows:

	Approximate area irrigated, 1961 in feddans
Dakhla Depression	30,000
Kharga Depression	20,000
Farafra Depression	100

The Parsons Engineering Company[12] reported in 1962 that the annual discharge from the Nubian Sandstones of the Western Desert was estimated to be 1850×10^6 cubic meters (1,500,000 acre-feet) per year. In terms of the probable total area under irrigation, this figure seems high, unless there is a considerable amount of uncontrolled flow or leakage. The Parsons report also estimated an annual recharge of about 355×10^6 cubic meters (290,000 acre-feet). Thus, according to these estimates, water was being "mined" at a rate of 1495×10^6 cubic meters (1,200,000 acre-feet) per year.

The Parsons report was based on limited exploration. It recommended further studies to determine the depletions more accurately, and also to determine the amount of water stored in the Nubian sandstones underlying the Western Desert. Studies have been conducted since that time on which final reports are not yet available. Measurements in the El Kharga Depression indicate a zone of saturation that extends from about 125 meters above sea level to 850 meters below sea level, with the depth below land surface to the artesian aquifer as shallow as 20 meters in topographic lows. Saturated thickness of the Nubian in El Kharga varies from about 300 meters near Baris to more than 800 meters between Ginah and El Kharga. Measurements in Dakhla have not delineated the water-bearing beds as well as has been done in the El Kharga Depression, but the deepest well drilled bottomed at 1,232 meters below land surface (1121 meters below sea level) and was still in water bearing Nubian Sandstone. The upper limit of these beds is at about 50 meters above sea level. and it is possible that a great potential water supply is in storage underground.

An extensive program of development has been carried on by the Egyptian government in Dakhla and El Kharga since about 1960. The rate of withdrawal of water from the Nubian formation in El Kharga increased by more than 150,000 cubic meters from over 100 new wells in the 1960–66 period. Similar development also has taken place in the Dakhla area, with about 100 new wells having been drilled there. With the new government wells being pumped (in both the El Kharga and Dakhla depressions), the older existing private flowing wells have experienced a decrease in supply and a number have ceased to flow, indicating a reduction in piezometric pressure in the aquifer. It is unfortunate that the governmental program since 1960 has appeared to concentrate on production, with only limited effort to utilize the drilled wells for evaluation of the potential water supply available in the Nubian formation. Based on the studies made to date, the following general conclusions related to groundwater development in Egypt are possible:

(1) The Nubian sandstones underlying the Western Desert contain water moving at a relatively low rate in a general northeasterly direction. The size and extent of these water-bearing beds is not known.

(2) Water is being discharged from the groundwater

---

[10] *Ibid.*
[11] Ray Roberts, *Reconnaissance Soil Survey, New Valley Western Desert, Egypt* (Cairo, 1962).

[12] Parsons Engineering Co., *Bahariya and Farafra Areas—New Valley Project, Western Desert of Egypt*, Final Report (November 1962).

in the Western Desert areas at a greater rate than it is being recharged, and increase in development of these waters will further deplete the amount of water in storage.

(3) The volume of water in storage underground which can be "mined" economically is not known. Some estimates described it as "extraordinarily large," but further detailed investigations are needed to determine the volumes and limits of possible development.

(4) In determining the extent of economical development of Western Desert lands by mining of the groundwater, it should be borne in mind that the soils are generally inferior to the Nile Valley soils, that there is an urgent need to provide drainage for any lands developed, and that such drainage may be difficult to provide since the better soils may often be located in the lower portions of the New Valley Depressions—sometimes at elevations below sea level.

The Tigris and Euphrates River Systems

General Description

The Tigris and Euphrates Rivers both have their headwaters in the mountains of the Kurdestan and Armenian areas of eastern Turkey (Figure B-2).

The combined basins occupy an area of over 784,000 square kilometers—359,000 in Iraq, 162,000 in Turkey, 71,000 in Syria, 146,000 in western Iran, and 46,000 in Saudi Arabia. The area within Iran is mostly in the Karun River Basin, which flows into the lower reaches of the Tigris Euphrates delta. The area within Saudi Arabia is all desert land. Neither of these latter two areas contributes to the more readily developable water of the basin; the following paragraphs therefore emphasize the impact of this river system on Iraq, Turkey, and Syria.

In the lower delta of the Mesopotamia valley of Iraq, the Tigris and Euphrates rivers merge before flowing into the Persian Gulf. These rivers originate in an area of mountain peaks reaching elevations greater than 12,000 feet above sea level, where precipitation often exceeds 1600 mm. (63 in.) annually. In their flow to the sea they irrigate deserts and provide water to sustain life for millions of people. In times of flood these rivers become raging torrents, causing great loss of life and property. During periods of low flow, the rivers fall far short of meeting the needs of those who are dependent upon them. There is pressing need for development of the waters of the Tigris and Euphrates Rivers.

Climate and Geology

The regions occupied by the basins of the Tigris and Euphrates can be roughly divided into three climatic and geologic zones: (1) the high and rugged mountains, in which the river headwaters are located; (2) the foothill zone, which forms the transition between the mountains and plains, or the mountains and the Syrian Plateau; and (3) the deep lacustrine and alluvial sediments filling the lower parts of a prominent geosynclinal trough and forming the zone in which the broad plains of Syria and Iraq are situated.

The mountain zone is characterized in the higher areas by Alpine topography—small, rocky fields and stony pastures; and vegetation—grasses, alpine flowers, mosses, and lichens, with about a three-month frost-free period. Below the high Alpine areas there are natural forest areas reaching down to the foothill zone. Precipitation in this zone is somewhat variable, depending on local topographic features. Much of the precipitation is in the form of snow, which occurs during the winter months and is usually melted by April or early May. All of the Tigris watershed in Turkey and a considerable part of the Euphrates watershed in Turkey are located in this zone.

The foothill zone is a zone of transition—from the standpoint of topography, climate, and geology. Only a relatively small area lies within this zone. The lower foothill area includes some gently sloping and rolling topography with relatively small amounts of precipitation limited to winter, spring, and early summer.

The plains zone is a large area, entirely in Syria and Iraq, which includes the desert lands in the west, the broad plateau of Syria, the great valley floor of Mesopotamia, and the swampy marsh lands of the river delta. A portion of this zone in western Syria and Iraq is underlain by limestone deposits, which in some locations represent good water-bearing aquifers. Quality of the groundwater is variable, and much of it is "bitter" because of the highly gypsiferous and saline deposits which overlay much of the limestone. Further to the east, and extending over a broad area, the underlying deposits are clastic, containing little water, and much of this is saline. These deposits cover a wide area under the plain of the Euphrates and the Tigris, interrupted only by a narrow belt of alluvial aquifers along the rivers themselves. In some relatively small areas of Eastern Syria, Turkey, and northern Iraq deposits of basalt occur which at some locations yield excellent groundwater. In lower Iraq, in the area of the broad delta formed at the mouth of the combined rivers, the land is poorly drained and highly saline. Among the salts in these soils, gypsum occurs in large quantity. There is also usually a positive hydrostatic pressure in the soil due to the water in the riverbed alluvium. Precipitation throughout this zone is very limited—averaging less than 200 mm. (8 in.) annually—and concentrated in the November to April period.

River Discharge Records

Records of streamflow are available from measurements made at various points along the Tigris, Euphrates, and their tributaries. These records provide a historical basis from which the total water available for development can be estimated. They also provide a basis for an evaluation of floods and measures which might be taken to reduce the loss of life and property from river flooding. Many studies[13] have been made relating to the development and control of the Tigris and Euphrates rivers. Many of these were available for use in the preparation of this report.

---

[13] S. W. Macksoud, in notes provided for this study, reports that ten different consultant groups had prepared (up to 1964) a total of 287 separate reports on water development in Iraq alone.

BLACK SEA
T U R K E Y
MED. SEA
Keban
KEBAN RES.
LAKE VAN
Birecik
Aleppo
Jerablus
Homs
S Y R I A
Khabur R.
Euphrates
River
Tigris River
Mosul
Khazir R.
Mangubak
Greater Zab R.
Eski Kelek
Lesser Zab R.
Altun Kupri
Hit
Adhaim R.
Injanah
Diyala R.
BAGHDAD
SAUDI ARABIA
I R A Q
I R A N
KUWAIT
PERSIAN GULF
RIVER BASINS OF THE
TIGRIS & EUPHRATES
0 50 100 150 200 250 300
Kilometers

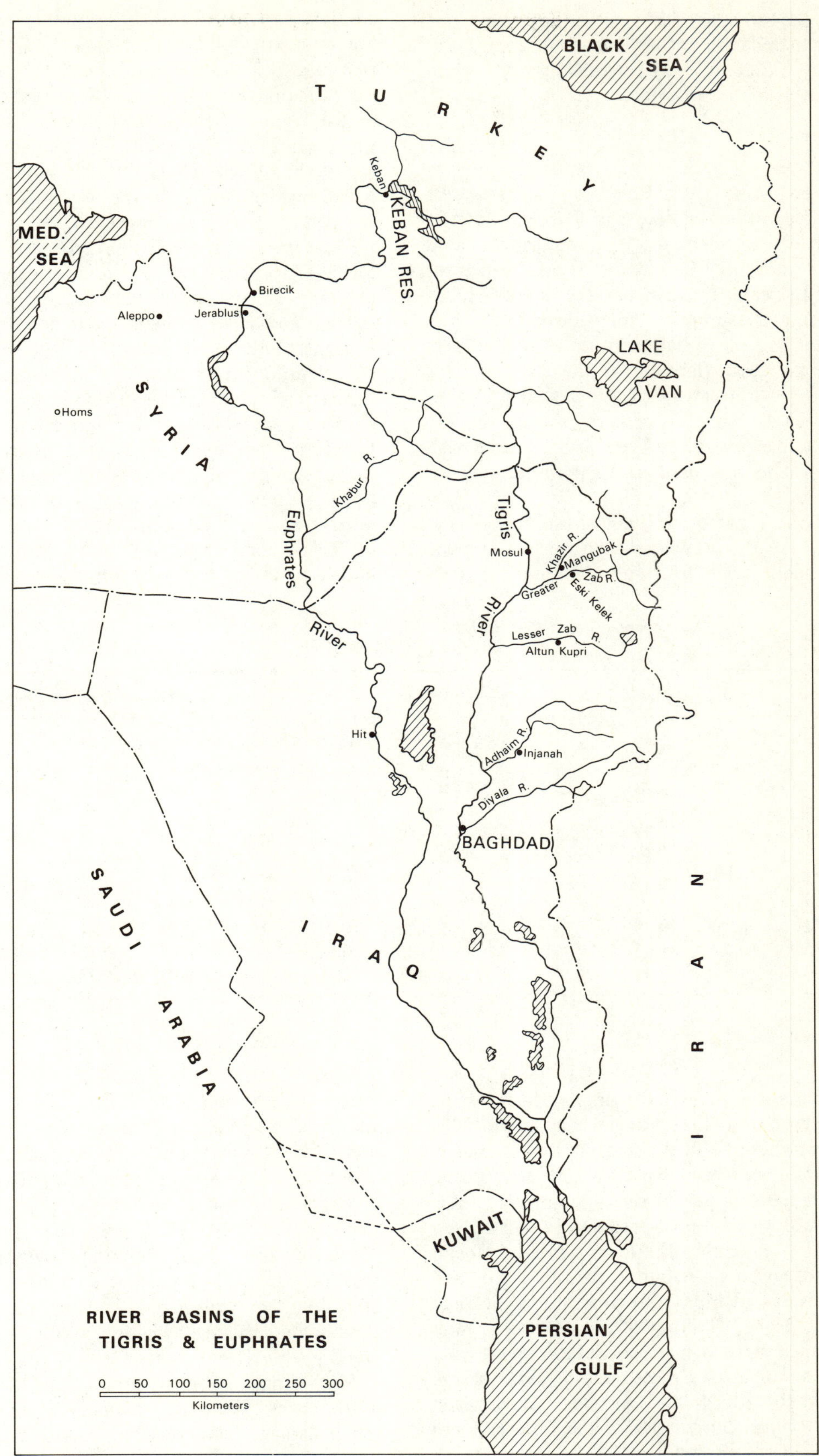

Fig. B-2

Because of this, the original discharge records were not examined and evaluated; but rather, the many reports, in the preparation of which this was presumably done, were used as a basis for this report.

The Tigris River

The upper reaches of the Tigris River are fed by waters originating in the high mountain areas of Turkey. Much of the runoff results from melting of snows accumulated during the winter months. The stream flow is somewhat typical of snowfed streams in other parts of the world, but the late winter and early spring precipitation is heavy. It is not uncommon for this precipitation to occur in the form of rain, which may cause flooding of the river in later winter. The Tigris flow is subject to rather violent floods, usually occurring in late winter or early spring. On the other hand, the upper river flows are usually quite well sustained in summer because of their origin in the snowfed mountain areas.

The portion of the Tigris Basin lying within Turkey is in rugged country where the growing season is short and the precipitation is usually ample to supply the limited agricul-

are measured at Altun Kupri and Injanah, respectively; and the summaries of measurements at these stations are given in Tables B-5 and B-6—the Lesser Zab from 1925—52 and the Adhaim from 1933—52. The last of the major tributaries of the Tigris is the Diyala River, which flows into the Tigris at Baghdad. Summary of the Diyala measurements (1924—52) is given in Table B-7. The Karkheh and Karun Rivers, which flow into the Tigris in the lower reaches of the delta, are located almost entirely in Iran. These streams provide water for navigation and saltwater repulsion in the lower Tigris. They are also used for irrigation of a considerable area of agricultural land in Iran, but are of relatively minor importance in the ultimate development of the Tigris River, and therefore are not considered important in the consideration at this time of development and control of the Tigris-Euphrates river system.

The combined flow of the Tigris River at Mosul and of the tributaries mentioned above—the Khazir, the Greater Zab, the Lesser Zab, the Adhaim and the Diyala—represent, for practical purposes, the total surface water available for development in the Tigris River basin. In order that the reader may be aware of the volume and seasonal distribution of these combined flows, they are tabulated below:

Average Volume of Runoff of Tigris River and Major Tributaries
(Milliards of Cubic Meters)

	Jan.	Feb.	Mar.	Apr.	May	June	July	Aug.	Sept.	Oct.	Nov.	Dec.	Total
Tigris at Mosul	1.27	1.87	2.59	3.84	3.53	1.69	0.86	0.47	0.37	0.44	0.65	0.92	18.44
Khazir at Mangubah	0.10	0.15	0.23	0.15	0.10	0.04	0.03	0.02	0.02	0.02	0.04	0.06	0.96
Gr. Zab at Eski-Kelek	0.69	0.98	1.57	2.57	2.71	1.56	0.79	0.43	0.32	0.32	0.43	0.47	12.84
Lesser Zab at Altun Kupri	0.64	0.99	1.37	1.35	0.96	0.50	0.27	0.15	0.11	0.12	0.19	0.33	6.98
Adhaim at Injanah	0.16	0.19	0.20	0.13	0.06	0.02	0.01	0.01	0.01	0.01	0.04	0.06	0.90
Diyalah River	0.51	0.72	1.05	1.05	0.66	0.25	0.14	0.10	0.09	0.11	0.18	0.30	5.16
Total	3.37	4.84	7.01	9.09	8.02	4.06	2.10	1.18	0.92	1.02	1.53	2.14	45.22

Source: Tables B-2 through B-7.

tural needs. Only a small part of the drainage area of the Tigris is within Syria, and there has been no appreciable water development there. The gauging station at Mosul, Iraq therefore should represent, for all practical purposes, the natural flow of the upper Tigris. A summary of the streamflow records at Mosul for the period 1919—52 is given in Table B-2 (end of Appendix B). Downstream of Mosul, there are many irrigation diversions from the Tigris, and there are also a number of important tributaries which add to the flow of the Tigris. The first of these is the Greater Zab River, combined with a major tributary, the Khazir River. Both of these streams are measured—the Greater Zab at Eski Kelek, and the Khazir at Mangubah—above diversions of any importance. Summaries of the Khazir record (1943—52) and the Greater Zab (1925—52) are presented in Tables B-3 and B-4. The Lesser Zab and the Adhaim Rivers are next in order moving downstream. These streams

The average annual total flow of this river system, 45.28 milliards of cubic meters, is equivalent to 36.7 million acre-feet of water—or about 5 million acre-feet more than the capacity of Lake Mead on the Colorado River.

The streamflow records of the Tigris River and its major tributaries through 1952 have been published in various forms by agencies of the Iraq Government. More recent data on stream flow at these stations are not readily available, but the published records are of sufficient length to permit their use in estimating total volumes of developable water in this river system in Iraq.

Many floods of major proportions have occurred on the Tigris River. Within the memory of many people of Iraq, much damage from flooding of the river has been experienced. The records of flood flow have been thoroughly studied by various engineering groups. Through the modern hydrologic techniques available, it has been possible to analyze the recorded flood flows and from these analyses to projected major floods which could and may occur in the future. This is done by projecting rates of precipitation coupled with other hydrologic events such as soil-moisture conditions, periods of sustained high temperatures during spring snowmelt periods, direction of storm movement— and projecting the flood runoff which could occur if all of the events were synchronized to produce a maximum result. Studies of this kind on the Tigris River indicate that, serious as past floods have been, a future flood considerably exceeding past records, could—and likely will—occur. A maximum possible flow of the Tigris at Baghdad of 34,000 cubic meters per second is predicted, yet the maximum recorded measurement of the river over a 57-year period only shows a maximum flow of 12,000 cubic meters per second. At the Eski Mosul dam site, the estimated probable maximum flood flow is 30,000 cubic meters per second, and the actual recorded maximum flow is 5,770 cubic meters per second. The reader may wonder at this great difference between the historical record and the predicted floods. There are many examples in the recorded river flows at many locations in the world where a "recorded" flood of forty or fifty years standing has later been exceeded as

such points have been made for a number of years; and some of the important Euphrates streamflow measurements have been summarized by Hathaway, Adams, and Clyde.[14] Records at Keban, Turkey, for the 1937—64 period tabulated from their report are presented in Table B-8. Measurements near the Syrian border at Birecik, Turkey, for the 1937—64 period are shown in Table B-9.

Measurements at Hit, Iraq, include the flow from the Khabur and Balikh rivers in Syria and other tributaries flowing into the Euphrates between Birecik and Hit. Flow at Hit is a good indicator of the flow crossing the Iraq-Syria border, since there are no important tributaries or diversions between the border and the station at Hit. A summary of the records at Hit for the period 1937—64 are presented in Table B-10.

None of the Euphrates records cited above represents the "natural" river flow. In order to estimate the average "natural" river flow, it is necessary to add the amounts of water diverted to the flow measured at some point below all major tributaries. Using figures from the report of Hathaway, Adams, and Clyde—in which estimates of irrigation diversions in Turkey, Syria, and Iraq are made—it is possible to calculate an estimated "natural" river flow as follows:

Calculations of Average "Natural" Flow of Euphrates River at Hit, Iraq[15]
(Milliards of Cubic Meters)

	Jan.	Feb.	Mar.	Apr.	May	June	July	Aug.	Sept.	Oct.	Nov.	Dec.	Total
Measured river at Hit (Iraq)	1.86	1.94	3.14	5.77	6.54	3.27	1.48	0.86	0.73	0.90	1.21	1.54	29.24
Diversions in Turkey	0	0	0	0.07	0.17	0.30	0.37	0.33	0.18	0.07	0	0	1.49
Diversions in Syria	0.05	0.07	0.24	0.24	0.37	0.46	0.52	0.44	0.29	0.14	0.09	0.05	2.96
Total "natural" river at Hit (Iraq)	1.91	2.01	3.38	6.08	7.08	4.03	2.37	1.63	1.20	1.11	1.30	1.59	33.69

much as threefold. On the Tigris River it may be expected that more serious floods may take place in the future than those that have occurred in the past. It is extremely important therefore that any development being considered for the Tigris must include flood control as one of the operational conditions.

The Euphrates River

The Euphrates River represents an important source of water for Turkey, Syria, and Iraq. Unlike the Tigris, whose present and foreseeable future development will be largely within Iraq, the orderly development of the Euphrates will require international agreements relating to the division of the water between these three countries. The total natural stream flow of the Euphrates is not as great as the Tigris. The quantity available at different points on the river and the natural flow within the boundaries of each country become important in considerations related to development of such international streams. Streamflow measurements at

The records of stream flow available to us do not give much information as to peak flood flows on the Euphrates River. In the streamflow records prior to 1952 which have been studied, the peak discharge of the Euphrates River at Hit, Iraq, was indicated to be 5,200 cubic meters per second on May 5, 1929. The extent of flood damage on the Euphrates in times of flooding is not known. Various reports contain general references to flood damages and the need for control of the river flooding. It is likely that the existing and proposed projects on the Euphrates will limit flooding considerably in the future. Keban Dam in Turkey and Tabqa Dam in Syria will provide flow regulation which

---

[14] Gail A. Hathaway, Harry W. Adams, and George D. Clyde, *Report on International Water Problems, Keban Dam-Euphrates River,* Report to International Bank for Reconstruction and Development (December 1965).

[15] Adapted from Hathaway, *et al., ibid.* Diversions were estimated as net diversions after taking into account "return flow" from irrigated lands whose areas and cropping patterns were made available to the authors.

will certainly have the effect of reducing floods on the river below these dams.

Groundwater in the Tigris-Euphrates Basin

Relatively little quantitative information is available on groundwater in this basin. Somewhat limited geologic information indicates that in some small areas of eastern Syria and northwestern Iraq, water-bearing basalt rocks occur at depths which can probably be economically developed. Recharge areas for these aquifers are likely the high mountain headwaters of the Tigris and Euphrates Rivers. In the limited areas where these formations occur, they should yield excellent groundwater supplies. Some of the water-bearing formations—especially in southwestern Syria—are overlain by highly gypsiferous formations, which may result in the yield of waters which are high in gypsum. The same might be predicted for waters from the chalky limestone and cavernous limestone formations of south-central Syria, where gypsiferous formations in some locations overlay the aquifers. Waters taken from these aquifers may not be of the desired quality. Along the river channels, alluvial aquifers exist which generally should yield well with good quality waters. It should be recognized that these aquifers usually are fed by waters directly related to the stream and would be expected to reflect quality of the stream. In eastern Syria for 100 kilometers or more on both sides of the Euphrates River (outside of the alluvial riverbed deposits), there exists a broad belt of clastic deposits which generally would be expected to yield only small quantities of water with considerably salinity. This same formation probably also extends southeastward into Iraq. Considerably more information is needed to delineate areas of possible groundwater development.

Jordan, Litani, and Orontes Rivers and Miscellaneous Sources

The Jordan River and its tributaries represent a very important water resource directly affecting Jordan, Israel, Syria, and Lebanon (Figure B-3). In the northern part of the river basin, the Bareighit, the Hasbani, the Dan, and the Banyas rivers combine to form the Upper Jordan above Lake Huleh. The Jordan flows from Lake Huleh to Lake Tiberias (the Sea of Galilee). Just below Lake Tiberias, the Yarmuk River, a major tributary, joins with the Jordan in its meandering path to the Dead Sea at an elevation of about 392 meters below sea level.

The valley of the Jordan River, including its northern tributaries, and the valley of the Litani River to the north and west are all in a down-faulted rift valley. This valley begins in the north between the Lebanon Mountains and the Anti-Lebanon Mountains, and extends southward to Lake Tiberias and the Dead Sea. Geologically the rift also extends southward from the Dead Sea, but the Dead Sea has no outlet and its waters are extremely salty. The watershed areas in the southern part of the Jordan Valley, between Lake Tiberias and the Dead Sea, contain a considerable accumulation of salt which is carried into the water in the natural runoff process. This salt accumulation makes the Jordan River waters below Lake Tiberias of questionable quality much of the time. Even Lake Tiberias, which is used as a storage reservoir of Jordan River waters, has been subject to considerable increase in salt content over the past decade. Measures are under consideration at the present time to remedy this situation by (1) construction of a bypass channel to prevent salty inflows from entering the lake or (2) intercepting salty mineral springs which now enter the lake and conducting the water to the river below the lake. Other possibilities also will undoubtedly be given consideration to protect the water quality in order that maximum development of the water resource may be achieved.

Full development of the Jordan River and its tributaries is complicated by the fact that several nations are riparian to the river. International agreements are necessary under these conditions to define the rights of the several countries.

In 1953, under the sponsorship of the United Nations, a report was prepared by Charles T. Main, Inc.,[16] which undertook to consider the most efficient development of the meager water supplies of the Jordan River—irrespective of national boundaries. In preparing their report this company used available records from a number of sources, and tabulated estimated available total water supply of the Jordan and tributaries at important points as follows:

	Estimated average yearly flow in millions of cubic meters (M.C.M.)
Hasbani River at proposed site of Hasbani Dam	130
Jordan River below Lake Huleh marshes (after correcting for marsh evaporation)	640
Flow into Lake Tiberias	838
Evaporation from Lake Tiberias	300
Net outflow from Tiberias	538
Yarmuk River at junction with Jordan	475
Beisan Springs	67
Wadi Zerka	45
Other inflow below Lake Tiberias	194
Flood flows developable	163

The reader should recognize that the first two figures given are only for the purpose of setting apart that portion of the supply which might be developed upstream from Lake Tiberias, and that these amounts are included in the figure of 538 M.C.M. shown as Lake Tiberias outflow (after taking into account the lake evaporation). Summing up the figures beyond that point it may be seen that the Main report estimated a total developable water supply equal to 1482 M.C.M. It further estimated the water already developed and used (in 1952) at 232 M.C.M., leaving a net supply of 1250 M.C.M. per year available for development. In the years that have elapsed since the Main report was

---

[16] Charles T. Main, Inc., *The Unified Development of the Water Resources of the Jordan Valley Region* (Boston, 1953).

MEDITERRANEAN SEA
LEBANON
SYRIA
ISRAEL
JORDAN
Beruit
Bekaa Valley
Damascus
Litani R.
Hasbani R.
Dan R.
Banyus R.
Tyre
L. Huleh
Jordan R.
Lake Tiberias
Yarmuk R.
Haifa
Beisan
Wadi Zerka
Tel Aviv
JORDAN RIVER
Amman
Jerusalem
DEAD SEA
NORTH
JORDAN and LITANI RIVERS
5 10   20   30
SCALE IN MILES

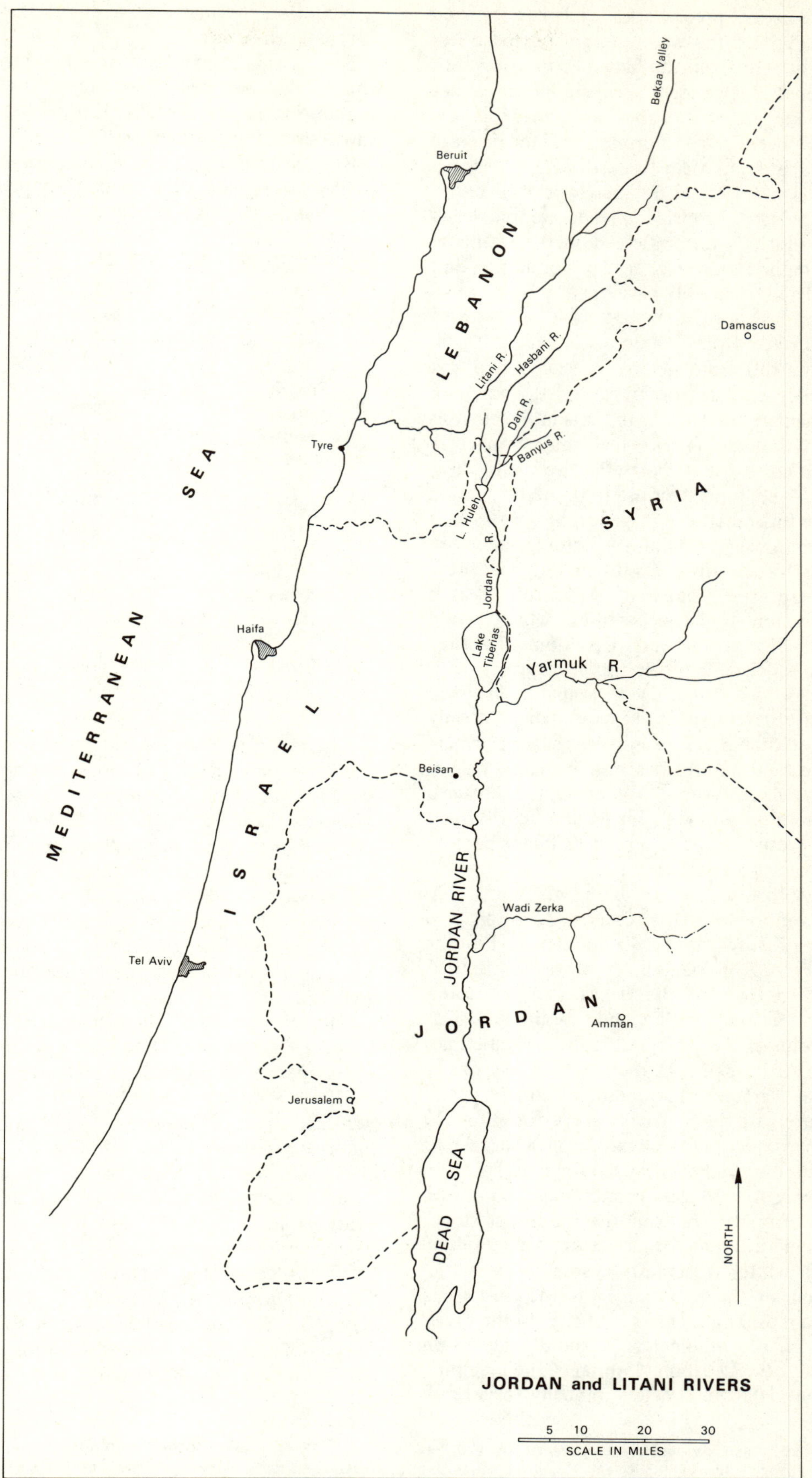

Fig. B-3

completed, a below-normal precipitation cycle has occurred in the Jordan Valley; and the water supply has been less than the averages quoted from the Main report. However, evaluation of the hydrological data available indicates that the precipitation and runoff of the past decade in the Jordan Valley has been below normal, and the average supplies indicated in the Main report still appear to represent reasonable estimates of the long-term availability. In the period since 1953, considerable development has been undertaken which is not reflected in the estimates given in the 1953 report; and the river may, for all practical purposes be considered to be fully developed.

Although there has been a considerable amount of groundwater use—mostly from shallow wells and springs in the Jordan Valley, and from nearby coastal and desert areas—there is a lack of information regarding the potential for such development in the future. Experts in the analysis of groundwater hydrologic factors feel that supplies in some areas are limited and that most effective use of these supplies would probably limit their use to domestic use and stock watering, or to limited use in providing irrigation at times of low streamflow and lack of other storage facilities.

Outside of the Jordan River Basin, in both southern Jordan and southern Israel, much of the desert areas is underlain by cavernous limestone aquifers which would have some potential for groundwater development if they were exposed to sufficient precipitation sources. It is apparent that there is limited water available in these aquifers since overdevelopment has probably already occurred in the coastal areas of southern Israel, where pumping from the sand aquifer has brought some fear of encouraging saltwater intrusion.[17] The geology of the area indicates some promise, although high pumping lifts and low water quality would likely be a limiting factor because of the limited recharge potential.

The Litani River has a drainage basin of about 2200 square kilometers, and is located entirely in Lebanon. The river has a length of only about 150 kilometers. For the first 60 kilometers of its length, it flows southward through the southern part of the Bekaa Valley, located between the Lebanon and Anti-Lebanon mountains. This valley is an extension of the geological fault line sometimes referred to as the African rift. At the southern end of the Bekaa Valley, near Karaoun, the mountains converge; and the river flows through a deep, narrow gorge for about 20 kilometers, then turns sharply westward, near Khardale, to flow into the Mediterranean Sea, near the town of Tyr.

In the coastal areas west of the Lebanon mountain range, the climate is subtropical; but on the mountain slopes above 600 meters elevation and in the Bekaa Valley (elevation, 850 meters), winter frosts impose some crop restrictions. Precipitation in the mountains east of the Bekaa Valley is only about 500 millimeters annually. In the valley itself, an average of 700 millimeters is found, whereas on the western slopes of the Lebanon Mountains, the precipitation averages about 1000 millimeters annually. Nearly all

precipitation occurs during the six winter months of November to April.

The Litani River flows result primarily from rainfall which is concentrated in the winter months; and the runoff therefore varies considerably throughout the year, with low flows during the summer months.

Average measured flow at Karaoun for the period from 1939 to 1952 and 1953 to 1967, expressed in millions of cubic meters, was as follows:

	Average flow in M.C.M.	
	1939-52	1953-67
January	72.6	42.6
February	89.3	68.8
March	88.8	68.4
April	63.5	47.9
May	40.7	26.6
June	21.6	9.3
July	14.7	5.0
August	11.7	3.7
September	11.9	5.7
October	15.1	9.1
November	18.1	12.8
December	30.0	25.6
Total	478.0	325.5

The record is analyzed in two 14-year periods because of the difference in flow during these two periods, part of which is the result of below normal precipitation and part of which is caused by the use of water for irrigation in the Bekaa Valley, which began in about 1953.

Depletion of Litani stream flow for irrigation in the Bekaa Valley is estimated at about 85 million cubic meters per year—all of this development having occurred since 1953. Streamflow records at Karaoun and any other stations downstream should be interpreted with this fact in mind.

The average annual flow measured at Khardale for the years 1939–52 and 1953–67 was 722 and 542 million cubic meters, respectively. Comparing this record with that at Karoun, we note that during the 1939–52 period there was an average increase of 244 million cubic meters between the two stations; but in the below-normal years 1953–67, the average annual increase was only 217 million cubic meters. The amounts of flow available at these two stations are important in any planning for development of water resources, since most of the power potential and part of the potential irrigated areas are downstream of Khardale.

The Orontes River heads in the Bekaa Valley of Lebanon, flowing northward from the area of the headwaters of the Litani River which runs southward from this valley. Burdon and Mazloum[18] report that the Orontes has

---

[17] See, for example, *Israel Economic Development—Past Progress and Plan for the Future* (Jerusalem: Isreal Economic Planning Authority, March 1968).

[18] David J. Burdon and Soubhi Mazloum, *Water Resources in the Mediterranean Zone for Pulp and Paper Processing* (UNESCO Middle East Science Cooperation Office, 1963).

an average annual flow of 500 million cubic meters. Most of the drainage basin for this river is located in Syria. Development of these waters has occurred in both Lebanon and Syria.

A large area of western Syria south and east of Damascus is reportedly underlain by an extensive basalt aquifer. These rocks have high surface porosity and may have considerable potential for development in areas where there is potential recharge of the aquifer. The large springs of Mzeirib and Zeizon issue from these basalt aquifers. Testing should be done to determine yield characteristics and potential for development. A rather extensive area underlain by cavernous limestone aquifers also exists in south-central Syria and under most of Lebanon. Wherever aquifers of this type have recharge areas of high rainfall which cause infiltration at exposed locations, they usually yield good water. Their potential should certainly be investigated. A large section running through central Syria all the way north to Turkey is underlain by a chalky limestone aquifer. This kind of aquifer often yields good water in large quantity but is more susceptible to contamination by dissolved impurities which may reduce the quality of the water—especially if recharge is limited. Investigation of groundwater potential, for domestic and irrigation use, is encouraged throughout most of Syria and Lebanon.

Table B-1
Discharge of Nile River at Aswan, 1871-1965
Milliards (Billions) of Cubic Meters, Rounded to Three Significant Figures

Year	Jan.	Feb.	Mar.	Apr.	May	June	July	Aug.	Sept.	Oct.	Nov.	Dec.	Annual total
1871	6.98	4.82	3.98	2.42	1.96	1.97	6.63	24.00	25.80	17.60	9.13	6.48	112.0
1872	4.48	2.59	1.95	1.57	1.50	2.28	8.36	23.60	27.30	21.30	12.90	8.31	116.0
1873	5.86	4.48	3.31	1.96	1.49	3.61	6.15	19.00	21.80	15.20	7.96	5.49	96.3
1874	3.59	2.05	1.78	1.34	1.25	2.50	8.20	28.90	30.60	21.60	11.10	7.78	121.0
1875	5.92	4.08	2.84	2.00	1.71	1.81	6.42	24.80	27.10	20.60	11.50	7.58	116.0
1876	5.67	4.49	3.86	2.45	1.94	2.30	8.32	23.60	28.70	18.10	9.82	6.76	116.0
1877	4.91	3.06	2.53	1.99	1.94	2.52	7.07	15.80	16.70	12.60	7.39	4.87	81.3
1878	3.68	2.29	1.82	1.41	1.25	1.32	5.72	22.10	32.00	27.40	14.40	10.10	123.0
1879	7.70	5.81	5.72	4.79	4.55	5.16	10.60	25.50	27.70	19.40	11.30	8.60	137.0
1880	7.03	5.68	4.87	3.45	2.86	2.96	9.50	22.50	23.30	16.80	8.46	6.35	114.0
1881	4.78	3.20	2.66	1.96	1.85	1.98	4.96	18.00	26.30	17.60	9.67	6.54	99.5
1882	4.90	3.25	2.37	1.70	1.48	1.39	4.08	18.20	22.60	15.50	11.00	6.92	93.5
1883	5.02	3.69	3.36	2.25	1.79	1.84	7.26	24.30	26.50	17.40	9.99	7.31	111.0
1884	5.58	4.40	3.87	2.67	2.28	2.28	4.75	17.80	21.20	17.00	10.70	6.94	99.4
1885	5.18	3.55	2.54	1.82	1.50	1.48	8.20	24.80	23.00	15.80	8.21	5.70	102.0
1886	4.02	2.47	2.18	1.80	1.79	2.00	5.07	20.20	25.30	15.80	8.51	6.91	96.0
1887	4.98	3.22	2.44	1.75	1.87	2.51	8.21	28.70	29.30	18.20	10.00	7.23	118.0
1888	5.53	3.38	2.54	1.96	1.77	1.84	4.23	17.80	18.90	11.70	5.88	4.35	79.9
1889	3.04	1.88	1.66	1.36	1.28	1.34	5.02	22.30	25.90	17.60	8.63	5.74	95.8
1890	4.39	2.56	1.96	1.43	1.27	1.60	5.87	25.80	27.40	21.40	12.30	7.86	114.0
1891	5.67	3.40	2.33	1.71	1.62	3.12	5.71	21.80	25.60	19.70	12.30	7.56	110.0
1892	5.20	3.37	2.31	1.59	1.34	1.38	5.64	23.80	31.60	23.70	12.40	8.36	121.0
1893	6.60	5.20	5.04	4.37	2.68	2.11	4.93	22.40	23.80	19.90	10.80	7.05	115.0
1894	5.39	3.51	2.46	1.85	1.85	2.15	7.65	26.10	28.70	24.20	12.40	8.72	125.0
1895	6.78	5.05	4.71	3.41	2.82	2.60	7.75	29.10	27.10	17.50	10.60	8.46	126.0
1896	6.42	4.83	3.77	2.67	2.38	2.32	7.10	21.90	27.90	18.80	13.00	11.00	122.0
1897	7.00	4.89	4.04	2.76	2.42	2.70	5.76	19.00	23.60	16.40	8.25	5.83	103.0
1898	4.65	3.29	2.43	1.80	1.59	1.52	4.00	24.30	27.00	20.10	11.30	8.14	110.0
1899	6.04	4.65	4.32	2.99	2.22	2.10	4.81	14.80	16.90	10.30	4.97	3.36	77.4
1900	2.14	1.47	1.25	1.02	1.06	1.41	3.86	22.20	21.50	15.80	7.22	5.04	84.0
1901	3.75	2.35	1.93	1.45	1.51	1.62	5.32	20.60	23.50	13.30	7.23	4.94	87.4
1902	3.30	1.98	1.67	1.40	1.33	1.46	3.42	11.10	18.40	14.20	6.81	4.40	69.4
1903	3.47	2.16	1.67	1.24	1.25	2.37	4.96	17.50	24.00	18.10	10.20	5.62	92.6
1904	4.19	3.23	2.38	1.67	1.79	2.92	5.11	18.90	19.00	13.00	6.10	4.51	82.7
1905	3.78	2.44	1.92	1.35	1.44	1.59	2.96	12.40	19.60	13.20	5.29	4.19	70.1
1906	3.60	2.63	2.09	1.86	1.80	1.97	3.83	17.80	24.80	17.50	8.68	5.17	91.7
1907	4.05	2.86	2.04	1.79	1.83	1.87	3.63	11.70	17.60	11.20	6.26	4.12	69.0
1908	3.52	2.17	1.77	1.55	1.48	1.52	3.47	21.60	27.90	20.20	9.32	5.32	99.9
1909	4.39	3.40	2.49	1.77	1.99	3.60	5.74	20.70	25.80	19.40	9.52	6.42	105.0
1910	4.26	3.79	3.12	1.83	1.71	2.20	3.55	16.30	24.50	18.90	10.70	5.91	96.8
1911	4.32	2.56	2.13	1.40	1.55	2.08	3.50	15.00	23.30	14.50	7.40	4.92	82.7
1912	3.38	2.34	1.78	1.48	1.48	1.53	3.44	18.60	18.20	10.30	4.43	4.03	70.9

Table B-1. Continued

Year	Jan.	Feb.	Mar.	Apr.	May	June	July	Aug.	Sept.	Oct.	Nov.	Dec.	Annual total
1913	3.16	1.93	1.56	1.64	1.87	1.96	2.48	6.55	12.20	7.52	2.65	2.00	45.5
1914	1.51	1.14	1.61	1.43	1.49	1.69	2.29	19.40	20.00	16.50	10.70	6.31	84.1
1915	3.36	2.57	2.05	1.71	1.79	2.26	3.44	10.70	14.90	14.70	7.75	4.07	69.2
1916	3.23	2.10	1.71	1.50	1.89	2.25	5.34	25.00	26.60	22.40	13.30	7.85	113.0
1917	4.13	3.03	3.34	2.22	2.20	3.20	5.47	17.50	27.80	22.80	11.60	7.34	111.0
1918	3.98	3.29	4.23	4.19	4.00	4.62	5.87	13.60	17.70	11.10	5.38	3.07	81.0
1919	3.34	2.09	2.01	2.10	2.21	2.31	4.10	16.60	20.80	12.60	5.83	3.00	77.1
1920	2.64	1.89	1.66	1.61	1.88	3.02	5.76	18.20	17.80	15.00	8.94	3.82	82.2
1921	3.00	2.14	1.91	1.79	1.89	2.11	3.44	15.40	19.60	14.80	7.09	3.22	76.5
1922	2.99	1.98	1.54	1.44	1.51	1.84	3.73	18.80	23.60	15.90	8.15	3.41	84.9
1923	2.71	2.22	1.58	1.52	2.08	3.30	4.41	20.10	21.40	16.20	7.14	3.86	86.5
1924	3.02	2.43	1.91	1.76	2.11	2.52	5.60	18.00	22.30	14.30	7.98	4.49	86.4
1925	2.49	2.40	1.89	1.72	2.07	2.60	4.14	13.30	16.50	12.40	4.63	3.49	67.7
1926	3.10	2.19	1.65	1.71	2.34	3.15	4.99	18.80	20.50	14.40	8.00	3.52	84.4
1927	2.80	2.75	1.96	1.80	2.25	2.64	4.51	15.30	18.30	13.30	4.42	2.91	72.9
1928	2.18	1.76	1.61	1.67	2.37	2.96	6.65	17.90	19.90	12.30	6.39	3.38	79.0
1929	2.77	2.23	1.97	1.87	2.45	3.72	9.42	22.90	24.40	17.80	9.59	3.51	103.0
1930	3.31	2.74	2.13	1.96	2.52	2.81	4.71	17.80	17.50	11.70	3.52	3.19	77.1
1931	2.37	1.76	1.94	1.71	1.78	2.02	3.08	15.50	21.20	14.80	7.84	3.10	77.1
1932	2.73	2.02	2.07	1.65	2.19	2.80	4.37	18.50	22.80	15.70	7.06	3.08	85.0
1933	3.51	3.29	2.70	1.99	2.58	3.03	3.69	12.70	22.10	15.10	7.34	4.31	82.4
1934	3.49	2.73	2.37	2.05	2.61	3.21	6.46	20.20	23.70	15.40	6.04	3.38	91.6
1935	3.51	2.73	2.41	2.19	3.04	4.19	7.87	21.00	23.90	17.40	5.00	3.58	96.9
1936	3.42	2.68	2.40	2.19	3.10	3.65	6.00	18.70	23.80	14.30	4.50	3.49	88.2
1937	2.95	2.25	2.28	1.99	2.43	3.17	4.67	19.50	22.10	12.40	3.19	3.62	80.6
1938	3.23	2.29	2.32	2.02	2.77	3.39	4.63	22.50	28.10	19.30	6.42	3.58	101.0
1939	3.45	3.05	2.51	2.21	3.26	4.11	5.11	13.20	18.20	10.50	5.30	4.15	75.0
1940	3.15	2.35	2.31	2.07	2.98	3.39	3.96	15.00	17.90	7.70	2.87	2.40	66.1
1941	2.19	2.13	2.08	1.95	2.64	3.61	4.69	11.90	14.20	8.56	5.41	4.09	63.4
1942	2.66	2.45	2.37	2.48	3.34	4.04	5.63	20.10	20.10	13.90	3.26	2.95	83.3
1943	2.50	2.49	2.21	2.38	3.10	3.38	3.96	14.60	24.40	12.00	4.60	3.70	79.3
1944	2.80	2.69	2.49	2.28	3.17	4.04	5.09	16.50	17.90	9.47	3.01	3.26	72.7
1945	2.36	2.56	2.12	2.22	2.79	3.96	4.55	14.80	18.80	14.80	4.82	4.70	78.5
1946	3.66	2.84	2.27	2.27	2.99	4.15	7.25	23.80	26.90	14.90	7.31	4.85	103.0
1947	3.89	3.58	2.92	2.66	3.91	4.66	5.01	14.80	22.30	12.10	4.07	4.18	84.1
1948	3.36	2.99	2.47	2.35	3.21	4.51	5.65	17.20	19.10	13.70	8.07	3.89	86.6
1949	2.50	3.06	2.45	2.48	3.19	4.49	5.47	17.80	20.10	12.20	4.76	4.23	82.8
1950	3.11	3.05	2.47	2.51	3.45	4.63	5.38	21.30	22.30	12.90	3.28	3.88	88.3
1951	2.75	2.92	2.41	2.43	2.92	3.66	3.83	14.90	17.50	10.10	4.35	4.30	72.3
1952	2.95	2.77	2.32	2.30	2.89	3.67	4.49	14.50	20.90	10.40	3.20	2.98	73.4
1953	2.21	2.45	2.22	2.20	2.70	3.91	5.23	21.50	20.60	12.20	2.85	3.28	81.3
1954	2.55	2.40	2.24	2.30	2.76	4.04	6.79	23.20	27.00	19.70	6.87	3.82	104.0
1955	3.39	2.59	2.47	2.48	3.29	4.80	5.70	18.80	22.50	16.50	4.03	3.70	90.2
1956	2.88	2.65	2.46	2.56	3.38	4.92	7.50	19.30	20.50	16.00	10.20	4.04	96.5
1957	3.51	2.98	2.50	2.90	3.82	5.30	6.21	17.20	21.10	7.44	2.18	2.23	77.4
1958	2.11	2.32	2.34	2.25	2.72	3.93	5.79	23.20	24.90	12.90	6.32	3.51	92.2
1959	2.87	2.40	2.34	2.42	3.22	4.46	4.88	18.10	27.90	15.30	6.62	3.78	94.3
1960	3.14	2.60	2.37	2.35	3.17	4.48	4.97	16.50	20.20	12.90	3.14	2.18	78.0
1961	2.24	2.17	2.32	2.40	3.08	4.22	6.50	21.90	26.10	17.90	5.91	4.59	98.4
1962	3.56	2.78	2.75	2.95	3.67	5.32	6.67	15.40	22.20	15.40	3.42	3.30	87.5
1963	2.90	2.71	2.77	2.59	3.86	5.42	7.04	19.80	21.30	9.42	3.80	4.42	86.0
1964	4.16	3.33	2.95	2.68	3.76	5.18	8.11	22.80	25.30	15.40	10.30	5.38	109.0
1965	5.07	5.31	4.63	4.26	5.57	6.36	6.38	9.35	12.60	10.80	7.76	4.68	82.7

Source: H. E. Hurst, R. P. Black, and Y. M. Simaika, *The Nile Basin* (Cairo: Ministry of Public Works, 1946, 1959, and 1966), Vols VII, IX, and X. No information is available to verify the accuracy of the measurements implied by the figures given.

Table B-2
Discharge of the Tigris River at Mosul, Iraq, 1919-52*

Year	Jan.	Feb.	Mar.	Apr.	May	June	July	Aug.	Sept.	Oct.	Nov.	Dec.	Annual average
1919	–	2070	1990	2060	1800	1150	659	420	340	340	392	541	–
1920	576	370	1190	1460	1120	690	300	197	170	168	341	187	565
1921	186	248	296	1170	830	551	225	144	118	118	133	288	359
1922	513	930	726	1280	948	561	252	153	130	129	284	534	537
1923	501	746	1450	1270	1590	852	322	178	141	139	143	167	625
1924	308	489	850	999	885	604	249	146	118	128	409	216	450
1925	179	154	673	721	576	306	159	126	98	162	146	439	312
1926	174	758	1430	1780	1420	730	407	252	195	128	148	329	696
1927	209	493	564	1290	1070	462	192	127	104	114	156	149	411
1928	192	598	709	1610	829	341	162	117	104	101	209	384	446
1929	306	752	1040	2110	1640	780	290	161	114	107	116	176	633
1930	165	300	315	396	376	216	114	91	92	103	165	336	222
1931	724	460	758	1480	1140	715	298	175	137	130	170	390	548
1932	218	476	823	875	1000	538	207	130	118	120	170	142	402
1933	172	282	614	843	1190	623	253	137	108	115	120	322	399
1934	232	418	578	1040	882	587	210	127	111	116	136	150	382
1935	477	1170	910	1310	972	411	174	115	103	111	242	679	556
1936	278	964	685	1370	1240	588	280	163	133	140	349	382	548
1937	251	641	712	1470	896	441	216	135	107	121	300	379	472
1938	716	955	808	2050	1730	752	340	177	143	143	274	264	696
1939	548	588	1420	2070	1720	647	283	190	166	170	228	402	703
1940	1140	1130	877	2140	1470	737	332	184	147	280	238	585	771
1941	882	1570	1960	1800	1530	617	293	178	147	163	178	213	794
1942	603	755	1640	1920	1750	823	328	187	152	233	1120	1040	879
1943	978	808	1080	1830	1940	787	382	228	185	193	221	262	741
1944	368	513	876	1530	1590	558	249	156	146	138	302	196	552
1945	758	593	720	1180	1110	615	277	165	145	140	180	340	518
1946	580	1060	1390	1910	2460	1050	468	265	223	372	247	232	835
1947	731	732	1070	959	700	405	198	159	98	92	351	392	491
1948	380	977	799	2260	2490	1220	476	237	156	166	177	263	801
1949	239	473	859	1970	1770	840	354	188	161	162	171	231	618
1950	374	474	1130	1510	2020	853	386	210	139	193	184	243	644
1951	606	486	727	1090	998	493	227	146	133	304	295	332	486
1952	–	1710	1270										
Mean	474	739	969	1480	1320	653	320	175	142	162	251	342	585
Milliards of cubic meters	1.27	1.81	2.59	3.84	3.53	1.69	0.86	0.47	0.37	0.44	0.65	0.92	

*Figures represent average flows in cubic meters per second (cumecs) unless noted otherwise.

Source: Statistical data released in various forms by the Government of Iraq.
Note: Maximum instantaneous discharge during period of record—6,100 cumecs on February 17, 1935.

Table B-3

Discharge of Khazir River at Mangubah, Iraq, 1943-52*

Year	Jan.	Feb.	Mar.	Apr.	May	June	July	Aug.	Sept.	Oct.	Nov.	Dec.	Annual average
1943	39	49	101	63	34	21	13	11	11	12	13	20	34
1944	45	45	50	49	18	8	5	6	6	6	22	14	23
1945	56	41	46	50	25	14	11	8	6	11	17	31	26
1946	33	91	160	75	53	23	19	14	14	15	17	28	44
1947	74	45	60	28	17	10	6	6	6	7	17	28	25
1948	17	23	34	68	49	10	6	5	4	5	7	11	20
1949	20	48	141	116	42	16	11	11	9	9	11	27	38
1950	61	96	193	68	90	18	12	9	9	12	13	13	49
1951	20	37	22	18	12	7	5	5	5	5	8	16	13
1952	15	123	57	–									
Mean	38	60	86	59	38	14	10	8	8	9	14	21	30
Milliards of cubic meters	0.10	0.15	0.23	0.15	0.10	0.04	0.03	0.02	0.02	0.02	0.04	0.06	–

* Figures represent average flows in cubic meters per second (cumecs) unless noted otherwise.

Source: Statistical data released in various forms by the Government of Iraq.
Note: Estimated maximum instantaneous discharge during period of record–3,000 cumecs on February 10, 1941.

Table B-4

Discharge of the Greater Zab River at Eski Kelek, Iraq, 1925-52*

Year	Jan.	Feb.	Mar.	Apr.	May	June	July	Aug.	Sept.	Oct.	Nov.	Dec.	Annual average
1925	85	74	246	429	345	184	99	62	42	39	58	42	142
1926	325	427	598	1070	1380	763	267	94	64	109	144	192	452
1927	142	480	354	546	1000	529	264	120	109	83	85	109	319
1928	93	188	462	947	773	364	173	128	112	105	169	129	304
1929	154	383	752	1260	1230	802	338	172	155	150	159	251	484
1930	250	190	292	377	384	240	128	83	72	71	82	136	192
1931	360	260	363	840	647	520	271	131	95	86	108	152	319
1932	135	454	656	813	908	608	253	130	96	81	129	93	363
1933	159	248	531	761	1020	577	295	137	93	84	79	277	355
1934	134	275	373	776	700	500	207	115	88	83	85	94	286
1935	177	414	445	771	749	418	176	103	84	83	145	148	310
1936	101	346	349	1030	993	505	251	122	93	85	352	214	370
1937	195	637	631	1310	857	692	215	115	90	95	161	128	427
1938	397	566	499	1600	1240	721	375	215	161	150	223	281	536
1939	312	444	754	1450	1530	855	395	198	158	144	169	288	558
1940	613	710	637	1590	1030	652	354	201	135	161	138	235	537
1941	411	990	944	1070	1090	549	292	157	121	120	115	137	499
1942	435	404	939	1260	1420	681	342	191	137	142	439	213	550
1943	194	241	419	810	1100	685	396	192	135	133	126	144	382
1944	261	265	664	1180	1080	594	324	179	155	122	291	166	443
1945	463	416	395	758	786	538	294	160	118	114	137	219	367
1946	244	514	1040	1450	1750	995	509	279	159	193	154	155	627
1947	397	461	763	764	564	321	171	146	152	156	269	250	367
1948	200	266	409	1050	1220	889	470	245	158	154	170	167	449
1949	201	309	733	1210	1550	937	477	302	199	163	153	169	535
1950	322	308	913	1090	1440	730	370	212	155	150	148	151	501
1951	200	299	405	557	564	382	220	185	185	174	188	251	301
1952	–	637	833	–									
Mean	258	400	586	991	1010	601	294	162	123	121	166	177	407
Milliards of cubic meters	0.69	0.95	1.57	2.57	2.71	1.56	0.79	0.43	0.32	0.32	0.43	0.47	–

* Figures represent average flows in cubic meters per second (cumecs) unless noted otherwise.

Source: Statistical data released in various forms by the Government of Iraq.
Notes: (1) Maximum instantaneous discharge during period of record–7,100 cumecs on February 10, 1941.
(2) Some discharge between 1925 and 1930 derived by correlation with other stream records.

Table B-5
Discharge of the Lesser Zab River at Altun Kupri, Iraq, 1925-52*

Year	Jan.	Feb.	Mar.	Apr.	May	June	July	Aug.	Sept.	Oct.	Nov.	Dec.	Annual average
1925	130	99	589	291	218	148	100	46	42	50	55	115	157
1926	280	504	792	656	318	270	165	105	80	90	120	200	298
1927	130	336	374	491	378	220	105	70	55	55	65	100	198
1928	120	380	342	582	335	170	100	65	60	60	110	170	208
1929	180	469	672	515	346	250	135	80	65	64	63	140	248
1930	110	271	198	231	204	95	48	35	29	32	34	85	114
1931	277	275	357	429	277	160	82	43	32	32	53	124	178
1932	135	324	452	293	220	143	64	38	31	32	98	72	159
1933	178	326	519	623	383	204	107	58	41	40	45	286	234
1934	194	277	252	552	432	231	97	49	34	33	44	82	190
1935	163	357	295	344	242	134	72	40	37	37	96	124	162
1936	70	462	332	522	336	160	97	54	44	42	138	197	205
1937	284	584	407	552	335	200	115	62	45	52	112	99	237
1938	360	568	461	658	424	194	104	60	50	52	74	201	267
1939	454	518	545	705	367	214	130	77	51	46	97	167	281
1940	535	932	648	690	414	271	178	66	49	61	66	165	340
1941	494	668	584	422	321	211	90	55	45	47	53	123	259
1942	263	338	759	464	335	188	104	62	52	61	120	91	236
1943	99	205	456	666	422	189	91	51	36	37	47	62	197
1944	230	220	310	405	248	121	53	33	27	29	108	62	154
1945	396	317	307	412	299	173	87	56	36	39	118	132	198
1946	267	554	939	826	684	302	176	111	88	88	93	103	352
1947	254	405	412	282	237	153	83	43	29	34	70	114	176
1948	131	146	267	361	367	216	108	46	27	29	39	76	155
1949	91	247	870	1040	636	241	49	12	8	7	9	73	273
1950	436	437	1010	770	706	277	107	45	28	28	31	36	325
1951	132	288	332	221	169	85	39	27	22	31	43	119	125
1952	–	808	841										
Mean	237	404	511	519	358	193	99	55	42	45	74	123	219
Milliards of cubic meters	0.64	0.99	1.37	1.35	0.96	0.50	0.27	0.15	0.11	0.12	0.19	0.33	–

* Figures represent average flows in cubic meters per second (cumecs) unless noted otherwise.

Source: Statistical data released in various forms by the Government of Iraq.

Notes: (1) Maximum instantaneous discharge during period of record–2,690 cumecs–February 11, 1941. (2) Some discharges prior to 1930 and during 1936-37 derived by correlation with other stream records.

Table B-6
Discharge of the Adhaim River at Injanah, Iraq, 1933-52*

Year	Jan.	Feb.	Mar.	Apr.	May	June	July	Aug.	Sept.	Oct.	Nov.	Dec.	Annual average
1933			206	167	15								
1934	75	146	81	203	215	82	12	0	0	0	1	8	68
1935	9	12	2	5	4	0	0	0	0	0	12	10	5
1936	0	193	32	36	18	0	0	0	0	0	65	65	34
1937	61	48	40	50	20	5	3	2	2	6	34	8	32
1938	145	130	88	25	11	1	0	0	0	0	4	93	41
1939	96	105	49	134	11	3	2	1	1	1	7	28	36
1940	107	249	74	33	13	6	3	2	2	3	27	17	45
1941	33	102	81	29	7	4	2	1	2	2	3	14	23
1942	36	35	48	6	1	0	0	0	0	0	8	2	11
1943	12	31	137	50	10	4	2	1	1	2	3	14	22
1944	43	25	25	12	4	2	1	1	1	2	18	5	12
1945	165	30	17	13	8	6	4	4	4	4	47	31	28
1946	148	111	219	90	30	4	1	0	0	0	1	12	51
1947	13	14	25	3	1	0	0	0	0	1	2	9	6
1948	3	4	3	32	20	9	0	0	0	0	0	7	7
1949	17	32	244	75	13	2	1	0	0	0	0	101	40
1950	125	106	77	12	21	2	0	0	1	1	2	2	29
1951	13	46	10	6	2	0	0	0	0	1	2	5	7
1952		17	25										
Mean	61	76	74	52	22	7	2	1	1	1	14	24	28
Milliards of cubic meters	0.16	0.19	0.20	0.13	0.06	0.02	0.01	0.01	0.01	0.01	0.04	0.06	

*Figures represent average flows in cubic meters per second (cumecs) unless noted otherwise.

Source: Statistical data released in various forms by the Government of Iraq.
Note: Maximum instantaneous discharge during period of record—1,260 cumecs on March 14, 1946.

Table B-7
Discharge of the Diyalah River at Discharge Site in Iraq, 1924-52*

Year	Jan.	Feb.	Mar.	Apr.	May	June	July	Aug.	Sept.	Oct.	Nov.	Dec.	Annual average
1924	153	252	414	318	198	98	52	40	40	52	68	165	154
1925	143	118	422	231	123	53	28	20	20	32	59	76	110
1926	243	602	734	543	322	131	64	48	45	48	116	150	254
1927	124	264	254	380	272	94	57	43	42	50	60	77	143
1928	101	318	263	444	189	68	32	30	26	33	83	77	139
1929	107	168	375	222	128	59	25	18	15	33	33	85	106
1930	122	261	184	215	181	69	27	21	20	25	31	70	102
1931	182	213	208	202	140	65	27	20	18	24	37	78	101
1932	154	210	274	203	156	78	23	18	17	22	37	51	103
1933	162	297	447	551	280	104	49	32	27	31	36	142	180
1934	202	258	199	401	355	149	54	35	30	31	38	70	152
1935	83	208	177	193	101	39	22	17	16	18	53	194	93
1936	61	365	338	575	343	139	58	42	39	41	79	250	194
1937	156	279	203	356	248	97	51	39	35	46	115	91	143
1938	476	522	385	561	331	126	73	52	48	50	71	210	242
1939	423	514	512	709	314	135	81	61	50	55	82	126	255
1940	310	757	567	489	250	113	121	52	45	66	106	101	248
1941	235	422	434	400	241	62	57	44	39	45	53	107	178
1942	164	229	502	312	147	75	43	32	31	61	111	99	150
1943	141	227	572	635	358	125	64	42	36	41	51	63	196
1944	154	169	267	221	114	52	30	24	23	27	131	71	107
1945	376	232	264	341	245	119	58	40	31	35	118	150	167
1946	359	367	921	754	439	138	95	69	62	62	70	77	298
1947	158	229	303	180	118	50	29	24	25	32	45	59	104
1948	85	94	166	250	196	62	28	20	19	21	29	93	89
1949	100	155	812	965	439	162	81	52	44	47	52	175	257
1950	304	292	644	512	499	210	113	78	63	66	73	96	246
1951	179	240	257	211	132	57	37	31	30	37	68	125	116
1952	49	270	311										
Mean	190	294	393	406	245	97	53	37	33	40	68	112	164
Milliards of cubic meters	0.51	0.72	1.05	1.05	0.66	0.25	0.14	0.10	0.09	0.11	0.18	0.30	

* Figures represent average flows in cubic meters per second (cumecs) unless noted otherwise.

Source: Statistical data released in various forms by the Government of Iraq.
Note: Maximum instantaneous discharge during period of record—3,600 cumecs on March 14, 1940.

Table B-8
Discharge of Euphrates River at Keban, Turkey, 1937-64*

Year	Jan.	Feb.	Mar.	Apr.	May	June	July	Aug.	Sept.	Oct.	Nov.	Dec.	Annual average
1937	280	399	859	2098	1376	719	405	285	258	284	445	490	658
1938	408	385	524	2682	2190	902	496	307	267	265	346	322	758
1939	323	366	673	1849	1701	617	373	268	244	242	288	315	605
1940	525	594	620	3466	2240	992	527	301	250	354	390	517	898
1941	439	883	1793	2616	1970	699	374	264	233	253	292	264	840
1942	285	339	657	2537	2365	825	346	248	223	277	701	601	784
1943	354	313	419	2297	2070	705	328	236	212	238	261	292	644
1944	260	421	1340	1805	2761	940	437	298	270	278	387	295	701
1945	324	304	420	1401	1479	717	278	207	183	181	204	263	497
1946	225	228	594	1806	2349	1040	443	306	235	480	321	281	692
1947	338	366	1116	1332	743	430	252	192	172	176	482	246	487
1948	242	371	340	2643	2793	1313	453	274	227	227	245	231	780
1949	200	227	383	1269	1796	659	268	212	190	193	196	193	482
1950	184	229	508	1641	1858	637	308	240	215	312	275	263	556
1951	286	283	772	1469	1110	554	273	207	213	315	334	306	510
1952	241	518	623	2854	1956	793	354	241	221	218	229	252	708
1953	256	382	452	2368	2166	1007	417	255	218	223	281	226	688
1954	234	266	759	2801	2440	1030	473	265	231	233	261	325	776
1955	286	314	493	883	1088	422	252	213	189	192	211	258	400
1956	241	289	414	2171	1853	898	378	257	233	237	242	239	621
1957	227	336	1077	1263	1885	853	352	232	206	207	230	247	593
1958	248	287	672	1431	1054	670	267	204	182	182	197	226	468
1959	210	203	417	1170	975	615	252	216	203	236	253	230	415
1960	270	389	648	2402	1741	677	355	252	229	227	244	228	638
1961	231	258	369	868	666	286	184	151	136	148	209	345	321
1962	229	357	942	1323	1058	504	266	172	153	161	186	345	475
1963	505	723	743	2724	2964	1942	739	386	298	337	367	325	1004
1964	228	268	1227	2228	1556	723	285	198	181				
Mean	289	368	709	1978	1757	792	362	246	215	249	302	303	
Milliards of cubic meters	0.78	0.90	1.91	5.12	4.71	2.05	0.97	0.66	0.56	0.67	0.78	0.81	

* Figures represent average flows in cubic meters per second (cumecs) unless noted otherwise.

Source: Adapted from Gail A. Hathaway, Harry W. Adams, and George D. Clyde, *Report on International Water Problems, Keban Dam-Euphrates River,* Report to International Bank for Reconstruction and Development, December 1965.

Table B-9

Discharge of Euphrates River at Birecik, Turkey, 1937-64*

Year	Jan.	Feb.	Mar.	Apr.	May	June	July	Aug.	Sept.	Oct.	Nov.	Dec.	Annual average
1937	496	667	1328	2644	1753	870	522	370	336	369	573	798	894
1938	680	647	873	3257	2686	1084	638	398	347	345	448	557	997
1939	558	620	1061	2382	2126	717	482	349	445	316	374	547	831
1940	848	947	985	4081	2742	1190	677	391	326	458	503	837	1165
1941	725	1362	2669	3188	2434	750	483	344	304	330	379	473	1120
1942	504	581	1038	3105	2886	994	448	323	292	360	898	957	1032
1943	603	544	696	2853	2548	853	421	308	278	311	340	514	856
1944	468	699	2019	2336	3339	1129	563	387	351	361	500	518	1056
1945	560	531	697	1911	1871	868	361	271	241	238	268	472	691
1946	417	422	947	2337	2868	1246	571	397	307	617	416	499	920
1947	580	620	1697	1839	1028	570	328	252	227	232	620	448	703
1948	442	627	583	3216	3376	1566	583	356	297	297	320	426	1007
1949	381	420	644	1773	2234	737	349	278	250	254	258	371	662
1950	359	423	824	2164	2305	727	399	313	282	405	358	472	753
1951	505	501	1203	1983	1449	676	355	271	279	408	432	534	716
1952	440	838	989	3438	2418	957	458	315	289	285	299	456	932
1953	462	643	743	2927	2658	1207	538	332	285	292	365	419	906
1954	430	476	1184	3382	2972	1234	609	345	302	304	340	561	1012
1955	505	545	802	1367	1424	560	328	279	249	252	277	465	588
1956	440	509	689	2720	2300	1080	488	335	304	309	316	437	827
1957	420	577	1641	1766	2336	1027	455	303	270	271	301	449	818
1958	450	506	1059	1943	1385	740	347	268	240	240	259	419	655
1959	369	315	754	1701	1351	718	302	249	239	261	301	327	574
1960	640	529	1055	3005	2098	740	400	297	272	252	297	326	826
1961	363	495	534	1409	1038	426	219	174	156	177	293	525	484
1962	434	805	1443	1710	1240	637	359	254	229	239	293	659	692
1963	1008	1118	1025	3291	4115	2321	951	508	410	510	507	512	1356
1964	369	540	1837	2528	1781	911	400	267	249				
Mean	516	625	1108	2509	2241	948	466	319	288	324	393	521	
Milliards of cubic meters	1.38	1.52	2.97	6.50	6.00	2.46	1.25	0.85	0.75	0.87	1.01	1.40	

*Figures represent average flows in cubic meters per second (cumecs) unless noted otherwise.

Source: Adapted from Gail A. Hathaway, Harry W. Adams, and George D. Clyde, *Report on International Water Problems, Keban Dam-Euphrates River*, Report to International Bank for Reconstruction and Development, December 1965 .

Table B-10
Discharge of Euphrates River at Hit, Iraq, 1937-64*

Year	Jan.	Feb.	Mar.	Apr.	May	June	July	Aug.	Sept.	Oct.	Nov.	Dec.	Annual average
1937	525	620	1070	2080	1800	1090	558	343	275	291	679	1090	862
1938	1130	971	966	2220	3200	1450	778	461	355	359	491	528	1076
1939	731	761	1120	2000	2530	1230	685	452	375	352	395	586	935
1940	1010	1080	1270	3060	2950	1330	700	418	343	407	708	908	1182
1941	922	1300	2700	2700	2420	1040	549	303	321	341	417	390	1117
1942	603	821	1220	2640	3030	1190	451	281	238	329	919	1210	1078
1943	1220	979	990	2350	2990	1190	573	376	309	330	483	485	1023
1944	666	754	1650	2250	3210	1400	622	394	359	379	634	513	1069
1945	904	726	847	1670	2120	1420	630	358	290	304	371	574	851
1946	580	656	1130	2160	3100	1660	765	463	376	591	612	473	1047
1947	870	900	1560	2080	1140	745	449	301	261	281	549	575	809
1948	554	1160	919	2560	3560	1950	749	408	349	356	368	495	1119
1949	407	542	585	1670	2200	1120	472	319	273	283	311	355	711
1950	448	354	1010	1970	2520	1130	494	311	264	315	401	373	799
1951	554	503	764	1870	1580	836	371	246	226	399	471	576	700
1952	451	1270	1140	2940	2350	1160	558	334	281	308	343	416	963
1953	537	1030	1310	3010	3110	1660	712	397	342	359	483	478	1119
1954	644	890	1630	3820	3380	1670	761	423	336	373	508	617	1254
1955	1090	706	899	1410	1720	777	340	228	228	284	318	521	710
1956	751	819	988	1750	2730	1230	558	314	269	328	370	402	876
1957	362	441	1580	1640	2690	1520	588	293	238	300	417	649	893
1958	679	644	1080	1820	1560	1140	414	219	196	307	373	495	744
1959	502	474	664	1672	1513	1029	364	209	194	290	390	360	638
1960	881	607	1298	2684	2766	1177	522	303	253	355	418	412	973
1961	494	571	466	1338	1209	475	197	94	99	191	377	905	535
1962	694	979	1338	1835	1454	883	298	153	317	248	297	491	749
1963	851	1300	1365	2585	4368	2819	931	422	311	451	630	505	1378
1964	398	468	1218	2621	1597	1075	373	168	172				
Mean	695	797	1171	2229	2457	1264	552	321	280	337	468	576	
Milliards of cubic meters	1.86	1.94	3.14	5.77	6.54	3.27	1.48	0.86	0.73	0.90	1.21	1.54	

* Figures represent average flows in cubic meters per second (cumecs) unless noted otherwise.

Source: Adapted from Gail A. Hathaway, Harry W. Adams, and George D. Clyde, *Report on International Water Problems, Keban Dam-Euphrates River,* Report to International Bank for Reconstruction and Development, December 1965.
Note: Other data available indicate a peak discharge of 5,200 cumecs occurring on May 5, 1929.

APPENDIX C

Statistical Tables

Notes

(1) This Appendix covers as many of the six countries as feasible in each instance. In a few instances (Tables C8-1 to 6, C9-1 to 3, and C11), easy availability has also led to the presentation of data for Saudi Arabia, but otherwise data for the latter are omitted as they are either unavailable or comparatively insignificant.

(2) Where the *FAO Production Yearbook* is cited without identification of year, it is usually the 1967 edition, supplemented by earlier editions for back years not covered in the 1967 edition.

(3) Attention is called to the conversion factors covered in the front matter of the book (pp. xix–xx).

(4) Original sources frequently do not make clear whether lack of entry in any given year signifies a "zero" or unavailability. Rather than guessing which applies, we have left blank spaces as found in the source.

List of Tables

Appendix Table C5-1

Syria: Agricultural Machines and Implements Sold, 1961—66

Agricultural Machines	1961	1962	1963	1964	1965	1966
Tractors	958	1,592	1,786	1,538	630	105
Harvesters	8	1	57	3	1	—
Combined Harvesters-Threshers	134	149	337	99	27	—
Threshers	4	41	16	44	42	13
Mouldboard Plows	115	414	123	108	66	52
Disc Harrows	450	731	966	682	272	51
Harrows	70	265	251	151	49	60
Seed Drills	44	65	57	49	16	7
Fertilizer Distributors	1	2	4	1	7	—
Tuber Harvesters	—	1	—	1	—	—
Irrigation Pumps	1,119	822	570	700	668	434
Motors (Stationary & mobile)	1,948	2,538	1,943	2,022	1,534	1,124
Sprayers, Dusters, etc.	160	35	11	658	421	706

Source: Syria, Ministry of Planning, *Statistical Abstract 1966* (Damascus: Government Press, 1967), p. 321.

Appendix Table C5-2

Iraq: Agricultural Machines and Implements Sold and Distributed During 1950—66

| Year | Tractors | Ploughs | | Culti-vators | Combines | | Disc harrows | Ditchers | Blade scrapers | Farm wagons | Agricultural pumping units | Binders | Roto-tillers | Grain drills | Disc drills | Others |
		Mould-board	Disc		Self-propelled	Horse-drawn										
1950	167	122	13	46	28	58	14	3	--	--	--	--	--	21	--	3
1951	54	62	14	43	43	54	21	--	--	--	--	--	--	3	--	--
1952	105	61	13	45	49	42	31	3	--	--	--	--	--	9	--	1
1953	343	143	44	109	151	243	48	6	11	6	--	2	--	8	--	8
1954	258	100	39	126	138	1	29	6	10	5	--	--	--	2	--	8
1955	253	130	49	94	41	26	25	15	12	--	--	--	--	--	--	12
1956	381	189	93	107	195	14	76	15	21	--	--	--	--	--	--	39
1957	377	136	86	138	209	123	53	16	16	--	--	--	--	--	--	63
1958	247	135	60	62	60	19	31	13	16	--	--	--	--	13	--	29
1959	129	61	38	62	10	171	42	3	9	--	--	--	--	4	--	24
1960	422	240	59	162	15	--	49	14	13	--	64	--	--	5	--	92
1961	743	←—530—→		324	←—343—→		76	23	17	--	189	--	--	--	--	--
1962	1,096	468	84	446	253	--	58	28	34	43	187	--	--	3	--	34
1963	842	491	105	337	129	81	144	12	18	25	218	--	--	10	--	30
1964	869	515	122	287	195	68	128	15	12	46	125	3	8	--	--	13
1965	981	588	82	305	250	130	92	32	17	30	192	2	3	18	--	4
1966	1,027	677	48	305	68	70	132	29	16	80	176	6	18	24	34	33

Source: Republic of Iraq, Ministry of Planning, Central Bureau of Statistics, *Statistical Abstract 1952—1966.*

Appendix Table C5-3

Iraq: Number of Pumps and Total Horsepower, 1946–65

Year	No. of Pumps	Horsepower (1000 HP)	Average HP per pump[a]
1946	2,836	99	35
1947	3,168	112	35
1948	3,311	118	36
1949	3,425	124	36
1950	3,593	131	36
1951	3,775	142	38
1952	4,068	157	39
1953	4,339	167	39
1954	4,555	175	38
1955	4,784	183	38
1956	5,025	192	38
1957	5,264	200	38
1958	5,444	206	38
1959	5,650	213	38
1960	5,796	217	37
1961	6,129	225	37
1962	6,654	233	35
1963	7,867	241	30
1964	8,512	264	31
1965	9,009	280	31

[a] Calculated before rounding figures in preceding column.

Source: Republic of Iraq, Ministry of Planning, Central Bureau of Statistics, *Statistical Abstracts 1956–1966*.

Appendix Table C5-4

Egypt: Chemical Fertilizers Consumed in Agriculture, 1952/53–1964/65

	Nitrogenous		Phosphates	
Years	Quantity (1000 tons)	Index (1952/53 = 100)	Quantity (1000 tons)	Index (1952/53 = 100)
1952/53	648	100	92	100
1953/54	679	105	105	114
1954/55	645	100	113	123
1955/56	632	98	156	170
1956/57	698	108	181	197
1957/58	694	107	179	195
1958/59	734	113	171	186
1959/60	763	118	177	192
1960/61	913	141	230	250
1961/62	939	145	245	266
1962/63	984	152	254	276
1963/64	1,157	179	293	318
1964/65	1,231	190	322	350

Source: UAR, Central Organization for General Mobilization and Statistics, *Statistical Indicators 1952–1966* (Cairo, 1967–in Arabic), p. 111.

Appendix Table C5-5

Egypt: Pesticides Consumed in Agriculture, 1952/53–1964/65

Years	Quantity of pesticides (tons)	Index (1952/53 = 100)
1952/53	2,143	100
1953/54	1,627	76
1954/55	8,871	414
1955/56	9,188	429
1956/57	10,489	489
1957/58	8,075	377
1958/59	15,078	704
1959/60	11,062	516
1960/61	23,398	1,092
1961/62	7,447	348
1962/63	12,552	586
1963/64	20,916	978
1964/65	20,450	954

Source: UAR, Central Organization for General Mobilization and Statistics, *Statistical Indicators 1952–66* (Cairo, 1967), p. 111.

Appendix Table C5-6

Iraq: Quantities of Seeds Distributed to Farmers by the Ministry of Agriculture, 1953–66

Year	Cotton (tons)	Wheat (tons)	Barley (tons)	Linseed (tons)	Tomatoes (kg.)
1953	333	182	333	22	67
1954	––	337	39	18	17
1955	––	283	48	33	33
1956	––	526	97	20	6
1957	––	422	11	––	28
1958	––	693	22	––	150
1959	54	598	65	3	24
1960	118	384	69	3	84
1961	369	300	50	4	11
1962	267	204	65	33	40
1963	244	905	93	––	26
1964	187	434	114	17	53
1965	352	449	104	2	90
1966	73	840	124	9	32

Source: Republic of Iraq, Ministry of Planning, *Statistical Abstract, 1966*, p. 153.

Appendix Table C7-1

Egypt: Distribution of Ownership of Landholdings by Size Groups, 1952, 1961, 1965

Size of holdings in hectares	1952[a]				1952[b]			
	No. of holdings (1000)	Area (1000 hectares)	Percent of total holdings	Percent of total area	No. of holdings (1000)	Area (1000 hectares)	Percent of total holdings	Percent of total area
Less than 2	2,642	891	94.3	35.4	2,841	1,168	94.4	46.6
2–4	79	221	2.8	8.8	79	221	2.6	8.8
4–8	47	268	1.7	10.7	47	268	1.6	10.7
8–21	22	275	0.8	10.9	30	343	1.0	13.6
21–42	6	181	0.2	7.2	6	181	0.2	7.2
42–84	3	184	0.1	7.3	3	184	0.1	7.2
42 only	–	–	–	–	–	–	–	–
84 only	–	–	–	–	2	149	0.1	5.9
84 & above	2	494	0.1	19.7	–	–	–	–
Total	2,801	2,514	100	100	3,008	2,514	100	100

Size of holdings in hectares	1961[c]				1965[d]			
	No. of holdings (1000)	Area (1000 hectares)	Percent of total holdings	Percent of total area	No. of holdings (1000)	Area (1000 hectares)	Percent of total holdings	Percent of total area
Less than 2	2,919	1,332	94.1	52.1	3,033	1,551	94.5	57.1
2–4	80	221	2.6	8.6	78	258	2.4	9.5
4–8	65	268	2.1	10.7	61	221	1.9	8.2
8–21	26	343	0.8	13.4	29	342	0.9	12.6
21–42	6	181	0.2	7.0	6	165	0.2	6.1
42–84	–	–	–	–	–	–	–	–
42 only	5	210	0.2	8.2	4	177	0.1	6.5
84 only	–	–	–	–	–	–	–	–
84 & above	–	–	–	–	–	–	–	–
Total	3,101	2,555	100	100	3,211	2,714	100	100

[a] Prior to the promulgation of the Agrarian Reform Law in 1952.
[b] After the Agrarian Reform Law, which set a maximum ownership of 84 hectares (200 feddans).
[c] After the new Agrarian Reform Law, which set a maximum ownership of 42 hectares (100 feddans).
[d] Excludes State owned lands and Agrarian Reform lands not yet distributed.

Source: UAR, Central Organization for General Mobilization and Statistics, *Annual Statistical Abstract 1952–1966* (Cairo, June 1967), pp. 50-54.

Appendix Table C7-2

Lebanon: Number and Area of Agricultural Holdings, by Size, 1961

	Holdings		Area	
Size Class	Number	%	Dunums	%
From 1 up to 5 dunums	44,510	35	113,361	4
From 5 up to 10 dunums	23,615	19	158,056	5
From 10 up to 20 dunums	23,708	19	324,722	11
From 20 up to 30 dunums	10,598	8	256,525	8
From 30 up to 40 dunums	6,706	5	223,301	7
From 40 up to 50 dunums	5,949	5	261,732	9
From 50 up to 100 dunums	7,277	6	491,949	16
From 100 up to 200 dunums	3,017	2	386,527	13
From 200 up to 500 dunums	1,304	1	364,345	12
From 500 up to 1,000 dunums	299	–	211,024	7
From 1,000 up to 2,500 dunums	119	–	144,612	5
From 2,500 dunums and more	21	–	103,989	3
Total	127,123	100	3,040,143	100

Source: Lebanon, Ministry of Agriculture, *Census of Agriculture, 1961* (Beirut, March 1965).

Appendix Table C7-3

Jordan: Number of Agricultural Holdings, by Size, 1965*

Size group in dunums	Number of holdings	Percent[a]
Less than 10	34,039	36.4
10–19	10,193	10.9
20–29	9,363	10.0
30–39	7,621	8.1
40–49	5,393	5.8
50–99	14,221	15.2
100–199	8,003	8.6
200–499	3,747	4.0
500–999	688	0.7
1,000–1,999	198	0.2
2,000–4,999	60	0.1
5,000–9,999	16	––
10,000 & over	2	––
Total	93,544	100.0

*The mean size of holdings is approximately 62 dunums or 6.2 hectares. The size of the median holding is 22 dunums or 2.2 hectares. On the assumption that the distribution of holdings is a unimodal one it would be highly skewed to the right.

[a]Percentages do not add up to total due to rounding.

Source: Jordan, Department of Statistics, *Report on Agricultural Census, 1965* (June 1967) and *Report on the Results of the Agricultural Sample Survey, 1966.*

Appendix Table C7-4

Jordan: Distribution of Land Ownership, by Size, in East Ghor Canal Project Area, July 1960

	Landowners		Area owned		Average area of ownership (ha.)
Size of ownership (hectares)	Number	Per-cent	Area (ha.)	Per-cent	
0.1–0.9	1,309	35.7	550	3.5	0.4
1.0–1.9	708	19.3	993	6.3	1.4
2.0–2.9	378	10.3	907	5.7	2.4
3.0–7.5	866	23.6	3,609	24.7	4.6
7.6–10.0	113	3.1	987	6.2	8.7
10.1–50.0	252	6.9	4,781	30.2	19.0
50.1–100.0	32	0.9	2,178	13.8	68.1
Over 100	10	0.3	1,524	9.6	152.4
Total/Average	3,668	100.0	15,830	100.0	4.3

Source: Mediterranean Development Project, *Jordan Country Report* (Rome: FAO 1967), p. 97.

Appendix Table C7-5

Jordan: Distribution of Land Ownership* in East Ghor Canal Project Area After the Reallocation of Land

	Landowners		Area Owned	
Size of Ownership (hectares)	Number	Percent	Area (ha.)	Percent
3.0–5.0	1,046	74.9	3,486	54.8
5.1–6.2	128	9.1	748	11.8
6.3–13.0	195	14.0	1,623	25.5
13.1–20.0	19	1.4	307	4.8
Over 20	18	0.6	197	3.1
Total	1,396	100.0	6,360	100.0

*This land ownership covers only 52% of the total project area of 12,000 hectares. The remaining area is almost completely allocated to new farmers but their ownership classifications were not available.

Source: Mediterranean Development Project, *Jordan Country Report* (Rome: FAO 1967), p. 98.

Appendix Table C7-6

Syria: Cumulative Percent Distribution of Farmers According to Size and Extent of Irrigation, 1962

Farm size (dunums)[a]	Completely unirrigated	Irrigated			Total
		Less than 50 percent	More than 50 but less than 100 percent	Completely	
0–5	6.2	4.5	–	8.9	6.4
5–10	12.4	12.7	5.8	19.2	13.7
10–20	20.4	27.6	21.5	39.0	25.2
20–30	27.5	33.4	35.7	56.8	34.7
30–40	34.1	40.8	44.9	70.1	42.9
40–50	39.6	47.3	55.2	77.8	49.0
50–100	57.2	61.5	70.8	92.3	65.6
100–200	76.5	75.4	85.5	98.2	81.2
200–500	93.8	95.4	92.0	99.5	95.1
500–1,000	97.6	99.1	96.1	99.6	98.1
1,000–2,000	99.1	99.2	96.9	–	99.1
2,000–5,000	99.9	99.5	100.0	–	99.8
Above 5,000	100.0	100.0	–	100.0	100.0
Number of farms	267,126	49,971	12,719	88,419	418,235

[a]Information as to exact size of Syrian dunum not available. Sources vary between 0.0919 and 0.1 hectares.

Source: Narindar S. Randhawa, *Preliminary Report on Appropriate Size of Farm in the Ghab,* U.N. Special Fund Ghab Development Project (Land Use and Production Economics Series No. 4 [Damascus: Syrian Arab Republic, July 1965]).

Appendix Table C7-7

Distribution of Holdings and Cultivated Areas by Size Groups in Iraq During 1958/59

Size group (in hectares)	Distribution of Holdings[a]		Distribution of Cultivated Units and Holdings[b]	
	Number	Area (1000 hectares)	Number	Area (1000 hectares)
Less than one	57,958	18.3	73,110	25.6
1 and under 5	45,539	107.5	70,906	163.5
5 and under 10	18,891	131.9	30,945	210.7
10 and under 15	10,802	130.5	17,460	207.4
15 and under 20	6,748	116.7	13.422	223.0
20 and under 25	4,864	107.0	8,675	187.9
25 and under 30	3,805	101.9	8,087	211.8
30 and under 50	7,659	292.0	12,913	485.4
50 and under 100	5,459	369.8	9,012	580.2
100 and under 150	1,693	203.2	2,453	289.5
150 and under 250	1,510	289.2	1,959	370.9
250 and under 500	1,395	503.2	1,832	640.0
500 and under 1,000	1,066	749.8	1,293	896.0
1,000 and under 2,500	682	1,019.6	835	1,241.6
2,500 and under 5,000	181	614.4	224	757.7
5,000 and under 25,000	89	857.7	120	1,181.1
25,000 and over	5	219.2	8	356.2
Total	168,346	5,831.8	253,254	8,038.7

[a] Includes all holdings except Miri Sirf, Waqf, and unsettled land. A holding is defined as a piece of Tapu, Lazma, or Mulk land owned and managed by one person. Miri Sirf land and Waqf land are excluded from this part of the table because the first is owned by the State and the latter belongs to the Department of Awqaf. Also, unsettled lands are excluded because their type of tenure is not decided and almost all of these lands belong to the State.

[b] This includes all types of land–i.e., it includes holdings as defined in (a) above and the cultivated units which are defined as a piece of Miri Sirf, Waqf or unsettled land cultivated and managed by one person. The distribution of cultivated units alone could be found approximately by subtracting the first two columns from their corresponding last two columns in the table.

Source: Republic of Iraq, Ministry of Planning, Central Bureau of Statistics, *Results of the Agricultural and Livestock Census in Iraq for the Year 1958/59* (Baghdad: Government Press, 1961), pp. 8, 9. All area measures were converted from mesharas to hectares (one meshara equals one fourth of a hectare).

Appendix Table C8-1
Wheat Area,* by Countries, Years of Record 1844–1966
(1,000 Acres*)

Year	Egypt	Israel	Lebanon	Jordan	Syria	Iraq	Saudi Arabia	Total reported
1966	1,495	188	168	529	2,720	4,292	247	9,039
1965	1,426	178	188	689	3,000	4,213	247	9,941
1964	1,344	138	173	734	3,647	4,020	210	10,266
1963	1,396	128	136	509	3,852	4,213	247	10,481
1962	1,510	119	170	704	3,501	3,931	222	10,157
1961	1,436	156	170	675	3,249	3,326	210	9,222
1960	1,512	146	161	247	3,830	3,141	N.A.	9,037
1959	1,532	153	178	420	3,514	3,682	N.A.	9,479
1958	1,480	146	168	741	3,610	3,788	111	10,044
1957	1,572	141	173	692	3,694	3,598	106	9,976
1956	1,631	141	173	803	3,783	3,247	106	9,884
1955	1,581	116	173	667	3,615	3,521	N.A.	9,673
1954	1,863	77	173	675	3,328	3,435	N.A.	9,551
1953	1,858	74	173	608	3,247	2,958	N.A.	8,918
1952	1,455	82	168	665	2,884	2,392	N.A.	7,646
1951	1,554	101	168	571	2,562	2,293	59	7,308
1950	1,432	91	178	423	2,451	2,348	30	6,944
1949	1,470	69	173	294	2,441	2,157		6,604
1948	1,574		173	294	1,947	2,380		
1947	1,690		173	173	2,086	2,357		
1946	1,646		161	346	2,002	2,760		
1945	1,710		173		1,853	1,920		
1944	1,715		131		1,690	1,799		
1943	1,989		168		1,517	1,680		
1942	1,636		161		1,441	2,100		
1941	1,559		198		1,063	1,920		
1940	1,562		173		1,201	1,433		
1935-39 av.	1,465							
1930-34 av.	1,562							
1925-29 av.	1,556							
1920-24 av.	1,425							
1915-19 av.	1,355							
1889	1,290							
1879	925							
1878	1,193							
1877	926							
1871	1,245							
1844	950							

[a] Figures are mostly harvested, but occasionally planted, acreage.

Sources: 1940–66: *FAO Production Yearbooks.* 1915–39: Donald C. Mead, *Growth and Structure Change in the Egyptian Economy* (Homewood, Ill.: Richard D. Irwin, 1967), see p. 321. 1844–89: Patrick O'Brien, "The Long-term Growth of Agricultural Production in Egypt: 1821–1962," in P. M. Holt, ed., *Political and Social Change in Modern Egypt* (London: Oxford University Press, 1968).

Appendix Table C8-2

Wheat Yields Per Hectare, by Countries, Years of Record 1879–1966
(100 Kilograms)

Year	Egypt[a]	Israel	Leba-non	Jordan	Syria	Iraq	Saudi Arabia
1966	26.8	13.2	10.3	4.7	6.5	4.8	14.9
1965	27.7	20.9	7.2	10.0	8.6	5.9	14.8
1964	27.6	22.7	8.5	9.9	7.5	5.0	14.7
1963	26.4	10.4	10.9	3.7	7.6	2.9	13.5
1962	26.1	10.9	11.0	3.9	9.7	6.8	14.4
1961	24.7	10.5	10.0	5.1	5.8	6.4	14.1
1960	24.5	6.9	6.2	4.4	3.6	4.7	N.A.
1959	23.3	11.9	9.3	6.1	4.4	3.8	N.A.
1958	23.6	10.5	7.1	2.2	3.8	4.9	9.8
1957	23.1	14.6	10.0	7.9	9.1	7.7	8.6
1956	23.4	13.0	8.9	7.4	6.8	5.9	8.7
1955	22.7	7.7	8.6	2.9	3.0	3.2	N.A.
1954	22.9	11.0	8.6	8.5	7.2	8.3	N.A.
1953	20.6	10.0	7.1	4.1	6.6	6.4	N.A.
1952	18.5	9.7	7.4	8.2	7.7	5.0	8.3
1951	19.2	3.4	7.4	3.0	4.9	5.3	9.2
1950	17.7	7.3	7.5	6.2	8.4	5.5	
1949	19.6	7.5	7.1	11.7	9.2	5.2	
1948	17.0		7.1	8.4	8.3	3.1	
1947	15.3		7.1	5.3	4.8	2.5	
1946	17.5		10.8	6.7	7.1	4.4	
1945	17.1		7.6		5.2	5.1	
1944	13.6		3.8		7.5	2.5	
1943	16.0		9.3		9.1	5.6	
1942	19.1		6.5		8.0	4.9	
1941	17.8		6.0		8.3	6.2	
1940	21.5		5.7		10.3	8.2	
1935–39 av.	21.2						
1930–34 av.	18.6						
1925–29 av.	17.4						
1920–24 av.	17.1						
1915–19 av.	16.7						
1911	18.2						
1910	18.0						
1909	21.6						
1908	18.7						
1907	20.8						
1906	19.3						
1905	21.0						
1904	22.9						
1903	20.6						
1902	17.9						
1901	20.5						
1900	18.4						
1899	19.9						
1898	22.9						
1897	15.9						
1896	14.9						
1895	17.8						
1894	17.5						
1893	18.5						
1892	13.0						
1891	17.1						
1890	15.2						
1889	14.3						
1888	13.9						

Appendix Table C8-2—*Continued*

Year	Egypt[a]	Israel	Leba-non	Jordan	Syria	Iraq	Saudi Arabia
1887	14.1						
1886	10.7						
1885	10.3						
1884	11.4						
1883	12.8						
1882	12.8						
1881	12.8						
1880	10.9						
1879	10.6						

[a]Data for 1915 to 1965 relate to all land growing wheat in Egypt; data for 1879 to 1911 relate to wheat grown in State domains.

Sources: 1940–66: *FAO Production Yearbooks.* 1915–39: Donald C. Mead, *Growth and Structural Change in the Egyptian Economy* (Homewood, Ill.: Richard D. Irwin, 1967), see p. 321. 1879–1911: Patrick O'Brien, "The Long-term Growth of Agricultural Production in Egypt: 1821–1962," in P. M. Holt, ed., *Political and Social Change in Modern Egypt* (London: Oxford University Press, 1968).

Appendix Table C8-3
Wheat Production, by Countries, Years of Record 1821–1967*
(1,000 Metric Tons)

Year	Egypt	Israel	Lebanon	Jordan	Syria	Iraq	Saudi Arabia	Total reported
1967	1,500	220	70	248	600	700	–	3,338
1966	1,465	101	60	132	400	630	149	2,897
1965	1,600	150	55	278	1,044	1,005	148	4,280
1964	1,500	127	60	295	1,100	807	125	4,014
1963	1,493	55	60	76	1,190	488	135	3,497
1962	1,593	52	75	112	1,374	1,085	130	4,421
1961	1,436	66	69	138	757	857	120	3,443
1960	1,499	41	40	44	555	592	N.A.	2,771
1959	1,443	74	67	103	632	564	N.A.	2,883
1958	1,412	62	48	66	562	757	N.A.	2,907
1957	1,467	83	70	220	1,354	1,118	44	4,356
1956	1,547	74	62	242	1,051	776	37	3,789
1955	1,451	36	60	79	438	453	35	2,552
1954	1,729	34	60	233	965	1,160	32	4,213
1953	1,547	30	50	100	870	762	N.A.	3,359
1952	1,089	32	50	220	900	480	N.A.	2,771
1951	1,209	14	50	69	510	488	20	2,360
1950	1,018	27	54	106	830	520	11	2,566
1949	1,167	21	50	139	909	450		2,736
1948	1,080	N.A.	50	100	657	301		2,188
1947	1,044	16	50	37	404	235		1,786
1946	1,163	16	70	94	577	495		2,415
1945	1,182		53		390	400		
1944	946		20		512	182		
1943	1,292		63		561	380		
1942	1,262		43		465	420		
1941	1,124		48		358	480		
1940	1,361		40		500	478		
1935–39 av.	1,248							
1930–34 av.	1,173							
1925–29 av.	1,090							
1920–24 av.	981							
1915–19 av.	913							
1886	600							
1875	1,000							
1844	380							
1835	142							
1833	201							
1832	150							
1821	173							

* USDA data for some years and some countries differ from the FAO data, sometimes by small amounts, sometimes by rather large ones.

Sources: 1967: USDA. 1940–66: *FAO Production Yearbook.* 1915–39: Donald C. Mead, *Growth and Structural Change in the Egyptian Economy* (Homewood, Ill.: Richard D. Irwin, 1967), see p. 32. 1821–86: Patrick O'Brien, "The Long-term Growth of Agricultural Production in Egypt: 1821–1962," in P. M. Holt, ed., *Political and Social Change in Modern Egypt* (London: Oxford University Press, 1968).

Appendix Table C8-4
Barley Area, by Countries, Years of Record 1844–1966
(1,000 Acres)

Year	Egypt	Israel	Lebanon	Jordan	Syria	Iraq	Saudi Arabia	Total reported
1966	104	121	35	161	830	2,889	62	4,202
1965	131	128	35	212	1,685	2,711	62	4,964
1964	126	180	37	225	1,890	2,713	62	5,233
1963	126	161	25	188	1,984	3,012	67	5,563
1962	136	161	35	260	1,787	2,938	67	5,384
1961	126	173	32	235	1,796	2,572	64	4,998
1960	153	148	44	84	1,834	2,565	N.A.	4,828
1959	146	141	44	124	1,796	2,696	N.A.	4,947
1958	141	143	47	285	1,900	2,859	N.A.	5,375
1957	138	128	52	235	2,009	3,064	57	5,683
1956	136	143	47	269	1,572	2,894	57	5,118
1955	141	153	49	250	1,517	2,978	N.A.	5,088
1954	126	193	49	257	1,342	2,772	N.A.	4,739
1953	121	173	49	227	1,085	2,708	N.A.	4,363
1952	141	200	42	232	981	2,179	N.A.	3,775
1951	124	151	49	203	850	2,132	42	3,551
1950	121	124	49	158	1,028	2,471	32	3,983
1949	175	37	49	89	860	2,355		3,565
1948	227		59	89	845	2,407		3,627
1947	247		37	37	902	2,592		3,815
1946	255		54	82	917	1,998		3,306
1945	373		49		860	2,041		
1944	344		37		788	1,920		
1943	435		77		709	1,799		
1942	334		64		768	1,013		
1941	264		136		640	2,281		
1940	279		124		689	2,399		
1935–39 av.	276							
1930–34 av.	319							
1925–29 av.	368							
1920–24 av.	377							
1915–19 av.	419							
1889	541							
1879	562							
1878	542							
1877	510							
1875	542							
1844	907							

Sources: 1940–66: *FAO Production Yearbooks.* 1915–39: Donald C. Mead, *Growth and Structural Change in the Egyptian Economy* (Homewood, Ill.: Richard D. Irwin, 1967), see p. 32. 1844–89: Patrick O'Brien, "The Long-term Growth of Agricultural Production in Egypt: 1821–1962," in P. M. Holt, ed., *Political and Social Change in Modern Egypt* (London: Oxford University Press, 1968), see p. 165.

Appendix Table C8-5
Barley Yields Per Hectare, by Countries, Years of Record 1879—1966
(100 Kilograms)

Year	Egypt[a]	Israel	Leba-non	Jordan	Syria	Iraq	Saudi Arabia
1966	24.0	4.3	9.0	3.5	6.0	7.1	13.4
1965	24.6	12.9	9.3	11.0	10.1	7.4	12.8
1964	27.6	15.9	10.0	10.6	8.3	5.7	13.2
1963	26.5	5.6	10.0	3.0	9.8	6.5	13.0
1962	26.6	7.4	9.7	3.4	11.0	9.5	12.6
1961	26.1	9.0	9.4	6.5	4.6	8.8	12.3
1960	25.0	4.5	6.1	3.8	2.1	7.7	N.A.
1959	24.1	11.4	13.9	5.2	3.0	6.6	N.A.
1958	23.7	9.1	9.5	2.3	3.0	8.2	N.A.
1957	23.4	14.2	11.4	8.5	8.9	10.5	10.0
1956	23.5	14.7	11.1	8.8	7.3	9.1	10.0
1955	22.3	6.8	13.0	2.5	2.2	6.3	N.A.
1954	22.7	11.5	13.5	10.0	11.7	11.0	N.A.
1953	21.0	9.1	13.0	4.7	10.8	10.1	N.A.
1952	20.7	11.5	15.9	9.8	11.8	7.4	N.A.
1951	20.0	4.6	13.0	3.7	4.5	9.7	8.8
1950	18.6	7.4	11.5	6.4	7.7	8.0	8.5
1949	19.4	13.3	13.5	15.6	10.3	7.9	
1948	18.2		9.2	11.4	8.9	5.9	
1947	17.0		11.3	5.3	4.6	4.8	
1946	17.3		11.8	11.3	7.6	7.7	
1945	17.4		13.3		7.1	7.9	
1944	16.3		4.9		8.7	4.9	
1943	17.9		6.2		10.8	8.2	
1942	20.5		5.5		7.6	16.0	
1941	19.6		9.1		8.0	10.3	
1940	21.4		8.8		9.9	10.3	
1935—39 av.	21.0						
1930—34 av.	17.2						
1925—29 av.	16.7						
1920—24 av.	16.2						
1915—19 av.	16.0						
1911	11.4						
1910	9.2						
1909	9.2						
1908	9.0						
1907	10.9						
1906	9.8						
1905	10.1						
1904	10.7						
1903	12.4						
1902	9.8						
1901	10.7						
1900	9.5						
1899	10.8						
1898	12.0						
1897	10.4						
1896	10.0						
1895	12.8						
1894	11.3						
1893	12.5						
1892	8.7						
1891	10.2						
1890	12.2						
1889	10.4						
1888	9.3						

Appendix Table C8-5—*Continued*

Year	Egypt[a]	Israel	Leba-non	Jordan	Syria	Iraq	Saudi Arabia
1887	8.7						
1886	6.9						
1885	5.9						
1884	6.3						
1883	6.4						
1882	7.2						
1881	5.7						
1880	7.6						
1879	5.4						

[a]Data for 1915 to 1965 relate to all land growing barley in Egypt; data for 1879 to 1911 relate to barley grown on state domains.

Sources: 1940—66: *FAO Production Yearbooks.* 1915—39: Donald C. Mead, *Growth and Structural Change in the Egyptian Economy* (Homewood, Ill.: Richard D. Irwin, 1967), see p. 321. 1879—1911: Patrick O'Brien, "The Long-term Growth of Agricultural Production in Egypt: 1821—1962," in P. M. Holt, ed., *Political and Social Change in Modern Egypt* (London: Oxford University Press, 1968), see p. 165.

Appendix Table C8-6

Barley Production,* by Countries, Years of Record 1821–1967

(1,000 Metric Tons)

Year	Egypt	Israel	Lebanon	Jordan	Syria	Iraq	Saudi Arabia	Total reported
1967	110	45	12	80	350	700		1,597
1966	102	21	10	32	150	700	34	1,049
1965	130	67	13	96	690	807	32	1,835
1964	141	117	15	97	637	623	33	1,663
1963	134	36	10	23	784	790	35	1,812
1962	146	48	13	36	798	1,125	34	2,200
1961	133	63	12	62	335	911	32	1,548
1960	155	27	11	13	156	804	N.A.	1,166
1959	142	65	25	26	218	725	N.A.	1,201
1958	135	53	18	17	228	954	N.A.	1,405
1957	131	74	24	81	721	1,305	N.A.	2,336
1956	129	85	21	96	462	1,066	23	1,882
1955	127	42	26	25	137	757	22	1,331
1954	116	90	27	104	635	1,239	20	2,231
1953	103	64	26	43	472	1,111	N.A.	1,819
1952	118	93	27	93	467	652	N.A.	1,450
1951	100	28	26	30	155	839	15	1,193
1950	91	37	23	41	322	801	11	1,326
1949	138	20	27	56	357	750		1,348
1948	167	9	22	41	305	570		1,114
1947	170	9	17	8	169	500		873
1946	178	15	26	37	282	620		1,158
1945	262		27		248	650		
1944	227		7		277	378		
1943	314		19		311	599		
1942	277		14		235	658		
1941	210		50		207	950		
1940	241		44		276	1,000		
1935–39 av.	233							
1930–34 av.	220							
1925–29 av.	247							
1920–24 av.	246							
1915–19 av.	271							
1875	372							
1844	373							
1835	67							
1833	71							
1832	108							
1821	72							

*USDA figures differ from FAO figures for some countries in some years.

Sources: 1967: USDA. 1940–66: *FAO Production Yearbooks.* 1915–39: Donald C. Mead, *Growth and Structural Change in the Egyptian Economy* (Homewood, Ill.: Richard D. Irwin, 1967), see p. 321. 1821–75: Patrick O'Brien, "The Long-term Growth of Agricultural Production in Egypt: 1821–1962," in P. M. Holt, ed., *Political and Social Change in Modern Egypt* (London: Oxford University Press, 1968), see p. 165.

Appendix Table C8-7

Cotton Area, by Countries, Years of Record 1871–1966

(1,000 Acres)

Year	Egypt	Israel	Lebanon	Syria	Iraq	Total reported
1966	1,930	54		633	96	2,713
1965	1,972	42		707	99	2,820
1964	1,671	32		709	99	2,511
1963	1,690	30		722	99	2,541
1962	1,720	42		746	84	2,592
1961	2,061	42		615	91	2,809
1960	1,945	27		524	77	2,573
1959	1,826	17		561	91	2,495
1958	1,977	17		645	138	2,777
1957	1,888	12	2	638	161	2,701
1956	1,715	15	2	672	143	2,547
1955	1,885	5	2	600	141	2,633
1954	1,638		2	462	138	2,240
1953	1,374		7	334	52	1,767
1952	2,041			467	126	2,634
1951	2,056		7	536	111	2,710
1950	2,048			193	82	2,323
1949	1,757			62	27	1,846
1948	1,497			59	17	1,573
1947	1,302			47	22	1,371
1946	1,258			49		1,307
1945	1,021			44		1,065
1944	885			42		927
1943	739			40	27	806
1942	734			37	27	798
1941	170			47	225	442
1940	175			79	148	402
1935–39 av.	1,820					
1930–34 av.	1,740					
1925–29 av.	1,825					
1920–24 av.	1,745					
1915–19 av.	1,535					
1889	1,097					
1886	1,090					
1879	985					
1871	725					

Source: 1940–66: *FAO Production Yearbooks.* 1915–39: Donald C. Mead, *Growth and Structural Change in the Egyptian Economy* (Homewood, Ill.: Richard D. Irwin, 1967), see p. 322. 1871–89: Patrick O'Brien, "The Long-term Growth of Agricultural Production in Egypt: 1821–1962," in P. M. Holt, ed., *Political and Social Change in Modern Egypt* (London: Oxford University Press, 1968), see p. 165.

Appendix Table C8-8

Cotton Yield Per Hectare, by Countries, Years of Record 1871–1966

(100 Kilograms)

Year	Egypt	Israel	Lebanon	Syria	Iraq
1966	5.9	11.4		5.3	2.5
1965	6.5	12.3		6.2	2.5
1964	7.4	12.2		6.2	2.5
1963	6.5	10.9		5.2	2.5
1962	6.6	9.7		5.0	2.4
1961	4.0	8.7		5.0	2.3
1960	6.1	10.0		5.2	2.4
1959	6.2	10.6	4.0	4.3	2.1
1958	5.6	7.3	2.5	3.7	1.2
1957	5.3	8.5	2.5	4.1	2.2
1956	4.7	5.8	2.0	3.4	1.3
1955	4.4	9.6	4.0	3.6	1.3
1954	5.2	8.9	4.0	4.3	1.2
1953	5.7		2.0	3.5	1.1
1952	5.4			2.4	0.7
1951	4.4			2.2	1.4
1950	4.6			4.5	2.5
1949	5.5			5.2	1.6
1948	6.6			3.8	0.6
1947	5.4			2.9	
1946	5.4			2.4	
1945	5.7			2.5	
1944	5.8			1.8	
1943	5.4			2.1	0.6
1942	6.4			2.0	0.8
1941	5.4			1.5	0.4
1940	5.8			1.1	0.8
1935–39 av.	5.5				
1930–34 av.	4.5				
1925–29 av.	4.7				
1920–24 av.	4.0				
1915–19 av.	3.8				
1889	3.2				
1886	3.0				
1879	3.6				
1871	3.1				

Sources: 1941–66: *FAO Production Yearbooks.* 1915–39: Donald C. Mead, *Growth and Structural Change in the Egyptian Economy* (Homewood, Ill.: Richard D. Irwin, 1967), see p. 322. 1871–89: Patrick O'Brien, "The Long-term Growth of Agricultural Production in Egypt: 1821–1962," in P. M. Holt, ed., *Political and Social Change in Modern Egypt* (London: Oxford University Press, 1968). Yield per hectare calculated from data on area and production.

Appendix Table C8-9

Cotton Production, by Countries, Years of Record 1821–1967

(1,000 Metric Tons Lint Cotton)

Year	Egypt	Israel	Lebanon	Syria	Iraq	Total reported
1967	435	30		111	10	586
1966	462	25		135	10	632
1965	520	22		178	10	730
1964	504	16		176	10	706
1963	442	13		153	10	618
1962	457	16		150	8	631
1961	336	15		124	9	484
1960	478	11		111	8	608
1959	457	7		97	8	569
1958	446	5		97	7	555
1957	405	4		107	14	530
1956	325	3		93	8	429
1955	335	2		87	8	432
1954	348	1		80	7	436
1953	318		1	47	2	368
1952	446		1	45	3	495
1951	363		1	48	6	418
1950	382			35	8	425
1949	391			13	2	406
1948	400			9		409
1947	286			5	1	292
1946	272			5	1	278
1945	235			4	1	240
1944	209			3	1	213
1943	161			3	1	165
1942	190			3	1	194
1941	376			3	4	383
1940	412			4	5	421
1935–39 av.	411					
1930–34 av.	321					
1925–29 av.	344					
1920–24 av.	278					
1915–19 av.	238					
1889	143					
1886	132					
1879	144					
1878	76					
1877	117					
1875	132					
1872	103					
1871	92					
1844	7					
1835	6					
1834	10					
1833	3					
1832	6					
1821	2					

Sources: 1967: USDA. 1940–66: *FAO Production Yearbooks.* 1915–39: Donald C. Mead, *Growth and Structural Change in the Egyptian Economy* (Homewood, Ill.: Richard D. Irwin, 1967), p. 322. 1821–89: Patrick O'Brien, "The Long-term Growth of Agricultural Production in Egypt: 1821–1962," in P. M. Holt, ed., *Politics and Social Change in Modern Egypt* (London: Oxford University Press, 1968), pp. 165, 179.

Appendix Table C8-10

Egypt: Areas Under Main Crops and Total Cropped Area, 1935–66*

(1000 Hectares)

	Total cropped area	Wheat	Maize	Millet	Barley	Rice	Cotton	Berseem	Beans[b]	Lentils	Onions[c]	Sugar-cane	Other
1935–39 average	3,477.6	592.2	646.8	151.2	113.4	189.0	735.0	688.8	163.8	33.6	16.8	29.4	117.6
1940–44 average	3,750.6	684.6	747.6	260.4	134.4	243.6	470.4	772.8	163.8	33.6	12.6	33.6	193.2
1945	3,872.4	693.0	789.6	285.6	151.2	264.6	411.6	789.6	163.8	33.6	12.6	42.0	235.2
1946	3,759.0	667.8	693.0	231.0	100.8	264.6	508.2	823.2	159.6	33.6	8.4	37.8	231.0
1947	3,851.4	684.6	676.2	226.8	100.8	327.6	525.0	835.8	159.6	29.4	12.6	37.8	235.2
1948	3,847.2	638.4	651.0	218.4	92.4	331.8	604.8	831.6	168.0	29.4	12.6	37.8	231.0
1949	3,851.4	596.4	625.8	210.0	71.4	294.0	709.8	852.6	176.4	29.4	16.8	37.8	231.0
1945–49 average	3,834.6	655.2	688.8	235.2	105.0	294.0	554.4	827.4	168.0	29.4	12.6	37.8	226.8
1950	3,876.6	575.4	609.0	163.8	50.4	294.0	831.6	915.6	151.2	33.6	16.8	33.6	201.6
1951	3,893.4	630.0	697.2	176.4	50.4	205.8	831.6	890.4	134.4	33.6	16.8	37.8	189.0
1952	3,910.2	588.0	714.0	180.6	58.8	155.4	827.4	924.0	151.2	25.2	12.6	37.8	235.2
1953	3,935.4	751.8	848.4	205.8	50.4	176.4	554.4	898.8	126.0	29.4	16.8	42.0	235.2
1954	4,153.8	751.8	798.0	193.2	50.4	256.2	663.6	953.4	130.2	37.8	16.8	50.4	252.0
1950–54 average	3,956.4	659.4	735.0	184.8	50.4	218.4	739.2	915.6	138.6	29.4	16.8	42.0	226.8
1955	4,187.4	638.4	768.6	184.8	58.8	252.0	764.4	987.0	151.2	33.6	21.0	46.2	281.4
1956	4,183.2	659.4	772.8	201.6	54.6	289.8	693.0	974.4	142.8	33.6	21.0	46.2	294.0
1957	4,242.0	634.2	743.4	189.0	54.6	306.6	764.4	991.2	151.2	33.6	21.0	46.2	306.6
1958	4,237.8	596.4	819.0	176.4	58.8	218.4	789.0	999.6	151.2	29.4	21.0	46.2	332.4
1959	4,313.4	621.6	781.2	197.4	58.8	306.6	739.2	1,008.0	147.0	33.6	25.2	46.2	348.6
1955–59 average	4,233.6	630.0	777.0	189.0	58.8	273.0	751.8	991.2	147.0	33.6	21.0	46.2	315.0
1960	4,355.4	613.2	764.4	189.0	63.0	298.2	785.4	1,012.2	159.6	33.6	21.0	46.2	369.6
1961	4,187.4	579.6	672.0	193.2	50.4	226.3	835.8	1,029.0	151.2	25.2	25.2	46.2	353.3
1962	4,355.5	613.2	768.6	189.0	54.6	348.6	697.2	1,024.8	159.6	33.6	21.0	50.4	394.5
1963	4,334.4	562.8	722.4	201.6	50.4	403.2	684.6	1,020.6	151.2	33.6	21.0	54.6	428.4
1964	4,351.2	546.0	697.2	205.8	50.4	403.2	676.2	1,104.6	172.2	33.6	21.0	54.6	386.4
1960–64 average	4,317.6	583.8	726.6	197.4	54.6	336.0	735.0	1,024.8	159.6	33.6	21.0	50.4	394.8
1965	4,347.0[a]	–	609.0	210.0	50.4	–	798.0	1,045.8	168.0	37.8	21.0	54.6	352.4
1966	–	–	659.4[a]	218.4[a]	42.0	–	781.2	1,062.6	168.0	33.6	21.0	–	

* For reasons not explained in sources, period averages shown may not agree precisely with averages computed from annual data.

[a] Provisional.

[b] Area includes green consumption as from 1960.

[c] Area excludes interplanted crop and is confined to winter crop as from 1960.

Source: National Bank of Egypt, *Economic Bulletin,* various issues, Cairo, Egypt; also, Central Bank of Egypt, *Economic Review,* VII, Nos. 1 & 2 (Cairo, 1967), pp. 106-107.

Appendix Table C8-11

Egypt: Average Yields of Major Crops, 1935–66*

(Kilograms Per Hectare)

Year	Wheat	Maize	Millet	Barley	Rice	Cotton	Berseem[a]	Beans[b]	Lentils[c]	Onions	Sugarcane
Average 1935/39	2,107	2,483	3,016	2,055	3,619	559	–	1,807	1,637	14,464	77,313
Average 1940/44	1,748	1,926	2,631	1,890	3,021	574	–	1,807	1,637	13,333	72,381
1945	1,706	2,149	2,570	1,733	3,273	568	–	1,880	1,488	14,682	62,976
1946	1,742	2,052	2,273	1,766	3,545	535	–	1,880	1,518	19,762	66,878
1947	1,525	2,072	2,548	1,686	3,898	545	–	1,635	1,633	18,730	71,614
1948	1,692	2,164	2,555	1,796	3,942	661	–	1,708	1,633	19,682	66,032
1949	1,957	1,997	2,700	1,933	3,973	551	–	1,820	1,599	19,583	59,735
Average 1945/49	1,720	2,085	2,517	1,743	3,779	572	50	1,756	1,667	18,492	66,878
1950	1,769	2,144	2,601	1,806	4,224	459	42	1,310	1,518	15,119	75,238
1951	1,919	2,038	2,931	1,984	3,013	436	40	1,726	1,399	18,333	74,233
1952	1,852	2,109	2,890	2,007	3,327	538	39	1,653	1,270	21,190	86,243
1953	2,058	2,184	2,828	2,044	3,696	574	44	1,659	1,599	18,036	87,762
1954	2,300	2,197	2,842	2,302	4,364	524	40	1,805	1,587	22,560	80,258
Average 1950/54	1,999	2,133	2,808	2,083	3,800	503	42	1,623	1,633	18,036	78,595
1955	2,273	2,230	2,906	2,160	4,936	438	38	1,733	1,458	19,714	89,719
1956	2,346	2,139	2,951	2,363	5,159	469	36	1,442	1,428	19,190	88,550
1957	2,313	2,011	2,995	2,399	5,297	531	38	1,680	1,577	22,952	89,545
1958	2,368	2,146	3,078	2,296	4,702	565	35	1,720	1,428	22,048	91,558
1959	2,321	1,920	3,191	2,398	5,010	620	37	1,415	1,428	22,143	93,701
Average 1955/59	2,324	2,090	3,037	2,245	5,073	524	37	1,619	1,428	22,095	90,606
1960	2,444	2,212	3,190	2,476	4,983	609	40	1,817	1,488	24,000	98,377
1961	2,478	2,406	3,266	2,639	5,046	402	38	1,065	1,349	18,611	90,606
1962	2,598	2,607	3,487	2,674	5,846	655	32	2,055	1,667	28,571	95,397
1963	2,653	2,584	3,616	2,659	5,503	646	26	1,739	1,399	31,381	94,377
1964	2,745	2,774	3,596	2,798	5,050	745	24	2,125	1,548	30,762	89,560
Average 1960/64	2,576	2,509	3,404	2,601	5,310	603	32	1,767	1,428	27,428	93,571
1965	–	3,516	3,838	2,579	–	653	27	2,048	1,614	31,905	86,795
1966	–	3,576[d]	3,947[d]	2,428	–	591[d]	32	2,268	1,310	–	–

* For reasons not explained in sources, period averages shown may not agree precisely with averages computed from annual data.

[a] Production of seeds from areas allotted to seed production alone.

[b] Area includes green consumption as from 1960.

[c] From 1945 to 1949 production estimates were calculated by multiplying annual area estimates by period average yields for 1945–49 and 1950–54, respectively.

[d] Provisional estimate.

Source: National Bank of Egypt, *Economic Bulletin,* various issues, Cairo, Egypt; also, Central Bank of Egypt, *Economic Review,* VII, Nos. 1 & 2 (Cairo, 1967), pp. 106 and 107.

Note: Production figures for 1945–54 were converted from ardebs, dariba (units of volume), and cantars (units of weight) into metric tons by applying conversion ratios for different crops as reported in NBE, *Economic Bulletin,* various issues, Cairo, Egypt.

Appendix Table C8-12

Egypt: Production of Main Crops, 1935–66*

(1000 Metric Tons)

Year	Wheat	Maize	Millet	Barley	Rice	Cotton	Berseem[a]	Beans[b]	Lentils[c]	Onions	Sugarcane
Average 1935/39	1,248	1,606	456	233	684	411	–	296	55	243	2,273
Average 1940/44	1,197	1,440	685	254	736	270	–	296	55	168	2,432
1945	1,182	1,697	734	262	866	234	–	308	50	185	2,645
1946	1,163	1,422	525	178	938	272	–	300	51	166	2,528
1947	1,044	1,401	578	170	1,277	286	–	261	48	236	2,707
1948	1,080	1,409	558	166	1,308	400	–	287	48	248	2,496
1949	1,167	1,250	567	138	1,168	391	–	321	47	329	2,258
Average 1945/49	1,127	1,436	592	183	1,111	317	41	295	49	233	2,528
1950	1,018	1,306	426	91	1,242	382	39	198	51	254	2,528
1951	1,209	1,421	517	100	620	363	36	232	47	308	2,806
1952	1,089	1,506	522	118	517	445	36	250	32	267	3,260
1953	1,547	1,853	582	103	652	318	40	209	47	303	3,686
1954	1,729	1,753	549	116	1,118	348	38	235	60	379	4,045
Average 1950/54	1,318	1,568	519	105	830	372	38	225	48	303	3,301
1955	1,451	1,714	537	127	1,244	335	38	262	49	414	4,145
1956	1,547	1,652	595	129	1,495	325	35	206	48	403	4,091
1957	1,467	1,495	566	131	1,624	406	38	254	53	482	4,137
1958	1,412	1,758	543	135	1,027	446	35	260	42	463	4,230
1959	1,443	1,500	630	141	1,536	458	37	208	48	558	4,329
Average 1955/59	1,464	1,624	574	132	1,385	394	37	238	48	464	4,186
1960	1,499	1,691	603	156	1,486	478	41	290	50	504	4,545
1961	1,436	1,617	631	133	1,142	336	39	161	34	469	4,186
1962	1,593	2,004	659	146	2,038	457	33	328	56	600	4,808
1963	1,493	1,867	729	134	2,219	442	27	263	47	659	5,153
1964	1,499	1,934	740	141	2,036	504	27	366	52	646	4,890
Average 1960/64	1,504	1,823	672	142	1,784	443	33	282	48	576	4,716
1965	–	2,141	806	130	–	521	28	344	61	670	4,739
1966	–	2,358[d]	862[d]	102	–	462[d]	34	381	44	–	–

* For reasons not explained in sources, period averages shown may not agree precisely with averages computed from annual data.

[a] Production of seeds from areas allotted to seed production alone.

[b] Area includes green consumption as from 1960.

[c] From 1945 to 1949 production estimates were calculated by multiplying annual area estimates by period average yields for 1945–49 and 1950–54, respectively.

[d] Provisional estimates.

Source: see Table C8-11.

Appendix Table C8-13

Egypt: Areas Cropped with Vegetables, 1959–66

(1000 Hectares)

Vegetables	1959	1960	1961	1962	1963	1964	1965	1966
Potatoes	27	28	28	28	29	28	29	26
Garlic	7	5	5	6	8	5	6	8
Tomatoes	54	57	61	66	71	78	80	86
Squash	11	11	13	12	13	13	13	15
Beans	2	3	3	3	5	9	8	8
Peas	3	4	3	3	5	5	5	5
Cabbages	10	10	10	9	10	11	10	10
Eggplant	5	6	6	7	8	8	8	9
Pepper	3	3	4	4	4	5	5	6
Okra	4	4	5	5	5	5	5	5
Moloukia	3	3	3	3	4	4	4	4
Red melons	26	26	26	27	31	33	38	43
Yellow melons	6	6	6	6	8	8	8	8
Cucumbers	14	15	14	18	17	17	15	15
Others	25	26	28	26	27	26	28	27
Total	200	207	215	223	245	255	262	275

Source: UAR, Central Organization for General Mobilization and Statistics, *General Statistics, 1952–66* (Cairo, 1967), pp. 28–30.

237

Appendix Table C8-14

Egypt: Area, Production, and Yields of Vegetables, 1964

Vegetables	Winter			Summer			Nili			Total		
	Area (hectares)	Average yield (ton/ha.)	Production (1000 tons)	Area (hectares)	Average yield (ton/ha.)	Production (1000 tons)	Area (hectares)	Average yield (ton/ha.)	Production (1000 tons)	Area (hectares)	Average yield (ton/ha.)	Production (1000 tons)
Potatoes	52	18.69	1	15,195	20.74	315	13,527	17.75	240	28,774	19.25	556
Garlic	5,301	12.54	66	–	–	–	–	–	–	5,301	12.54	66
Tomatoes	29,577	13.26	393	21,251	17.40	370	27,369	15.71	430	78,198	15.25	1,193
Squash	5,215	13.02	68	4,182	20.28	85	4,035	18.38	74	13,432	16.90	227
Green beans	87	6.95	1	1,341	9.30	12	1,916	7.42	14	3,345	8.16	27
Dry beans	127	1.78	–	2,959	1.71	5	2,906	1.61	5	5,993	1.66	10
Green French beans	5	7.14	–	785	7.85	6	668	7.00	5	1,458	7.45	11
Dry French beans	34	2.14	–	2,773	7.00	5	1,320	1.96	2	4,128	1.69	7
Green peas	3,027	7.07	21	–	–	–	14	2.97	–	3,040	7.04	21
Dry peas	737	1.47	1	–	–	–	48	2.38	–	785	1.52	1
Broad beans	746	9.16	7	–	–	–	–	–	–	746	9.16	7
Cabbages	6,328	25.78	163	226	20.26	5	3,956	24.11	95	10,510	25.04	263
Cauliflower	1,465	25.80	38	51	16.76	1	813	20.42	16	2,329	23.71	55
Eggplant	49	16.90	1	4,343	23.73	103	3,613	19.50	70	8,005	21.78	174
Pepper	54	8.97	1	3,163	15.35	48	1,432	13.21	19	4,649	14.61	68
Okra	52	4.52	–	5,311	12.40	66	21	7.42	–	5,384	12.30	66
Moloukia	396	10.83	4	3,344	15.92	53	275	13.52	4	4,015	15.26	61
Spinach	1,389	16.35	22	–	–	–	35	20.23	1	1,424	16.45	23
Kobbayzé	1,082	30.50	33	–	–	–	–	–	–	1,082	30.50	33
Artichoke	4	14.54	–	–	–	–	759	19.26	15	762	19.23	15
Radish	1,156	13.04	15	1,029	10.50	11	639	10.45	7	2,824	11.54	33
Turnip	2,528	18.02	46	6	13.09	–	69	18.85	1	2,603	18.04	47
Lettuce	1,724	21.11	37	33	33.54	1	–	–	–	1,757	21.35	38
Carrots	1,335	19.71	26	–	–	–	35	19.04	1	1,370	19.69	27
Parsley	200	28.73	6	205	22.14	5	82	28.11	2	487	25.85	13
Rocket (Jarjir)	179	20.40	4	293	23.19	7	148	23.80	3	620	22.52	14
Carrat Masri	178	31.80	6	331	36.11	12	112	41.85	4	621	35.92	22
Red melons	–	–	–	32,457	26.42	857	162	22.02	4	32,619	26.39	861
Yellow melons	–	–	–	7,539	27.73	209	87	17.61	2	7,626	27.61	211
Cucumbers	828	2.40	2	9,988	13.61	136	3,595	11.69	42	14,411	12.47	180
Snake cucumbers	58	21.76	1	2,327	14.33	34	12	20.07	–	2,397	14.54	35
Others	851	–	15	3,461	21.32	76	418	–	6	4,730	20.45	97
Total	64,764	–	978	122,594	–	2,422	68,068	–	1,062	255,428	–	4,462

Source: UAR Ministry of Agriculture, *Agricultural Economy*, bulletin published by the Department of Agricultural Economics and Statistics (1966), pp. 371-72.

Appendix Table C8-15
Egypt: Gross Value of Fruit, 1965

Fruit	Area (hectares)	Output (tons)	Average yield (kg./ha.)	Value (£E 1000)	Gross return per hectare (£E)
Oranges	36,433	341,683	9.378	6,801	187
Mandarines	4,298	60,574	14,094	744	180
Sour lemons	3.987	80,139	20,100	662	166
Sweet lemons	248	3,090	12,460	59	238
Bitter oranges	114	919	8,061	10	88
Lemons	249	22,036	8,177	47	189
Other citrus	205	5,579	2,824	11	54
Apricots	1,016	5,940	5,846	317	312
Damson plums	286	2,163	7,563	231	808
Apples	449	6,375	14,198	417	929
Pears	1,136	16,095	14,168	800	704
Plums	791	5,172	6,539	282	357
Mango	9,734	78,546	8,069	4,837	497
Bananas	3,522	64,005	18,173	2,372	673
Figs	843	4,296	5,096	86	102
Pomegranates	731	7,625	10,431	209	300
Olives	1,041	10,455	10,043	307	295
Guava	2,653	33,539	12,642	826	311
Grapes	10,014	92,453	9,232	2,690	269
Dates	—	388,385	—	10,075	—
Other palm	—	—	—	824	—
Other	582	26,197	45,012	519	892
Total	78,332	1,230,266	15,706	33,136	423

Appendix Table C8-16

**Israel: Area of Field Crops, by Unirrigated and Irrigated, Winter and Summer,
Years of Record 1948/49 to 1966/67***

(1,000 Hectares)

	1948/49	1952/53	1954/55	1955/56	1956/57	1958/59	1959/60
Unirrigated							
Winter crops:							
Hay	12	38	35	31	34	39	36
Fodder, pasture, silage	1	4	4	4	5	4	4
Green manure	1	8	12	10	10	10	8
Pulses for grain	4	13	13	12	11	7	7
Barley	16	83	62	58	52	57	60
Wheat	30	35	47	57	57	62	59
Oats and rye	2	2	1	1	1	1	1
Sugar beets	–	@	@	@	–	–	–
Peas for canning	–	1	1	1	1	1	1
Safflower, winter	–	–	2	3	2	1	1
Miscellaneous	–	@	@	@	1	@	1
Subtotal	66	184	179	177	174	183	177
Summer crops:							
Hay, fodder, silage	3	7	3	3	3	3	1
Pulses for grain	@	1	2	2	2	3	2
Maize for grain	7	5	3	3	1	1	1
Sorghum for grain	4	17	12	16	22	22	6
Sunflowers, sesame, safflower	4	3	3	5	6	5	6
Tobacco	1	4	4	3	3	4	4
Cotton	–	–	–	–	–	1	1
Melons	4	8	5	6	6	7	4
Cultivated fallow	17	15	23	22	20	20	40
Miscellaneous'	@	@	@	@	@	@	@
Subtotal	41	61	55	60	62	67	65
Irrigated							
Winter crops:							
Sugar beets	–	@	1	1	1	3	4
Green fodder	3	4	5	6	8	11	11
Subtotal	3	5	6	8	10	15	15
Summer crops:							
Hay	–	–	–	1	1	2	1
Pulses for grain	–	–	@	@	@	@	@
Maize for grain	@	2	6	4	7	2	1
Sorghum for grain	@	@	1	1	1	1	1
Cotton	–	–	2	6	5	6	10
Green fodder	3	6	9	9	10	12	13
Pasture and silage	@	2	2	2	2	2	2
Miscellaneous	@	@	@	@	@	1	1
Subtotal	4	11	20	23	27	27	28
Total field crop area	113	261	260	236	236	291	285

Appendix Table C8-16 *—continued*

	1960/61	1961/62	1962/63	1963/64	1964/65	1965/66	1966/67
Unirrigated							
Winter crops:							
Hay	33	30	35	30	26	36	30
Fodder, pasture, silage	4	4	5	5	4	4	5
Green manure	6	6	5	4	4	3	2
Pulses for grain	6	6	5	6	5	3	4
Barley	70	65	65	73	52	49	43
Wheat	63	48	52	56	72	76	94
Oats and rye	1	1	1	1	@	@	@
Sugar beets	@	@	@	@	2	1	2
Peas for canning	1	1	1	1	1	1	1
Safflower, winter	1	1	@	1	3	@	1
Miscellaneous	1	1	@	@	@	@	@
Subtotal	185	162	170	177	168	174	183
Summer crops:							
Hay, fodder, silage	1	1	1	1	1	1	1
Pulses for grain	2	1	1	1	2	2	2
Maize for grain	1	1	@	1	@	@	@
Sorghum for grain	18	14	8	25	27	3	9
Sunflowers, sesame, safflower	9	5	5	5	6	4	3
Tobacco	4	4	@	2	3	4	4
Cotton	3	2	@	1	1	2	5
Melons	5	6	7	7	7	6	7
Cultivated fallow	14	30	21	10	13	21	10
Miscellaneous	@	@	@	@	@	@	@
Subtotal	58	64	45	52	60	42	40
Irrigated							
Winter crops:							
Sugar beets	6	5	5	5	6	5	5
Green fodder	10	11	12	11	9	8	8
Subtotal	17	16	17	16	15	13	13
Summer crops:							
Hay	1	1	1	@	1	2	1
Pulses for grain	@	@	@	@	@	@	@
Maize for grain	1	1	1	1	1	@	1
Sorghum for grain	1	2	4	5	3	1	1
Cotton	10	15	12	12	16	20	25
Green fodder	13	12	12	11	10	10	10
Pasture and silage	2	2	2	1	1	2	2
Miscellaneous	1	@	1	@	1	1	1
Subtotal	30	33	32	30	33	36	39
Total field crop area	290	275	264	276	276	265	278

* Israel economic data are compiled on the basis of the Hebrew calendar; the year begins in early fall (September, usually) and extends for about twelve months. No data available for 1949/50 to 1951/52, 1953/54, and 1957/58.

@ = Less than 500 hectares.

Sources: Various issues of *Statistical Abstract of Israel,* published by the Central Bureau of Statistics.

Appendix Table C8-17
Israel: Cropped Area, Total and Irrigated, 1948/49 to 1967/68[a] *
(1,000 Hectares)

	1948/49	1949/50	1950/51	1951/52	1952/53	1953/54	1954/55	1955/56	1956/57	1957/58
All cropped area:										
Field crops	113	185	223	243	261	254	260	267	273	279
Area in preparation[b]	–	–	44	30	10	10	2	3	3	5
Vegetables, potatoes, and groundnuts	7	13	16	20	24	27	27	26	28	26
Fruit plantations	36	38	39	41	43	48	52	54	60	64
Fish ponds	2	2	3	3	4	4	4	4	4	4
Miscellaneous	8	9	11	11	12	14	14	15	15	15
Subtotal	165	248	335	348	355	356	359	368	382	394
Irrigated cropped land:[c]										
Field crops	6	8	12	13	16	19	26	30	37	41
Vegetables, potatoes, and groundnuts	5	9	12	16	21	24	24	24	25	24
Fruit plantations	15	16	17	18	19	24	28	31	36	41
Fish ponds	2	2	3	3	4	4	4	4	4	4
Miscellaneous	2	3	4	5	5	6	7	7	8	8
Subtotal	30	38	47	54	65	76	89	96	110	118

	1958/59	1959/60	1960/61	1961/62	1962/63	1963/64	1964/65	1965/66	1966/67	1967/68
All cropped area:										
Field crops	292	285	290	275	264	276	276	265	276	278
Area in preparation[b]	5	2	3	2	2	2	2	2	2	2
Vegetables, potatoes, and groundnuts	26	27	26	26	27	26	27	28	28	28
Fruit plantations	68	72	75	81	84	86	87	87	88	88
Fish ponds	5	5	5	6	6	6	6	6	6	6
Miscellaneous	16	16	16	16	17	18	17	17	17	18
Subtotal	411	408	415	406	397	414	416	406	416	419
Irrigated cropped land:[c]										
Field crops	42	43	47	49	49	47	48	49	52	54
Vegetables, potatoes, and groundnuts	24	25	24	24	25	23	25	27	26	26
Fruit plantations	45	49	52	58	62	65	66	66	68	68
Fish ponds	5	5	5	6	6	6	6	6	6	6
Miscellaneous	8	8	8	8	9	9	9	9	9	9
Subtotal	124	130	136	144	150	150	154	157	162	164

* Israel economic data are compiled on the basis of the Hebrew calendar; the year begins in early fall (September, usually) and extends for about twelve months.

[a] Israel statistics call this "cultivated area," but "an area is included as many times as it is sown," hence this conforms to "cropped area" as shown in other appendix tables.

[b] New cultivated areas which were prepared for sowing in the following year.

[c] Does not include area under auxiliary irrigation.

Source: Various issues of *Statistical Abstract of Israel*, published by the Central Bureau of Statistics.

Appendix Table C8-18

Israel: Area of Fruit Plantations, 1948/49 to 1967/68*

(1,000 Hectares)

	1948/49	1949/50	1950/51	1951/52	1952/53	1953/54	1954/55
Citrus	12	13	13	14	14	16	20
Table grapes	} 5	6	7	8	9	10	10
Wine grapes							
Pome and stone fruits[a]	2	3	3	3	3	4	4
Olives	14	14	14	14	14	14	13
Bananas	1	1	1	1	1	1	1
Other	2	2	2	2	2	2	3
Total	36	38	39	41	43	48	52

	1955/56	1956/57	1957/58	1958/59	1959/60	1960/61	1961/62
Citrus	21	25	28	30	33	34	39
Table grapes	7	7	7	7	7	8	8
Wine grapes	4	4	4	4	4	4	4
Pome and stone fruits[a]	4	6	6	7	8	10	10
Olives	13	13	13	13	12	12	12
Bananas	2	2	2	2	2	2	2
Other	3	4	4	5	5	5	5
Total	54	60	64	68	72	75	81

	1962/63	1963/64	1964/65	1965/66	1966/67	1967/68	
Citrus	42	44	44	44	44	45	
Table grapes	7	7	6	6	6	5	
Wine grapes	4	5	5	6	6	6	
Pome and stone fruits[a]	10	11	11	11	12	12	
Olives	12	12	11	11	11	11	
Bananas	2	2	2	2	2	2	
Other	6	6	6	6	6	7	
Total	84	86	87	88	88	88	

* Israel economic data are based upon the Hebrew calendar; the year begins in early fall (September, usually) and extends for about twelve months.

[a] Including almonds.

Source: Various issues of *Statistical Abstract of Israel,* published by Central Bureau of Statistics.

Appendix Table C8-19

Israel: Area of Vegetables and Potatoes, Years of Record 1948/49 to 1966/67*

(100 Hectares)

	1949/50	1950/51	1951/52	1952/53	1953/54	1954/55	1955/56	1956/57	1958/59
Potatoes	23	24	28	38	53	51	47	51	45
Tomatoes	32	31	39	39	44	30	42	36	36
Cucumbers	11	15	19	19	19	17	21	18	22
Carrots	6	9	7	12	9	10	11	12	10
Cabbage and cauliflower	9	10	11	17	13	12	12	16	13
Marrows	4	4	6	6	5	4	5	6	7
Onions, green and dry	7	9	19	22	20	32	17	19	16
Peppers and gamba	4	5	6	6	6	5	7	7	7
Beans and peas[a]	9	8	10	12	13	10	11	13	13
Melons, irrigated	1	3	3	5	5	10	14	11	17
Groundnuts	2	7	10	21	41	54			
Other	25	32	40	47	41	36	37	32	26
Total	133	157	198	244	269	271	224	221	212

	1959/60	1960/61	1961/62	1962/63	1963/64	1964/65	1965/66	1966/67
Potatoes	43	45	53	52	45	50	42	44
Tomatoes	33	33	32	32	31	28	30	31
Cucumbers	25	25	30	24	28	27	31	31
Carrots	11	11	11	12	13	11	12	13
Cabbage and cauliflower	14	11	12	10	12	13	14	12
Marrows	8	9	9	8	9	8	9	9
Onions, green and dry	18	15	18	24	25	19	23	20
Peppers and gamba	8	9	9	10	10	11	13	13
Beans and peas[a]	14	16	12	12	12	11	12	15
Melons, irrigated	18	10	11	14	8	15	17	12
Groundnuts	38	55	38	47	51	41	35	37
Other	27	30	29	35	39	39	39	39
Total	219	213	226	233	232	231	243	239

* Israel economic data are based on Hebrew calendar; year begins in early fall (September, usually) and extends for about twelve months. Data for 1957/1958 not available.

[a] Beans, lubia, horse beans, and peas.

Source: Various issues of *Statistical Abstract of Israel,* published by the Central Bureau of Statistics.

Appendix Table C8-20.
Israel: Average Yield on Main Field Crops, 1948/49 to 1966/67[a]*
(Metric Tons Per Hectare)

	1948/49	1949/50	1950/51	1951/52	1952/53	1953/54	1954/55	1955/56	1956/57	1957/58
Wheat	.75	1.05	.45	1.30	1.10	1.40	.90	1.50	1.70	1.35
Barley	1.40	1.10	.60	1.30	.85	1.25	.80	1.75	1.70	1.25
Maize										
unirrigated	1.10	1.15	.30	.75	1.20	1.40	.75	1.40	1.50	1.45
irrigated	–	2.90	2.50	2.80	2.90	3.30	4.30	4.50	5.10	5.45
Sorghum										
unirrigated	.60	.75	.25	.60	.80	.90	.85	1.50	1.55	1.90
irrigated	–	–	–	–	–	3.00	2.70	2.60	3.50	3.60
Hay	3.2	2.3	1.5	3.15	2.9	3.2	2.3	3.1	3.1	2.75
Green fodder										
unirrigated	9.5	9.0	6.0	12.5	8.5	13.0	10.5	15.5	25.0	14.0
irrigated	55	60	47	48	51	49	51	52	58	55.5
Sugar beets	–	–	–	7.6	12.05	16.95	36.8	35.9	41.8	40.6
Cotton, lint										
unirrigated	–	–	–	–	–	–	–	–	–	–
irrigated	–	–	–	–	–	.90	.95	.80	.85	.85

	1958/59	1959/60	1960/61	1961/62	1962/63	1963/64	1964/65	1965/66	1966/67
Wheat	1.50	.80	1.25	1.50	1.30	2.75	2.45	1.55	2.05
Barley	1.65	.65	1.40	1.35	.85	2.30	2.15	.9	2.40
Maize									
unirrigated	1.75	1.70	2.20	2.00	2.65	2.10	2.00	2.45	2.40
irrigated	5.80	5.80	5.55	5.00	4.70	4.50	4.70	4.25	4.15
Sorghum									
unirrigated	1.75	2.35	1.70	2.50	2.50	2.15	1.70	2.5	2.45
irrigated	3.80	4.85	4.80	5.45	5.55	5.25	5.10	5.6	5.30
Hay	2.55	1.7	2.9	3.0	2.65	3.85	3.65	2.4	3.75
Green fodder									
unirrigated	13.5	11.5	13.0	13.5	14.0	15.0	17.0	16.5	18.15
irrigated	60	64	67	67	68	73	72	69	70.55
Sugar beets	39.65	44.5	42.7	46.4	53.9	46.35	42.75	52	47.25
Cotton, lint									
unirrigated	–	–	.45	.45	.45	.50	.50	.45	.9
irrigated	1.05	1.00	1.00	1.10	1.10	1.25	1.30	1.2	1.8

* Israel economic data are based upon the Hebrew calendar; the year begins in early fall (September, usually) and extends for about twelve months.

[a] These are yields reported in cited source, for Jewish farming only; not calculated by present authors from area and production data.

Source: Various issues of *Statistical Abstract of Israel,* published by the Central Bureau of Statistics.

Appendix Table C8–21

Israel: Volume of Agricultural Production (Including Intermediate Products), 1948/49 to 1966/67[a*]

(1,000 Metric Tons, Except as Noted)

	1948/49	1949/50	1950/51	1951/52	1952/53	1953/54	1954/55	1955/56	1956/57	1957/58
Wheat	21	27	14	31	30	34	36	74	83	62
Barley	20	37	28	93	64	90	42	85	74	53
Sorghum for grain	3	4	1	6	14	20	11	26	38	34
Other cereals	8	17	4	10	13	24	27	25	40	32
Pulses	2	2	1	6	4	8	6	8	10	8
Hay	41	53	55	120	110	126	91	109	123	111
Green fodder and silage	373	491	498	569	642	672	866	940	1,218	1,301
Groundnuts	@	1	3	3	8	15	19	14	18	13
Sunflower, safflower, sesame	2	3	1	2	2	3	2	4	5	3
Cotton, lint	–	–	–	@	@	@	2	3	4	5
Cottonseed	–	–	–	@	@	@	4	5	7	8
Tobacco	1	1	2	3	2	3	2	1	2	1
Sugar beets	–	–	1	4	8	7	21	28	56	94
Melons and pumpkins	13	46	14	41	60	42	38	58	69	61
Vegetables	80	126	142	179	203	205	209	231	242	265
Potatoes	26	35	37	46	55	79	82	92	93	98
Citrus fruit	273	270	353	350	352	470	392	452	439	436
Table grapes	11	9	7	12	13	18	14	18	23	26
Wine grapes	7	7	6	8	10	13	11	14	21	22
Olives	11	4	3	14	14	22	3	25	7	18
Bananas	4	2	6	8	11	11	17	24	20	28
Deciduous fruit	5	6	4	7	6	11	7	12	17	20
Other fruit	2	5	6	6	5	5	6	9	10	11
Cow milk (million litres)	79	92	103	119	128	147	159	172	184	216
Other milk (million litres)	7	13	15	18	24	30	33	38	41	42
Eggs–millions	242	330	392	366	369	414	504	510	630	886
Beef and veal (live weight)	2	2	2	2	3	4	6	8	9	10
Sheep and goat meat (live weight)	1	@	1	1	1	2	3	3	4	5
Poultry meat (live weight)	5	7	7	7	8	9	16	23	22	34
Other meat (live weight)	–	–	–	–	–	–	2	2	3	4
Fish	4	6	7	8	8	9	11	11	11	12

Appendix Table C8-21—*continued*

	1958/59	1959/60	1960/61	1961/62	1962/63	1963/64	1964/65	1965/66	1966/67
Wheat	74	41	66	52	55	126	150	101	222
Barley	65	27	65	48	36	117	67	21	56
Sorghum for grain	41	16	34	43	40	80	67	13	24
Other cereals	17	10	9	10	8	7	4	2	4
Pulses	7	3	4	4	4	7	7	4	6
Hay	123	78	109	104	109	123	109	110	137
Green fodder and silage	1,534	1,596	1,559	1,616	1,756	1,714	1,538	1,423	1,400
Groundnuts	15	17	14	12	13	10	14	13	13
Sunflower, safflower, sesame	4	4	6	3	3	4	6	4	3
Cotton, lint	7	11	14	16	13	16	22	25	28
Cottonseed	12	17	23	26	22	24	35	40	48
Tobacco	2	2	2	2	@	1	2	2	2
Sugar beets	122	169	245	221	250	256	295	282	239
Melons and pumpkins	83	52	70	88	93	92	102	84	92
Vegetables	270	296	277	280	297	317	307	344	342
Potatoes	88	82	85	111	109	107	109	104	93
Citrus fruit	588	610	516	532	736	839	878	905	1,082
Table grapes	32	31	37	42	34	49	46	43	42
Wine grapes	26	22	27	30	26	34	34	33	40
Olives	8	7	21	5	13	21	10	11	24
Bananas	32	34	44	49	51	50	44	54	50
Deciduous fruit	30	38	52	72	68	105	101	117	134
Other fruit	12	14	17	16	17	14	14	16	18
Cow milk (million litres)	260	277	284	316	297	305	323	349	380
Other milk (million litres)	41	40	44	43	43	44	45	45	48
Eggs—millions	1,027	1,114	1,270	1,273	1,113	1,278	1,296	1,233	1,402
Beef and veal (live weight)	19	25	23	22	31	36	31	28	30
Sheep and goat meat (live weight)	5	5	5	6	6	7	7	7	8
Poultry meat (live weight)	41	46	55	66	67	74	74	82	89
Other meat (live weight)	6	6	6	6	6	4	5	5	6
Fish	13	14	15	16	16	19	19	23	23

* Israel economic data are based upon the Hebrew calendar; the year begins in early fall (September, usually) and extends for about twelve months.

[a] Excludes some minor products.

@ = Less than 500.

Source: Various issues of *Statistical Abstract of Israel,* published by the Central Bureau of Statistics.

Lebanon: Areas Under Crops, 1956–67

(1000 Hectares)

Grains	1956	1957	1958	1959	1960	1961	1962	1963	1964	1965	1966	1967
Wheat	70.0	70.0	68.0	72.2	65.0	68.8	68.6	55.0	70.0	76.4	68.1	66.5
Sorghum	5.5	5.0	3.5	3.0	2.0	1.5	1.2	0.8	1.2	1.4	1.4	1.3
Maize	8.0	10.0	8.0	11.0	6.5	6.0	7.0	6.0	5.5	4.9	4.8	3.0
Barley	19.0	21.0	19.0	18.0	18.0	12.8	13.5	10.0	15.0	13.5	14.1	13.4
Total	102.5	106.0	98.5	104.2	91.5	89.1	90.3	71.8	91.7	96.2	88.4	84.1
Legumes												
Drybeans	2.5	2.6	2.4	2.7	2.5	2.0	1.5	2.0	2.2	2.5	2.4	1.1
Horsebeans	2.0	2.0	2.0	2.2	2.0	2.0	1.2	1.0	1.3	1.5	1.5	0.6
Lentils	2.1	2.5	2.3	2.2	2.0	1.6	1.6	1.6	2.8	2.8	4.6	2.9
Chickpeas	1.8	2.1	2.3	2.5	2.0	1.8	1.2	1.0	1.5	1.6	3.3	2.9
Vetch[a]	3.2	3.0	2.8	2.6	2.2	2.0	1.4	1.3	2.0	2.7	2.4	5.4
Others[a]	2.6	2.9	3.8	3.5	3.0	2.5	2.1	1.8	2.9	3.2	3.4	0.9
Total	14.2	15.1	15.6	15.7	13.7	11.9	9.0	8.7	12.7	14.3	17.6	13.8
Vegetables												
Potatoes	4.5	5.0	4.7	4.8	4.5	6.3	6.0	6.5	8.0	4.5	6.2	6.9
Onions	2.5	2.3	2.5	2.4	2.1	1.2	2.2	2.5	1.0	2.2	2.6	3.0
Watermelons	1.5	1.7	2.8	2.6	2.5	2.5	2.7	3.0	3.2	3.2	2.7	2.0
Squash	–	–	–	–	–	–	1.1	1.2	1.4	1.2	1.1	1.0
Cucumbers	1.0	1.3	1.8	2.0	2.1	2.0	2.2	3.5	3.7	3.2	3.2	2.6
Tomatoes	1.5	1.7	2.3	2.5	2.0	2.3	2.4	2.4	2.9	4.3	4.6	4.5
Others	6.2	6.4	6.7	6.6	5.9	5.9	6.4	7.4	12.6	11.9	11.2	10.5
Total	17.2	18.4	20.8	20.9	19.1	20.2	23.0	26.5	32.8	30.5	31.6	30.5
Fruits												
Citrus	7.8	7.9	8.0	8.0	8.2	8.5	8.5	9.0	10.0	12.2	12.3	11.0
Apples	7.4	8.2	9.1	9.5	10.0	11.3	11.3	11.5	11.4	11.0	11.0	10.9
Figs	2.5	2.5	2.5	2.5	2.6	2.8	3.9	3.9	3.5	3.1	1.9	1.5
Almonds	1.4	1.4	1.4	1.4	1.5	1.7	2.4	2.4	1.1	1.9	2.0	5.2
Bananas	2.0	2.1	2.2	2.3	2.4	2.6	2.6	3.0	3.4	2.5	2.5	2.5
Grapes[a]	22.0	22.2	22.5	23.0	24.0	24.2	24.5	25.5	23.8	18.1	18.7	15.2
Olives[a]	21.5	22.0	22.5	22.6	27.0	27.0	27.0	27.5	28.6	27.3	26.0	26.8
Others	8.1	8.4	8.6	8.8	9.0	8.9	8.6	6.4	4.3	3.9	3.9	1.5
Total	72.7	74.7	76.8	78.1	84.7	87.0	88.8	89.2	86.1	80.0	78.3	74.6
Industrial Crops												
Sugar cane	0.3	0.2	0.2	0.1	0.1	0.1	0.1	0.1	0.1	–	–	–
Sugar beets	0.6	0.6	0.6	0.9	0.8	0.8	1.0	0.9	1.8	1.5	2.2	2.1
Tobacco	3.2	4.0	4.0	4.0	4.0	4.0	4.2	4.7	6.8	6.6	6.6	6.8
Ground nut	1.6	1.5	1.4	1.2	0.6	1.5	2.5	3.0	3.0	2.4	2.5	2.5
Others	2.3	1.6	0.9	0.9	0.6	0.4	0.5	0.4	9.2	6.2	6.3	1.8
Total	8.0	7.9	7.1	7.1	6.1	6.8	8.3	9.1	20.9	16.7	17.6	13.2
All Crops												
Grand Total	214.6	222.1	218.8	226.0	215.1	215.0	219.4	205.3	244.2	225.9	223.5	216.2

[a]Area under vetch in 1967 includes more than one variety of vetches that were included with "other legumes" in previous years. Area of "other legumes" in 1967 is unreasonable if compared with production of other legumes in 1967. Figures for grapes and olives in 1965–67 are revised on basis of new surveys for these crops. All figures are rounded.

Source: 1956–63 figures: Lebanon, Central Bureau of Statistics, *Statistical Abstract 1963* (Beirut, n.d.); 1964–67 figures: Lebanon, Ministry of Agriculture, *Agricultural Statistics* for 1964, 1965, 1966, 1967. Also, Lebanon Ministry of Agriculture, *Agricultural Statistics 1954–66* (Beirut, March 1968).

Appendix Table C8-23
Lebanon: Crop Yields, 1956—67
(Metric Tons Per Hectare)

Crop	1956	1957	1958	1959	1960	1961	1962	1963	1964	1965	1966	1967
Grains												
Wheat	0.9	1.0	0.7	0.9	0.6	1.0	0.9	1.1	0.8	0.7	1.0	1.0
Sorghum	1.0	1.0	1.0	1.1	1.2	1.1	1.1	1.2	1.2	1.6	1.0	0.9
Maize	1.8	1.7	1.8	1.1	1.5	2.1	2.1	2.0	2.0	1.8	1.9	1.8
Barley	1.1	1.1	1.0	1.4	0.6	0.9	1.0	1.0	1.0	0.9	0.9	1.2
Total	1.0	1.1	0.8	1.0	0.7	1.1	1.2	1.2	0.9	0.8	1.1	1.1
Legumes												
Dry beans	2.0	2.0	2.0	2.0	1.6	1.5	2.3	1.4	1.4	1.4	1.4	1.4
Horse beans	0.8	0.7	0.8	0.8	0.8	0.7	1.3	0.9	0.9	1.0	1.1	1.3
Lentils	1.0	1.1	1.1	0.7	0.5	0.7	0.6	1.2	1.2	1.2	0.4	1.0
Chickpeas	1.1	1.2	1.0	1.6	1.0	0.4	1.5	0.9	1.0	0.8	0.7	1.2
Vetches	0.7	0.7	0.7	0.7	0.7	0.8	0.7	0.9	0.9	0.8	0.6	1.0
Others	1.2	1.5	1.5	1.5	1.4	1.6	2.2	4.2	3.0	5.9	5.1	—
Total	1.1	1.2	1.2	1.3	1.0	1.0	1.5	1.8	1.5	2.3	1.6	2.1
Vegetables												
Potatoes	8.0	5.6	8.1	8.8	6.2	8.0	10.0	10.8	10.0	12.0	12.7	11.8
Onions	16.0	13.5	9.0	13.8	11.0	20.8	18.6	13.2	12.0	11.5	12.4	14.6
Watermelons	16.0	17.1	9.3	10.4	8.8	10.0	12.9	7.0	8 5	10.3	5.7	8.9
Squash	12.0	12.9	12.4	13.2	13.0	14.2	16.4	11.7	11.6	11.6	6.8	8.1
Cucumbers	9.5	10.0	10.0	10.5	10.1	10.0	11.4	5.7	5.1	5.5	6.4	8.6
Tomatoes	15.3	14.7	12.2	12.4	12.5	13.0	13.3	14.2	14.2	10.6	11.6	13.2
Others	12.9	12.7	12.9	13.1	13.0	13.8	13.9	12.0	12.3	15.4	12.3	11.0
Total	12.3	11.3	11.2	11.5	10.4	11.5	13.0	10.6	11.0	12.1	11.0	11.4
Fruits												
Citrus	13.9	14.7	16.5	20.0	18.9	23.6	23.5	23.3	21.0	19.0	20.3	20.7
Apples	3.9	4.5	4.6	6.8	5.3	7.5	7.1	6.5	10.9	10.5	9.5	14.4
Figs	7.6	7.8	7.2	6.8	5.4	6.1	6.7	6.4	6.9	4.7	6.2	8.9
Almonds	1.9	1.9	1.8	1.9	1.3	1.5	1.2	1.2	1.6	1.3	1.2	1.1
Bananas	13.5	12.4	12.8	9.1	11.1	10.0	9.6	9.3	6.6	10.0	11.7	10.9
Grapes	3.6	3.4	5.6	3.7	2.9	3.7	3.5	3.6	4.2	4.6	4.1	5.8
Olives	2.6	0.6	2.0	0.8	1.1	2.4	1.7	2.2	1.1	1.9	1.8	2.5
Others	6.4	5.8	5.4	5.9	4.7	6.7	5.6	7.3	13.8	11.9	9.7	—
Total	5.1	4.5	5.1	5.4	4.6	6.3	5.4	6.0	6.8	7.0	6.9	8.7
Industrial												
Sugar Cane	33.5	33.2	33.3	20.0	33.8	40.0	40.0	40.0	40.0	33.6	33.5	25.3
Sugar Beets	16.7	16.4	14.7	26.4	30.3	31.2	34.4	39.8	41.0	45.9	52.3	52.4
Tobacco	0.8	0.9	0.8	0.9	0.8	1.0	1.0	0.9	0.9	0.9	0.9	0.9
Groundnut	1.2	1.3	1.4	1.4	1.3	1.3	1.4	1.7	1.7	2.2	1.6	1.3

Source: Lebanon, Ministry of Agriculture, *Agricultural Statistics, 1967* (Beirut, May 1968), pp. 12–20. All figures are rounded.

Appendix Table C8-24
Lebanon: Crop Production, 1956–67
(1000 Metric Tons)

Crop	1956	1957	1958	1959	1960	1961	1962	1963	1964	1965	1966	1967
Grains												
Wheat	62.0	70.0	48.0	67.2	40.0	68.5	75.2	60.0	59.5	55.0	70.0	67.8
Sorghum	5.6	5.0	3.5	3.2	2.3	1.6	1.3	0.9	1.4	2.2	1.4	1.2
Maize	14.2	17.0	14.0	12.0	13.0	12.5	15.0	12.0	11.0	9.1	8.8	5.2
Barley	21.0	23.5	18.5	25.0	11.0	12.0	13.1	10.0	15.0	12.6	12.7	15.8
Total	102.8	115.5	84.0	107.4	66.3	94.6	104.6	82.9	86.9	78.9	92.9	90.0
Legumes												
Dry beans	5.0	5.3	4.8	5.5	4.0	3.0	3.5	2.8	3.0	3.4	3.4	1.6
Horsebeans	1.6	1.5	1.7	1.8	1.7	1.5	1.6	0.9	1.2	4.5	1.6	0.8
Lentils	2.2	2.8	2.5	1.6	1.0	1.1	1.0	2.0	3.4	3.2	1.8	2.8
Chickpeas	2.0	2.5	2.2	3.9	2.0	0.8	1.8	0.9	1.4	1.4	2.4	3.3
Vetch	2.3	2.1	1.9	1.8	1.5	1.5	1.0	1.3	1.8	2.2	1.4	5.1
Others	3.1	4.3	5.6	5.4	4.1	3.9	4.7	7.5	8.7	18.9	17.5	15.6
Total	16.2	18.5	18.7	20.0	14.3	11.8	13.6	15.4	19.5	33.6	28.1	29.2
Vegetables												
Potatoes	36.0	28.0	38.0	42.0	28.0	50.0	60.0	70.0	80.0	53.6	79.1	80.9
Onions	40.0	31.0	36.0	33.0	23.0	25.0	41.0	33.0	24.0	25.9	32.4	44.0
Watermelons	24.0	29.0	26.0	27.0	22.0	25.0	35.0	21.0	27.3	32.5	15.2	18.7
Squash	–	–	–	–	–	–	18.0	14.0	15.7	13.5	7.4	8.0
Cucumbers	9.5	13.0	18.0	21.0	23.0	20.0	25.0	20.0	18.9	16.5	20.6	21.9
Tomatoes	23.0	25.0	28.0	31.0	25.0	30.0	32.0	34.0	41.8	45.3	53.8	59.2
Others	79.8	81.4	86.4	86.3	76.7	81.6	89.2	88.7	154.8	183.1	137.8	115.3
Total	212.3	207.4	232.4	240.3	197.7	231.6	300.2	280.7	362.5	370.4	346.3	348.0
Fruits												
Citrus	108.5	116.0	131.0	160.0	155.0	200.0	200.0	225.0	225.0	231.6	249.9	228.1
Apples	29.0	37.0	42.0	65.0	53.0	85.0	80.0	75.0	125.0	115.0	104.0	157.0
Figs	19.0	19.5	18.0	17.0	14.0	17.0	25.0	25.0	24.0	14.7	11.7	13.4
Almonds	2.7	2.8	2.6	2.6	2.0	2.5	2.8	3.0	1.8	2.4	2.4	5.9
Bananas	27.0	26.0	27.5	21.0	26.0	26.0	25.0	28.0	22.0	19.5	29.8	27.2
Grapes	80.0	75.0	80.0	85.0	70.0	90.0	85.0	90.0	100.0	83.8	76.0	88.3
Olives	55.0	13.0	44.0	18.0	30.0	65.0	16.0	60.0	30.0	49.0	29.4	67.8
Others	51.5	48.5	46.4	52.0	42.6	59.7	48.1	47.0	59.5	46.3	37.9	62.0
Total	372.7	338.0	391.5	420.6	392.6	545.2	482.0	553.0	587.3	562.4	541.1	649.7
Industrial												
Sugar cane	8.7	8.3	8.0	2.0	2.7	2.8	3.0	2.0	2.0	0.8	0.8	0.2
Sugar beets	10.0	9.0	8.1	12.5	18.0	21.9	30.0	31.0	72.7	68.8	115.0	110.0
Tobacco	2.4	3.5	3.0	3.7	3.3	4.0	4.2	4.4	6.0	5.8	6.3	6.4
Groundnuts	2.0	2.0	1.9	1.8	0.9	2.0	3.5	5.2	5.0	5.2	4.1	3.2
Others	1.8	1.2	0.9	0.9	0.5	0.4	0.6	0.4	6.6	4.8	5.0	0.8
Total	24.9	24.0	21.9	20.9	25.4	31.1	41.3	43.0	92.3	85.4	131.2	120.6
All crops												
Grand total	729.1	703.4	748.7	809.4	696.5	914.5	941.8	975.1	1,148.5	1,130.7	1,139.6	1,237.5

Source: 1956–63 figures: Lebanon, Central Bureau of Statistics, *Statistical Abstract 1963* (Beirut, n.d.); 1964–67 figures: Lebanon, Ministry of Agriculture, *Agricultural Statistics* for 1964, 1965, 1966, 1967. Also, Lebanon, Ministry of Agriculture, *Agricultural Statistics 1954–66* (Beirut, March 1968). All figures are rounded.

Appendix Table C8-25
Lebanon: Area, Yields, Production, and Value of Vegetables, 1967

Crop	Area (hectares)	Yield (kg. per ha.)	Production (metric tons)	Average farm price (£L per metric ton)	Total value of production (£L 000's)	Gross return (£L per ha.)
Green peas	370	3,750	1,389	340	473	1,278
Green broad beans	1,260	8,290	10,448	290	3,068	2,435
Colocase	52	10,000	520	400	208	4,000
Watermelon	2,015	8,860	18,737	150	2,821	1,400
Melon	236	7,940	1,869	180	329	1,394
Eggplant	1,526	13,010	19,848	220	4,464	2,925
Okra	725	3,000	2,173	540	1,170	1,614
Onions	3,018	14,570	43,978	125	5,520	1,829
Potatoes	6,889	11,750	80,945	190	15,289	2,219
Radishes	190	18,960	3,603	120	435	2,289
Tomatoes	4,489	13,180	59,176	190	11,090	2,470
Garlic	569	6,540	3,723	660	2,442	4,292
Carrots	567	22,040	12,488	150	1,821	3,212
Lettuce	638	20,920	13,340	130	1,682	2,636
Cucumbers	2,550	8,590	21,904	266	5,820	2,282
Spinach	176	9,730	1,716	300	515	2,926
Chenpodia	172	16,040	2,762	120	326	1,895
Artichokes	45	16,000	720	550	396	8,800
Squash	988	8,120	8,025	200	1,600	1,619
Turnips	159	17,030	2,704	150	402	2,528
Green beans	1,444	4,340	6,260	390	2,412	1,670
Cabbage and cauliflower	2,049	14,050	28,796	190	5,351	2,612
Others	351	8,120	2,850	390	1,125	3,205
Total vegetables	30,476		347,974		68,759	2,256

Source: Lebanon, Ministry of Agriculture, Statistics Section, *Agricultural Statistics 1967* (A. Ec. No. 11, May 1968), pp. 7–11. See note of Appendix Table C8-27.

Appendix Table C8-26

Lebanon: Area, Number, and Production of Fruit Trees, 1967

Crop	(1) Total area under fruits (hectares)	(2) Productive area under fruits (hectares)	(3) Nonproductive area under fruits[a] (hectares)	(4) Number of scattered fruit trees (1000)	(5) Average yield of fruits[b] (tons per ha.)	(6) Average yield of scattered trees (kg./tree)	(7) Production of fruits[c] (metric tons)	(8) Average farm price (£L/ton)	(9) Total value (£L 1000)	(10) Gross return per hectare of productive area (£L)[d]
Pears	944	687	257	181	10.7	22	11,272	410	4,669	4,370
Loquats	184	170	13	84	12.1	30	4,605	410	1,903	4,969
Oranges	7,869	6,692	1,177	76	22.6	33	153,848	240	37,172	5,426
Tangerines	1,064	976	88	32	13.1	30	13,775	360	4,926	4,723
Grapefruits	130	115	16	5	28.7	56	3,601	290	1,057	8,326
Lemons	1,965	1,747	218	76	30.6	46	56,910	240	13,829	7,334
Total citrus	11,028	9,530	1,498	190			228,134	–	56,984	–
Apples	10,850	9,793	1,057	83	15.8	29	157,021	180	27,869	2,842
Stone Pine	–	–	–	–	–	–	235	7,000	1,645	–
Figs	1,507	1,115	392	141	7.6	35	13,447	240	3,191	1,822
Walnuts	68	8	60	28	4.2	26	757	1,410	1,071	5,880
Plums	364	318	47	91	8.9	36	6,576	300	1,982	2,673
Peaches	727	621	106	87	11.8	31	10,041	285	2,862	3,374
Pomegranates	277	217	60	97	9.1	27	4,562	280	1,298	2,548
Olives	26,837	22,162	4,675	102	3.0	16	67,773	510	34,618	1,530
Quinces	221	90	130	27	10.3	21	1,520	285	434	2,944
Grapes	15,222	12,689	2,533	323	5.8	44	88,321	220	19,518	1,287
Strawberries	96	96	–	–	4.6	–	445	1,170	521	5,417
Cherries	1,842	993	849	79	8.8	25	10,010	655	6,552	5,777
Almonds	515	436	79	140	6.1	23	5,948	960	5,723	5,856
Apricots	1,186	668	518	52	9.3	37	8,120	300	2,471	2,781
Bananas	2,457	2,454	3	23	11.0	13	27,157	540	14,615	5,913
Others	265	225	40	28	10.2	52	3,781	590	2,227	6,012
Total fruits	74,591	62,272	12,319	1,758	–	–	649,725	–	190,153	–

[a]Representing area of fruits not yet of productive age.
[b]Average yields are calculated over productive areas only.
[c]Include production from productive areas plus production from scattered fruit trees.
[d]This column is a product of columns (5) and (8).

Source: Lebanon, Ministry of Agriculture, *Agricultural Statistics 1967* (Beirut, May 1968), pp. 12–20.

Appendix Table C8-27
Lebanon: Areas and Production of Grains, Legumes, and Industrial Crops, 1967

Crop	Area (hectares)	Average yield (kg. per ha.)	Production (metric tons)	Average farm price (£L per metric ton)	Total value of production (£L 1000)	Gross return (£L per ha.)
Grains						
Wheat	66,522	1,020	67,852	270	18,320	275
Barley	13,396	1,180	15,807	210	3,319	248
Maize	2,950	1,760	5,192	290	1,506	510
Sorghum	1,267	930	1,178	250	294	231
Total	84,135		90,029		23,439	279
Legumes						
Vetches	5,453	940	5,126	260	1,333	244
Dry broadbeans	631	1,330	839	350	294	466
Alfalfa	246	59,187	14,560	70	1,019	4,142
Dry peas	370	2,176	805	410	330	892
Lentils	2,858	972	2,779	680	1,890	661
Chickpeas	2,885	1,150	3,311	440	1,457	505
Dry beans	1,117	1,450	1,620	770	1,247	1,116
Lupins	203	910	185	480	89	438
Total	13,763		29,225		7,659	556
Industrial crops						
Groundnut	2,528	1,280	3,239	470	1,524	603
Sesame	146	750	109	860	93	637
Anis	30	900	27	1,100	30	1,000
Sugar beets	2,100	52,380	110,000	60	6,600	3,143
Silk cocoon	–	–	96.3	6,000	5,778	–
Sugarcane	8	25,320	195	200	40	5,000
Sunflower	1,494	350	530	750	398	266
Tobacco	6,778	940	6,378	4,360	27,818	4,104
Flowers	128				960	
Total	13,212		120,574		43,241	7,500

Source: Lebanon, Ministry of Agriculture, Statistics Section, *Agricultural Statistics 1967* (A. Ec. No. 11, May 1968), pp. 21–25.

Note: The above estimates are based on a stratified random sample of 532 villages in the four provinces of Lebanon. The sample was selected from two strata in which all the villages of Lebanon were divided. The first stratum included all the important agricultural villages which had reported substantial area under one or more crops in 1965. These villages were all taken in the sample. The remaining villages constituted the second stratum, from which 10% were selected at random. Yield figures are weighted averages, the weights being the cropped areas in the four provinces.

Appendix Table C8-28

Lebanon: Area and Number of Trees of Fruit Crops, 1961

Crop	Area (dunums)	Number of trees (1000)			Number of trees per dunum	Productive trees	
		Total	Scattered	Compact		Number (1000)	% of total trees
(1)	(2)	(3)	(4)	(5)	(6)	(7)	(8)
Citrus	76,399	3,254	13	3,241	43	1,965	60
Apples	103,505	5,087	18	5,069	49	3,858	76
Pears	6,146	367	1	366	60	284	77
Quinces	2,140	77	1	76	36	69	89
Loquats	2,234	91	3	88	39	53	59
Apricots	6,793	178	8	170	25	124	70
Peaches	3,615	242	1	241	67	225	93
Prunes	3,547	252	4	248	70	239	95
Figs	28,707	884	27	868	30	696	78
Pomegranates	1,618	72	9	63	39	70	97
Almonds	9,255	436	11	425	46	319	73
Bananas	30,368	3,331	24	3,307	109	2,951	89
Grapes	126,334	13,443	69	13,373	106	10,782	80
Olives	144,715	3,246	46	3,200	22	2,621	81
Others	11,980	1,325	9	1,316	110	1,159	87
Total	557,356	32,295	245	32,050	57	25,415	78

Source: Lebanon, Ministry of Agriculture, *Census of Agriculture 1961* (Beirut, March 1966).

Appendix Table C8-29
Jordan: Area Under Principal Grains, Vegetables, and Fruits, 1960–66
(1000 Dunums)

Crop	1960	1961	1962	1963	1964	1965	1966
Wheat	2,510.3	2,732.3	2,848.1	2,057.2	2,966.7	2,788.7	2,138.9
Barley	746.6	949.7	1,050.0	757.3	914.7	859.5	645.2
Lentils	138.6	165.5	208.0	157.2	258.8	286.5	194.8
Vetch (Kersenneh)	148.6	160.7	188.7	148.8	164.0	246.4	144.2
Dry broad beans	26.0	28.8	32.0	22.2	30.6	39.0	35.8
Chick peas	40.9	54.1	56.6	31.1	56.7	85.0	34.7
Sorghum	108.7	95.1	107.4	75.9	81.9	69.5	51.7
Sesame	68.7	51.7	50.8	45.1	35.7	35.2	22.0
Other grains	6.7	9.2	14.5	20.0	21.4	9.9	12.3
Total grains[a]	3,794.1	4,247.1	4,506.1	3,314.8	4,530.5	4,419.7	3,279.6
Tomatoes	139.3	190.4	210.1	205.3	239.2	205.9	167.7
Eggplants	30.7	36.3	38.2	37.7	34.9	38.2	33.1
Onions & garlic	25.2	28.7	33.7	32.6	34.2	31.2	23.6
Cauliflower & cabbages	24.3	26.2	21.5	21.4	25.8	26.0	14.5
Watermelons & melons	142.0	223.3	225.6	128.9	175.1	217.5	91.5
Cucumbers				71.3	94.4	86.9	68.2
Green broad beans	17.6	19.3	17.0	20.3	24.9	23.0	16.5
Potatoes	16.8	17.1	17.4	15.7	16.1	16.4	18.8
Radishes, turnips & carrots	3.8	4.0	3.9	3.3	5.0	5.7	3.2
Other vegetables	27.2	33.3	42.9	35.6	41.6	38.9	26.6
Total vegetables	427.0	578.6	610.3	572.1	691.2	689.7	463.7
Olives	–	–	–	–	562.6	601.9	602.0
Apples & pears	–	14.7	15.3	17.7	19.1	19.6	19.4
Plums & peaches	–	9.9	10.9	11.6	13.3	13.2	14.5
Almonds	–	19.9	21.9	23.6	26.3	30.1	30.5
Apricots	–	6.5	6.7	6.7	6.2	6.7	6.7
Figs	–	80.2	81.2	75.4	80.0	76.0	72.2
Bananas	–	8.8	9.3	7.3	5.7	10.5	10.8
Citrus fruits	–	16.5	18.6	18.0	18.9	23.8	25.1
Pomegranates	–	6.8	6.4	6.1	6.0	6.2	5.1
Vines	–	199.5	197.3	193.2	194.5	198.0	200.6
Other fruit trees	–	5.2	6.0	5.3	5.3	5.8	6.0
Total fruit trees		368.0	373.6	364.9	937.9	991.8	992.9
Total area of crops	4,221.1	5,193.7	5,490.0	4,251.8	6,159.6	6,101.2	4,736.2

[a]Includes winter and summer crops, whether irrigated or not.

Source: Jordan, Department of Statistics, *Statistical Yearbook, 1966* (No. 17 [Amman]), pp. 124, 127.

Appendix Table C8-30

Jordan: Yield Per Dunum of Principal Grains, Vegetables, and Fruits, 1960–66[a]

(kilograms)

Crop	1960	1961	1962	1963	1964	1965	1966
Grains							
Wheat	17.4	50.6	39.3	36.9	99.3	99.7	47.3
Barley	17.8	65.0	34.0	30.2	106.3	110.3	35.3
Lentils	13.0	38.7	62.5	26.0	97.0	101.6	56.5
Vetch (kersenneh)	23.6	39.8	42.5	24.2	80.5	63.7	47.5
Dry broad beans	19.2	41.7	46.9	63.1	68.6	69.2	55.8
Chick peas	26.9	70.2	53.0	46.7	74.1	76.5	42.7
Sorghum	27.6	85.2	49.3	44.3	106.2	119.4	50.2
Sesame	26.2	42.6	37.4	36.6	47.6	39.8	36.4
Vegetables							
Tomatoes	1,117.0	1,122.9	806.3	1,045.3	952.3	917.4	862.3
Eggplants	1,677.5	1,427.0	1,377.0	1,273.2	1,194.8	1,246.0	1,522.6
Onions & garlic	646.8	508.7	537.1	582.4	617.0	602.6	919.5
Cauliflower & cabbages	1,177.0	1,244.3	1,386.0	1,275.4	1,391.5	1,480.8	1,489.6
Watermelons & melons ⎱	638.7	730.0	733.6	933.6	912.6	736.1	521.3
Cucumbers ⎰				634.5	681.1	698.5	580.6
Green broad beans	812.5	848.7	764.7	821.9	650.6	600.0	648.9
Potatoes	1,000.0	789.5	626.4	838.5	677.0	963.4	1,191.4
Radishes, turnips & carrots	1,184.2	1,150.0	1,051.3	1,471.4	1,180.0	1,386.0	1,500.0
Fruit trees							
Olives	–	–	–	–	172.6	62.1	54.3
Apples and pears	–	224.5	268.0	203.4	288.0	260.2	350.5
Plums and peaches	–	333.3	348.6	293.1	300.7	310.6	324.1
Almonds	–	110.5	132.4	110.2	125.5	123.0	114.7
Apricots	–	400.0	283.6	358.2	338.7	373.1	358.2
Figs	–	260.6	259.8	250.7	263.7	253.9	227.1
Bananas	–	1,568.2	1,849.5	1,315.1	1,438.6	1,466.7	1,555.5
Citrus fruits	–	987.9	1,150.5	2,155.5	1,962.9	1,974.8	2,278.8
Pomegranates	–	602.9	593.7	491.8	416.7	387.1	313.7
Grapes	–	393.0	400.4	303.8	395.4	400.0	308.6

[a]Including winter and summer crops whether irrigated or not.

Source: Jordan, Department of Statistics, *Statistical Yearbook, 1966* (No. 17 [Amman]), pp. 126, 127.

Appendix Table C8-31
Jordan: Production of Principal Grains, Vegetables, and Fruits, 1960–66[a]
(1000 tons)

Crop	1960	1961	1962	1963	1964	1965	1966
Grains							
Wheat	43.6	138.2	111.9	75.8	294.7	277.9	101.1
Barley	13.3	61.7	35.7	23.0	97.2	94.8	22.8
Lentils	1.8	6.4	13.0	4.1	25.1	29.1	11.0
Vetch (kersenneh)	3.5	6.4	8.0	3.6	13.2	15.7	6.9
Dry broad beans	0.5	1.2	1.5	1.4	2.1	2.7	2.0
Chick peas	1.1	3.8	3.0	1.5	4.2	6.5	1.5
Sorghum	3.0	8.1	5.3	3.4	8.7	8.3	2.6
Sesame	1.8	2.2	1.9	1.7	1.7	1.4	0.8
Other grains	0.2	0.4	1.2	2.1	2.3	1.4	1.0
Vegetables							
Tomatoes	155.6	213.8	169.4	214.6	227.8	188.9	144.6
Eggplants	51.5	51.8	52.6	48.0	41.7	47.6	50.4
Onions and garlic	16.3	14.6	18.1	19.0	21.1	18.8	21.7
Cauliflower & cabbages	28.6	32.6	29.8	27.3	35.9	38.5	21.6
Watermelons and melons	90.7	163.0	165.5	120.4	159.8	160.1	47.7
Cucumbers				45.3	64.3	60.7	39.6
Green broad beans	14.3	16.4	13.0	16.7	16.2	13.8	10.7
Potatoes	16.8	13.5	10.9	12.4	10.9	15.8	22.4
Radishes, turnips & carrots	4.5	4.6	4.1	4.9	5.9	7.9	4.8
Other vegetables	20.3	22.6	31.1	28.9	28.1	30.4	19.0
Fruit trees							
Olives	–	114.4	7.4	38.7	97.1	37.4	32.7
Apples and pears	–	3.3	4.1	3.6	5.5	5.1	6.8
Plums and peaches	–	3.3	3.8	3.4	4.0	4.1	4.7
Almonds	–	2.2	2.9	2.6	3.3	3.7	3.5
Apricots	–	2.6	1.9	2.4	2.1	2.5	2.4
Figs	–	20.9	21.1	18.9	21.1	19.3	16.4
Bananas	–	13.8	17.2	9.6	8.2	15.4	16.8
Citrus fruits	–	16.3	21.4	38.8	37.1	47.0	57.2
Pomegranates	–	4.1	3.8	3.0	2.5	2.4	1.6
Grapes	–	78.4	79.0	58.7	76.9	79.2	61.9
Other fruits	–	2.4	2.8	3.3	2.1	2.7	2.4

[a] Including winter and summer crops whether irrigated or not.

Source: Jordan, Department of Statistics, *Statistical Yearbook, 1966* (No. 17 [Amman]), pp. 125, 127.

Appendix Table C8-32

Jordan: Vegetable Yields, 1963/64

(Kilograms Per Dunum)*

Crop	National average[a]	East Ghor[b]	West Ghor[b]
Tomatoes	1,000	1,500	4,000
Eggplants	1,230	1,650	4,000
Cauliflower	1,330[c]	1,650	3,500
Cucumber	–	750	1,250
Marrow	–	600	1,500
Green pepper	–	900	1,500
Onions	600	1,500	2,500
Lettuce	–	1,000	1,600
Okra	–	500	800
Potato	740	600	2,000
Cabbage	–	2,000	3,400

* Ten dunums equal one hectare.

[a] Table C8-30.

[b] Estimated by Agriculture Section of U. S. Embassy, Amman, June 1968.

[c] Includes cabbage.

Source: Agriculture Section of the U.S. Embassy in Amman, June 1968.

Appendix Table C8-33

Jordan: Index of Average Rainfall and Indices of Yields of Major Crops, 1954–66

(Average 1954–66 = 100)

Year	Index of rainfall (1)	(2)	Wheat	Barley	Lentils	Tomatoes	Watermelons & cucumbers	Olives	Grapes	Citrus	Bananas
1954	116		135	141	130	103	127	132	130	84	81
1955	66		47	35	33	98	86	25	81	90	118
1956	118		118	124	138	99	86	143	89	125	101
1957	112		124	120	75	113	88	21	112	94	56
1958	85		35	23	27	140	85	94	95	84	91
1959	79		63	46	32	159	104	19	116	105	94
1960	51	47	28	25	20	216	109	26	91	76	68
1961	96	89	80	92	58	215	124	171	154	107	112
1962	93	95	62	48	94	155	125	11	157	125	132
1963	56	64	59	43	39	201	133	53	119	234	94
1964		128	157	150	146	183	135	128	155	213	102
1965		132	158	140	152	177	122	48	157	214	104
1966		81	56	50	91	166	94	40	121	247	111

(1) Average 1953/54 to 1955/56 seasons - 100. The average seasonal rainfall is for all reading stations.

(2) Average for 17 main stations. The index is linked to the first series by using the ratio of 1959/60–1962/63 averages in the two series as a factor for inflating the second series.

Source: FAO, Mediterranean Development Project, *Jordan Country Report* (Rome, 1967), p. 52.

Appendix Table C8-34
Syria: Production of Cereals, Legumes, and Industrial Crops, 1967

Crops	Area (1000 hectares)	Average yield (kg./ha.)	Production (1000 metric tons)	Average farm price in 1965 (£S./ton)	Estimated value of production (£S. million)	Gross return (£S./ha.)
Cereals						
Wheat	1,201	874	1,049	185	194.1	162
Barley	646	913	590	145	85.5	132
Maize	6	1,611	9	250	2.2	404
Millet	39	1,016	40	109	4.3	111
Oats	3	836	2	150	0.3	128
Rice	1	2,329	2	333	0.7	777
Total	1,896	893	1,692	170	287.1	151
Legumes						
Lentils	77	1,086	84	218	18.2	237
Chick peas	52	1,233	64	334	21.2	412
Beans	12	1,197	14	305	4.3	365
Peas	1	637	[a]	643	0.3	402
Rambling vetch	26	1,027	26	310	8.2	318
Bitter vetch	20	1,091	22	233	5.1	254
Flowering sern	5	1,252	6	238	1.4	299
Haricot beans	1	1,629	2	860	2.0	1,413
Total	193	1,131	218	278	60.7	314
Industrial						
Cotton	239	1,374	329	720[b]	236.9	991
Sesame	12	781	9	825	7.7	645
Tobacco	11	541	6	480	2.9	260
Sugar beets	7	23,455	154	185	28.5	4,318
Cumin	16	229	4	2,000	7.2	453
Nuts	8	1,602	14	600	8.1	964
Others	2	1,524	3	1,400	4.5	2,133
Total	295	–	519	–	295.8	1,002

[a] Less than half a ton.
[b] Farm price of seed cotton at Aleppo.

Source: Syria, Ministry of Agriculture and Agrarian Reform, *Annual Statistical Bulletin, 1967* (Damascus).
Note: All figures are rounded.

Syria: Area of Main Crops, 1946—67

(1000 Hectares)

Main crops	1946	1947	1948	1949	1950	1951	1952	1953	1954	1955	1956	1957
Cereals												
Wheat	810	843	788	988	992	1,037	1,167	1,314	1,347	1,463	1,537	1,495
Barley	371	365	342	348	416	344	397	439	543	614	636	813
Maize	23.7	22.1	33.1	25.9	24.8	16.0	19.2	18.1	17.3	13.0	10.1	9.7
Rice	7.6	8.2	7.5	7.6	9.4	0.5	2.8	5.3	6.0	4.0	2.4	1.0
Millet	95.0	104.0	90.0	92.9	94.0	71.4	114.9	97.6	102.4	69.9	86.7	70.3
Oats	7.8	7.0	10.2	9.6	7.9	6.4	7.7	6.1	6.5	4.2	5.5	5.9
Total cereals	1,315.1	1,349.3	1,270.8	1,472.0	1,544.1	1,475.3	1,708.6	1,880.1	2,022.2	2,168.1	2,277.7	2,394.9
Dry Legumes												
Lentils	42.6	49.6	46.2	51.4	59.2	65.7	66.2	70.9	76.4	77.5	85.1	93.1
Chick peas	34.2	31.9	32.0	26.2	33.1	27.4	28.4	22.5	27.4	25.2	25.4	29.2
Peas	0.2	0.3	0.4	0.2	0.2	0.2	0.2	0.2	0.4	0.2	0.2	0.2
Rambling vetch	49.4	44.8	53.8	60.3	62.9	57.3	58.3	59.1	61.3	61.6	49.2	59.2
Broad beans	16.0	18.3	19.3	22.6	23.8	20.3	15.7	16.9	16.4	15.2	11.6	12.7
Haricot beans	1.0	1.1	1.1	1.3	1.8	1.5	1.4	1.6	1.8	1.5	1.6	1.5
Bitter vetch	13.7	13.2	13.6	13.5	12.9	15.5	15.9	13.5	11.9	15.2	15.8	20.2
Flowering sern	5.5	7.8	3.4	3.5	4.4	3.4	3.7	2.5	3.6	3.2	5.5	6.0
Total dry legumes	162.6	167.0	169.8	179.0	198.3	191.3	189.8	187.2	199.2	199.6	194.4	222.1
Vegetables												
Potatoes	3.3	3.1	3.4	3.4	4.2	3.7	3.9	3.1	2.7	2.1	2.6	3.1
Garlic	1.6	1.5	1.0	1.1	1.1	1.1	1.0	1.2	1.2	1.0	0.9	0.8
Tomatoes	5.9	5.7	6.1	7.3	7.1	6.3	7.7	8.8	8.8	8.9	10.3	11.4
Onions	4.3	4.4	3.7	4.5	4.2	4.0	4.2	4.5	4.4	3.6	3.7	3.3
Eggplant	2.5	2.3	2.5	3.0	3.2	2.7	3.7	3.7	3.5	3.0	3.5	4.0
Onion sets	0.7	0.7	0.5	0.6	0.5	0.6	0.7	0.8	0.8	0.7	0.6	0.7
Total vegetables	18.3	17.7	17.2	19.9	20.3	18.4	21.2	22.1	21.4	19.3	21.6	23.3
Industrial Crops												
Cotton	19.8	19.3	24.0	25.3	78.0	217.4	189.3	127.6	187.3	248.8	272.2	258.3
Sesame	6.2	7.0	8.4	8.2	10.9	5.9	35.8	22.3	22.6	20.9	22.6	6.9
Sugarcane	0.1	0.1	0.1	0.1	0.2	0.1	0.1	0.1	0.1	0.2	0.1	0.1
Tobacco	7.7	5.6	4.7	5.4	7.5	8.0	7.3	5.4	6.4	6.8	6.6	6.6
Tombac	0.5	0.5	0.5	0.6	0.4	0.4	0.4	0.3	0.4	0.3	0.2	0.2
Linseed	0.2	0.2	0.2	0.2	0.2	0.2	0.2	0.5	0.6	0.4	0.4	0.4
Hemp	4.7	4.9	4.7	4.7	4.6	3.3	4.3	4.0	3.8	3.2	3.5	3.7
Sugar beets	–	–	–	–	0.1	0.2	2.6	4.3	4.6	3.0	3.0	3.4
Total ind. crops	39.2	37.6	42.6	44.5	101.9	235.5	240.0	164.5	225.8	283.6	308.6	279.6
Fruits												
Olives	78.8	79.9	79.9	80	81	80	82	86	93	98	105	109
Grapes	59	61	64	66	68	67	68	70	70	71	72	69
Apricots	7	7	7	7	7	7	8	8	7	8	8	8
Apples	2	2	2	3	3	3	3	4	4	4	4	4
Pears	1.5	1.5	1.5	1.5	1.6	1.6	1.7	1.8	1.9	1.9	1.8	1.9
Plums	0.6	0.6	0.7	0.7	0.8	0.9	1.0	1.0	1.1	1.1	1.1	1.2
Peaches	0.6	0.6	0.5	0.6	0.6	0.7	0.7	0.7	0.9	0.9	0.9	0.8
Nuts	6.5	6.6	6.0	6.1	6.5	6.6	6.8	6.9	6.8	6.8	6.7	6.7
Pomegranates	1.9	2.1	1.7	1.8	2.0	2.0	1.9	2.0	2.1	2.2	2.2	2.2
Figs	11	12	15	12	14	14	15	15	15	15	18	18
Almonds	3.1	3.2	3.1	3.1	3.3	2.3	2.4	2.4	2.4	2.5	2.3	1.9
Cherries	1.1	1.1	0.9	0.9	1.1	1.2	1.2	1.3	1.3	1.3	1.3	1.3
Quince	0.5	0.6	0.6	0.6	0.6	0.6	0.6	0.7	0.6	0.7	0.7	0.6
Pistachio of Aleppo	2.5	2.7	2.6	2.6	2.8	3.1	3.2	3.4	2.4	3.4	3.1	3.1
Total fruits	175.3	180.0	184.6	185.9	192.3	190.0	195.5	203.2	208.5	216.8	227.1	227.7
Total main crops	1,710.5	1,751.6	1,685.0	1,901.3	2,056.9	2,110.5	2,355.1	2,457.1	2,677.1	2,887.4	3,029.4	3,147.6

Main crops	1958	1959	1960	1961	1962	1963	1964	1965	1966	1967
Cereals										
Wheat	1,461.0	1,550.0	1,550	1,315.0	1,417.0	1,559.0	1,476.0	1,214.0	854.0	1,200.8
Barley	769.0	727.0	742.0	727.0	723.0	804.0	765.0	683.0	336.0	645.6
Maize	8.5	8.5	6.8	9.7	7.5	7.2	5.1	4.9	4.9	5.5
Rice	0.4	0.8	0.2	0.2	0.4	0.5	0.6	1.1	0.9	0.9
Millet	57.4	57.7	52.4	60.5	68.7	57.9	49.8	45.7	27.8	38.8
Oats	5.2	5.4	3.0	2.9	3.2	3.2	2.8	3.1	2.7	2.7
Total cereals	2,301.5	2,221.4	2,354.4	2,115.3	2,219.8	2,431.8	2,299.3	1,951.8	1,226.3	1,894.3
Dry Legumes										
Lentils	119.3	83.7	55.8	57.6	81.2	75.2	90.0	92.7	59.7	77.0
Chick peas	35.5	28.6	11.7	24.8	40.1	31.2	41.2	65.7	25.2	51.6
Peas	0.2	0.3	0.4	0.6	0.4	0.4	1.1	0.9	0.7	0.8
Rambling vetch	62.2	42.5	56.4	23.7	28.9	32.5	33.0	38.0	31.9	25.7
Broad beans	12.5	11.3	8.9	8.2	9.8	9.5	10.2	11.7	12.6	11.7
Haricot beans	1.5	1.6	1.1	1.2	1.6	1.6	1.4	1.7	1.5	1.4
Bitter vetch	28.6	19.0	18.9	14.6	25.2	19.4	24.0	28.4	13.3	20.2
Flowering sern	9.3	4.5	6.0	7.0	9.0	7.1	8.9	7.8	7.1	4.7
Total dry legumes	269.1	191.5	159.2	137.7	196.2	176.9	209.8	246.9	152.0	193.1
Vegetables										
Potatoes	2.2	1.6	2.6	2.5	3.1	3.5	4.8	4.4	4.5	3.5
Garlic	0.8	0.8	0.4	0.5	2.0	2.1	2.6	1.9	1.9	1.3
Tomatoes	10.3	11.3	8.5	10.5	13.6	23.5	18.7	16.5	14.6	18.8
Onions	3.4	3.7	2.7	3.2	3.9	4.9	4.2	4.3	4.0	4.6
Watermelons	30.0	30.0	29.0	41.0	49.0	70.0	48.0	38.3	26.1	62.5
Yellow melons	18.0	21.0	14.0	20.0	26.0	37.0	22.0	22.0	16.2	27.0
Eggplant	4.0	4.0	2.0	3.0	3.0	5.0	4.0	3.9	3.9	4.4
Total vegetables	68.7	72.4	59.2	80.7	100.6	146.0	104.3	91.3	71.2	122.1
Industrial Crops										
Cotton	260.8	227.2	212.3	249.1	302.4	291.7	286.5	285.7	255.1	239.4
Sesame	9.1	11.7	7.1	7.2	8.3	7.5	9.7	7.0	6.4	11.9
Sugarcane	0.1	0.1	—	0.1	0.1	0.1	0.1	0.1	—	—
Tobacco	7.0	8.7	9.4	9.7	9.9	1.4	15.3	16.3	15.6	11.6
Hemp	3.4	2.3	1.1	0.5	0.4	0.5	0.4	0.3	0.5	0.3
Sugar beets	2.2	5.0	5.2	4.4	4.0	4.0	7.0	8.6	8.5	6.6
Total ind. crops	282.6	255.0	235.1	271.0	325.1	305.2	319.0	318.0	286.1	269.8
Fruits										
Olives	110.6	110.8	124.0	134.5	132.5	111.2	112.2	116.4	117.7	141.7
Grapes	70.6	68.2	66.0	69.4	69.6	70.5	69.8	70.2	70.2	68.6
Apricots	7.8	7.8	7.3	8.1	8.1	8.7	8.8	9.8	9.5	9.9
Apples	4.4	4.5	5.7	5.9	6.4	6.6	7.0	7.1	7.5	7.0
Pears	2.0	2.0	1.8	2.0	2.0	2.1	2.3	2.5	2.5	2.6
Plums	1.2	1.3	0.9	0.9	1.0	1.1	1.3	1.3	1.1	—
Peaches	0.9	0.9	0.9	1.0	1.1	1.2	1.4	1.5	1.7	1.7
Nuts	6.8	6.6	4.6	4.1	4.1	4.3	4.6	4.7	4.9	8.4
Pomegranates	2.3	2.4	2.2	2.9	2.5	2.6	2.7	2.7	2.8	2.9
Figs	18.0	17.7	14.9	18.8	20.7	21.3	21.4	22.4	22.5	22.4
Almonds	2.0	2.0	1.9	2.2	2.4	2.4	2.5	2.5	2.6	2.5
Cherries	1.4	1.5	1.4	1.2	0.8	1.0	1.1	1.2	1.4	1.5
Pistachio of Aleppo	3.2	3.3	3.4	3.9	5.0	5.1	5.5	5.7	5.9	6.3
Orange	0.7	0.7	0.6	0.8	0.9	1.0	1.1	1.2	1.2	1.1
Other citrus	0.4	0.4	0.3	0.5	0.6	0.6	0.6	0.8	0.8	0.8
Other fruits	0.7	0.6	0.8	1.2	2.2	1.2	1.5	6.5	2.1	—
Total fruits	233.0	230.7	236.7	257.4	259.9	240.9	243.8	256.5	254.4	277.4
Total main crops	3,154.9	2,971.0	3,044.6	2,862.1	3,101.6	3,300.8	3,176.2	2,864.5	1,990.0	2,756.7

Source: Syria, Ministry of Planning, *Statistical Abstract*, various issues, Damascus.

Syria: Yield of Main Crops, 1946–67

(Metric Tons Per Hectare)

Main crops	1946	1947	1948	1949	1950	1951	1952	1953	1954	1955	1956	1957
Cereals												
Wheat	0.7	0.5	0.8	1.2	0.8	0.5	0.8	0.7	0.7	0.3	0.7	0.9
Barley	0.8	0.5	0.9	1.0	0.8	0.5	1.2	1.1	1.2	0.2	0.7	0.9
Maize	1.2	1.3	1.1	1.5	1.4	1.4	1.2	1.2	1.4	1.4	1.5	1.6
Rice	1.6	2.6	4.1	2.6	1.9	1.8	2.1	3.0	3.2	2.6	2.4	2.2
Millet	0.6	0.6	0.7	0.6	0.8	0.3	0.9	1.3	1.1	1.0	0.9	0.7
Oats	1.2	0.7	0.5	0.6	0.8	0.9	0.9	0.8	0.8	0.9	0.7	1.1
Average all cereals	0.7	0.5	0.9	0.9	0.8	0.5	0.9	0.8	0.9	0.3	0.7	0.9
Dry Legumes												
Lentils	0.5	0.8	0.7	1.1	0.5	0.5	0.8	0.9	0.8	0.5	0.9	0.8
Chick peas	0.5	0.4	0.4	0.8	0.5	0.3	0.9	0.5	0.8	0.5	0.6	0.4
Peas	0.8	1.3	0.2	0.7	1.1	0.8	1.2	0.8	1.0	0.9	1.0	1.0
Rambling vetch	1.9	1.3	0.7	1.0	0.5	0.5	1.8	1.0	0.8	0.4	1.0	0.9
Broad beans	0.9	1.6	1.1	1.6	0.8	1.3	0.8	0.7	0.7	0.6	0.6	0.8
Haricot beans	1.0	0.9	1.1	0.9	0.8	0.7	0.7	0.9	0.8	0.5	0.6	0.6
Bitter vetch	0.6	0.4	0.7	0.7	0.6	0.4	0.7	0.9	1.0	0.6	0.8	0.8
Flowering sern	0.4	0.3	0.6	0.5	0.6	0.6	0.8	0.9	0.7	0.4	1.0	0.7
Average dry legumes	0.6	0.9	0.7	1.0	0.5	0.5	0.8	0.9	0.8	0.5	0.8	0.8
Vegetables												
Potatoes	4.5	4.8	8.7	8.8	8.1	9.9	9.8	12.2	10.0	11.2	9.7	10.5
Garlic	5.5	4.8	5.4	5.6	3.8	4.0	4.1	3.6	3.5	3.5	4.2	4.4
Tomatoes	7.0	6.2	9.6	7.8	7.2	10.6	8.7	7.9	7.9	7.2	7.9	8.4
Onions	10.6	9.8	12.1	10.2	10.2	9.8	8.0	8.3	8.4	6.9	8.3	10.0
Eggplant	4.7	8.3	9.5	8.8	8.3	12.8	9.5	7.6	7.8	9.8	10.8	10.0
Onion sets	12.7	15.1	9.2	14.4	13.9	1.8	3.2	4.9	4.7	2.3	2.2	2.4
Average vegetables	7.1	7.3	9.7	8.7	8.1	10.0	8.5	8.2	7.9	7.6	8.3	8.9
Industrial Crops												
Cotton	0.7	0.8	1.1	1.5	1.3	0.8	0.9	1.0	1.2	0.9	0.9	1.1
Sesame	0.5	0.5	0.3	0.8	0.8	0.6	0.6	0.5	0.6	0.6	0.5	0.4
Sugarcane	23.0	19.0	28.5	27.0	32.7	38.9	41.9	29.1	30.9	30.0	27.0	19.4
Tobacco	0.8	0.8	0.8	0.6	1.0	0.9	0.9	0.8	0.8	0.7	0.8	0.9
Tombac	0.8	0.8	0.8	0.8	1.0	1.0	1.0	0.8	0.9	1.0	1.0	1.1
Linseed	1.0	0.9	1.0	1.0	1.0	1.0	1.0	1.0	1.0	0.8	1.0	1.0
Hemp	0.6	0.5	0.6	0.8	0.8	1.0	1.0	0.5	0.6	0.4	0.5	0.7
Sugar beets	–	–	–	–	18.7	13.0	21.1	11.6	11.0	11.6	15.1	18.1
Average ind. crops	0.7	0.7	0.9	1.2	1.2	0.8	1.1	1.2	1.3	1.0	1.0	1.3
Fruits												
Olives	0.7	1.0	1.3	1.2	0.2	0.5	0.4	0.6	0.4	0.3	0.8	0.4
Grapes	2.9	1.8	2.0	2.1	3.6	2.8	2.5	3.5	3.7	2.9	2.7	3.5
Apricots	3.0	2.3	3.4	3.4	3.0	3.3	1.9	2.9	3.6	1.5	1.3	3.0
Apples	2.9	2.8	2.9	1.7	2.1	2.4	2.1	1.9	2.6	2.0	2.5	2.7
Pears	1.0	0.9	1.3	1.5	1.3	1.0	0.8	1.1	1.2	0.9	1.3	1.4
Plums	2.8	2.7	2.6	2.9	2.5	2.3	2.0	2.4	2.5	2.0	2.4	2.5
Peaches	2.0	1.8	2.2	2.2	2.0	2.1	2.0	2.0	1.8	1.2	1.6	1.8
Nuts	0.9	0.5	0.6	0.7	0.6	0.8	0.5	0.9	0.9	0.8	0.7	0.7
Pomegranates	7.1	7.6	5.8	7.0	3.9	1.9	2.6	2.7	3.4	2.9	3.0	2.1
Figs	4.2	3.3	2.7	3.8	3.0	2.4	3.2	3.5	3.7	3.3	2.7	2.8
Almonds	0.7	0.7	0.9	0.8	1.2	1.0	0.9	1.1	1.2	0.9	0.8	1.3
Cherries	1.8	1.7	1.7	2.3	1.5	1.8	1.9	2.0	1.8	1.5	1.3	1.2
Quinces	1.4	1.0	1.0	1.2	2.0	2.8	2.0	1.9	2.2	1.4	1.4	1.5
Pistachio of Aleppo	0.5	0.2	0.2	0.6	1.6	0.2	0.2	0.1	0.4	0.4	0.5	0.4
Average fruits	1.9	1.5	1.8	1.8	1.9	1.6	1.5	2.0	2.0	1.5	1.6	1.7

Main crops	1958	1959	1960	1961	1962	1963	1964	1965	1966	1967
Cereals										
Wheat	0.4	0.4	0.4	0.6	1.0	0.8	0.7	0.9	0.7	0.9
Barley	0.3	0.3	0.2	0.5	1.1	1.0	0.8	1.0	0.6	0.9
Maize	1,1	1.5	1.1	1.0	0.9	1.1	1.2	1.1	1.5	1.6
Rice	1.6	2.1	1.6	1.6	2.6	2.4	2.0	2.0	2.4	2.3
Millet	0.9	0.7	0.6	0.6	0.8	0.9	0.9	0.9	0.5	1.0
Oats	1.4	0.7	0.9	1.0	0.8	0.5	0.8	0.9	0.8	0.8
Average all cereals	0.37	0.41	0.32	0.54	1.01	0.84	0.78	0.92	0.64	0.89
Dry Legumes										
Lentils	0.3	0.4	0.2	0.6	0.9	0.8	1.0	0.7	0.4	1.1
Chick peas	0.2	0.2	0.3	0.6	0.6	0.5	0.7	0.7	0.6	1.2
Peas	0.7	0.7	0.4	0.7	1.0	1.1	0.4	1.3	1.1	0.6
Rambling vetch	0.3	0.5	0.2	0.6	0.9	0.7	0.9	0.9	0.4	1.0
Broad beans	0.5	0.7	0.6	0.9	1.2	1.1	1.2	1.0	1.0	1.2
Haricot beans	0.6	0.9	0.8	1.3	1.1	0.9	1.0	0.9	1.0	1.6
Bitter vetch	0.3	0.4	0.2	0.5	0.7	0.5	0.9	0.8	0.4	1.1
Flowering sern	0.2	0.3	0.2	0.6	0.8	0.7	1.1	0.8	0.6	1.2
Average dry legumes	0.31	0.40	0.21	0.61	0.80	0.72	0.93	0.76	0.49	1.13
Vegetables										
Potatoes	9.6	8.9	10.7	11.8	10.9	9.0	9.9	11.1	9.1	11.2
Garlic	4.3	3.7	5.4	5.8	7.1	2.7	1.8	3.5	4.0	3.5
Tomatoes	6.9	8.6	8.8	8.8	8.5	7.1	8.2	8.2	8.6	8.8
Onions	9.3	7.7	11.7	11.3	10.0	7.4	8.1	7.7	8.1	9.5
Red melons	3.9	5.4	4.6	5.0	2.4	4.4	5.5	5.0	4.0	6.6
Yellow melons	5.7	5.6	8.0	4.6	8.0	3.5	5.3	6.5	8.0	5.1
Eggplant	9.3	10.0	14.5	10.7	13.3	7.2	8.3	8.8	9.2	10.3
Average vegetables	5.6	6.4	6.9	6.1	6.9	4.9	6.3	6.5	6.7	6.9
Industrial crops										
Cotton	1.0	1.2	1.3	1.3	1.3	1.4	1.6	1.7	1.5	1.4
Sesame	0.4	0.6	0.5	0.6	0.7	0.7	0.7	0.7	0.9	0.8
Sugarcane	28.4	26.9	—	39.0	29.4	66.1	30.1	27.5	21.8	—
Tobacco	0.9	0.8	0.6	0.7	0.7	0.5	0.7	0.7	0.6	0.5
Hemp	0.6	1.0	1.0	0.8	1.0	1.5	0.9	1.6	0.8	1.2
Sugar beets	14.2	18.2	23.6	19.6	20.3	21.5	24.5	20.0	22.2	23.4
Average ind. crops	1.0	1.5	1.7	1.6	1.5	1.7	2.1	2.1	2.0	1.8
Fruits										
Olives	0.58	0.25	0.43	0.62	0.65	0.61	1.10	0.56	0.99	0.8
Grapes	2.82	3.20	3.00	3.52	3.64	2.27	3.28	2.94	2.88	3.1
Apricots	2.42	2.03	2.57	2.25	5.18	3.11	3.22	0.90	1.38	2.2
Apples	2.02	1.50	1.72	2.04	3.77	4.01	3.50	3.04	3.73	4.0
Pears	1.10	1.20	0.72	0.90	3.55	3.33	2.43	2.16	2.20	2.3
Plums	2.33	2.46	2.22	2.44	3.50	2.18	2.54	1.85	3.00	—
Peaches	1.56	1.67	1.33	1.60	6.18	3.33	3.14	3.00	2.88	3.7
Nuts	0.50	0.56	0.54	1.93	2.29	1.58	1.56	0.96	1.08	1.3
Pomegranates	2.04	2.25	2.64	2.14	3.16	3.00	4.22	3.85	4.58	4.6
Figs	2.72	2.22	2.00	1.47	2.62	2.09	2.52	3.09	2.35	2.3
Almonds	0.80	1.10	0.89	0.77	1.00	0.83	0.84	0.48	1.46	3.5
Cherries	0.57	1.07	1.86	0.58	1.00	0.60	1.18	1.08	1.31	1.0
Pistachio of Aleppo	0.12	0.39	0.47	0.51	0.18	0.18	0.24	0.10	0.19	0.2
Oranges	4.0	7.3	6.0	4.4	3.6	2.7	2.1	3.0	3.8	4.3
Other citrus	5.0	7.5	6.3	3.0	2.3	2.7	3.7	3.0	3.9	3.9
Other fruits	2.4	15.8	1.1	1.8	2.1	2.7	1.2	0.4	1.9	—
Average fruits	1.6	1.5	1.4	1.6	2.0	1.6	2.1	1.5	1.8	1.8
Average main crops	0.6	0.7	0.6	0.9	1.3	1.1	1.2	1.3	1.2	1.4

Source: Syria, Ministry of Planning, *Statistical Abstract,* various issues, Damascus.

Syria: Production of Main Crops, 1946–67

(1000 Metric Tons)

Main crops	1946	1947	1948	1949	1950	1951	1952	1953	1954	1955	1956	1957
Cereals												
Wheat	578	404	657	909	830	510	900	870	965	438	1,051	1,354
Barley	284	169	305	357	322	155	467	472	635	137	462	721
Maize	27.5	28.6	37.6	39.1	35.7	21.8	22.6	22.2	23.8	18.5	14.8	15.1
Rice	12.0	21.6	19.1	20.0	18.0	0.9	5.8	16.2	19.0	10.5	5.8	2.2
Millet	58.9	67.6	63.4	57.3	75.7	22.2	106.1	123.7	114.3	71.4	75.4	51.6
Oats	9.0	4.8	5.0	5.7	6.1	5.6	6.7	4.6	5.0	3.8	4.1	6.6
Total cereals	969.4	695.6	1,087.1	1,388.1	1,287.5	715.5	1,508.2	1,508.7	1,762.1	679.2	1,613.1	2,150.5
Dry legumes												
Lentils	19.5	39.2	33.1	54.1	26.6	30.2	50.6	62.8	58.3	36.7	75.0	77.2
Chick peas	17.0	12.3	12.7	20.3	15.7	8.6	24.2	10.6	22.6	13.6	14.1	12.4
Peas	0.2	0.3	0.3	0.1	0.2	0.2	0.3	0.2	0.3	0.2	0.2	0.2
Rambling vetch	36.7	57.0	37.1	62.0	34.2	26.7	45.3	58.8	48.3	26.3	48.8	55.0
Broad beans	15.3	28.6	21.0	35.1	19.0	25.5	12.0	11.2	11.7	8.8	6.6	10.2
Haricot beans	0.9	1.0	1.2	1.2	1.4	1.0	1.0	1.4	1.4	0.8	1.0	0.8
Bitter vetch	˙7.6	5.2	9.8	9.3	7.1	5.9	11.6	12.2	12.2	9.8	13.3	16.0
Flowering sern	2.4	2.4	2.0	1.8	2.7	1.9	2.9	2.2	2.7	1.2	5.3	4.3
Total dry legumes	99.6	146.0	117.2	183.9	106.9	100.0	147.9	159.4	157.5	97.4	164.3	176.1
Vegetables												
Potatoes	15.0	15.1	29.7	29.8	34.2	36.6	38.2	37.7	27.2	23.7	25.2	32.4
Garlic	8.8	7.3	5.3	6.3	4.4	4.6	4.0	4.3	4.2	3.6	3.8	3.7
Tomatoes	40.9	35.2	58.4	56.6	50.9	67.4	66.7	70.0	69.1	64.2	81.0	95.5
Onions	45.0	42.7	44.9	45.7	42.4	39.8	33.1	36.8	36.7	25.0	30.8	33.6
Eggplant	11.6	19.5	23.3	26.7	26.1	35.0	34.9	28.4	27.1	29.3	37.9	39.6
Onion sets	8.5	10.0	4.9	8.4	7.2	1.1	2.3	3.7	3.7	1.6	1.3	1.6
Total vegetables	129.8	129.8	166.5	173.5	165.2	184.5	179.2	180.9	168.0	147.4	180.0	206.4
Industrial crops												
Cotton	14.1	15.5	25.6	38.1	100.3	175.5	176.4	126.0	220.8	233.3	252.5	291.5
Sesame	3.0	3.3	2.8	6.7	8.4	3.6	21.2	10.7	14.2	13.1	10.4	3.0
Sugarcane	2.1	1.6	2.9	2.7	4.9	0.4	5.0	4.1	3.1	5.9	2.7	1.9
Tobacco	6.0	4.3	3.5	3.3	7.1	7.3	6.2	4.5	5.3	5.0	5.2	5.7
Tombac	0.4	0.4	0.4	0.4	0.4	0.4	0.4	0.3	0.4	0.3	0.2	0.3
Linseed	0.2	0.2	0.2	0.2	0.2	0.2	0.2	0.5	0.6	0.3	0.4	0.4
Hemp	2.7	2.6	2.7	2.9	3.0	2.8	4.1	2.2	2.4	1.3	1.6	2.6
Sugar beets	–	–	–	–	2.4	2.7	55.3	49.6	50.6	34.5	45.3	61.0
Total ind. crops	28.5	27.9	38.1	54.3	126.7	192.9	268.8	197.9	297.4	293.7	318.3	366.4
Fruits												
Olives	53	79	103	92	19	40	33	49	36	29	78	38
Grapes	171	108	126	137	245	185	167	244	257	206	194	241
Apricots	21	16	24	24	21	23	15	23	25	12	10	24
Apples	5.7	5.6	5.7	5.1	6.4	7.1	6.4	7.6	10.3	8.1	9.9	10.7
Pears	1.5	1.3	1.9	2.3	2.0	1.6	1.4	1.9	2.2	1.8	2.3	2.6
Plums	1.7	1.6	1.8	2.0	2.0	2.1	2.0	2.4	2.8	2.2	2.6	3.0
Peaches	1.2	1.1	1.1	1.3	1.2	1.5	1.4	1.4	1.6	1.1	1.4	1.4
Nuts	5.8	3.4	3.7	4.4	3.7	5.0	3.6	6.0	6.1	5.6	4.4	4.5
Pomegranates	13.4	16.0	9.8	12.6	7.8	3.7	4.9	5.4	7.2	6.4	6.6	4.6
Figs	46	40	41	46	42	34	48	53	56	50	48	51
Almonds	2.1	2.2	2.7	2.5	4.0	2.4	2.2	2.7	2.9	2.3	1.9	2.5
Cherries	2.0	1.9	1.5	2.1	1.7	2.1	2.3	2.6	2.4	2.0	1.8	1.6
Quince	0.7	0.6	0.6	0.7	1.2	1.7	1.2	1.3	1.3	1.0	1.0	0.9
Pistachio of Aleppo	1.2	0.5	0.6	1.5	4.4	0.5	0.6	0.5	0.9	1.5	1.5	1.3
Total fruits	326.3	277.2	323.4	333.5	361.4	309.7	289.0	400.8	411.7	329.0	363.4	387.1
Total main crops	1,553.6	1,276.5	1,732.3	2,133.3	2,047.7	1,502.6	2,393.1	2,447.7	2,732.3	1,546.7	2,639.1	3,286.5

Main crops	1958	1959	1960	1961	1962	1963	1964	1965	1966	1967
Cereals										
Wheat	562.0	632.0	555.0	757.0	1,374.0	1,190.0	1,100.0	1,044.0	559.0	1,049.0
Barley	228.0	218.0	156.0	335.0	798.0	784.0	637.0	690.0	203.0	589.6
Maize	9.6	12.4	7.2	9.9	6.6	7.6	6.3	5.6	7.5	8.9
Rice	0.7	1.7	0.4	0.5	1.1	1.1	1.2	2.2	2.2	2.1
Millet	50.4	41.7	30.7	37.9	56.5	49.5	44.5	44.0	15.1	39.5
Oats	7.0	3.9	2.6	2.7	2.6	1.7	2.2	3.0	2.2	2.3
Total cereals	857.7	909.7	751.9	1,143.0	2,238.8	2,033.9	1,791.2	1,788.8	789.0	1,691.4
Dry legumes										
Lentils	35.8	31.6	10.4	34.4	69.5	58.5	89.6	65.7	22.2	83.6
Chick peas	7.0	5.9	3.4	14.1	24.1	16.8	28.6	45.7	15.7	63.6
Peas	0.2	0.2	0.2	0.4	0.4	0.4	0.5	1.2	0.9	0.5
Rambling vetch	21.8	21.6	8.6	14.1	24.7	23.3	30.4	32.5	12.2	26.4
Broad beans	6.7	7.5	5.3	7.8	12.1	10.9	11.8	12.0	12.5	14.0
Haricot beans	1.0	1.3	0.9	1.5	1.9	1.4	1.5	1.5	1.5	2.3
Bitter vetch	7.7	7.0	3.6	7.5	17.6	10.4	22.0	22.6	5.1	22.0
Flowering sern	2.1	1.3	1.5	4.4	6.8	5.3	10.0	6.0	4.6	5.9
Total dry legumes	82.3	76.4	33.9	84.2	157.1	127.0	194.4	187.2	74.7	218.3
Vegetables										
Potatoes	21.1	14.3	27.9	30.1	33.7	31.8	47.7	48.9	40.8	39.7
Garlic	3.4	3.1	2.3	3.0	13.9	5.7	4.8	6.7	7.2	4.5
Tomatoes	71.1	98.1	74.6	93.2	115.9	166.1	153.2	135.4	125.9	161.7
Onions	32.0	28.4	31.9	36.3	38.9	36.1	33.9	32.2	32.3	44.0
Red melons	116.0	163.0	132.0	205.0	245.0	307.0	265.0	193.3	104.2	415.8
Yellow melons	102.0	118.0	112.0	92.0	208.0	128.0	117.0	142.4	129.2	138.0
Eggplant	37.0	40.0	29.0	32.0	40.0	36.0	33.0	34.5	36.0	42.5
Total vegetables	382.6	464.9	409.7	491.6	695.4	710.7	654.6	593.4	475.6	846.2
Industrial crops										
Cotton	249.8	265.0	278.7	324.9	403.9	410.0	470.1	472.7	375.3	329.1
Sesame	4.0	7.0	3.7	4.2	5.9	5.3	6.5	4.9	5.7	9.3
Sugarcane	2.4	2.4	1.7	3.1	2.1	5.2	2.4	1.4	0.9	—
Tobacco	6.6	7.4	6.0	7.2	6.6	0.7	11.2	11.9	9.8	—
Hemp	2.1	2.3	1.1	0.4	0.4	0.7	0.4	0.6	0.4	0.4
Sugar beets	31.6	91.6	122.4	85.7	81.0	86.9	171.3	171.4	189.1	154.4
Total ind. crops	296.5	375.7	413.6	425.5	499.9	508.8	661.9	662.9	581.2	493.2
Fruits										
Olives	64.7	27.7	52.9	83.2	87.0	77.2	122.6	65.6	116.3	113.1
Grapes	198.7	217.6	197.8	242.8	255.0	159.2	230.3	205.7	206.0	213.5
Apricots	28.5	15.7	17.8	18.4	43.4	28.0	29.0	9.3	13.2	22.2
Apples	8.1	7.5	8.6	10.2	22.6	28.1	24.5	21.4	26.1	27.9
Pears	2.2	2.4	1.3	1.8	7.1	7.0	5.6	5.3	5.6	6.1
Plums	2.8	3.2	2.0	2.2	3.5	2.4	3.3	2.4	3.3	—
Peaches	1.4	1.6	1.2	1.6	6.8	4.0	4.4	4.5	4.9	6.3
Nuts	3.4	3.7	2.5	7.9	9.4	6.8	7.2	4.5	5.2	10.5
Pomegranates	4.7	5.4	5.8	6.2	7.9	7.8	11.4	10.4	11.8	13.2
Figs	49.0	40.0	30.0	28.0	55.0	44.0	52.8	54.6	53.8	52.4
Almonds	1.6	2.2	1.7	1.7	2.4	2.0	2.1	1.2	3.9	8.7
Cherries	0.8	1.6	1.2	0.7	0.8	0.6	1.3	1.2	1.8	1.5
Pistachio of Aleppo	1.4	1.3	1.6	2.0	0.9	0.9	1.3	0.6	1.5	1.4
Oranges	2.8	5.1	3.6	3.5	3.2	2.7	2.3	3.6	4.6	4.7
Other citrus	2.0	3.0	1.9	1.5	1.4	1.6	2.2	2.4	3.1	3.1
Other fruits	1.7	9.5	0.9	2.2	4.6	3.2	1.8	2.5	3.9	12.1
Total fruits	373.8	347.5	330.8	413.9	511.0	375.5	503.1	395.2	465.0	496.7
Total main crops	1,992.9	2,174.2	1,939.9	2,558.2	4,102.2	3,755.9	3,805.2	3,627.5	2,385.5	3,745.8

Source: Syria, Ministry of Planning, *Statistical Abstract,* various issues, Damascus.

Appendix Table C8-38

Syria: Area, Production, and Yield of Vegetables, 1967

Vegetables	Area (hectares)	Production (tons)	Average yield (kg./ha.)	Average farm price in 1965 (£S./ton)	Estimated value of production at 1965 prices (£S.000's)	Gross return per hectare (£S.)
Tomatoes	18,263	161,682	8,852	160	25,869	1,416
Eggplant	4,120	42,482	10,311	150	6,372	1,547
Watermelon	62,493	415,829	6,654	70	29,108	466
Melons	27,070	138,359	5,111	400	55,344	2,044
Potatoes	3,547	39,723	11,199	235	9,335	2,632
Squash	3,811	41,879	10,988	126	5,277	1,385
Green beans	2,007	11,755	5,857	275	3,233	1,611
Kidney beans	1,343	2,845	2,118	150	427	318
Dry onions	4,613	43,962	9,530	129	5,671	1,229
Garlic	1,275	4,458	3,496	293	1,306	1,024
Cucumbers	8,876	63,992	7,209	210	13,438	1,514
Okra	2,625	10,293	3,921	240	2,470	941
Green peppers	3,981	14,781	7,462	183	2,705	680
Green peas	649	1,583	2,439	407	644	993
Green broad beans	4,137	17,706	4,279	225	3,984	963
Green onions	1,140	9,552	8,378	200	1,910	1,676
Other vegetables	6,435	67,270	10,454	350	23,545	3,659
Total vegetables	156,385	1,088,151	6,958	–	190,638	1,219

Source: Syria, Ministry of Agriculture and Agrarian Reform, *Annual Statistical Bulletin 1967* (Damascus).

Appendix Table C8-39

Syria: Fruit Production, 1967

Fruits	Area (1000 hectares)	Number of trees — Total (1000)	Number of trees — Fruit bearing (1000)	Average yield — Per hectare (kg.)	Average yield — Per bearing tree (kg.)	Production (1000 tons)	Average farm price in 1965 (£S./ton)	Estimated value of production (£S. 1000)	Gross return per hectare (£S.)
Olives	142	18,618	13,876	798	8.2	113	750	84,825	599
Grapes	69	67,681	54,586	3,112	3.9	214	227	48,465	71
Figs	22	4,975	4,368	2,339	12.0	52	225	11,790	526
Apricots	10	2,562	2,069	2,242	10.7	22	400	8,880	897
Walnut	5	427	345	2,100	30.4	10	1,500	15,750	3,150
Apples	7	2,121	1,518	3,986	18.4	28	320	8,928	1,275
Pears	3	771	574	2,346	10.6	6	370	2,257	868
Plums	2	532	405	4,800	17.8	7	300	2,160	1,440
Peaches	2	593	452	3,706	13.9	6	460	2,898	1,705
Cherries	2	379	224	1,000	6.7	2	1,000	1,500	1,000
Almonds	2	714	562	3,480	15.5	9	3,450	30,015	12,006
Pomegranates	3	1,405	974	4,552	13.6	13	500	6,600	2,276
Quince	1	154	132	3,600	13.7	2	640	1,152	2,304
Janark	1	356	238	2,308	12.6	3	500	1,500	1,154
Pistachio	6	1,364	710	222	2.0	1	3,000	4,200	667
Oranges	1	446	282	4,273	16.7	5	330	1,551	1,410
Lemon	–	115	66	4,000	18.2	1	250	300	1,000
Other citrus	1	199	111	3,600	16.2	2	400	720	1,440
Loquats	1	16	10	2,000	10.0	–	146	25	292
Bananas	–	8	7	6,000	8.5	–	580	35	348
Total fruits	280	103,436	81,509	1,771	–	496		233,541	834

Source: Syria, Ministry of Agriculture and Agrarian Reform, *Annual Statistical Bulletin 1967.*

Appendix Table C8-40
Syria: Number and Production of Main Fruit Trees, 1958−60 and 1964−66

Fruit	1958−60				1964−66			
	Total number (1000)	Fruit bearing (1000)	Production (1000 tons)	Average yield per bearing tree (kg.)	Total number (1000)	Fruit bearing (1000)	Production (1000 tons)	Average yield per bearing tree (kg.)
Olives	15,369	8,814	49	5.5	15,808	11,580	102	8.8
Grapes	58,463	51,323	205	4.0	60,990	51,469	213	4.1
Apricots	2,206	1,829	24	13.1	2,433	1,896	18	9.5
Apples	1,505	1,017	8.1	8.0	2,180	1,568	24.0	15.3
Pears	614	407	2.0	4.9	722	529	5.5	10.4
Plums	379	291	2.7	9.3	485	346	3.0	8.7
Peaches	282	221	1.4	6.3	543	382	4.6	12.0
Nuts	406	324	3.2	9.9	421	321	5.7	17.8
Pomegranates	958	677	5.3	7.8	1,283	881	11.2	12.7
Figs	3,417	2,782	40	14.4	4,859	3,976	58	14.6
Almonds	542	395	1.8	4.6	717	487	2.4	4.9
Cherries	275	213	1.2	5.6	323	179	1.4	7.8
Quince	168	138	0.8	5.8	157	132	0.8	6.1
Pistachio of Aleppo	601	231	1.1	4.8	1,262	498	1.1	2.2

Source: Syria, Ministry of Planning, *Statistical Abstract, 1966* (Damascus: Government Press, 1967), pp. 298−311.

Appendix Table C8-41
Syria: Tobacco Crop, Area, Production, and Value, 1958−66

Year	Cultivated area (ha.)	Production (ton)	Ex-farm value (1000 £S.)
1958	7,047	6,600	14,079
1959	8,749	7,422	14,670
1960	9,360	6,011	13,818
1961	9,727	7,203	16,485
1962	9,858	6,645	17,261
1963	1,385	743	2,809
1964	15,283	11,221	30,054
1965	16,341	12,211	33,714
1966	15,623	9,747	26,742

Source: Syria, Ministry of Planning, *Statistical Abstract, 1966* (Damascus: Government Press, 1967), p. 314.

Appendix Table C8-42

Iraq: Area of Principal Crops and Vegetables, 1951—66*

(1000 Hectares)

Crops & vegetables	1951	1952	1953	1954	1955	1956	1957	1958	1959	1960	1961	1962	1963	1964	1965	1966
Field crops																
Wheat	928	968	1,198	1,390	1,425	1,314	1,456	1,533	1,490	1,271	1,346	1,591	1,705	1,627	1,703	1,737
Barley	917	882	1,096	1,122	1,205	1,160	1,240	1,157	1,091	1,038	1,041	1,190	1,219	1,098	1,097	1,169
Rice	61	75	95	120	54	70	91	89	63	77	64	91	108	110	116	111
Sesame	13	21	25	30	21	19	24	22	11	10	10	10	10	14	15	16
Maize	9	10	8	6	8	8	6	6	4	5	3	3	3	3	4	4
Millet	15	11	22	19	8	9	10	7	5	5	6	5	5	6	7	7
Cotton	113	51	21	56	58	58	65	56	37	31	37	34	25	40	34	33
Other field crops	42	37	40	49	43	53	54	61	56	54	60	59	60	64	75	75
Total field crops	2,098	2,055	2,505	2,792	2,822	2,691	2,946	2,931	2,757	2,491	2,567	2,983	3,135	2,962	3,051	3,152
Winter vegetables																
Green peas	–	–	–	–	–	–	–	–	1	13	10	9	10	9	10	–
Green onions	–	–	–	–	–	–	–	–	5	5	5	5	5	6	7	–
Turnip	–	–	–	–	–	–	–	–	2	3	3	3	3	3	3	–
Lettuce	–	–	–	–	–	–	–	–	1	1	1	1	2	2	2	–
Radish	–	–	–	–	–	–	–	–	1	1	1	2	2	2	2	–
Other winter vegetables	–	–	–	–	–	–	–	–	3	4	5	4	5	5	7	–
Total winter vegetables	–	–	–	–	–	–	–	–	13	27	25	24	27	27	31	–
Summer vegetables																
Tomatoes	–	–	–	–	–	–	–	–	12	16	20	22	23	28	28	–
Okra	–	–	–	–	–	–	–	–	5	8	8	9	9	12	13	–
Eggplant	–	–	–	–	–	–	–	–	4	5	6	7	8	9	9	–
Cucumbers	–	–	–	–	–	–	–	–	5	8	9	10	11	13	14	–
Watermelons	–	–	–	–	–	–	–	–	17	19	21	24	25	32	34	–
Red melons	–	–	–	–	–	–	–	–	6	10	11	11	12	15	16	–
Green french beans	–	–	–	–	–	–	–	–	2	4	5	5	6	7	8	–
Other summer vegetables	–	–	–	–	–	–	–	–	3	4	6	5	6	8	8	–
Total summer vegetables	–	–	–	–	–	–	–	–	54	74	86	93	100	124	130	–
Total vegetables	–	–	–	–	–	–	–	–	67	101	111	117	127	151	161	–
Dates	–	–	–	–	–	–	–	–	112	112	112	112	112	112	112	112
Tobacco & Tombac	–	–	–	–	–	–	–	–	10	13	13	13	13	15	–	–
Total	2,098	2,055	2,505	2,792	2,822	2,691	2,946	2,931	2,946	2,717	2,803	3,225	3,387	3,240	3,324	

* Statistics of vegetables for 1951 to 1958 and for 1966 are not available.

Source: Iraq, Ministry of Planning, *Statistical Abstract*, various issues, 1951 to 1966; Ministry of Agriculture, *Year Book of Agricultural Statistics, 1964* (Baghdad), pp. 206, 220.

Appendix Table C8-43
Iraq: Yield of Principal Crops and Vegetables, 1951–66*
(Metric Tons Per Hectare)

Crops & Vegetables	1951	1952	1953	1954	1955	1956	1957	1958	1959	1960	1961	1962	1963	1964	1965	1966
Field crops																
Wheat	0.5	0.5	0.6	0.8	0.3	0.6	0.8	0.5	0.4	0.5	0.6	0.7	0.3	0.5	0.6	0.5
Barley	0.9	0.7	1.0	0.2	0.6	0.9	1.1	0.8	0.7	0.8	0.9	0.9	0.6	0.6	0.7	0.7
Rice	1.2	1.7	1.7	1.5	1.5	1.6	1.7	1.5	1.5	1.5	1.1	1.2	1.3	1.7	1.7	1.6
Sesame	0.7	0.6	0.6	0.5	0.6	0.7	0.6	0.6	0.6	0.6	0.5	0.5	0.6	0.6	0.7	0.7
Maize	1.0	0.8	1.0	0.7	0.9	0.8	0.8	0.8	0.5	0.6	0.7	0.7	0.7	1.0	0.8	1.0
Millet	0.5	0.6	1.3	1.0	0.8	0.7	0.7	0.7	0.8	0.6	0.5	0.6	0.8	0.8	0.7	0.9
Cotton	0.1	–	0.1	0.1	0.1	0.1	0.2	0.2	0.2	0.3	0.2	0.2	0.2	0.3	0.3	0.2
Other field crops	0.6	0.7	0.8	0.7	0.6	0.8	0.8	0.6	0.7	0.7	0.7	0.8	0.7	0.8	0.8	0.9
Total field crops	0.7	0.6	0.8	1.0	0.5	0.7	0.9	0.7	0.5	0.6	0.7	0.8	0.5	0.6	0.7	0.6
Winter vegetables[a]																
Green peas	–	–	–	–	–	–	–	–	27.0	2.5	4.4	5.0	5.2	3.3	5.7	–
Green onions	–	–	–	–	–	–	–	–	3.8	4.8	5.8	6.8	5.6	6.3	8.4	–
Turnip	–	–	–	–	–	–	–	–	6.5	7.0	7.3	9.3	10.3	7.7	11.3	–
Lettuce	–	–	–	–	–	–	–	–	6.0	8.0	8.0	17.0	12.0	6.0	10.0	–
Radish	–	–	–	–	–	–	–	–	5.0	8.0	10.0	7.0	9.0	6.5	10.0	–
Other winter vegetables	–	–	–	–	–	–	–	–	8.7	6.0	7.2	8.5	8.8	7.0	8.7	–
Total winter vegetables	–	–	–	–	–	–	–	–	7.4	4.3	6.0	7.2	7.3	5.6	8.1	–
Summer vegetables[a]																
Tomatoes	–	–	–	–	–	–	–	–	7.3	8.9	7.0	6.4	6.2	6.8	7.0	–
Okra	–	–	–	–	–	–	–	–	5.6	6.0	5.5	5.9	6.0	5.8	5.5	–
Eggplant	–	–	–	–	–	–	–	–	10.0	11.0	10.7	11.4	11.0	14.9	15.2	–
Cucumbers	–	–	–	–	–	–	–	–	5.8	5.3	5.3	6.2	6.5	7.1	6.6	–
Watermelons	–	–	–	–	–	–	–	–	10.4	10.0	8.9	26.0	11.4	11.3	11.3	–
Red melons	–	–	–	–	–	–	–	–	8.3	7.7	7.5	7.6	7.7	8.0	8.9	–
Green french beans	–	–	–	–	–	–	–	–	6.5	5.8	5.2	5.4	5.5	5.4	5.0	–
Other summer vegetables	–	–	–	–	–	–	–	–	9.0	9.5	7.3	9.4	8.7	8.4	10.0	–
Total summer vegetables	–	–	–	–	–	–	–	–	8.3	8.3	7.4	8.2	8.2	8.6	8.8	–
Total vegetables	–	–	–	–	–	–	–	–	8.1	7.3	7.0	7.9	8.0	8.1	8.7	–
Dates	–	–	–	–	–	–	–	–	4.0	2.4	2.7	3.1	3.8	2.9	2.8	2.5
Tobacco & tombac	–	–	–	–	–	–	–	–	1.1	0.9	0.8	0.7	0.4	0.9	–	

* Statistics on vegetables for 1951 to 1958 and for 1966 are not available.

[a] Yield figures are derived from estimates of area and production. All figures are rounded.

Source: Iraq, Ministry of Planning, *Statistical Abstract,* various issues, 1951 to 1966; Ministry of Agriculture, *Year Book of Agricultural Statistics, 1964* (Baghdad), pp. 206, 220.

Appendix Table C8-44

Iraq: Production of Principal Crops and Vegetables, 1951–66*

(Metric Tons)

Crops & Vegetables	1951	1952	1953	1954	1955	1956	1957	1958	1959	1960	1961	1962	1963	1964	1965	1966
Field crops																
Wheat	488	480	762	1,160	453	776	1,118	757	564	592	857	1,085	488	807	1,006	826
Barley	839	652	1,111	1,239	757	1,016	1,305	954	725	804	911	1,125	790	623	806	832
Rice	84	127	163	180	83	111	154	137	94	118	69	113	143	184	198	182
Sesame	9	12	16	16	12	13	15	14	7	6	5	5	6	8	10	11
Maize	9	8	8	4	7	6	5	5	2	3	2	2	2	3	3	4
Millet	8	7	28	20	6	6	7	5	4	3	3	3	4	5	5	6
Cotton	8	1	2	7	7	8	14	11	8	8	9	8	5	10	10	7
Other field crops	24	25	30	32	25	42	42	39	39	39	44	47	44	48	63	66
Total field crops	1,469	1,312	2,120	2,658	1,350	1,978	2,660	1,922	1,443	1,573	1,900	2,388	1,482	1,688	2,101	1,934
Winter vegetables																
Green peas	–	–	–	–	–	–	–	–	27	32	44	45	52	30	57	–
Green onions	–	–	–	–	–	–	–	–	19	24	29	34	28	38	59	–
Turnip	–	–	–	–	–	–	–	–	13	21	22	28	31	23	34	–
Lettuce	–	–	–	–	–	–	–	–	6	8	8	17	24	12	20	–
Radish	–	–	–	–	–	–	–	–	5	8	10	14	18	13	20	–
Other winter vegetables	–	–	–	–	–	–	–	–	26	24	36	34	44	35	61	–
Total winter vegetables	–	–	–	–	–	–	–	–	96	117	149	172	197	151	251	–
Summer vegetables																
Tomatoes	–	–	–	–	–	–	–	–	87	143	139	140	143	189	196	–
Okra	–	–	–	–	–	–	–	–	28	48	44	53	54	69	72	–
Eggplant	–	–	–	–	–	–	–	–	40	55	64	80	88	134	137	–
Cucumber	–	–	–	–	–	–	–	–	29	42	48	62	72	92	93	–
Watermelons	–	–	–	–	–	–	–	–	176	190	186	265	285	363	383	–
Red melons	–	–	–	–	–	–	–	–	50	77	83	84	92	120	142	–
Green french beans	–	–	–	–	–	–	–	–	13	23	26	27	33	38	40	–
Other summer vegetables	–	–	–	–	–	–	–	–	27	38	44	47	52	67	80	–
Total summer vegetables	–	–	–	–	–	–	–	–	450	616	634	758	819	1,072	1,143	–
Total vegetables	–	–	–	–	–	–	–	–	546	733	783	930	1,016	1,223	1,394	–
Dates	–	–	–	–	–	–	–	–	450	270	300	350	420	320	310	280
Tobacco & tombac	1.85	4.59	7.36	9.01	5.31	5.40	5.10	5.42	11.31	12.29	10.70	8.50	4.74	13.45	–	–

* Statistics of vegetables for 1951 to 1958 and for 1966 are not available.

Source: Iraq, Ministry of Planning, *Statistical Abstract,* various issues, 1951 to 1966; Ministry of Agriculture, *Year Book of Agricultural Statistics, 1964.*

Appendix Table C9-1
Cattle, by Countries, Years of Record 1904—65
(1,000)

Year[a]	Egypt	Israel	Leb-anon	Jordan	Syria	Iraq	Saudi Arabia	Total reported
1965	1,630	202	105	73	524	1,455	105	4,094
1964	1,608	217	104	65	508	1,455	70	4,027
1963	1,587	216	92	61	463	1,550	66	4,035
1962	1,566	243	110	60	393	1,550	65	3,987
1961	1,715	235	95	45	433	1,550	62	4,135
1960	1,867	229	85	62	432	1,550	63	4,288
1959		224	70		535	1,535		
1958	1,499	192	60	100	596			
1957	1,390	152	88	116	609	1,500		3,855
1956		125	95	64	574	1,535		
1955		109	96	67	545			
1954	1,362	102	90	99	552			
1953	1,344	94	30	52	509			
1952		88	30	32	487	721		
1951	1,356	83	35	42	477	1,510		3,503
1950		70		81	429			
1949		37	20		368			
1948		33	22	64	369	822		
1947		36	22	59	354			
1946	1,321	33	22	60	371			
1945			22	76	354			
1944	1,265		22		391	866		
1943			20		418	704		
1942	1,202		25		413	613		
1941			44		456			
1940	991		45		368			
1939	1,230							
1938			47	53	357	250		
1937	983							
1927	740							
1917	566							
1912	620							
1904	605							

[a] Year beginning in October.

Sources: 1938—65: *FAO Production Yearbook;* 1904—37: Donald C. Mead, *Growth and Structural Change in the Egyptian Economy* (Homewood, Ill.: Richard D. Irwin, 1967), p. 332.

Appendix Table C9-2
Sheep, by Countries, Years of Record 1917—65
(1,000)

Year[a]	Egypt	Israel	Leb-anon	Jordan	Syria	Iraq	Saudi Arabia	Total reported
1965	1,947	195	90	987	5,422	11,040	3,500	23,781
1964	1,855	192	87	803	5,075	11,040	2,670	20,922
1963	1,770	194	84	803	4,524	9,450	2,750	19,575
1962	1,691	192	82	741	4,297	9,450	2,830	19,283
1961	1,650	189	80	528	3,823	9,450	2,900	18,620
1960	1,600	182	77	609	3,503	9,450	3,000	18,421
1959		192	75	621	4,740	9,221		
1958	1,418	191	75		5,912	9,221		
1957	1,259	168	70	689	5,466			
1956		129	72	453	4,703	9,221		
1955		118	74	494	4,340			
1954	1,237	102	60	515	3,955			
1953	1,216	78	60	364	3,746			
1952		74	60	223	3,560	4,942		
1951	1,254	69	25	274	3,085	10,000		
1950		77		226	2,930			
1949		50	25		2,750	7,490		
1948		30	20	113	2,935	7,055		
1947		23	20	125	3,176			
1946	1,868		21	239	3,260	8,000		
1945			20	265	3,504	7,424		
1944	1,385		12		3,080	6,526		
1943			10		2,482	6,125		
1942	1,423		20		2,322	7,500		
1941			34		2,976			
1940	1,242		36		3,123			
1939	1,897		38	200	3,100	5,525		
1937	1,919							
1927	1,232							
1917	808							

[a] Year beginning in October.

Sources: 1939—65: *FAO Production Yearbook;* 1917—37: Donald C. Mead, *Growth and Structural Change in the Egyptian Economy* (Homewood, Ill.: Richard D. Irwin, 1967), p. 332.

Appendix Table C9-3

Goats, by Countries, Years of Record 1917–65
(1,000)

Year[a]	Egypt	Israel	Leb-anon	Jordan	Syria	Iraq	Saudi Arabia	Total reported
1965	790	152	442	759	832	1,845	2,500	7,320
1964	787	153	442	651	818	1,845	2,341	7,037
1963	784	157	471	650	790	2,600	2,306	7,758
1962	780	164	450	565	581	2,600	2,272	7,012
1961	797	155	470	451	751	2,600	2,238	7,462
1960	815	150	480	513	1,273	2,600	2,205	8,036
1959		168	500	454	1,223	2,639		
1958	778	169	500		1,645	1,733		
1957	723	162	380	584	1,803	2,500		
1956		140	400	541	1,741	2,639		
1955		140	420	666	1,690	1,618		
1954	744	135	450	626	1,652			
1953		115	500	545	1,614			
1952	735	100	500	348	1,572	1,689		
1951	703	95	450	393	1,434	3,000		
1950		150		358	1,230			
1949		70	390		1,196			
1948		70	450	332	1,220	1,849		
1947		23	450	280	1,185			
1946	1,474		450	304	1,257	2,250		
1945			550	381	1,412	1,947		
1944	732		450		1,467	1,868		
1943			400		1,173	1,902		
1942	760		500		1,369			
1941			543		1,570			
1940	671		551		1,800			
1939	1,088		550		1,275	2,224		
1938				394				
1937	1,311							
1927	622							
1917	308							

[a] Year beginning in October.

Sources: 1938–65: *FAO Production Yearbook;* 1917–37: Donald C. Mead, *Growth and Structural Change in the Egyptian Economy* (Homewood, Ill.: Richard D. Irwin, 1967), p. 332.

Israel: Number of Livestock and Poultry, 1948–67

(1,000)

Livestock and poultry	1948	1949	1950	1951	1952	1953	1954	1955	1956	1957	1958	1959	1960	1961	1962	1963	1964	1965	1966	1967
Cattle:																				
on Jewish farms,																				
dairy	34	38	47	60	64	69	74	73	77	89	113	130	128	136	143	122	126	131	129	139
beef		c	c	1	2	3	5	11	15	27	39	50	58	57	59	68	65	57	55	57
on other farms		12	18	20	21	22	24			36			44		40	26	26	25	25	26
Total[a]		50	65	81	87	94	103			150			230		242	216	217	213	209	221
Poultry:																				
Chickens[b]	1,676	2,848	3,162	2,893	2,720	3,050	3,550	3,350	3,750	4,250	5,750	6,750	7,750	7,750	7,550	7,050	7,050	7,050	6,850	6,950
Geese, ducks, and turkeys	21	19	15	13	12	62	133	170	120	250	400	600	800	900	950	1,000	1,050	1,100	1,300	1,400
Sheep:																				
on Jewish farms	22	29	37	44	49	61	75	90	102	112	117	115	117	121	125	125	122	124	126	127
on other farms		10	12	12	15	18	23			56			65		67	69	70	71	69	72
Total[a]		39	49	56	64	79	98			168			182		192	194	192	195	195	219
Goats, purebred:																				
Jewish farms	5	11	15	20	25	40	45	45	45	45	40	35	32	32	30	27	30	28	30	29
other farms																		4	4	6
local breeding		33	55	60	70	75	85			117			118		134	130	120	120	116	109
Total[a]																		152	150	144
Horses and mules:																				
on Jewish farms	5	7	9	10	11	12	12	14	14	15	16	16	16	16	16	15	14	14	13	12
on other farms		3	4	4	4	5	6			6			6		6	4	4	3	3	3
Total[a]		10	13	14	15	17	18			21			23		22	19	18	17	16	15
Asses:																				
on Jewish farms	2	4	6	6	6	6	5	5	5	5	5	5	5	5	4					
on other farms		6	14	13	15	15	15			16			16		14					
Camels			4	4	4	4	4			10			9		10	10	10	10	10	11

[a] Total shown only where all parts recorded.
[b] Laying hens, including an estimate of 250,000 each year on non-Jewish farms.
[c] @ = Less than 500.

Source: Various issues of *Statistical Abstract of Israel,* published by the Central Bureau of Statistics.

Appendix Table C9-5
Israel: Production of Cows Milk, Eggs, and Fish, 1948/49 to 1966/67

Year[a]	Cows milk (million litres)	Eggs (millions)		Fish (tons)			
		Total	Exported[b]	Ponds	Lakes	Marine	Total
1948/49	79	242	–	2,510	360	630	3,500
1949/50	92	330					6,150
1950/51	103	392	–	4,350	835	2,115	7,300
1951/52	119	366					7,500
1952/53	128	369	–	4,710	760	2,130	7,600
1953/54	147	414					9,000
1954/55	159	504	4	7,295	878	2,827	11,000
1955/56	172	510		7,385	945	2,570	10,900
1956/57	184	630	21	7,310	1,300	2,790	11,400
1957/58				7,850	1,005	3,495	12,350
1958/59	260	1,027	280	7,816	917	4,467	13,200
1959/60	277	1,114	342	8,692	1,313	3,895	13,900
1960/61	284	1,290	459	9,300	1,350	4,000	14,650
1961/62	316	1,273	425	9,440	1,210	5,750	16,400
1962/63	315	1,113	128	10,030	1,265	5,155	16,450
1963/64	305	1,278	210	10,675	1,375	6,850	18,900
1964/65	323	1,296	222	10,300	1,260	7,740	19,300
1965/66	349	1,233	94	9,905	1,615	11,030	22,550
1966/67	380	1,402	195	8,735	1,850	12,365	22,950

[a] Most Israel economic data are on the basis of the Hebrew calendar; the year begins in early fall (September, usually) and extends for about twelve months.
[b] Including hatching eggs exported.

Source: Various issues of the *Statistical Abstract of Israel*, published by the Central Bureau of Statistics.

Appendix Table C9-6
Lebanon: Livestock Population, 1966, 1967*
(1000)

Type of animal	1966	1967
Milk cows		
Under 15 months	19.2	14.7
15 months & over	37.3	31.1
Total milk cows	56.4	45.8
Working cattle		
Male	(included with	14.6
Female	other cattle)	13.6
Total working cattle		28.1
Other cattle		
Under 15 months	20.3	14.6
15 months & over	28.1	8.4
Total other cattle	48.4	23.0
Total cattle	104.8	96.9
Sheep		
Female	172.6	168.0
Male	40.6	29.8
Total sheep	213.2	197.8
Goats		
Female	354.2	350.3
Male	88.1	81.0
Total goats	442.3	431.3
Poultry		
Layers	2,908.2	2,730
Broilers	14,317.7	12,250
Other livestock		
Pigs	9.6	8.8
Horses	3.2	2.9
Mules	4.7	3.5
Donkeys	37.0	30.3
Camels	0.8	0.5
Rabbits	24.4	21.5
Beehives	37.1	36.3

* Columns do not in all instances add to totals due to rounding.

Source: Lebanon, Ministry of Agriculture, *Agricultural Statistics 1966, 1967* (Beirut).

Appendix Table C9-7
Lebanon: Production of Honey, 1967

Number of beehives	Average yield per beehive (kg.)	Total output (metric tons)	Farm gate price (£L per ton)	Farm gate value (£L 1000)
36,334	4.7	172	6,420	1,104

Appendix Table C9-8
Lebanon: Total Meat and Fish Production, 1967

Meat and fish	Production (tons)	Farm gate price (£L per ton)	Farm gate value (£L 1000)
Beef	2,017	3,240	6,532
Mutton	1,662	3,400	5,649
Goat meat	2,760	3,170	8,759
Offals	1,288	980	1,264
Poultry meat	16,177	2,410	38,917
Pork	411	3,190	1,312
Fish	1,923	2,820	5,422
Total meat	26,238		67,855

Appendix Table C9-9
Lebanon: Other Animal Products, 1967

Product	Output unit	Output quantity	Farm gate price unit	Farm gate price (£L)	Farm gate value (£L 1000)
Hides	number	22,410	pieces	8.26	185
Skins	number	251,656	pieces	3.50	882
Wool	tons	494	tons	2,000	988
Manures	bags	5,230	bags	1.34	7,003
Total					9,058

Source for C9-7, 8, 9: Lebanon, Ministry of Agriculture, *Agricultural Statistics 1967* (Beirut, May 1968), pp. 27–30.

Appendix Table C9-10
Lebanon: Milk Production From Locally Produced Stock, 1967

Type of animal	Number	Lactation period (days/ year)	Average yield per animal per year (kg.)	Production (metric tons)	Farm gate price (£L per ton)	Farm gate price (£L 1000)
Milk cows	27,219	213	1,749	56,666	360	20,437
Sheep	134,265	129	77	10,375	470	4,804
Goats	248,830	138	112	27,808	400	11,208
Total				94,849		36,449

Appendix Table C9-11
Lebanon: Locally Produced Livestock and Poultry Slaughtered, 1967

Livestock/Poultry	Unit	Total number
Cattle	number	22,410
Sheep	number	79,140
Goats	number	172,516
Pigs	number	7,486
Poultry	1000	14,706

Appendix Table C9-12
Lebanon: Production of Eggs, 1967

Number of layers (1000)	Average production per layer	Total production (1000)	Farm gate price (£L per 1000)	Total farm gate value (£L 1000)
2,730	210	573,300	99	56,931

Source for C9-10, 11, 12: Lebanon, Ministry of Agriculture, *Agricultural Statistics 1967* (Beirut, May 1968), pp. 27–30.

Appendix Table C9-13
Jordan: Livestock Population and Composition, 1953–66
(in 1000)

Year	Sheep	Goats	Cattle	Camels	Total
1953	226	334	71	15	646
1954	464	545	52	19	1,080
1955	515	626	99	23	1,263
1956	494	669	67	26	1,256
1957	453	541	64	13	1,071
1958	689	584	116	20	1,409
1959	621	454	100	23	1,198
1960	609	513	62	20	1,204
1961	528	451	45	19	1,043
1962	702	537	60	12	1,311
1963	741	565	61	13	1,380
1964	803	650	65	19	1,537
1965	987	759	73	19	1,838
1966	1,136	766	78	17	1,997

Source: Jordan, Department of Statistics, *Statistical Yearbook,* various years.

Appendix Table C9-14

Syria: Number of Livestock, Poultry, and Beehives, 1957–66

(1000)

Year	Sheep	Goats	Camels	Buffalo	Pigs	Cows	Oxen	Horses	Mules	Asses	Poultry	Calves	Beehives
1957			79	4.3	0.3	304	204	101	81	239	2,973	97	67
1958	5,912	1,645	63	3.6	0.3	298	200	97	77	227	2,821	95	74
1959	4,756	1,227	25	2.9	0.4	266	183	84	71	206	3,107	83	69
1960	3,649	660	25	2.4	0.3	250	130	63	67	189	2,860	73	58
1961	2,901	439	13	1.4	0.1	230	96	66	72	186	3,296	86	46
1962	3,223	535	12	1.7	0.1	231	112	67	70	196	3,867	91	58
1963	3,926	690	11	1.6	0.2	244	113	67	68	202	3,736	92	71
1964	4,753	800	11	1.5	0.2	252	112	67	65	200	4,675	99	78
1965	5,373	877	8	1.3	0.1	274	111	70	65	190	4,599	121	87
1966	5,682	910	7	1.8	0.1	293	108	67	66	199	4,090	124	83

Source: Syrian Arab Republic, Ministry of Planning, Directorate of Statistics, *Statistical Abstract, 1966*, pp. 315–316.

Appendix Table C9-15

Syria: Number of Animals Slaughtered and Inspected, 1956–66

(1000)

Year	Sheep	Lambs	Goats	Kids	Cows	Calves	Camels	Pigs	Buffalo
1956	365	375	129	54	14	13	6	0.1	0.2
1957	419	344	145	45	17	10	9	0.09	0.2
1958	533	454	129	37	14	7	12	0.05	0.2
1959	692	368	160	32	19	8	17	0.1	0.1
1960	789	240	82	38	13	9	19	0.1	0.4
1961	646	367	61	32	14	6	12	0.1	0.02
1962	503	491	38	52	19	8	10	0.08	0.1
1963	523	533	47	41	23	13	9	0.2	0.5
1964	663	493	56	35	23	21	8	0.1	0.06
1965	698	514	61	37	23	20	8	0.1	0.6
1966	690	776	77	41	24	18	7	0.05	0.3

Source: Syria, Ministry of Planning, *Statistical Abstract, 1966* (Damascus, Government Press, 1967), p. 317.

Appendix Table C9-16

Syria: Animal Products, 1957–66

Year	Milk (1000 tons)	Ghee sanmeh (tons)	Cheese (tons)	Butter (tons)	Wool, washed (tons)	Goat hair (tons)	Eggs (1000)	Honey (tons)	Silk cocoons (tons)
1957	501	8,716	14,867	856	5,238	1,514	146	211	424
1958	256	5,749	11,058	558	4,838	1,259	138	150	284
1959	159	3,185	7,365	420	5,335	1,120	105	182	364
1960	133	2,072	4,201	419	3,963	361	147	74	507
1961	161	2,542	4,962	496	3,161	840	168	104	329
1962	265	5,766	12,766	438	3,989	400	234	117	264
1963	539	11,168	47,691	2,471	4,705	786	286	207	306
1964	539	12,876	28,379	1,729	5,562	519	293	210	390
1965	600	15,330	30,892	2,094	6,592	490	306	233	281
1966	604	15,439	32,379	1,801	5,649	506	222	169	281

Source: Syria, Ministry of Planning, *Statistical Abstract, 1966* (Damascus: Government Press, 1967), p. 320.

Appendix Table C9-17
Iraq: Livestock Population, 1964/65

Kind	Number (1000)	Percent of total
Sheep	11,040	71
Goats	1,846	12
Cattle	1,455	9
Buffalo	235	2
Camels	202	1
Horses	122	1
Mules	72	1
Donkeys	542	34
Total	15,514	100
Hens	5,494	

Source: Republic of Iraq, Ministry of Agriculture, *Annual Agricultural Statistics Bulletin for 1964* (in Arabic) (Baghdad, n.d.).

Appendix Table C9-18
Iraq: Animals Slaughtered in Abattoirs, 1952–66
(1000)

Year	Sheep	Goats	Cows	Buffalo	Camels
1952	720	218	86	9	2
1953	957	361	110	11	2
1954	1,199	423	149	12	3
1955	1,381	429	171	11	2
1956	1,483	511	188	13	2
1957	1,549	620	206	13	4
1958	1,712	532	200	13	6
1959	1,758	610	216	17	5
1960	1,905	545	222	18	13
1961	1,886	618	216	18	16
1962	1,820	682	211	13	11
1963	1,926	503	226	19	11
1964	1,896	502	248	20	11
1965	1,844	517	276	22	13
1966	2,105	542	263	19	14

Source: Republic of Iraq, Ministry of Planning, Central Bureau of Statistics, *Statistical Abstract*, various issues.

Appendix Table C10-1

Egypt: Gross Value of Agricultural Production, Value of Inputs and Value Added, 1952–65

(Million Egyptian pounds, at current prices)

Item	1952	1953	1954	1955	1956	1957	1958	1959	1960	1961	1962	1963	1964	1965
Winter field crops	104	116	118	104	116	113	114	119	127	127	138	132	146	204
Summer field crops	146	119	146	148	183	194	182	199	215	156	213	215	244	291
Nili field crops	31	27	35	44	48	41	47	39	43	38	39	35	29	19
Total field crops	281	272	298	296	347	348	343	358	385	322	390	382	428	513
Vegetable products	16	18	19	19	20	23	26	31	31	38	39	56	54	82
Fruit products	15	15	15	14	14	14	17	15	19	17	21	21	31	33
Wood chopped trees	1	1	1	1	2	2	2	2	2	2	3	3	4	5
Bulbs and flowers	0.03	0.02	0.02	0.03	0.1	0.09	0.1	0.2	0.2	0.2	0.2	0.2	0.2	1
Medical plants	0.03	0.03	0.02	0.04	0.07	0.1	0.1	0.1	0.1	0.1	0.1	0.1	0.1	0.1
Fruit nurseries	0.09	0.09	0.08	0.1	0.1	0.2	0.2	0.3	0.3	0.4	0.4	0.2	0.3	0.3
Total other crops	32	34	35	34	36	39	45	49	53	58	63	82	90	122
Total value of production	312	305	333	330	383	387	388	406	438	380	454	464	519	636
Cattle meat	36	34	33	34	32	36	38	38	37	38	39	67	81	89
Dairy	26	23	25	26	28	31	32	33	32	34	38	49	54	56
Wool	1	1	1	0.8	0.8	0.8	0.9	1	1	1	1	0.7	0.8	2
Poultry meat	11	9	9	11	11	11	11	13	13	13	14	19	24	35
Eggs	3	3	3	4	4	5	5	5	5	6	6	7	11	11
Fish and sea animals	7	7	7	8	8	11	13	15	15	13	12	18	20	21
Honey and wax	0.4	0.3	0.2	0.3	0.3	0.3	0.3	0.3	0.3	0.3	0.4	0.5	0.9	0.8
Cocoons	0.01	0.01	0.01	0.02	0.03	0.02	0.01	0.01	0.01	0.03	0.03	0.03	0.05	0.04
Natural fertilizers	22	23	23	25	28	30	31	33	42	39	46	54	55	56
Total value of animal and fish production	106	99	101	110	112	125	131	138	146	145	156	215	246	271
Total value of agricultural production	418	405	434	440	496	512	519	544	584	525	609	679	765	906
Value of production inputs	148	129	125	127	138	146	150	154	166	173	183	199	229	291
Value added	270	276	209	314	358	367	369	391	418	352	426	480	536	615

Source: UAR, Central Organization for General Mobilization and Statistics, *Estimates of National Income from the Agricultural Sector, 1965* (Cairo, 1967), p. 5. Figures may not add to totals, due to rounding.

Appendix Table C10-2

Egypt: Gross Value of Agricultural Production, Value of Inputs and Value Added, 1952–65

(Million Egyptian pounds, 1954 prices)

Item	1952	1953	1954	1955	1956	1957	1958	1959	1960	1961	1962	1963	1964	1965
Field products	297	279	298	292	301	319	327	331	341	274	242	332	344	341
Vegetable products	16	17	19	20	21	22	25	28	30	31	34	39	39	41
Fruit products	14	14	15	17	15	15	18	17	19	18	21	22	21	20
Trees and nurseries	0.8	0.7	1	4	4	4	4	4	4	5	5	6	6	7
Dairy	25	25	25	26	25	24	27	29	26	27	27	27	28	28
Cattle meat	33	32	33	24	35	36	38	33	33	33	33	35	35	36
Poultry meat and eggs	11	12	12	15	16	16	16	17	17	18	18	20	21	22
Natural fertilizers	22	23	23	24	24	25	24	26	26	27	27	27	28	28
Fish	8	8	7	11	9	8	11	11	12	12	12	12	11	9
Other animal products	0.9	0.8	0.9	1	4	3	3	1	1	1	1	0.8	0.9	1
Total value of agricultural production	428	411	434	444	454	473	493	497	510	444	521	521	533	535
Production inputs	96	98	125	127	124	136	128	115	115	118	121	119	129	129
Value added	332	313	309	317	330	337	365	381	394	326	400	402	405	405

Source: UAR, Central Organization for General Mobilization and Statistics, *Estimates of National Income from the Agricultural Sector, 1965* (Cairo, 1967), p. 7.

Appendix Table C10-3

Egypt: Gross Value of Agricultural Production, Cost of Inputs and Value Added, 1961–65

(Thousand Egyptian pounds, 1959 prices)

Item	1961	1962	1963	1964	1965
I. *Agricultural Production*					
A. Plant Production					
Field products	301,561	378,543	369,000	384,986	383,889
Vegetable products	34,878	38,041	42,905	34,168	45,489
Fruit products	14,610	18,595	19,852	18,509	17,943
Trees and nurseries	3,367	2,951	3,623	3,695	4,653
Total	354,416	438,130	435,380	450,358	451,974
B. Animal Production					
Dairy	40,024	40,598	41,173	41,748	42,322
Cattle meat	37,824	38,826	40,127	40,912	41,741
Poultry meat and eggs	18,394	19,196	21,222	22,296	24,051
Natural fertilizers	33,729	34,325	34,949	35,569	36,196
Fish	15,000	15,411	15,426	13,845	12,271
Other animal production	1,436	1,576	971	1,048	1,435
Total	146,407	149,932	153,868	155,418	158,016
Total value of agricultural production	500,823	588,062	589,248	605,776	609,990
II. *Input Requirements*					
'Seeds	19,362	20,895	19,883	19,862	20,679
Chemical fertilizers	30,085	31,032	32,484	38,095	40,610
Natural fertilizers	33,729	34,325	34,949	35,569	36,196
Insecticides	3,651	1,251	2,516	5,053	4,569
Fodder	56,277	60,236	53,624	56,163	53,532
Fuel, oil, and lubricants	10,002	10,410	10,646	11,053	11,038
Maintenance	1,221	1,249	1,274	1,366	1,446
Depreciation	2,156	2,259	2,537	2,689	2,471
Eggs for hatcheries	1,174	1,244	1,384	1,764	1,608
Total value of the requirements of agricultural production	157,657	162,901	159,297	171,614	172,149
Value added	343,166	425,161	429,951	434,162	437,841

Source: UAR, Central Organization for General Mobilization and Statistics, *Estimates of National Income from the Agricultural Sector, 1965* (Cairo, 1964), p. 9.

Appendix Table C10-4

Egypt: Index Numbers of the Gross Value of Agricultural Production, Its Requirements, and Value Added, 1952–65

Item	1952	1953	1954	1955	1956	1957	1958	1959	1960	1961	1962	1963	1964	1965
Field crops	82	97	110	99	117	100	98	104	108	84	121	98	112	120
Fruits, vegetables, and nursery	103	106	104	99	104	108	117	107	108	110	109	129	111	135
All crops	83	98	109	99	116	101	100	105	108	87	119	102	113	123
Livestock	92	94	102	109	102	112	104	105	106	99	107	138	115	110
Total value of agricultural production	85	97	108	101	113	103	101	105	107	90	116	111	113	118
Total value of inputs	105	87	97	101	109	106	103	103	108	104	106	109	115	127
Value added	77	102	112	102	114	102	101	106	107	84	121	113	113	115

Source: UAR, Central Organization for General Mobilization and Statistics, *Estimates of National Income from the Agricultural Sector, 1965* (Cairo, 1967), p. 10. Index numbers are calculated on a moving base = 100.

Appendix Table C10-5

Egypt: Gross Value of Inputs, 1952–65

(Thousand Egyptian pounds, at current prices)

Item	1952	1953	1954	1955	1956	1957	1958	1959	1960	1961	1962	1963	1964	1965
I. *Seeds*														
A. Field products	21,850	15,733	14,236	14,259	18,108	16,100	16,737	16,970	16,702	18,218	19,248	17,756	17,420	22,950
B. Vegetable products	1,428	1,144	1,314	1,225	1,372	1,894	2,234	2,327	2,489	1,780	1,610	1,858	1,703	2,163
C. Fruit products	107	108	91	141	129	211	168	2,271	329	394	395	196	285	290
Total	23,385	16,985	15,641	15,625	19,609	19,005	19,139	19,568	19,520	20,392	21,253	19,810	19,408	25,403
II. *Fertilizers*														
A. Field products	20,777	18,486	17,840	18,314	17,428	21,319	20,908	21,265	22,146	23,535	23,325	25,620	29,009	33,032
B. Vegetable products	1,299	1,312	1,338	1,455	1,448	1,709	1,793	1,848	2,220	2,276	2,408	2,905	3,480	2,761
C. Fruit products	821	733	743	689	603	750	774	888	971	968	961	1,127	1,447	2,351
D. Natural fertilizers	22,461	23,080	22,956	25,470	28,084	29,567	31,083	32,662	41,730	39,401	45,546	53,656	54,613	55,581
Total	45,358	43,611	42,877	45,925	47,563	53,345	54,558	56,663	67,067	66,180	72,240	83,308	88,549	93,725
III. *Chemical insecticides*	117	109	188	816	965	1,334	1,302	1,913	1,784	4,462	4,775	7,404	12,093	10,404
IV. *Fodder*	69,888	58,294	56,554	52,032	57,383	58,297	59,975	59,828	62,751	67,092	69,287	72,551	91,698	143,670
V. *Fuel and lubricants*														
A. Irrigation equipment	3,333	3,636	3,555	2,245	2,154	2,403	2,555	2,541	2,360	2,419	2,401	2,440	3,137	3,984
B. Tractors	2,495	2,603	2,493	5,311	5,094	6,446	7,615	8,194	7,686	7,990	8,429	8,693	8,557	8,412
Total	5,828	6,239	6,048	7,556	7,248	8,849	10,170	10,735	10,046	10,409	10,830	11,133	11,694	12,396
VI. *Depreciation*														
A. Irrigation equipment	331	338	338	342	357	363	357	385	387	474	473	455	589	672
B. Tractors	252	266	296	554	599	626	610	782	802	658	744	777	799	763
C. Threshing, harvesting and other agricultural machines	1,669	1,669	1,669	1,704	1,717	1,735	1,534	1,353	1,248	1,078	1,112	1,371	1,211	1,075
Total	2,252	2,273	2,303	2,600	2,673	2,724	2,501	2,520	2,437	2,210	2,329	2,603	2,599	2,510
VII. *Maintenance*														
A. Irrigation equipment	435	444	444	450	470	480	470	508	510	514	520	517	620	737
B. Tractors	269	283	316	591	575	601	585	651	655	601	638	654	606	597
C. Other agricultural machines	183	183	183	185	186	187	177	168	172	154	156	169	161	154
Total	887	910	943	1,226	1,631	1,268	1,232	1,327	1,337	1,269	1,314	1,340	1,387	1,488
VIII. *Eggs for hatcheries*	602	673	674	815	907	881	946	1,030	1,101	1,123	1,285	1,192	1,640	1,871
Total inputs	148,217	129,094	125,228	126,595	137,579	145,703	149,823	153,584	166,043	173,147	183,313	199,341	229,068	291,467

Source: UAR, Central Organization for General Mobilization and Statistics, *Estimates of National Income from the Agricultural Sector, 1965* (Cairo, 1967), p. 6.

Appendix Table C10-6

Egypt: Gross Value of Agricultural Inputs, 1952−65

(Thousands of Egyptian pounds, at 1954 prices)

Item	1952	1953	1954	1955	1956	1957	1958	1959	1960	1961	1962	1963	1964	1965
Seeds	14,166	14,914	15,641	15,692	15,680	15,993	15,775	16,975	16,628	16,379	17,524	16,595	16,497	17,102
Chemical fertilizers	19,459	18,395	19,921	18,878	18,788	20,592	20,706	23,874	23,524	27,275	28,209	29,596	34,743	37,052
Natural fertilizers	22,461	23,080	22,956	23,639	24,316	24,769	24,452	25,691	26,037	26,511	26,977	27,466	27,951	28,460
Insecticides	151	171	188	832	862	1,130	757	1,414	1,037	2,194	698	1,177	1,961	1,918
Fodder	30,137	31,723	56,554	54,474	50,830	59,549	52,316	33,215	33,984	31,726	33,753	30,108	31,672	29,009
Fuel and lubricants	5,639	5,778	6,048	8,452	8,742	8,849	9,220	9,178	9,205	8,909	9,085	9,191	9,912	10,015
Depreciation	2,256	2,272	2,303	2,590	2,625	2,724	2,680	2,675	2,680	2,645	2,665	2,678	2,748	2,754
Maintenance	892	910	943	1,256	1,297	1,268	1,358	1,355	1,360	1,324	1,346	1,359	1,453	1,467
Eggs for hatcheries	602	674	674	812	887	834	858	922	986	1,051	1,113	1,239	1,579	1,439
Total inputs	95,763	97,917	125,228	126,625	124,027	135,708	128,122	115,299	115,441	118,014	121,370	119,409	128,516	129,216

Source: UAR, Central Organization for General Mobilization and Statistics, *Estimates of National Income from the Agricultural Sector, 1965* (Cairo, 1967), p.8.

Appendix Table C10-7

Egypt: Agricultural Income, 1951−65

(Million Egyptian Pounds)

	Agricultural Output			Costs of Agricultural Production									Value added
Year	Crops	Animal	Total	Seeds	Fertilizers	Insecticides	Fodder	Fuel	Depreciation	Maintenance	Others	Total	Value added
1951	375.0	93.0	468.0	17.5	19.1	0.2	72.8	5.5	2.2	0.9	−	118.2	349.8
1952	312.6	105.7	418.3	23.4	45.4	0.1	69.9	5.8	2.3	0.9	0.5	148.3	270.0
1953	305.5	99.2	404.7	17.0	43.6	0.1	58.3	6.2	2.3	0.9	0.7	129.1	275.6
1954	333.1	101.0	434.1	15.6	42.9	0.2	56.6	6.0	2.3	0.9	0.7	125.2	308.9
1955	330.7	109.8	440.5	15.6	45.9	0.8	52.0	7.6	2.6	1.2	0.9	126.6	313.9
1956	383.5	112.2	495.7	19.6	47.6	1.0	57.4	7.2	2.7	1.2	0.9	137.6	358.1
1957	387.2	125.1	512.3	19.0	53.3	1.3	58.3	8.8	2.7	1.3	1.0	145.7	366.6
1958	388.5	130.7	519.2	19.1	54.6	1.3	60.0	10.2	2.5	1.2	0.9	149.8	369.3
1959	406.5	137.6	544.1	19.6	56.7	1.9	59.8	10.7	2.5	1.3	1.0	153.5	390.5
1960	438.1	145.9	584.0	19.5	67.1	1.8	62.8	10.0	2.4	1.3	1.1	166.0	418.0
1961	379.9	145.1	525.0	20.4	66.2	4.5	67.1	10.4	2.2	1.3	1.1	173.2	351.9
1962	453.6	155.6	609.2	21.3	72.2	4.8	69.3	10.8	2.3	1.3	1.3	183.3	425.9
1963	464.5	214.6	679.1	19.8	83.3	7.4	72.6	11.1	2.6	1.3	1.2	199.3	479.7
1964	518.7	246.0	764.7	19.4	88.5	12.1	91.7	11.7	2.6	1.4	1.7	229.1	535.6
1965	635.5	270.6	906.1	25.4	93.7	10.4	143.7	12.4	2.5	1.5	1.9	291.5	614.7

Source: N.B.E., *Economic Bulletin*, XXI, No. 1 (Cairo, 1968). Also, Central Agency for General Mobilization and Statistics.

Appendix Table C10-8

Egypt: Breakdown of Production Costs of Main Crops by Labor, Material Inputs, and Rent, 1964

(Egyptian Pounds Per Feddan)

				Variable costs							Percentage distribution				
Crop	Labor	Working cattle	Seeds	Natural ferti-lizers	Chemical ferti-lizers	Irri-gation	Other	Total variable costs	Rent	Total direct costs	Cost of labor	Cost of cattle	Other costs	Rent	Total
Cotton	17.680	4.242	0.998	1.372	4.241	1.773	4.857	35.163	22.518	57.681	31	7	23	39	100
Sugarcane	16.366	4.221	4.013	0.640	15.241	7.003	2.797	50.281	24.804	75.084	22	6	39	33	100
Rice	11.376	5.758	3.016	1.072	3.747	–	0.391	25.360	9.695	35.055	33	16	23	28	100
Millet	7.083	1.778	0.309	0.679	4.791	5.546	0.473	20.659	8.445	29.104	24	6	41	29	100
Sesame	5.752	2.124	0.329	1.114	1.782	2.251	0.512	13.864	7.894	21.758	26	10	28	36	100
Peanut	9.162	3.105	1.766	2.415	1.167	1.324	1.604	20.543	10.572	31.115	29	10	27	34	100
Maize	6.068	2.428	1.053	3.738	3.571	0.131	0.971	17.960	8.657	26.617	23	9	36	32	100
Wheat	4.390	3.880	2.720	1.290	3.590	0.580	0.460	16.910	13.980	30.890	14	13	28	45	100
Barley	3.440	3.160	2.060	–	2.040	0.540	0.380	11.620	10.530	22.150	16	14	23	47	100
Broad beans	3.760	2.550	4.560	–	0.860	0.500	0.740	12.970	12.940	25.910	14	10	26	50	100
Lentil	3.570	2.560	5.570	–	–	–	0.460	12.160	13.950	26.110	14	10	23	53	100
Onions	8.480	2.380	12.760	2.870	3.430	0.830	0.780	31.530	15.000	46.530	18	5	45	32	100
Chick peas	3.940	2.270	6.440	–	–	–	0.560	13.210	14.080	27.290	14	8	26	52	100

Source: UAR, Ministry of Agriculture, *Agricultural Economy,* monthly bulletin published by the Department of Agricultural Economics and Statistics (Cairo, 1966), pp. 476–479.

Appendix Table C10-9

Egypt: Variable Cost of Agricultural Production, by Operations, 1964

(Egyptian Pounds Per Feddan)

Crop	Land preparation	Seeds and sowing	Irrigation	Fertilizing	Harvesting, threshing, etc.	Weeding	Miscellaneous	Total
Wheat	1.860	2.780	1.460	5.050	5.760	–	–	16.910
Barley	1.630	2.130	1.350	2.090	4.420	–	–	11.620
Broad beans	1.470	4.910	1.120	0.910	4.560	–	–	12.970
Lentils	0.890	6.110	–	–	5.160	–	–	12.160
Chick peas	0.410	7.420	–	–	5.380	–	–	13.210
Onions	2.060	14.980	1.120	6.560	5.170	1.640	–	31.530
Flax (linen)	2.720	4.510	1.490	4.420	5.380	1.290	–	19.810
Fenugreek	0.360	3.360	0.690	–	4.550	–	–	8.960
Lupins	0.570	3.380	0.460	–	3.700	–	–	8.110
Cotton	3.935	1.421	4.376	6.054	–	–	19.377	35.163
Sugarcane	2.873	1.181	–	–	–	–	46.227	50.281
Rice	2.140	3.016	4.355	4.965	5.460	1.745	3.679	25.360
Millet	2.238	0.911	6.402	5.728	3.748	–	11.632	20.659
Sesame	2.235	0.454	3.597	3.108	3.363	1.107	–	13.864
Peanuts	2.467	2.402	4.824	3.824	4.218	–	2.808	20.543
Maize	2.095	1.310	2.265	7.816	2.361	–	2.113	17.960

Source: UAR, Ministry of Agriculture, *Agricultural Economy,* monthly bulletin published by the Department of Agricultural Economics and Statistics (Cairo, 1966), pp. 465–473.

Appendix Table C10-9a

**Israel: Net Domestic Product (Value Added), Value of Output, and Value of Purchased Inputs
in Agriculture, Years of Record 1952–1966/67***

(Million Israel Pounds, Current Prices)

Item	1952	1955	1956/57	1957/58	1958/59	1959/60	1960/61	1961/62	1962/63	1963/64	1964/65	1965/66	1966/67
Crop products	79.2	169.9	246.7	278.3	296.5	319.3	366.7	421.9	568.4	578.1	673.9	725.5	833.7
Animals & products	61.3	142.2	211.5	271.6	304.1	330.2	372.5	413.8	456.2	510.4	530.7	569.7	636.0
Production for investment	16.1	41.4	65.7	85.8	74.2	60.6	62.8	61.9	62.1	67.7	64.3	57.5	64.0
Total	156.5	353.5	523.9	635.7	674.8	710.1	802.0	897.6	1,086.7	1,156.2	1,268.5	1,352.7	1,533.7
Purchased inputs	60.3	154.8	224.5	268.3	297.0	337.4	349.1	410.4	473.1	489.3	568.0	643.6	690.1
Net domestic product	96.2	198.7	229.4	367.4	377.8	372.7	452.9	487.2	613.6	666.9	700.5	709.1	843.6
Special subsidies	0.3	–		1.2		23.2	4.5	8.0	12.7	2.6	0.2	13.8	2.5
Income originating in agriculture	96.5	198.7		368.6		395.9	457.4	495.2	626.3	669.5	700.7	722.9	846.1

* When a single figure is given, year is calendar year; when a dual figure is given, year is agricultural year, October-September.
Source: Various issues of *Statistical Abstract of Israel*, published by the Central Bureau of Statistics.

Appendix Table C10-10

Israel: Value of Agricultural Production (Including Intermediate Products) at Current Prices, 1948/49 to 1966/67*

(Thousand Israel Pounds)

Item	1948/49	1949/50	1950/51	1951/52	1952/53	1953/54	1954/55	1955/56	1956/57	1957/58
Wheat	1,134	1,271	885	3,430	4,868	7,650	7,740	17,460	19,180	14,213
Barley	742	1,282	1,381	8,114	8,640	15,733	8,209	15,643	14,062	10,434
Sorghum for grain	124	177	140	832	2,055	3,607	1,998	5,090	7,024	6,017
Other cereals	369	865	253	1,413	2,278	5,498	5,551	5,681	8,936	7,079
Pulses	247	270	185	1,785	1,465	2,877	1,866	2,918	3,860	3,620
Hay	1,015	1,596	2,074	5,705	6,716	7,566	6,193	7,975	8,978	8,640
Green fodder and silage	1,731	2,538	4,241	6,114	8,432	13,021	16,885	18,586	26,495	27,165
Other roughage	–	13	326	991	1,083	2,251	2,372	2,361	2,677	2,382
Groundnuts	69	223	1,016	2,295	4,030	7,925	9,115	7,455	9,205	7,075
Sunflower, safflower, sesame	256	292	152	698	1,030	1,572	1,174	2,455	3,032	2,329
Cotton lint	–	–	–	–	47	562	4,400	6,639	10,096	11,888
Cottonseed	–	–	–	–	–	100	700	811	1,485	2,166
Tobacco	210	689	1,688	3,255	2,408	3,805	3,155	2,011	2,653	2,274
Sugar beets	–	–	5	193	391	388	1,176	1,674	3,073	5,048
Other industrial crops	–	–	52	45	567	342	292	480	729	332
Melons and pumpkins	220	1,318	738	4,511	4,502	4,063	4,775	6,643	6,600	7,054
Green manure and straw	581	673	497	2,358	2,716	3,707	3,515	6,177	6,895	6,178
Vegetables	4,195	5,696	8,813	19,237	24,132	28,436	33,077	39,886	45,077	50,734
Potatoes	1,143	1,265	1,535	4,237	7,830	10,727	12,106	14,120	14,646	14,745
Citrus fruit	6,924	7,470	8,648	16,040	31,099	67,319	62,599	75,394	87,654	100,046
Table grapes	831	811	919	2,996	4,366	5,697	6,991	7,530	9,275	10,582
Wine grapes	396	415	637	1,745	1,897	2,141	2,773	3,720	5,453	6,362
Olives	832	475	615	2,680	2,886	4,653	1,243	7,579	2,833	5,783
Bananas	477	286	913	2,023	3,943	6,056	8,600	10,143	10,562	12,912
Deciduous fruit	546	761	623	3,879	3,948	6,077	7,180	9,993	11,053	17,976
Other fruit	170	356	590	1,503	1,895	1,671	2,514	3,095	3,200	4,308
Cow milk	6,626	7,288	9,196	17,639	27,448	34,529	38,222	46,227	51,913	62,672
Other milk	587	923	1,266	2,570	4,822	6,671	8,265	9,670	10,438	11,390
Eggs	6,663	7,882	9,331	13,283	20,235	26,729	36,300	39,604	58,755	84,429
Honey	123	217	281	704	1,040	517	624	1,442	922	1,023
Changes in livestock inventory	1,433	1,886	2,445	3,668	4,118	5,285	3,511	9,451	15,170	25,232
Meat:										
cattle	545	375	563	1,160	1,896	2,828	10,998	16,491	22,570	27,897
sheep and goats	140	220	168	476	1,212	2,946	4,175	5,196	6,740	8,923
poultry	3,090	4,137	4,034	7,001	12,068	18,208	30,236	40,268	46,214	63,645
other	–	–	–	–	–	–	3,150	5,188	7,197	10,553
Fish	1,584	2,313	3,050	5,710	7,861	9,400	11,566	11,995	14,494	17,471
Miscellaneous[a]	1,410	2,320	3,710	6,683	9,378	11,206	14,159	15,652	16,941	20,904
Total	44,413	56,303	70,970	154,973	223,302	331,763	377,405	482,703	576,607	681,481

Appendix Table C10-10—*Continued*

Item	1958/59	1959/60	1960/61	1961/62	1962/63	1963/64	1964/65	1965/66	1966/67
Wheat	16,012	8,960	14,663	12,661	14,499	33,532	41,585	28,791	63,405
Barley	11,255	4,599	11,672	10,799	7,540	22,504	13,541	4,787	12,488
Sorghum for grain	6,521	2,722	6,687	9,544	8,003	15,110	13,393	2,630	5,366
Other cereals	3,506	2,402	2,467	3,047	2,895	2,730	1,417	1,193	1,706
Pulses	2,906	1,709	2,829	2,759	3,483	5,639	4,263	3,325	5,287
Hay	8,985	6,216	7,876	9,845	11,435	9,378	8,914	10,443	14,709
Green fodder and silage	30,583	31,922	32,737	36,367	40,420	40,435	39,885	38,915	38,388
Other roughage	2,243	2,074	1,904	2,039	1,869	2,032	1,788	2,803	2,848
Groundnuts	8,865	9,836	8,410	8,556	11,830	9,595	14,715	14,190	14,061
Sunflower, safflower, sesame	2,889	3,421	4,295	3,680	4,470	4,736	5,182	3,712	4,078
Cotton lint	17,903	23,001	30,659	32,373	30,131	36,103	49,235	56,138	64,438
Cottonseed	1,880	2,597	3,590	4,539	3,909	4,410	6,372	7,236	9,600
Tobacco	3,601	2,795	4,091	4,777	331	2,889	5,404	6,599	7,588
Sugar beets	5,930	8,851	12,858	12,627	14,753	16,209	18,962	19,389	18,448
Other industrial crops	606	1,131	1,576	1,075	1,078	997	1,896	2,637	1,852
Melons and pumpkins	8,351	7,443	8,755	11,350	12,197	15,157	17,098	21,208	19,694
Green manure and straw	6,329	4,853	6,258	6,035	6,094	9,163	9,282	6,386	11,135
Vegetables	45,964	48,810	55,823	65,142	80,308	91,568	99,080	108,798	113,942
Potatoes	12,766	11,822	13,734	17,259	18,598	23,596	22,464	23,844	25,137
Citrus fruit	106,166	110,690	111,552	144,595	245,584	195,412	240,320	268,628	317,163
Table grapes	9,282	12,347	14,701	15,399	17,204	17,736	19,501	20,017	17,082
Wine grapes	6,329	5,684	6,992	8,101	9,074	10,360	8,880	9,316	11,907
Olives	3,589	4,163	8,333	3,831	9,141	13,464	7,852	8,898	17,998
Bananas	12,173	14,325	18,186	18,825	20,959	19,581	24,933	22,626	22,189
Deciduous fruit	23,356	36,986	43,840	52,302	65,938	76,077	81,517	91,717	87,356
Other fruit	3,959	5,197	6,445	6,369	9,419	9,198	11,654	13,643	17,608
Cow milk	67,808	72,394	81,470	93,491	97,774	105,613	114,404	130,835	142,055
Other milk	10,776	12,105	13,881	14,062	14,791	16,215	17,141	19,305	21,673
Eggs	85,176	93,712	109,278	108,543	115,955	131,001	130,508	135,191	159,239
Honey	2,250	2,069	2,190	2,307	2,191	3,354	3,630	4,895	5,375
Changes in livestock inventory	15,125	5,392	8,444	9,030	0	2,367	2,133	1,220	12,996
Meat:									
cattle	43,103	54,531	55,030	56,601	78,269	92,463	87,041	83,636	87,409
sheep and goats	8,716	9,265	11,001	13,009	15,089	16,528	18,243	19,456	21,763
poultry	66,352	71,825	86,230	101,809	115,346	126,498	135,883	148,424	161,910
other	12,636	10,795	11,420	12,840	13,020	11,070	11,700	11,458	11,020
Fish	17,320	18,749	19,682	22,546	27,330	32,527	33,799	41,528	39,542
Miscellaneous[a]	21,711	24,371	25,473	27,046	30,400	29,128	31,198	32,433	34,559
Total	713,012	749,764	865,032	965,160	1,161,327	1,256,071	1,355,000	1,423,307	1,622,996

* Israel economic data are based upon the Hebrew calendar; the year begins in early fall (September, usually) and extends for about twelve months.

[a] Flowers, seedlings, and ornamental plants; organic manure; vegetable seeds and others.

Source: Various issues of *Statistical Abstract of Israel,* published by the Central Bureau of Statistics.

Appendix Table C10-10a

Israel: Value of Purchased Agricultural Inputs, by Product Purchased, 1952–67*

(Million Israel Pounds, at Current Prices)

	1952	1953	1954	1955	1956	1956/57	1957/58	1958/59	1959/60	1960/61	1961/62	1962/63	1963/64	1964/65	1965/66	1966/67
PURCHASED INPUTS	60.3	92.2	130.5	154.8	193.8	224.5	268.3	297.0	337.4	349.1	412.8	473.1	489.2	568.0	643.6	690.1
Feedingstuffs	10.5	24.1	32.4	45.6	63.1	75.0	99.6	112.0	140.9	138.0	151.8	167.2	157.2	195.6	230.7	245.0
Grains, imported	2.6	8.9	13.6	22.1	27.6	34.9	43.4	51.5	80.7	69.4	75.1	–	–	–	–	–
Milling by-products, etc.	2.6	6.6	6.7	6.7	7.6	7.7	7.2	8.1	7.7	9.8	9.4	–	–	–	–	–
Oil cake	2.8	3.7	4.8	7.2	14.3	15.3	21.0	25.7	25.8	31.7	37.0	–	–	–	–	–
Protein concentrates	1.0	1.6	2.8	3.7	5.3	5.7	13.5	11.6	11.5	11.0	11.6	–	–	–	–	–
Minerals	0.3	0.5	1.0	1.1	1.3	1.8	2.7	3.3	4.0	2.8	5.2	–	–	–	–	–
Vitamins and antibiotics	–	0.1	0.5	0.9	2.0	3.5	5.5	4.2	4.6	5.3	5.3	–	–	–	–	–
Miscellaneous	1.2	2.7	3.0	3.9	5.0	6.1	6.3	7.6	6.6	8.0	8.2	–	–	–	–	–
Water	6.7	9.4	12.8	15.5	18.3	22.3	27.5	30.0	34.4	33.6	42.5	45.2	39.4	46.8	56.7	51.0
Packing Materials	5.1	9.1	18.7	16.2	19.5	20.3	20.4	24.1	24.7	22.1	30.9	42.1	46.3	55.1	61.9	71.0
Citrus packing materials	3.1	6.7	15.3	12.6	15.2	15.6	15.3	17.6	17.8	14.4	21.6	–	–	–	–	–
Other packing materials	2.0	2.4	3.4	3.6	4.3	4.7	5.1	6.5	6.9	7.7	9.3	–	–	–	–	–
Fertilizers	5.0	5.0	9.3	9.8	12.4	13.3	15.7	16.3	16.9	16.3	19.9	22.8	23.2	28.0	30.3	32.9
Nitrogen fertilizers	2.0	2.3	5.0	4.7	6.7	7.5	9.0	9.6	10.2	10.1	12.6	–	–	–	–	–
Phosphate fertilizers	3.0	2.5	3.9	4.8	5.2	5.3	6.0	5.9	5.8	5.1	6.3	–	–	–	–	–
Potash and other fertilizers	–	0.2	0.4	0.3	0.5	0.5	0.7	0.8	0.9	1.1	1.0	–	–	–	–	–
Transport	4.4	5.5	7.5	9.1	11.0	12.7	14.7	16.4	16.9	18.8	20.9	33.6	37.4	42.3	46.8	52.8
Spare parts, repairs and tools	4.0	5.4	6.3	7.4	9.7	11.3	13.5	15.1	14.4	18.6	24.7	23.6	31.5	33.2	34.5	36.0
Fuel, lubricants and electricity	2.6	3.5	4.4	5.3	5.6	7.3	7.9	8.8	9.3	10.6	11.5	12.7	16.6	17.6	18.8	20.8
Plant and animal protection materials	1.6	2.2	3.1	3.6	5.1	7.9	9.1	9.6	10.8	13.2	17.1	17.3	16.5	16.6	18.1	21.7
Government	2.6	3.3	3.8	4.9	6.9	7.1	7.7	9.2	10.7	12.9	14.6	16.3	15.3	15.3	16.5	17.0
Seeds and seed cleaning	1.2	2.4	3.3	4.3	4.0	3.7	3.7	4.2	4.1	4.5	5.4	4.7	3.8	3.4	3.6	3.6
Services, miscellaneous	1.5	2.0	2.7	3.1	3.6	4.0	4.9	5.2	6.0	6.4	7.4	10.4	10.4	12.5	13.9	16.2
Depreciation	13.5	18.5	24.3	27.7	32.1	37.0	40.3	42.7	45.6	51.0	62.8	71.3	83.9	95.2	103.4	109.8
Miscellaneous expenses	1.6	1.8	1.9	2.3	2.5	2.6	3.3	3.4	2.7	3.1	3.3	6.3	7.1	6.4	8.4	12.3

* Calendar years, through 1956; agricultural years (October to September) beginning in 1957.
Sources: Israel, Central Bureau of Statistics, *National Income Originating in Israel's Agriculture, 1952–1963; Statistical Abstract of Israel.*

Appendix Table C10-11

Israel: Value of Agricultural Production (Excluding Intermediate Products) at Current Prices, 1960/61 to 1966/67*

(Thousand Israel Pounds)

Item	1960/61	1961/62	1962/63	1963/64	1964/65	1965/66	1966/67
Wheat	8,934	7,459	11,076	27,447	35,001	23,959	58,097
Barley	95	112	84	98	103	66	1,226
Sorghum for grain	0	0	0	0	0	0	0
Other cereals	340	376	422	532	492	467	450
Pulses	701	764	832	783	1,669	1,820	1,898
Hay	0	0	0	0	0	0	0
Green fodder and silage	8	205	191	18	14	25	182
Other roughage	0	0	0	0	0	0	0
Groundnuts	8,005	8,142	8,742	6,675	14,061	13,662	13,516
Sunflower, safflower, sesame	4,181	3,531	4,324	4,594	5,048	3,575	3,956
Cotton lint	30,659	32,373	30,131	36,103	49,235	56,138	64,438
Cottonseed	3,590	4,539	3,909	4,410	4,797	4,707	6,640
Tobacco	4,091	4,777	331	2,889	5,404	6,599	7,588
Sugar beets	12,858	12,627	14,753	16,209	18,962	19,389	18,448
Other industrial crops	1,576	1,075	1,078	992	1,888	2,500	1,478
Melons and pumpkins	8,570	11,091	11,883	14,813	16,721	20,659	19,501
Green manure and straw	0	0	0	0	0	0	0
Vegetables	53,333	62,820	77,614	87,994	98,683	107,949	113,455
Potatoes	12,520	15,897	17,224	21,936	20,910	22,237	23,247
Citrus fruit	111,552	144,595	234,014	185,414	237,389	268,619	317,163
Table grapes	14,701	15,399	17,204	17,736	19,501	20,017	17,082
Wine grapes	6,992	8,101	9,074	12,056	11,953	9,316	11,907
Olives	8,333	3,831	9,141	13,464	7,852	8,898	17,998
Bananas	18,186	18,825	20,959	19,581	24,933	22,626	22,189
Deciduous fruit	43,840	52,302	65,938	75,916	81,517	91,717	87,356
Other fruit	6,293	6,162	9,165	9,011	11,482	13,411	17,327
Cow milk	78,683	90,634	99,727	108,149	111,922	128,532	139,454
Other milk	10,394	10,542	10,895	11,799	12,179	13,953	15,283
Eggs	93,389	92,523	96,944	108,502	111,215	112,771	137,161
Honey	2,190	2,307	2,191	3,354	3,630	4,895	5,375
Changes in livestock inventory	8,444	9,030	0	2,367	2,133	1,220	12,996
Meat:							
cattle	55,030	56,601	78,269	92,463	87,041	83,636	87,409
sheep and goats	11,001	13,009	15,089	16,528	18,246	19,456	21,763
poultry	86,230	101,809	115,346	126,498	135,883	148,424	161,910
other	11,420	12,840	13,020	11,070	11,700	11,458	11,020
Fish	19,682	22,546	27,330	32,527	33,799	41,528	39,542
Miscellaneous	5,287	6,618	8,159	8,113	8,978	10,971	12,400
Total	741,108	833,462	1,015,059	1,080,041	1,204,341	1,295,200	1,469,696

* Israel economic data are based upon the Hebrew calendar; the year begins in early fall (September, usually) and extends for about twelve months.

Source: Various issues of *Statistical Abstract of Israel,* published by the Central Bureau of Statistics.

Appendix Table C10-12

Israel: Value of Agricultural Production (Including Intermediate Products) at 1948/49 Prices, 1948/49 to 1966/67*

(Thousand Israel Pounds)

Item	1948/49	1949/50	1950/51	1951/52	1952/53	1953/54	1954/55	1955/56	1956/57	1957/58
Field crops	6,698	10,396	7,951	16,133	16,730	21,650	21,678	27,571	34,251	30,822
Vegetables and potatoes	5,338	8,041	9,105	11,353	13,085	14,303	14,772	16,245	17,023	18,528
Citrus fruit	6,924	7,459	8,402	8,011	9,507	12,820	11,451	12,770	13,032	13,452
Other fruits	3,252	2,814	2,661	5,130	5,358	6,845	5,655	9,393	9,112	11,881
Milk	7,213	8,768	9,900	11,549	12,735	14,658	16,236	17,093	18,333	21,114
Eggs	6,663	9,018	10,766	10,079	10,147	11,421	14,159	14,549	17,730	24,820
Honey	123	98	96	139	167	110	120	241	145	145
Changes in livestock inventory	1,433	1,592	1,411	1,094	1,012	982	651	1,525	2,346	4,442
Meat	3,775	5,152	5,144	5,201	6,016	7,188	12,808	17,853	17,821	25,910
Fish	1,584	2,744	3,303	3,552	3,439	4,073	4,941	4,988	5,151	5,629
Miscellaneous	1,410	2,070	2,450	3,006	3,250	3,683	4,072	4,512	4,891	5,835
Total	44,413	58,152	61,189	75,247	81,446	97,733	106,543	126,730	139,835	162,578

Item	1958/59	1959/60	1960/61	1961/62	1962/63	1963/64	1964/65	1965/66	1966/67
Field crops	36,693	33,414	42,741	41,703	40,011	51,423	50,937	52,512	67,740
Vegetables and potatoes	18,391	19,503	18,555	19,814	20,519	19,561	23,087	25,093	24,089
Citrus fruit	16,532	17,467	14,772	15,362	21,343	20,657	21,896	23,210	26,692
Other fruits	13,541	14,415	19,760	23,960	24,224	31,132	29,499	35,015	40,968
Milk	24,667	25,973	26,964	31,267	31,071	31,109	32,887	34,570	27,681
Eggs	29,337	31,466	37,181	36,988	32,947	37,448	37,815	36,207	41,276
Honey	234	192	214	211	202	286	356	522	585
Changes in livestock inventory	2,937	1,316	1,136	1,532	0	106	154	175	1,860
Meat	32,766	37,358	42,436	49,839	52,582	58,060	38,716	62,964	68,631
Fish	5,954	6,345	6,817	7,716	7,753	9,028	9,475	10,906	11,015
Miscellaneous	6,828	7,477	7,571	7,803	7,885	7,176	8,145	8,848	9,556
Total	187,880	194,926	218,147	236,195	238,537	265,986	272,967	290,022	330,093

* Israel economic data are based upon the Hebrew calendar; the year begins in early fall (September, usually) and extends for about twelve months.

Source: Various issues of *Statistical Abstract of Israel,* published by the Central Bureau of Statistics.

Appendix Table C10-13

Lebanon: Value of Gross Agricultural Production at Constant Prices,* 1954/56 to 1964/66

(£L. Million)

Product	1954/56	1959/61	1964	1965	1966	1964/66
Cereals	24.2	21.2	20.7	18.9	22.6	20.7
Pulses	7.8	6.2	6.0	6.9	6.2	6.4
Fruits	83.0	114.8	153.2	142.8	140.8	145.6
Vegetables	29.0	33.3	47.7	43.9	48.4	46.6
Oil & oilseeds	27.1	29.3	25.1	39.4	24.2	29.6
Sugar	2.8	1.8	5.2	4.7	6.2	5.4
Tobacco	6.3	10.2	16.5	24.2	17.3	19.3
Total crops	180.2	216.8	274.4	280.8	265.7	273.6

Appendix Table C10-13–*Continued*

Product	1954/56	1959/61	1964	1965	1966	1964/66
Dairy products	10.5	19.0	34.9	35.0	33.8	34.6
Eggs	3.4	3.8	18.0	26.9	43.7	29.5
Meat: poultry	2.8	8.1	21.9	26.3	32.2	26.8
Other meat	18.5	20.7	22.7	23.7	25.3	23.9
Total meat	21.3	28.8	44.6	50.0	57.5	50.7
Other animal products	6.7	6.8	9.6	8.9	8.8	9.2
Total animal products	41.9	58.4	107.1	120.8	143.8	124.0
Total agricultural products:						
Food	209.1	258.2	355.4	368.5	383.4	369.1
Nonfood	13.0	17.0	26.1	33.1	26.1	28.5
Total	222.1	275.2	381.5	401.6	409.5	397.6

* At average farm gate prices during 1964–66.

Source: Rural Economics Institute, *Agriculture in the Lebanese Economy* (A. Ec. No. 9 [Beirut, February 1968]), p. 54.

Appendix Table C10-14
Jordan: Agriculture Income and Expenditure in Current Prices, 1959–66
(JD Million)

	1959	1960	1961	1962	1963	1964	1965	1966
Income								
1. Grains and legumes	3.10	2.10	5.65	4.50	3.06	10.84	10.08	4.24
2. Vegetables	5.22	5.71	7.87	6.87	7.36	8.57	9.01	5.73
3. Tobacco	0.36	0.11	0.39	0.29	0.10	0.41	0.40	0.31
4. Fruits, vines, and olives	2.64	2.75	9.00	3.44	5.32	7.93	5.65	7.51
5. Forest products	0.31	0.20	0.20	0.16	0.14	0.13	0.14	0.14
6. Sales of animals	4.20	3.36	2.48	2.30	3.37	3.36	4.22	6.15
7. Animal products	3.77	2.84	2.63	2.71	3.12	2.93	3.53	4.01
8. Poultry and game	1.03	1.14	1.26	1.57	1.70	2.03	2.62	2.79
9. Honey	0.02	0.02	0.04	0.05	0.03	0.04	0.04	0.05
10. Fish	0.01	0.02	0.03	0.04	0.04	0.04	0.04	0.04
11. Construction on farms (labor)	0.10	0.08	0.16	0.14	0.29	0.15	0.22	0.54
12. Increase in livestock numbers	–	–	–	2.02	0.84	1.43	2.77	1.15
Total	20.76	18.34	29.71	24.09	25.37	37.86	38.72	32.66
Expenditure								
1. Seeds	1.10	1.10	1.51	1.33	1.09	1.11	1.02	0.96
2. Machinery expenses	0.14	0.18	0.25	0.31	0.35	0.42	0.55	0.60
3. Fertilizers	0.14	0.16	0.11	0.19	0.23	0.26	0.41	0.47
4. Insecticides	0.05	0.04	0.05	0.06	0.09	0.09	0.11	0.12
5. Animal feed	0.89	0.82	0.74	0.92	1.06	1.26	1.90	2.33
6. Seedlings	–	0.01	0.01	0.01	0.01	0.01	0.02	0.01
7. Irrigation expenses	0.19	0.21	0.31	0.32	0.37	0.37	0.34	0.33
8. Imported chicks	–	0.01	0.04	0.05	0.09	0.20	0.26	0.19
9. Reduction in livestock numbers	3.17	1.19	1.39	–	–	–	–	–
Total	5.68	3.72	4.41	3.19	3.29	3.72	4.61	5.01
Net farm income	15.08	14.62	25.30	20.90	22.08	34.14	34.11	27.65

Source: Jordan, Department of Statistics, *The National Accounts 1959–66* (Amman, n.d.), p. 27.

Appendix Table C10-15
Syria: Index Numbers of Agricultural Production, 1952–66
(1956 = 100)

	Crop Production						Animal Production					Total agricultural production
Year	Cereals	Legumes	Vegetables	Industrial	Fruits	Total	Milk products	Meat & hides	Wool & hair	Others	Total	
1952	92	92	90	73	72	84	63	74	95	111	71	77
1953	91	97	142	54	100	86	96	80	84	111	92	86
1954	103	97	146	88	100	101	96	83	90	111	93	97
1955	42	61	73	91	80	63	82	98	96	100	88	76
1956	100	100	100	100	100	100	100	100	100	100	100	100
1957	132	107	106	113	95	119	88	105	106	96	95	107
1958	53	51	82	100	99	74	55	123	95	93	78	76
1959	57	47	110	118	85	79	34	138	99	71	68	77
1960	48	23	97	116	86	72	23	134	64	95	56	68
1961	71	53	115	130	113	94	28	123	63	103	57	86
1962	137	99	195	158	142	144	59	122	65	131	78	130
1963	123	79	167	156	106	129	173	132	84	155	119	127
1964	110	122	160	186	147	138	131	143	90	176	132	137
1965	108	112	152	196	106	133	150	149	104	181	147	136
1966	50	48	111	161	102	89	128	170	91	133	136	99

Source: Syria, Ministry of Planning, *Statistical Abstract* (Damascus), various issues.

Appendix Table C10-16

Syria: National Income Estimates, 1956–64, at 1963 prices

(In Million Syrian Pounds)

Sector	1956	1957	1958	1959	1960	1961	1962	1963	1964
National income	2,556.8	2,752.5	2,383.2	2,426.9	2,401.8	2,666.7	3,376.6	3,115.4	3,288.6
Agriculture	981.0	1,118.3	745.6	765.2	676.9	843.7	1,265.5	1,037.8	1,055.6
Industry	284.6	293.1	298.8	301.7	310.2	315.9	327.3	347.2	365.1
Construction	83.8	63.7	77.1	67.9	85.5	85.5	86.3	86.5	107.7
Transpt. & comm.	237.3	225.4	230.2	237.3	258.7	232.5	270.5	270.0	300.2
Trade	543.0	559.3	483.3	461.5	445.3	488.7	608.2	457.6	452.7
Finance	46.8	72.5	61.3	52.4	52.4	34.2	78.2	104.1	100.0
Rent	125.5	143.1	173.2	194.5	212.1	234.7	248.5	270.6	266.4
Public administration	79.7	100.4	129.1	158.6	166.6	224.7	278.9	320.7	386.1
Services	158.6	160.2	168.1	171.3	177.6	190.3	196.7	204.5	243.7
Rest of world	16.5	16.5	16.5	16.5	16.5	16.5	16.5	16.5	11.5

Source: A. Jano, "National Income of Syria," in *Al Iktisad,* bi-monthly in Arabic (Damascus, July 16, 1968), p. 15.

Appendix Table C10-17

Iraq: Agricultural Income, 1962–66

(Million Iraq Pounds)

Item	At current prices					At constant prices of 1962				
	1962	1963	1964	1965	1966	1962	1963	1964	1965	1966
Value of crop products										
Field crops	57.8	41.9	50.4	52.8	55.2	57.8	39.1	43.4	50.1	45.7
Vegetable crops	20.0	23.4	24.6	29.5	35.3	20.0	21.9	26.2	29.4	31.1
Fruit crops	13.6	12.8	12.2	11.0	13.3	13.6	10.4	10.7	11.5	11.6
Total crops	91.4	78.2	87.3	93.4	103.8	91.4	71.4	80.3	91.0	88.4
Value of inputs	7.4	7.1	7.3	8.6	7.9	7.4	6.1	6.3	8.4	7.1
Value added of crops	84.0	71.0	80.0	84.8	95.9	84.0	65.3	74.0	82.5	81.3
Value of animal products										
Change in livestock inventory	5.0	5.2	5.8	7.3	8.2	5.0	5.6	6.2	6.9	7.6
Milk production	23.2	23.0	24.2	29.2	29.8	23.2	23.6	24.7	29.3	29.7
Meat production	20.7	23.3	25.3	27.6	28.5	20.7	21.3	21.8	21.5	23.2
Wool and hair	2.2	2.0	3.6	3.3	3.5	2.2	2.0	2.1	2.2	2.2
Hides and skins	1.2	1.2	1.7	1.1	2.1	1.2	1.5	1.7	1.4	1.5
Eggs	2.5	2.9	3.1	2.9	4.0	2.5	2.7	2.8	2.6	3.9
Other animal products	7.4	8.0	8.6	7.9	7.9	7.4	8.7	9.0	7.1	7.5
Total animal production	62.3	65.6	72.2	79.3	84.0	62.3	65.4	68.4	71.0	75.6
Value of inputs	13.2	11.7	13.4	14.1	15.4	13.2	11.7	13.4	14.7	11.4
Value added of animal products	49.1	53.9	58.8	65.3	68.6	49.1	53.7	55.0	56.4	64.2
Value added, other products										
Forest products	2.5	2.4	3.0	2.8	2.1	2.5	2.9	3.2	2.7	2.0
Fish products	3.3	4.2	4.6	2.0	1.6	3.3	3.4	3.3	2.5	2.1
Chopped straw	9.6	6.0	6.5	8.4	6.1	9.6	6.0	6.5	8.4	6.1
Total	15.4	12.6	14.1	13.2	9.8	15.4	12.3	13.0	13.6	10.2
Total value added of all agricultural production	148.6	137.6	152.9	163.2	174.2	148.6	131.3	142.1	151.5	155.6

Source: Preliminary draft of National Income Estimates, Central Bureau of Statistics, Ministry of Planning, Republic of Iraq.

Appendix Table C10-18
Iraq: Breakdown of Agricultural Inputs 1962–66, at Current and Constant Prices

Inputs	At current prices					At constant prices of 1962				
	1962	1963	1964	1965	1966	1962	1963	1964	1965	1966
Fertilizers	1,043	1,349	1,527	1,442	1,136	1,043	1,153	1,305	1,641	970
Pesticides	193	217	471	371	380	193	209	453	359	346
Machinery	1,980	1,811	1,190	1,585	1,546	1,980	1,548	1,017	1,340	1,445
Animal feed	7,386	6,393	6,970	6,086	7,823	7,386	6,393	6,970	6,570	5,795
Land tax	799	797	706	600	600	799	681	604	600	740
Fuel and oils	1,135	933	1,001	1,604	1,042	1,135	798	856	856	1,050
Seeds	510	517	689	808	–	510	442	588	781	–
Waste	1,701	1,523	1,720	2,192	3,140	1,701	1,309	1,470	2,900	2,545
Other requirements of animal production	5,800	5,300	6,400	8,000	7,600	5,800	5,300	6,400	8,100	5,600
Total	20,547	18,840	20,674	22,688	23,267	20,547	17,833	19,663	23,147	18,491

Source: Central Bureau of Statistics, Iraq, September 1968.

Appendix Table C10-19
Iraq: Gross National Product and National Income at Current and Constant Prices, 1962–66

Sector	At current prices					Distribution		At constant prices of 1962					Distribution	
	1962	1963	1964	1965	1966	1962	1966	1962	1963	1964	1965	1966	1962	1966
	(million Iraqi pounds)					(%)		(million Iraqi pounds)					(%)	
Agriculture & fisheries	149	138	153	163	174	25.8	23.7	149	131	142	151	156	25.9	22.0
Mining & quarrying	215	243	276	285	304	37.4	41.3	215	244	276	283	302	37.3	42.6
Manufacturing	54	52	53	61	63	9.3	8.6	54	53	52	60	61	9.4	8.7
Building & construction	16	16	17	18	19	2.7	2.6	16	16	17	18	19	2.8	2.7
Water & electricity	7	7	8	10	12	1.2	1.6	7	7	8	10	12	1.2	1.7
Trade	40	39	46	53	57	7.0	7.7	40	38	44	52	55	6.9	7.8
Transport & comm.	52	53	56	60	59	9.0	8.0	52	53	56	60	57	9.0	8.1
Banking & insurance	8	8	7	8	12	1.4	1.6	8	8	7	7	12	1.4	1.6
Property income	15	15	15	16	16	2.6	2.2	15	16	17	18	18	2.6	2.6
Public administration	70	77	86	89	97	12.2	13.1	70	77	86	90	96	12.1	13.6
Services	44	43	54	59	62	7.6	8.4	44	42	48	55	57	7.6	8.1
Total value of domestic product at factor cost	669	692	772	822	874	116.2	118.8	669	684	754	806	846	116.2	119.5
Less factor cost paid abroad	93	–	–	–	–	16.2	18.8	93	108	125	129	138	16.2	–19.5
Gross national product at factor cost—national income	576	584	649	692	736	100	100	576	576	630	677	708	100	100

Source: Central Bureau of Statistics, Ministry of Planning, Iraq, August 1968. Figures may not add to totals due to rounding.

Appendix Table C11-1

Egypt: Commodity Balances, 1960–62

(1,000 Metric Tons)

Commodity	Production	Imports	Supply	Local utilization Food	Other	Exports	Imports As percent of supply	Exports
Wheat	1,554[a]	1,450	3,004	2,803	195	6	48	—
Rice (paddy)	1,678[b]	—	1,678	1,196	84	398	—	24
Coarse grains	2,919	179	3,098	2,420	675	3	6	—
Starchy roots	465[c]	17	482	292	100	90	4	19
Sugar (raw equivalent)	375	65	440	384	4	52	15	12
Pulses	367	14	381	267	90	24	4	6
Vegetables, inc. melons	3,681	—	3,681	3,121	328	232	—	6
Fruits, excl. melons	1,103	32	1,135	1,035	88	12	3	1
Meat (excl. offal)	184	15	199	199	—	—	8	—
Eggs	34	—	34	30	4	—	—	—
Fish & fish products (fish equiv.)	129	14	143	139	—	4	10	3
Milk products (liquid equiv.)	1,140	62	1,202	1,141	58	3	5	—
Oilseed & veg. oils (oil equiv.)	125	35	160	155	—	5	22	3
Cotton & products (raw cotton equiv.)	412	—	412	—	91	321	—	78

[a] Including 45,000 m.t. from stocks.
[b] Including 122,000 m.t. from stocks
[c] Including 10,000 m.t. from stocks.

Source: Indicative World Plan, Near East (Rome: FAO, 1966), II.

Appendix Table C11-2

Israel: Commodity Balances, 1965–66

(1,000 Metric Tons)

Commodity	Production	Imports	Supply[a]	Local utilization Food	Other	Exports	Imports As percent of supply	Exports
Wheat	101	317	384	338	45	1	83	—
Rice (milled)		17	17	16	1	—	100	—
Coarse grains[b]	112	483	595	—	595	—	81	—
Potatoes	104	6	115	85	18	12	5	10
Sugar (refined)	36	77	110	88	13	9	70	8
Pulses and groundnuts	12	7	18	14	1	3	39	17
Vegetables (incl. melons)	437	4	438	382	28	28	1	6
Fruits (excl. melons)	1,120	2	1,116	287	52	777	—	70
Meat, incl. poultry (carcass weight)	98	41	130	130	—	—	32	—
Eggs	69	—	69	60	6	3	—	4
Fish[c]	23	7	27	27	—	—	26	—
Milk & products[c][d]	461	6	468	237	231	—	1	—
Oils & fats[c]	68	—	68	50	15	3	—	4

[a] Adjusted for changes in stocks, except for coarse grains.
[b] Average of 1965 and 1966 crop years.
[c] Aggregate product tonnage.
[d] Includes milk used as livestock feed.

Source: Coarse grains: *Israel Economic Development, Past Progress and Plans for the Future* (Jerusalem: Prime Minister's Office, Economic Planning Authority, March 1968); all other: *Statistical Abstract of Israel, 1967.*

Appendix Table C11-3
Lebanon: Commodity Balances, 1960—62
(1,000 Metric Tons)

Commodity	Production	Imports	Supply	Local utilization Food	Other	Exports	Imports As percent of supply	Exports As percent of supply
Wheat	61	221	282	260	19	3	78	1
Rice (paddy)	–	26	26	25	1	–	100	–
Coarse grains	27	63	90	15	66	9	70	10
Starchy roots	46	12	58	29	6	23	21	40
Sugar (raw equivalent)	3	41	44	42	2	–	93	–
Pulses	8	26	34	19	3	12	76	35
Vegetables, incl. melons	237	50	287	246	25	16	17	6
Fruits, excl. melons	471	23	494	247	98	149	5	30
Meat, excl. offal	22	27	49	49	–	–	55	–
Eggs	4	2	6	5	1	–	33	–
Fish & fish products (fish equiv.)	2.4	6.7	9.1	9.1	–	–	74	–
Milk products (liquid equiv.)	80	68	148	144	4	–	46	–
Oilseed & veg. oils (oil equiv.)	7	10	17	17	–	–	59	–
Cotton & products (raw cotton equiv.)	–	9	9	–	7	2	100	22

Source: Indicative World Plan, Near East (Rome: FAO, 1966), II.

Appendix Table C11-4
Jordan: Commodity Balances, 1960—62
(1,000 Metric Tons)

Commodity	Production	Imports	Supply	Local utilization Food	Other	Exports	Imports As percent of supply	Exports As percent of supply
Wheat	98	169	267	237	25	5	63	2
Rice (paddy)	–	27	27	27	–	–	100	–
Coarse grains	43	18	61	22	38	1	30	2
Starchy roots	13	11	24	18	5	1	46	4
Sugar (raw equivalent)	–	47	47	42	5	–	100	–
Pulses	17	6	23	12	7	4	26	17
Vegetables, incl. melons	392	9	401	267	49	85	2	21
Fruits, excl. melons	176	27	203	130	62	11	13	5
Meat, excl. offal	11	7	18	18	–	–	39	–
Eggs	3	–	3	3	–	–	–	–
Fish & fish products (fish equiv.)	0.2	3.3	3.5	3.5	–	–	94	–
Milk products (liquid equiv.)	48	27	74	73	2	–	36	–
Oilseed & veg. oils (oil equiv.)	10	9	19	18	–	–	47	–
Cotton & products (raw cotton equiv.)	–	3	3	–	3	–	100	–

Source: Indicative World Plan, Near East (Rome: FAO, 1966), II.

Appendix Table C11-5

Syria: Commodity Balances, 1960–62

(1,000 Metric Tons)

Commodity	Production	Imports	Supply	Local utilization		Exports	Imports	Exports
				Food	Other		As percent of supply	
Wheat	895	270	1,165	649	445[a]	71	23	6
Rice (paddy)	–	46	46	45	1	–	100	–
Coarse grains	482	51	533	185	201	147	10	28
Starchy roots	31	20	51	42	8	1	39	2
Sugar (raw equivalent)	10	69	79	79	–	–	87	–
Pulses	91	1	92	50	25	17	1	18
Vegetables, incl. melons	588	27	615	527	58	30	4	5
Fruits, excl. melons	409	47	456	345	105	6	10	1
Meat, excl. offal	55	7	62	53	–	–	11	–
Eggs	8	–	8	7	1	–	–	–
Fish & fish products (fish equiv.)	2.1	5.7	7.8	6.4	–	1.4	73	18
Milk products (liquid equiv.)	315	15	330	309	16	5	5	2
Oilseed & veg. oils (oil equiv.)	50	4	54	38	7	9	7	17
Cotton & products (raw cotton equiv.)	142	1	143	–	23[b]	120	1	84

[a] Incl. 300,000 from stock.
[b] Incl. 8,000 from stock.

Source: Indicative World Plan, Near East (Rome: FAO, 1966), II.

Appendix Table C11-6

Iraq: Commodity Balances, 1960–62

(1,000 Metric Tons)

Commodity	Production	Imports	Supply	Local utilization		Exports	Imports	Exports
				Food	Other		As percent of supply	
Wheat	845	207	1,052	715	324[a]	13	20	1
Rice (paddy)	100	98	198	184	14	–	49	–
Coarse grains	957	1	958	121	707	130	–	14
Starchy roots	11	15	26	22	4	–	58	–
Sugar (raw equivalent)	2	240	242	207	35[b]	–	99	–
Pulses	36	5	41	34	4	3	12	7
Vegetables, incl. melons	544	25	569	516	51	2	4	–
Fruits, excl. melons	544	30	574	313	39	222	5	39
Meat, excl. offal	114	–	114	112	–	2	–	2
Eggs	8	–	8	7	1	–	–	–
Fish & fish products (fish equiv.)	16	0.5	16.5	16.5	–	–	3	–
Milk products (liquid equiv.)	501	10	511	486	25	–	2	–
Oilseed & veg. oils (oil equiv.)	5	11	16	16	–	–	69	–
Cotton & products (raw cotton equiv.)	9	8	17	–	16[c]	–	47	–

[a] Incl. 150,000 from stock.
[b] All from stock.
[c] 2,000 m.t. to stocks.

Source: Indicative World Plan, Near East (Rome: FAO, 1966), II.

Appendix Table C11-7
Saudi Arabia: Commodity Balances, 1962
(1,000 Metric Tons)

Commodity	Production	Imports	Supply	Local utilization Food	Local utilization Other	Exports	Imports As percent of supply	Exports As percent of supply
Wheat	129	138	267	246	21	—	52	—
Rice (paddy)	4	130	134	127	7	—	97	—
Coarse grains	103	52	155	145	10	—	34	—
Starchy roots	—	5	5	5	—	—	100	—
Sugar (raw equivalent)	—	60	60	60	—	—	100	—
Pulses	7	—	7	7	—	—	—	—
Vegetables, incl. melons	344	36	380	343	37	—	9	—
Fruits, excl. melons	344	38	382	344	38	—	10	—
Meat, excl. offal	38	17	55	55	—	—	31	—
Eggs	6	1	7	7	—	—	14	—
Fish & fish products (fish equiv.)	5	3.1	8.1	8.1	—	—	38	—
Milk products (liquid equiv.)	169	44	213	204	9	—	21	—
Oilseed & veg. oils (oil equiv.)	2	7	9	9	—	—	78	—
Cotton & products (raw cotton equiv.)	negligible	8[a]	8	—	8	—	100	—

[a] Including cotton textiles.

Saudi Arabia

SAUDI ARABIA is geographically, culturally, and economically part of the "Middle East," as that term is used in this book. It would have been desirable to have included this country in the detailed analysis throughout; indeed, we attempted to do so in the first draft of this report. Data about agricultural resources, land and water use, agricultural production, agricultural labor force, agricultural income, and other aspects of agriculture are so limited, however, that in many of the tables and much of the discussion in the main part of this book, it was not possible to consider Saudi Arabia on the same basis as the other countries. As a result, the decision was reached to omit it from the body of this book, and to include here, in an appendix, what is known or estimated about Saudi Arabian agriculture.

Saudi Arabia is a very large country, with 550 million acres of land; it is nearly as large as all of Western Europe or is more than one-fourth the area of the forty-eight contiguous states of the United States. It is roughly 750 by 1,200 miles in extent, the latter measured on its long axis from northwest to southeast and the former at right angles to this axis. There is a rough mountainous area along the western and southwestern border, of ancient basement rocks, rising in the south to elevations where rainfall is higher; and some precarious crop production without irrigation is possible. Most of the country slopes off toward the northeast, with such runoff as there is going predominantly in that direction. An immense area toward the southeast is labeled on most maps as the "empty quarter," with neither permanent settlements nor nomadic agriculture—or, indeed, clearly established national boundaries. On the southeastern edge of the Arabian Peninsula there are some other countries (Yemen, Aden, etc.) and on the northeast near Iraq there is Kuwait; none of these are included in the discussion of this appendix.

Saudi Arabia is an extremely arid country; average annual precipitation for the whole country probably does not exceed four inches and may be less. There is a lack of detailed and precise weather data. It is also a very hot country, with summer temperatures often well above 100 degrees Fahrenheit for extended periods of time. The combination of scant precipitation and high temperatures results in very sparse vegetation, especially on the slopes.

Much of the water which does fall on Saudi Arabia is wasted, as far as agricultural output is concerned. Rain is likely to come in infrequent but rather heavy storms; and the runoff is very high, approaching 100 percent in many cases. Under these circumstances, it is possible to have destructive floods in areas characterized by extreme aridity; a rain produces a heavy downpour, nearly all of which runs off, frequently in a high-velocity damaging torrent. The flashy runoff from wadis accumulates in shallow desert basins (or wadi bottoms), and from these basins it is evaporated quickly.

Much of the well water, whether pumped or artesian, and flowing springs at oases, is also wasted; flowing wells are not checked during the nonuse seasons. Most of the oases have areas of waterlogged land, or extremely saline lands, or both, where the waters have accumulated and the salts evaporated. For many oases, the amount of water available is wholly insufficient for flushing salts to the sea. Cropping has often been carried out too near the low point of the natural basis, so that salts accumulated in the farmed areas; in the future, cropping may have to be moved to higher ground, and the bottoms of the basins converted to salt sinks.

The groundwater situation in Saudi Arabia is complex.[1] Some water may be quite old, accumulated in past geologic periods. In many situations, the infiltrated water enters permeable strata overlain by tighter strata, so that artesian or flowing wells or natural oases develop, often at long distances from the infiltration areas. Some of the oases are very old, and water has been flowing naturally for many centuries. In addition, some of the confined artesian water reaches the surface in the Arabian Gulf, in submarine springs of fresh water.

The total groundwater resources of Saudi Arabia are poorly known, but have become the subject of intensive study during recent years. Within the next few years, it should be possible to form a much more accurate estimate of the total volume of water, according to its source, age, and probable sustained yield or output. It is known now, however, that the total volume of water will be very small in relation to the area of potentially irrigable soils, which in turn are very limited in extent. It is also known that much of the water that can be recovered is highly saline; salt concentrations of more than 3,000 parts per million are common, and little water of less than 1,000 parts per million is available. While water of such relatively high salt content can be used to grow many crops, the flushing of salts becomes extremely important with such highly saline water. Burdon[2] also points out that the full cost of well water, considering the frequently high cost of drilling and the low yield of wells, is likely to be on the order of $100 per acre-foot annually or more.

In the more mountainous southwestern part of the country, where precipitation is higher and runoff in defined streams is greater, it is possible to construct dams and store surface water. However, evaporation is very high, and the

---

[1] David J. Burdon and Galip Otkun, "Hydrogeological Control of Development in Saudi Arabia," *XXIII International Geological Congress, 1968.*

[2] *Ibid.*

net yield of reservoirs will be much below the amounts put into storage.

The soils of Saudi Arabia are not well known, but are now under more intensive study. The low and variable precipitation, the high temperatures which would quickly burn up any organic matter that did grow, and the sparse natural vegetation have combined to prevent the development of mature soils. For the most part, the soils of Saudi Arabia that have potential for irrigated agriculture are found in the wadi bottoms or outlets. These soils have clay and silt that have been brought to the low landscape position by flood waters. They have, thus, better physical properties for plant growth under irrigated agriculture than do the dominantly sandier soils of the higher lying ground from which they were washed. These are Alluvial soils, usually with a large component of soluble salts that must be washed out before they can produce most crops profitably. A large proportion of these Alluvial soils are in closed depressions in desert areas. Ultimate disposal of drainage water from the irrigated areas thus presents a problem; indeed, many oases already suffer from areas of poorly drained land where salt accumulations are unusually high, and drainage may be a necessary condition to better use of the existing soil and water resources. However, because of the relatively large area of arable soils in relation to probable water supply on a sustained basis, and because of the very high evaporation rates, the capacity of the natural depressions to serve as salt sinks may be adequate for the irrigation water available.

In the recent past, approximately 660,000 acres of crops have been grown in Saudi Arabia.[3] It is further estimated that the cultivated acreage was just over 600,000 acres (or 0.13 percent of the total land area), giving a cropping ratio of 1.09:1, and that the latter area was only 30 percent of the total cultivable lands in the country; limitations of water prevent the whole from being cropped. These are rough estimates. Studies now under way should in time provide more accurate data, but one can exclude the possibility that they will upset the major characteristic of the area: a very small area cropped in relation to total area of the country.

In past decades, the agriculture of Saudi Arabia has fallen into two major groupings: irrigated oases and grazing of desert areas, with relatively little interconnection between the two. In addition to the irrigated oases, some grain has been grown in the mountainous southwest without irrigation. In the recent past, of the 660,000 acres cropped, nearly two-thirds was in cereals, principally wheat, followed by millet, barley, sorghum, and rice. Yields (for irrigated and unirrigated lands in unspecified mixture) were about 21 bushels per acre for wheat and 31 bushels per acre for barley—not high, assuming that most of these acreages were irrigated. A variety of vegetables and melons was also grown. Dates, a staple article in the diet of a major part of

the population in the past, have been a traditional crop in the irrigated oases.

Farms in the irrigated areas have been small, averaging 5½ acres in four provinces—with 55 percent just over 1 acre each, 80 percent less than 3¾ acres, and only 4 percent 25 acres or more. Costs of production, including an allowance for the labor of the farmer, have been very high; in an effort to protect the domestic farm producer, while at the same time keeping food prices to the consumer as low as possible, most imported crops have been subsidized up to 17 percent of their value in recent years, and a customs tariff of 10 percent has been levied on imports of fresh vegetables.

Nomadic herds of livestock have been grazed in Saudi Arabia for centuries; as elsewhere, the nomads have followed the forage, which in turn has meant following the rain. When severe droughts hit, livestock losses have often been most severe, since there were no reserves of feed which could be drawn upon. In severe droughts of the early 1960's, for example, it has been estimated that in the Northern Region as many as 80 percent of the animals died or were moved out of the country.[4] The types of forage available have been best suited for goats, sheep, and camels, and hence these have been the kinds kept; a few cattle and horses are found, but chiefly within the irrigated oases. Goats, sheep, and camels are kept for milk as well as for meat; in many areas, the sheep are rarely shorn for wool. The camel was long the major means of transportation, but is increasingly being replaced by mechanized transport.

There has never been a detailed and accurate census of livestock numbers in Saudi Arabia or of total area grazed; given the nomadic character of livestock production, the relectance of nomads to reveal information about their operations, and the variations in livestock numbers as forage conditions change from year to year, it would be difficult indeed to count livestock in a meaningful way. An estimate by the Ministry of Agriculture in 1963 reported 2,800,000 sheep, 1,400,000 goats, 600,000 camels, and 270,000 cattle.[5] An FAO expert in 1963 made an independent estimate, based on travel around the country, of 4,158,000 sheep and goats (nearly identical with the one made by the Ministry), 1,004,000 camels, and 75,000 donkeys.[6]

In the mountainous areas of western Arabia, extending from the Yemen border north to Medina or further, there was for centuries a system of grazing reserves referred to as "Hema" or "Ahmia."[7] The origins of such reserves antedates Muhammed by several centuries. Their purposes seem

---

[3] Edmond Y. Asfour, with the collaboration of George S. Medawar, Hisham M. Al Kaylani, and Leila Takieddine, *Saudi Arabia— Long-Term Projections of Supply Of and Demand For Agricultural Products*, Economic Research Institute, American University of Beirut, 1965.

[4] Marvin Klemme, *Report to the Government of Saudi Arabia— Pasture Development and Range Management* (FAO No. 1993 [Rome: FAO, 1965]), p. 17.

[5] Reported in Lyle E. Moe, *Saudia Arabia: Supply and Demand Projections for Farm Products to 1975, with Implications for U. S. Exports,* (ERS-Foreign 168, Economic Research Service, U. S. Department of Agriculture [Washington, December 1966]), p. 10.

[6] Harold F. Heady, *Report to the Government of Saudi Arabia on Grazing Resources and Problems* (ETAP Report No. 1614, Project SAU/TE/PL [Rome: FAO, 1963]), pp. 5–7.

[7] Klemme, *op. cit.* (above, n. 4).

to have varied: a place for the military to hold their live-stock, a place established by local communities for their livestock, and others. At any rate, grazing on these areas was controlled and limited, and a degree of range conservation was undoubtedly practiced. In 1953, by royal decree, these lands were thrown open to any users, on a first-come-first-served basis; and the inevitable overgrazing quickly reduced the condition of these formerly protected and managed lands to that of the unrestricted use lands in others areas.

Elsewhere in Saudi Arabia, as in most of the Middle East, nomadic livestock have grazed on any forage available, without control or limit. The inevitable result nearly every-where has been severe overgrazing, with consequent reductions in forage output. FAO (in an unpublished report) estimates range condition as follows:

Best or excellent	5%
Good (50 to 75% of original productive plants present)	10%
Fair (25 to 50% of best forage plants still present)	25%
Poor (24% or less of better plants present)	60%

The various studies by range-management experts empha-size that more forage could be produced on these lands, by careful livestock management programs and by various range improvements. However, this would involve a major readjustment in traditional ways of using range lands, which would not be easy to accomplish. The Government has been making serious efforts to induce nomads to settle down in recent years; perhaps more effective has been the pull of better paying jobs in the cities and oil fields, which are drawing the young men away from the nomadic life. Regardless of such efforts, it is doubtful if much of the grazing resource of Saudi Arabia could be harvested in any way except by nomadic livestock.

Prior to World War II, Saudi Arabia was primarily a pastoral country, with little economic development not based upon that basic activity; it is the location of important religious centers, to which pilgrims have come annually for centuries, and this has provided a degree of economic support. But the modern attributes of an economically advanced nation were lacking at that time and to a degree still are. For instance, there has never been a systematic census of population in Saudi Arabia; published official estimates of total population in 1962 were 6.4 million, whereas most recent estimates are but half that, or 3.2 million.[8] While data are available on imports, which are probably fairly accurate, data on agricultural output and consumption of agricultural commodities are either only for recent years or are likely to have high margins of error or both.

Prior to World War II, average per capita incomes must have been low: a major part of the population was in agriculture and grazing, on a largely subsistence basis—little was produced for sale and little was consumed that had not been produced by the family. Oil has largely changed this. National income has risen greatly, and so has average income per capita. At the same time such income has filtered down to the average citizen slowly. Although accurate data are lacking, it is estimated that 60 percent of the population is still in agriculture but contributes only 15 percent of total GNP.[9] The government is fully aware of this situation, and has taken steps to increase agricultural output, of which the survey of land and water resources now under way is a basic one. At the same time, many people from nomadic groups and from oases populations have sought employment in the oil industry or in its services and have moved to cities and larger towns. Major social and economic changes are under way in Saudi Arabia, the full consequences and significance of which are beyond the scope of this appendix.

It seems highly probable that the demand for agricultural commodities will rise over the next decade or two. Asfour and associates estimate an increase in population of a third between 1961/62 (the base year for their study) and 1975; they estimate that GNP will increase two and one-half times in the same period.[10] The combination of more people and higher average incomes will surely mean a greatly increased demand for agricultural commodities. Because incomes are now relatively low, at least for much of the population, the income elasticity of demand for food is relatively high, according to the best estimates available. Without adequate domestic production, imports of food products into the country have grown greatly in recent years: in 1961/62, more than a third of the total food demand was met by imports. The subsidy on such imports, while making life easier for the low-income groups, has been an obstacle to increasing domestic agricultural output.

The prospect of increased demand for food lends emphasis to the desire of the government and of agricultural producers to increase agricultural output; but it seems highly likely that limited resources of water, arable land, and natural forage will be serious constraints on increasing agricultural output. At the same time, as in other countries of the region, there are considerable possibilities of adopting improved agricultural technologies and better resource management programs, though the information base known to us is insufficient to construct even the tenous kind of "potential" estimate that has been presented for the other countries. Saudi Arabia, however, is likely in the future to import an increasing proportion of its rising total demand for agricultural commodities; in a regional analysis, it is one area where the other countries of the region might look—and are looking—for markets for any surplus production. Almost all the major commodities must be imported in the future, according to the Asfour analysis, thus offering a market for a wide range of products from other countries. If Saudi Arabia seeks the cheapest sources of supply for its food imports, then the ability of other Arab countries of the Middle East to meet world prices competitively may be decisive; their relative nearness may be a major asset in the market for fresh fruits and vegetables but relatively unimportant for cereals and other staples. On the other hand, being a member of the Arab Common Market may make it an easier export target for its neighbors.

---

[8] Asfour, *op. cit.* (above, n. 3), p. 17.

[9] FAO unpublished report.

[10] Asfour, *op. cit.* (above, n. 3).

Author Index

Numbers in parentheses indicate the numbers of the references when these are cited in the text without the names of the authors.

Numbers set in *italics* designate the page numbers on which the complete literature citations are given.

Subject Index

---

* n = text footnote.

Becker, Abraham S. *Soviet National Income 1958-1964.* University of California Press, Berkeley and Los Angeles. 1969.

Bergson, A. *The Real National Income of Soviet Russia Since 1928.* Harvard University Press, Cambridge, Massachusetts. 1961.

Bergson, Abram, and Hans Heymann, Jr. *Soviet National Income and Product,* 1940-48. Columbia University Press, New York. 1954.

Chapman, Janet G. *Real Wages in Soviet Russia Since 1928.* Harvard University Press, Cambridge, Massachusetts. 1963.

Downs, Anthony. *Inside Bureaucracy.* Little Brown and Company, Boston, Massachusetts. 1967.

Gurtov, Melvin. *Southeast Asia Tomorrow: Problems and Prospects For U.S. Policy.* Johns Hopkins Press, Baltimore, Maryland. 1970.

Hirshleifer, Jack, James C. DeHaven, and Jerome W. Milliman. *Water Supply: Economics, Technology, and Policy.* The University of Chicago, Chicago, Illinois. 1960.

Hitch, Charles J., and Roland McKean. *The Economics of Defense in the Nuclear Age.* Harvard University Press, Cambridge, Massachusetts. 1960.

Hoeffding, Oleg. *Soviet National Income and Product in 1928.* Columbia University Press, New York. 1954.

Johnson, William A. *The Steel Industry of India.* Harvard University Press, Cambridge, Massachusetts. 1966.

Johnstone, William C. *Burma's Foreign Policy: A Study in Neutralism.* Harvard University Press, Cambridge, Massachusetts. 1963.

Kershaw, Joseph A., and Roland N. McKean. *Teacher Shortages and Salary Schedules.* McGraw-Hill Book Company. Inc., New York. 1962.

Leites, Nathan, and C. Wolf. *Rebellion and Authority.* Markham Publishing Company, Chicago, Illinois. 1970.

Liu, Ta-Chung, and Kung-Chia Yeh. *The Economy of the Chinese Mainland: National Income and Economic Development, 1933-1959.* Princeton University Press, Princeton, New Jersey. 1965.

Lubell, Harold. *Middle East Oil Crises and Western Europe's Energy Supplies.* The Johns Hopkins Press, Baltimore, Maryland. 1963.

McKean, Roland N. *Efficiency in Government through Systems Analysis with Emphasis on Water Resource Development.* John Wiley and Sons, Inc., New York. 1958.

Marschak, Thomas, Thomas K. Glennan Jr., and Robert Summers. *Strategy for R & D,* Springer-Verlag, Inc., Berlin, Heidelberg, New York, 1967.

Moorsteen, Richard. *Prices and Production of Machinery in the Soviet Union, 1928-1958.* Harvard University Press, Cambridge, Massachusetts. 1962.

Nelson, Richard R., Merton J. Peck, and Edward D. Kalachek. *Technology, Economic Growth and Public Policy.* The Brookings Institution, Washington, D. C. 1967.

Pascal, Anthony. *Thinking About Cities: New Perspectives on Urban Problems.* Dickenson Publishing Company, Belmont, California. 1970.

Pincus, John A. *Economic Aid and International Cost Sharing.* The Johns Hopkins Press, Baltimore, Maryland. 1965.

Rosen, George. *Democracy and Economic Change in India.* University of California Press, Berkeley and Los Angeles, California. 1966.

Wolf, Charles, Jr. *Foreign Aid: Theory and Practice in Southern Asia.* Princeton University Press, Princeton, New Jersey. 1960.

Selected Resources for the Future *Books*

Grunwald, Joseph, and Philip Musgrove. *Natural Resources in Latin American Development.* 1970. 512 pp. $20.00

Mikesell, Raymond F., and Associates. *Foreign Investment in the Petroleum and Mineral Industries: Case Studies of Investor–Host Country Relations.* 1970. 460 pp. $15.00.

Herfindahl, Orris C. *Natural Resource Information for Economic Development.* 1969. 232 pp. $7.00.

Christy, Francis T., Jr. and Anthony Scott. *The Common Wealth in Ocean Fisheries: Some Problems of Growth and Economic Allocation.* 1966. 296 pp. $6.00.

Kneese, Allen V., and Blair T. Bower. *Managing Water Quality: Economics, Technology, Institutions.* 1968. 338 pp. $8.95.

Landsberg, Hans H., Leonard L. Fischman, and Joseph L. Fisher. *Resources in America's Future: Patterns of Requirements and Availabilities, 1960-2000.* 1963. 1,040 pp. $15.00.

Barnett, Harold J., and Chandler Morse. *Scarcity and Growth: The Economics of Natural Resource Availability.* 1963. 304 pp. $9.00. Paper 1969, $2.25.

Crosson, Pierre R. *Agricultural Development and Productivity: Lessons from the Chilean Experience.* 1970. 208 pp. $7.00.

Brubaker, Sterling. *Trends in the World Aluminum Industry.* 1967. 276 pp. $6.95.

Campbell, Robert W. *The Economics of Soviet Oil and Gas.* 1968. 294 pp. $6.95.

Clawson, Marion, ed. *Natural Resources and International Development: Essays Based on the RFF Forum Lectures of 1963.* 1964. 474 pp. $10.00.

Clawson, Marion. *Policy Directions for U.S. Agriculture: Long-Range Choices in Farming and Rural Living.* 1968. 416 pp. $10.00.

Clawson, Marion, and Jack L. Knetsch. *Economics of Outdoor Recreation.* 1966. 348 pp. $8.50.

Arrow, Kenneth J., and Mordecai Kurz. *Public Investment, the Rate of Return, and Optimal Fiscal Policy.* 1970. 288 pp. $9.00.

(All titles cited above were published for RFF by The Johns Hopkins Press, Baltimore, Maryland 21218.)